Electrical

Level Three

ELEVENTH EDITION

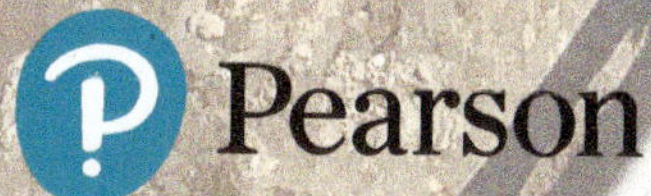

National Center for Construction Education and Research

Composition: NCCER
Content Technologies: Gnostyx
Printer/Binder: Lakeside Book Company
Cover Printer: Lakeside Book Company

Cover Image
Cover photo provided by: Elm Electrical, Inc..

10 9 8 7 6 5 4 3 2 1

1 2023

Paperback:
ISBN-13: 978-0-13-817488-0

PREFACE

To the Trainee

Welcome to the NCCER *Electrical* craft training curriculum! Electricians play a critical role in society because electricity is an integral part of our daily lives. It powers applications that increase productivity and enhances our standard of living. As a result, the demand for skilled electricians remains high and opportunities in the electrical field continue to grow. Whether people are at home or at work, an electrician's skills help ensure their safety as well as the reliability of equipment and systems they use every day.

According to the US Bureau of Labor Statistics, the demand for skilled electricians is expected to continue its growth, providing numerous job opportunities for trained professionals in the electrical field. To become a skilled electrician, individuals are trained through craft and apprenticeship programs. Apprenticeship programs offer a combination of classroom instruction and on-the-job learning with experienced electricians.

NCCER's craft training tools give students the knowledge and skills they need to be successful in today's competitive market. Whether they are learning to read construction drawings, developing an understanding of safety codes, pulling and terminating conductors, or installing electrical fixtures in a building, students using NCCER's *Electrical* curriculum are well equipped to meet today's increased demands. Ultimately, trainees will have the skills they need to install, maintain, and repair electrical components and equipment on residential, commercial, and industrial jobsites.

NCCER provides you with comprehensive training tools that combine classroom instruction with hands-on experience. The result is an approach that gives students the skills and confidence they need to succeed in a demanding and rewarding electrical career. We are committed to helping you achieve your goals, and we hope this curriculum will serve as a valuable resource as you pursue your passion for the electrical craft.

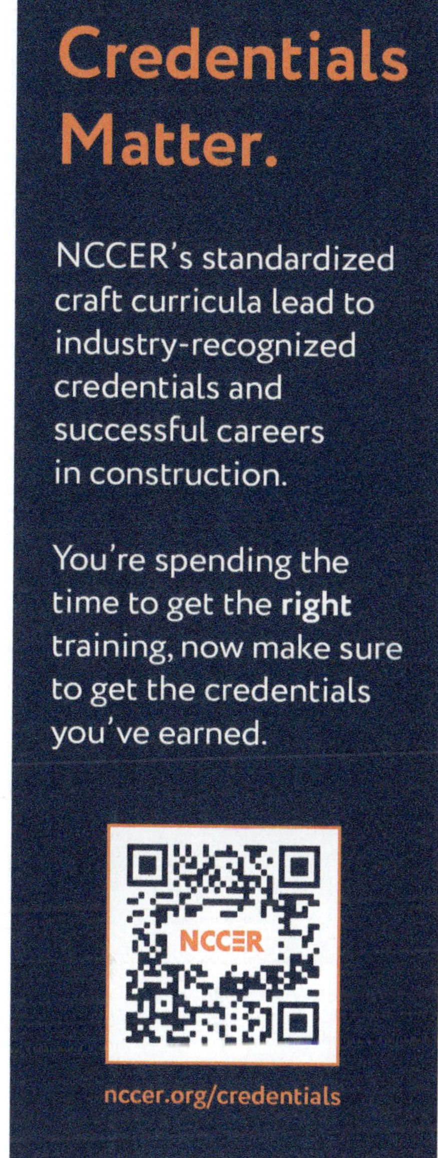

New with *Electrical Level Three*

NCCER and Pearson are pleased to present *Electrical Level Three*, which has been reformatted to provide a better experience to both Electrical trainees and instructors. *Electrical Level Three* has been updated to meet the 2023 *National Electrical Code®* along with updated art throughout.

In addition to the 2023 code changes, this edition of *Electrical Level Three* features several further updated modules. NCCER Module 26301-23, *Load Calculations – Branch and Feeder Circuits*, has been revised to improved clarity of math problems to aid the trainee in learning more complex concepts. NCCER Module 26307-23, *Transformers*, has been revised to simplify the presentation of complex materials and enhance the presentation of testing. Figures and tables throughout the modules have been updated to reflect the new code and new technology.

We wish you success as you progress through this training program. If you have any comments on how NCCER might improve upon this textbook, please complete the User Update form using the QR code in the frontmatter on this page. NCCER appreciates and welcomes its customers' feedback. You may submit yours by emailing **support@nccer.org**. When doing so, please identify feedback on this title by listing *#ElectricalL3* in the subject line.

Our website, **www.nccer.org**, has information on the latest product releases and training.

NCCER Standardized Curricula

NCCER is a not-for-profit 501(c)(3) education foundation established in 1996 by the world's largest and most progressive construction companies and national construction associations. It was founded to address the severe workforce shortage facing the industry and to develop a standardized training process and curricula. Today, NCCER is supported by hundreds of leading construction and maintenance companies, manufacturers, and national associations. The NCCER Standardized Curricula was developed by NCCER in partnership with Pearson, the world's largest educational publisher.

Some features of the NCCER Standardized Curricula are as follows:

- An industry-proven record of success
- Curricula developed by the industry, for the industry
- National standardization providing portability of learned job skills and educational credits
- Compliance with the Office of Apprenticeship requirements for related classroom training (*CFR 29:29*)
- Well-illustrated, up-to-date, and practical information

NCCER maintains a secure online database that provides certificates, digital badges, transcripts, and wallet cards to individuals who successfully complete programs under an NCCER-accredited organization or through one of NCCER's self-paced, online programs. This system also allows individuals and employers to track and verify industry-recognized credentials and certifications in real time.

For information on NCCER's credentials, contact NCCER Customer Service at 1-888-622-3720 or visit **https://www.nccer.org**.

Digital Credentials

Show off your industry-recognized credentials online with NCCER's digital credentials.

NCCER is now providing online credentials. Transform your knowledge, skills, and achievements into digital credentials that you can share across social media platforms, send to your network, and add to your resume. For more information, visit **www.nccer.org**.

Cover Image

Construction of a Final Conditioning Facility for a renewable energy power plant manufacturer in the Northeast. Elm Electrical, Inc. acted as the EPC contractor for the engineering, procurement, and construction of a 3,000 square feet pre-engineered Butler Building which housed the electrical room, mechanical room with boilers / air compressors / water treatment systems, and conditioning facility operator room. The outdoor conditioning facility included the installation of a 60-ton overhead crane, associated foundations, along with a 20,000 square feet radiant heated concrete pad to house (4) proprietary conditioning plants. The installation of this facility allowed this customer to scale to the demand for the production of their power plants that are helping drive the Electrification of our world and reducing our carbon footprint. For more information, visit **www.elmelec.com**.

DESIGN FEATURES

Content is organized and presented in a functional structure that allows trainees to access the information where they need it.

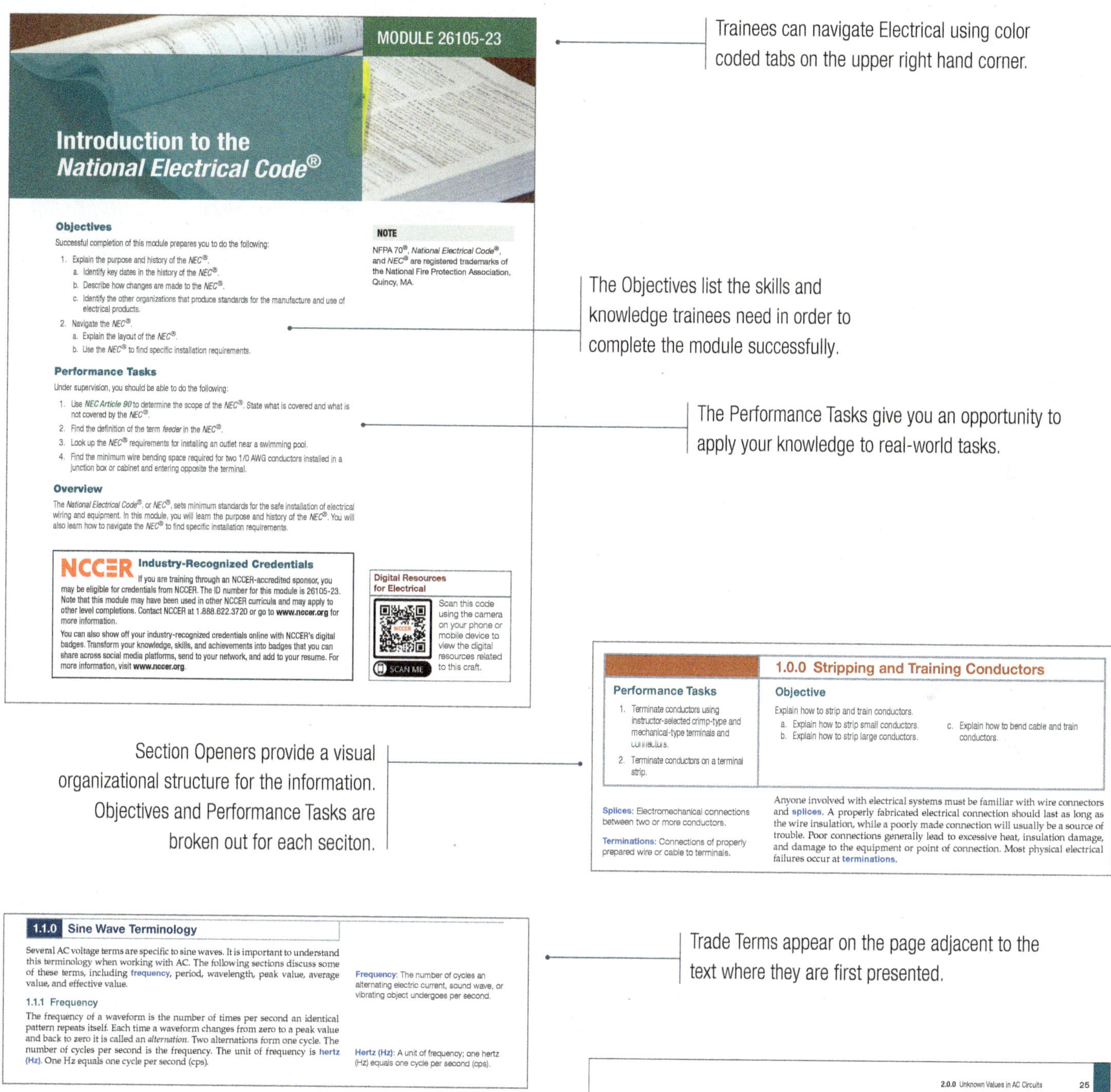

Trainees can navigate Electrical using color coded tabs on the upper right hand corner.

The Objectives list the skills and knowledge trainees need in order to complete the module successfully.

The Performance Tasks give you an opportunity to apply your knowledge to real-world tasks.

Section Openers provide a visual organizational structure for the information. Objectives and Performance Tasks are broken out for each seciton.

Trade Terms appear on the page adjacent to the text where they are first presented.

Step-by-step math equations help make the concepts clear and easy to grasp.

2.0.0 Unknown Values in AC Circuits 25

Solve for X_L as follows:

Step 1 Compute the value of X_L.

$X_L = 2\pi fL$

$X_L = 6.28 \times 100\ \text{Hz} \times 4\text{H}$

$X_L = 2{,}512\Omega$

Step 2 Draw vectors R and X_L as shown in *Figure 19*. R is drawn horizontally because the circuit current and voltage across vector R are in phase. It therefore becomes the reference line from which other angles are measured. X_L is drawn upward 90° from R because the voltage across X_L leads the circuit current through R.

Step 3 Compute the value of circuit impedance Z, which is equal to the vector sum of X_L and R.

$\tan = \frac{X_L}{R} = \frac{2{,}512}{1{,}500} = 1.67$

$\arctan(1.67) = 59.1°$

$\cos(59.1°) = 0.5135$

Find Z using the cosine function.

$\cos = \frac{R}{Z}$

$Z = \frac{R}{\cos}$

$Z = \frac{1{,}500}{0.5135} = 2{,}921\Omega$

Step 4 Compute the circuit current using Ohm's law for AC circuits.

$I = \frac{E}{Z} = \frac{100V}{2{,}921\Omega} = 0.034A$

QR codes link trainees directly to videos that highlight current content.

SUCCESS STORIES: *Credentials Matter: Start Building Your Career Today*

Scan this code using the camera on your phone or mobile device to view this video.

Important information is highlighted, illustrated, and presented to facilitate learning.

Placement of images near the text description and details such as callouts and labels help trainees absorb information.

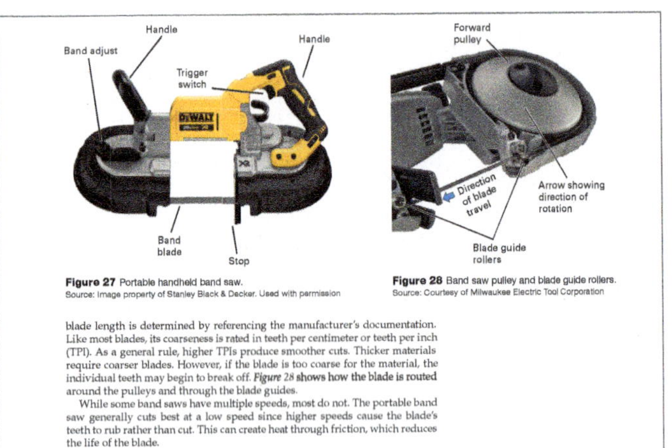

Figure 27 Portable handheld band saw.
Source: Image property of Stanley Black & Decker. Used with permission

Figure 28 Band saw pulley and blade guide rollers.
Source: Courtesy of Milwaukee Electric Tool Corporation

blade length is determined by referencing the manufacturer's documentation. Like most blades, its coarseness is rated in teeth per centimeter or teeth per inch (TPI). As a general rule, higher TPIs produce smoother cuts. Thicker materials require coarser blades. However, if the blade is too coarse for the material, the individual teeth may begin to break off. *Figure 28* **shows how the blade is routed** around the pulleys and through the blade guides.

While some band saws have multiple speeds, most do not. The portable band saw generally cuts best at a low speed since higher speeds cause the blade's teeth to rub rather than cut. This can create heat through friction, which reduces the life of the blade.

Preparing Drills with Keyless Chucks

Most cordless drills use a keyless chuck. While the steps for preparing a cordless drill are similar, there are some small differences. Follow the steps below when preparing to use drills with keyless chucks:

Step 1 Disconnect the drill from its power source by removing the battery pack before loading a bit.

Step 2 As shown in (*Figure 7A*), open the chuck by turning it counterclockwise until the jaws are wide enough to insert the bit shank.

Step 3 Insert the bit shank into the chuck opening (*Figure 7B*). Keeping the bit centered in the opening, turn the chuck by hand until the jaws grip the bit shank.

Step 4 Tighten the chuck securely with your hand so that the bit does not move (*Figure 7C*). You are now ready to use the cordless drill.

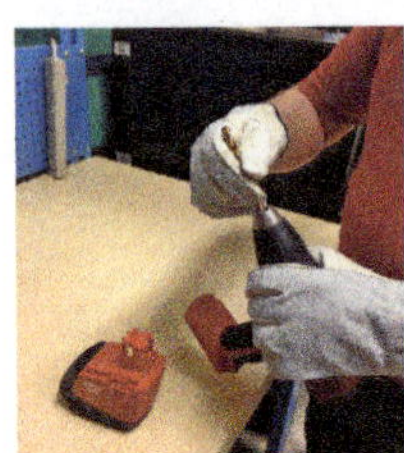

(A) Insert the Bit Shank.

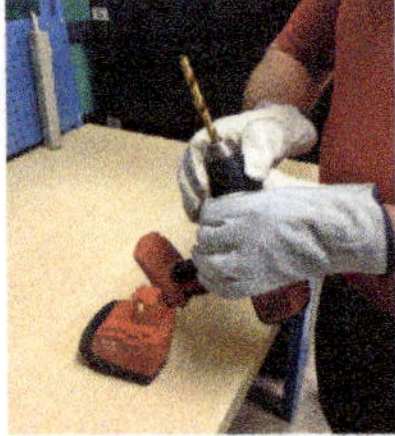

(B) Keep Bit Straight and Partially Tighten the Chuck.

(C) Tighten the Chuck Securely.

Figure 7 Loading the bit on a keyless chuck.
Source: Cianbro Corporation

New boxes highlight safety and other important information for trainees. Warning boxes stress potentially dangerous situations, while Caution boxes alert trainees to dangers that may cause damage to equipment. Note boxes provide additional information on a topic.

WARNING!

A portable band saw always cuts in the direction of the user. For that reason, workers must be especially careful to avoid injury when using this type of saw. Always wear appropriate PPE and stay focused on the work.

CAUTION

Never assume anything. It never hurts to ask questions, but disaster can result if you don't ask. For example, do not assume that an electrical power source is turned off. First ask whether the power is turned off, then check it yourself to be completely safe.

NOTE

This training alone does not provide any level of certification in the use of fall arrest or fall restraint equipment. Trainees should not assume that the knowledge gained in this module is sufficient to certify them to use fall arrest equipment in the field.

Did You Know?

Lightning Rods

An interesting fact about grounding is that a lightning rod (air terminal) isn't meant to bring a bolt of lightning to ground. To do this, its conductors would have to be several feet (1 m or more) in diameter. The purpose of the rod is to dissipate the negative static charge that would cause the positive lightning charge to strike the house.

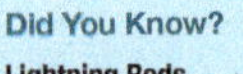

These boxed features provide additional information that enhances the text.

Orthographic Drawings

Orthographic drawings are used for elevation drawings. They show straight-on views of the different sides of an object with dimensions that are proportional to the actual physical dimensions. In orthographic drawings, the designer draws lines that are scaled-down representations of real dimensions. Every 12 inches, for example, may be represented by ¼ inch on the drawing. Similarly, in an example using metric measurements with a ratio of 1:2, every 30 millimeters may be represented by 15 millimeters on the drawing.

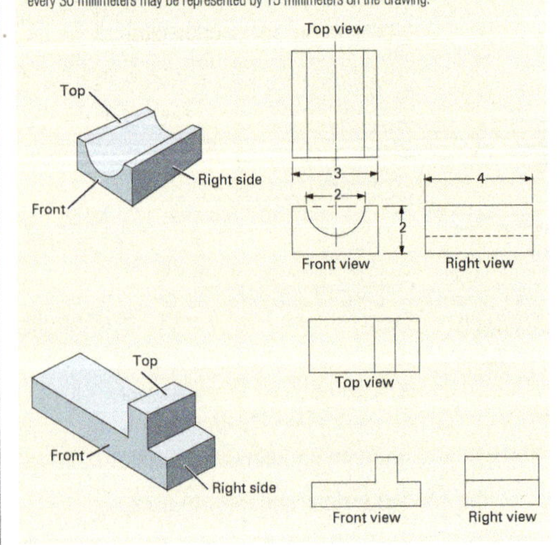

Around The World

Predicting Weather

Barometers measure atmospheric air pressure. Generally, an increased pressure indicates that the weather is pleasant. A reduced pressure indicates that the weather is cloudy and/or rainy.

Changes in barometric pressure are used to predict changes in weather conditions. For example, if it is raining outside, but the barometer is rising, it indicates that better weather is on the way. In the midst of a hurricane, the barometric pressure is an indicator of intensity. Extremely low pressures exist in powerful hurricanes. While traditional barometers work very well, sophisticated weather stations capture far more data and can transmit the information wirelessly.

Going Green

Residential Solar

As residential customers look for ways to reduce reliance on fossil fuels like coal and oil, they are turning to solar energy. Usually, they want to install photovoltaic (PV) panels on the roofs of their homes, or in their yards. A residential PV system usually includes the PV panels, a mounting structure that can support the panels, an inverter to convert the direct current (DC) electricity generated by solar photovoltaic modules into alternating current (AC) electricity, and a battery (or batteries) for storing the energy. Most states require a licensed electrician for solar energy system installation, but you should check the local requirements. (Source: **https://www.energy.gov/eere/solar/how-does-solar-work**)

Going Green looks at ways to preserve the environment, save energy, and make good choices regarding the health of the planet.

Cornerstone of Craftsmanship

John Lupacchino
Senior Design Engineer
Gaylor Electric, Inc.

How did you choose a career in the industry?
I knew that this was what I wanted to do since I was in 7th grade.

Who inspired you to enter the industry?
My grandfather, who was an electrician, inspired me to also become an electrician.

What types of training have you been through?
I went to vocational technical school for high school and studied electrical. Then I went on to complete an apprenticeship through the state of Connecticut.

How important is education and training in construction?
Training and education are extremely important in the construction industry. All craftworkers need to participate in continuing education in order to stay up with the advancements in technology.

How important are NCCER credentials to your career?
NCCER credentials are very important because they allow you to showcase your skills and abilities in a standardized way.

What kinds of work have you done in your career?
I have worked in all areas of electrical—residential, commercial, and industrial. Some of the types of facilities I have worked in include houses, stores, factories, warehouses, hospitals, prisons, steel mills, solar installations, cement plants, and many others.

Tell us about your current job.
In my present job as senior design engineer, I am responsible for estimating, designing, and managing electrical construction projects. I am also responsible for electrical code interpretation and compliance.

What do you enjoy most about your job?
My job is challenging and ever changing. It never gets boring. For me, it is great to be a part of building something that benefits others.

What factors have contributed most to your success?
I take advantage of the opportunities that come up, especially the training that is available. Applying my skills and putting in the effort required has definitely contributed to my success.

Would you suggest construction as a career to others? Why?
Yes! The construction industry has limitless opportunities. There will always be a need for building and maintaining facilities, which means there will always be a need for craftworkers.

What advice would you give to those new to the field?
Take advantage of any opportunities for training that you have. Show up to work on time with a "Get It Done" mentality. Do all you can to be the best you can be.

What is an interesting career-related fact or accomplishment?
I have been able to acquire licenses all over the country to enable my employer to work in various locations. I have also had the opportunity to be an instructor in the local ABC apprenticeship program for over 20 years.

How do you define craftsmanship?
Craftsmanship is the quality that comes from creating with passion, care, and attention to detail.

Cornerstone of Craftsmanship boxes feature career stories from people working in related fields.

Know the Code

Electrical Continuity of Metal Raceways, Cable Armor, and Enclosures
NEC Section 300.10

Know the Code boxes feature references to the *National Electrical Code®*.

Review questions at the end of each section and module allow trainees to measure their progress.

2.0.0 Section Review

1. A straight pull contains two raceways. One of the raceways has a trade size of 3" and one has a trade size of 2". The length of the box *must* be _____.
   a. 16"
   b. 24"
   c. 26"
   d. 32"
2. Where angle or U pulls are made, to determine the distance between raceway entries enclosing the same conductor, you need to multiply the trade size of the largest raceway by _____.
   a. two
   b. four
   c. six
   d. eight
3. If possible, pull boxes should be installed _____.
   a. high on the wall for security
   b. at a height/location that makes it easy to pull conductors
   c. behind wall coverings for a neater look
   d. with extra knockouts left open for air flow

Section Review questions can be found at the end of each section to test your knowledge of the content. Review Questions at the end of each module are provided to reinforce the knowledge you have gained.

Module 26207-23 Review Questions

1. In which *NEC®* article is cable tray installation primarily addressed?
   a. *NEC Article 300*
   b. *NEC Article 392*
   c. *NEC Article 517*
   d. *NEC Article 550*
2. Which type of tray is *best* suited for corrosive areas and atmospheres?
   a. Stainless steel cable tray
   b. Coated aluminum cable tray
   c. Nonmetallic cable tray
   d. Basket tray
3. When a cable tray has a solid bottom, it is referred to as _____.
   a. basket tray
   b. ladder tray
   c. trough
   d. raceway
4. Cable tray is generally manufactured in _____.
   a. 2' and 4' lengths
   b. 6' and 8' lengths
   c. 6' and 12' lengths
   d. 12' and 24' lengths

Module 26204-23 Supplemental Exercises

1. In which chapter of the *NEC®* will you find the installation requirements for conduit?
   a. *Chapter 2*
   b. *Chapter 3*
   c. *Chapter 4*
   d. *Chapter 5*
2. The maximum number of degrees allowed between pull points in a conduit run is _____.
3. Would 4" RMC be bent using a mechanical bender?
   ______
   ______
4. True or False? The one-shot shoes used with hydraulic benders can make 90° bends in a single operation.

Supplemental Exercises provide additional practice for strengthening your comprehension of the material.

NCCERCONNECT

This interactive online course is a unique web-based supplement that provides a range of visual, auditory, and interactive elements to enhance training. Also included is a full eText.

Visit **www.nccerconnect.com** for more information!

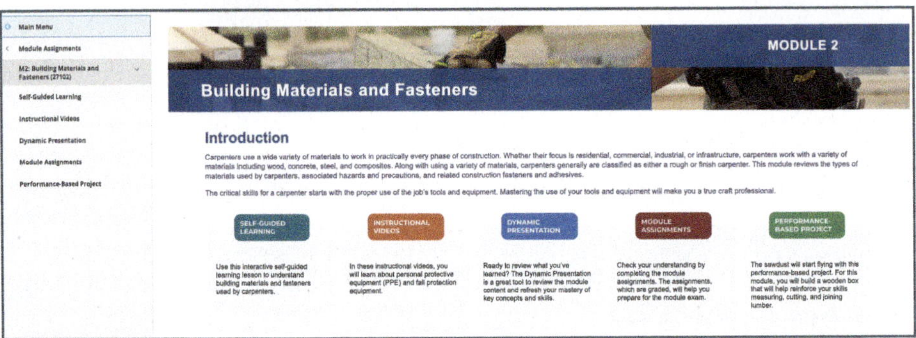

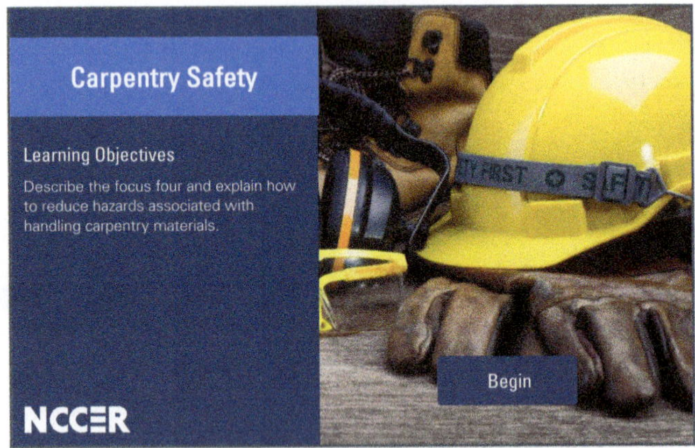

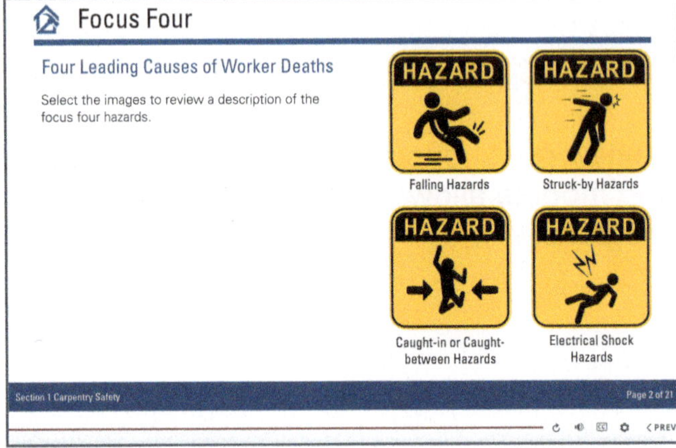

Use the interactive self-guided learning lesson to understand key concepts and terms needed for a career in construction.

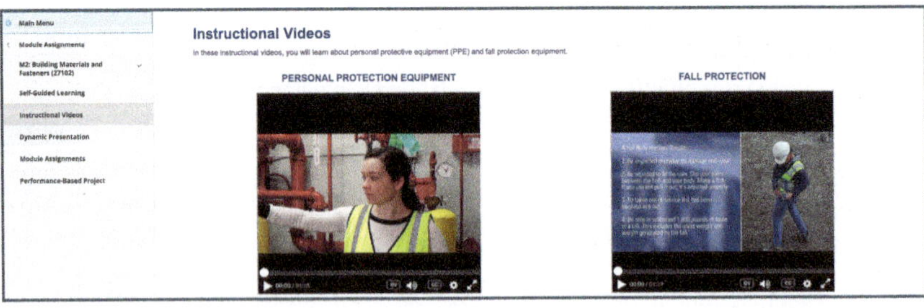

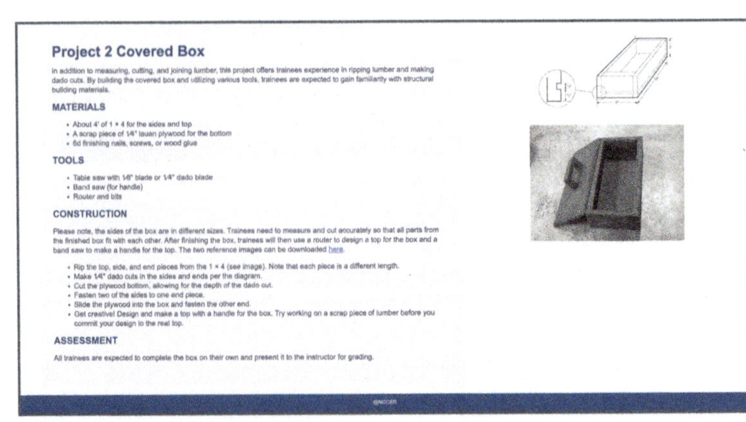

ACKNOWLEDGMENTS

This curriculum was revised as a result of the vision and leadership of the following sponsors:

Burkhead Electric
Carter Electric, Inc.
Cianbro Corporation
Construction Industry Training Council of Washington
Delta Electrical, Inc.
Elm Electrical, Inc.
Gaylor Electric, Inc.
Interstates, Inc.
Kansas City Veterans Administration Medical Center
M. Davis & Sons, Inc.
MMR Constructors, Inc.
Virginia Technical Academy
Willmar Electric Service Corp

This curriculum would not exist were it not for the dedication and unselfish energy of those volunteers who served on the Authoring Team. A sincere thanks is extended to the following:

Paul Asselin
Timothy Burkhead
Jason Canarr
Tim Dean
David Gillespie
Kenneth Hill
Justin Johnson
John Lupacchino II
Patrick Martin
Larry Meeks
Nicholas Mobley
John Mueller
Roy Oliver Jr.
Lowell Reith
Scott Robertson
Karl J Segner
Richard St. Vincent

NCCER PARTNERS

To see a full list of NCCER Partners, please visit:

www.nccer.org/about-us/partners.

You can also scan this code using the camera on your phone or mobile device to view these partnering organizations.

CONTENTS

Module 26301-23 Load Calculations – Branch and Feeder Circuits

Module 26302-23 Conductor Selection and Calculations

Module 26303-23 Practical Applications of Lighting

Module 26304-23 Hazardous Locations

Module 26305-23 Overcurrent Protection

Module 26306-23 Distribution Equipment

Module 26307-23 Transformers

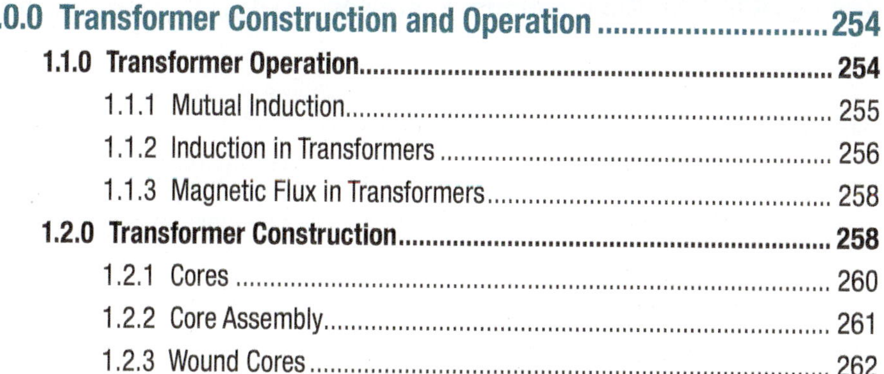

Module 26308-23 Commercial Electrical Services

Module 26309-23 Motor Calculations

Module 26310-23 Voice, Data, and Video

Module 26311-23 Motor Controls

Electrical Appendixes

MODULE 26301-23

Load Calculations – Branch and Feeder Circuits

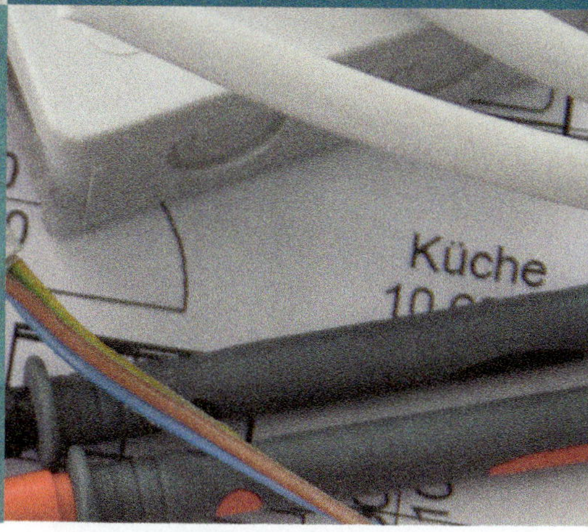

Source: iStock@Mark Hochleitner

Objectives

Successful completion of this module prepares you to do the following:

1. Calculate branch circuit electrical loads.
   a. Calculate branch circuit electrical ratings.
   b. Apply derating factors to the ampacity of conductors.
   c. Calculate branch circuit ampacity.
2. Identify residential branch circuit load requirements.
   a. Calculate lighting loads.
   b. Calculate receptacle loads.
   c. Calculate small appliance loads.
   d. Calculate laundry circuit loads.
   e. Calculate cooking appliance loads.
   f. Calculate water heater loads.
   g. Calculate electric heating loads.
   h. Calculate air conditioning loads.
3. Calculate commercial electrical loads.
   a. Calculate the loads on multioutlet assemblies.
   b. Calculate show window loads.
   c. Calculate sign loads.
   d. Calculate loads for heavy-duty lamp holder outlets.
   e. Calculate commercial kitchen equipment loads.
   f. Calculate motor loads.
   g. Calculate welding machine loads.

Performance Tasks

This is a knowledge-based module. There are no Performance Tasks.

Overview

Calculations to determine the electrical load for branch and feeder circuits are a necessary part of every electrical system design process. The information is used to determine the size of the conductors and branch circuit overcurrent protective devices needed in accordance with *NEC*® requirements. Properly sized conductors and protective devices ensure a safe and reliable electrical system. This module explains how branch and feeder circuit load calculations typical of residential and commercial applications are done, and the opportunity for extensive practice.

NOTE

Digital Resources for Electrical

Scan this code using the camera on your phone or mobile device to view the digital resources related to this craft.

NCCER Industry-Recognized Credentials

If you are training through an NCCER-accredited sponsor, you may be eligible for credentials from NCCER. The ID number for this module is 26301-23. Note that this module may have been used in other NCCER curricula and may apply to other level completions. Contact NCCER at 1.888.622.3720 or go to **www.nccer.org** for more information.

You can also show off your industry-recognized credentials online with NCCER's digital badges. Transform your knowledge, skills, and achievements into badges that you can share across social media platforms, send to your network, and add to your resume. For more information, visit **www.nccer.org**.

1.0.0 Branch Circuit Loads

Performance Tasks

There are no Performance Tasks in this section.

Objective

Calculate branch circuit electrical loads.

a. Calculate branch circuit electrical ratings.
b. Apply derating factors to the ampacity of conductors.
c. Calculate branch circuit ampacity.

Branch circuit: The circuit conductors connecting the final overcurrent protective device and the outlets or other loads it protects.

Overcurrent: Current that exceeds the rating of a conductor or a connected load, resulting from a ground fault, short circuit, or an overload condition.

Utilization equipment: Equipment that consumes electrical energy for a wide variety of purposes.

Device: Components other than conductors with the primary purpose of carrying or controlling electrical power, but that do not consume significant electrical energy.

Individual branch circuits: Branch circuits that power only one electrical load, such as a clothes dryer or motor.

Ampacity: The maximum current that a conductor can carry continuously without warming beyond its temperature rating.

Outlet: A point in a branch circuit to which a load device can be connected.

Receptacles: Points of connection installed at outlets that allow for the quick connection of an electrical load device with a compatible configuration. A *duplex receptacle* contains two such receptacles.

The purpose of **branch circuit** load calculations is to determine the correct size of branch circuit **overcurrent** protection and conductors based on the *NEC®* requirements. Branch circuits supply **utilization equipment**. Although it sounds intimidating, the term is just a formal name for equipment that is electrically powered.

NEC Article 210 covers branch circuits other than those that supply motors alone. *NEC Section 210.3* provides a listing of other code articles for specific-purpose branch circuits.

Per *NEC Section 210.18*, branch circuits are rated by the maximum capacity of the overcurrent protective **device** or by the highest setting if it is adjustable. Except for **individual branch circuits**, branch circuits are rated at 15A, 20A, 30A, 40A, and 50A. Again, this is based on the value of the overcurrent protective device. Circuits powering individual branch circuits have no **ampacity** restrictions.

Per *NEC Section 210.19(A)(2)*, branch circuit conductors must have an ampacity rating no less than the maximum load they serve, although there are some valid exceptions. Branch circuit overcurrent protection is required to be rated (or set) at a value less than or equal to the conductor rating specified in *NEC Section 240.4*. This is also specified in *NEC Section 240.3* for equipment and in *NEC Section 210.21* for **outlet** devices such as lamp holders and **receptacles**. *NEC Article 430* applies to branch circuits used for motor loads only, while *NEC Article 440* addresses branch circuits serving air conditioning and/or refrigerating equipment.

Think About It

Load Calculations

Which factors need to be taken into consideration when adding a large load to an existing service?

Branch circuit conductors must be selected for an ampacity equal to or greater than 125% of the **continuous load** plus any noncontinuous loads on the circuit. This is before the application of any adjustment or correction factors per *NEC Section 210.19(A) and (B)*, which can change the ampacity rating.

NEC Sections 210.23(A) through (E) define the permissible loads for multiple-outlet branch circuits. This is important information because it lists the types of loads that may be served based on the rating of the branch circuit. *NEC Section 210.24*, along with *NEC Tables 210.24(1) and 210.24(2)*, summarize the branch circuit requirements.

NEC Article 220 provides the requirements for determining the number of branch circuits needed and those related to computing the branch circuit, feeder, and service loads. Note that branch circuits may supply noncontinuous loads, continuous loads, or a combination of the two. *NEC Section 210.20(A)* indicates that the rating of a branch circuit overcurrent protective device shall not be less than 125% of the continuous load plus any noncontinuous loads. *NEC Table 220.42(A)* provides general lighting loads listed for various nonresidential buildings. The general lighting loads are expressed in volt-amperes per square foot (VA/ft^2) or in VA per square meter (VA/m^2).

For example, the minimum design lighting load for a school is 3VA/ft^2, a retail location is 1.9VA/ft^2, and a warehouse is 1.2VA/ft^2. *NEC Sections 220.14(A) through 220.14(K)* list minimum design loads for outlets used in all types of buildings. In this case, the outlets include general-use receptacles and other outlets not used for lighting.

NOTE

To find 125% of a load, convert 125% into a multiplier by dividing the percentage by 100. So, 125% divided by 100 becomes 1.25. Then, simply multiply the load by 1.25.

For example, 125% of a 14A load is 17.5A, since 14A × 1.25 = 17.5A.

Continuous load: A load where the maximum anticipated current is expected to continue for at least three hours.

NOTE

Local codes may require different values than the minimum *NEC*® values. For instance, local codes may limit the number of outlets on a branch circuit to less than the calculated value, or they may require dedicated circuits other than those that are listed in the *NEC*®. Always check local codes before beginning any installation.

1.1.0 Branch Circuit Ratings

The maximum load that a single-phase branch circuit can support is determined by multiplying the rating of the overcurrent protective device (circuit breaker or fuse) by the circuit voltage. For example, the maximum load that can be supported by a 20A, two-wire, 120V circuit is calculated as follows:

$$20A \times 120V = 2{,}400VA$$

The maximum load that can be supplied by a 20A, three-wire, 120/240V circuit is determined by multiplying by the higher voltage:

$$20A \times 240V = 4{,}800VA$$

The maximum load that can be supplied by a 20A, 208V three-phase circuit is determined by multiplying 20A by 208V, then multiplying the result by the square root of 3. The square root of 3 is approximately 1.732; you may wish to commit this to memory for use in future calculations. The maximum load would be calculated as follows:

$$20A \times 208V \times 1.732 = 7{,}205VA$$

(answer has been rounded down to the nearest whole number)

Branch circuits may supply noncontinuous loads, continuous loads, or a combination of the two. Per *NEC Section 210.20(A)*, the rating of a branch circuit overcurrent protective device cannot be less than 125% of the continuous load plus any noncontinuous loads.

However, the exceptions to *NEC Section 210.19(A)(1) and 210.20(A)* allow the ampacity of the branch circuit conductors to be equal to or greater than the sum of the continuous and noncontinuous loads if the overcurrent protective device is listed for operation at 100% of its rating. This allows a lower rating for the overcurrent protective device.

For 15A and 20A branch circuits, *NEC Section 210.23(B)(2)* allows fastened-in-place utilization equipment (loads) to be connected in the same circuit with lighting units and cord- connected loads not fastened in place. In this situation, the fastened-in-place loads cannot exceed 50% of the branch circuit current rating.

Know the Code

Overcurrent Protection
NEC Section 210.20

Know the Code

Conductors — Minimum Ampacity and Size
NEC Section 210.19

Know the Code

Permissible Loads, Multiple-Outlet Branch Circuits
NEC Section 210.23

Let's look at a few examples of branch circuit load calculations.

Example 1:

A store has fluorescent lighting fixtures consisting of nine fluorescent ballasts rated at 1.5A at 120V. The fixtures will operate continuously during normal business hours from 9:00 a.m. until 9:00 p.m. daily. What is the minimum circuit breaker rating required for the branch circuit that will serve this load?

Solution:

Determine the branch circuit load:

$$9 \times 1.5A = 13.5A$$

Determine the continuous duty load. This is considered a continuous duty load because the lighting fixtures will stay on for more than three hours at a time. Thus, we need a value that is 125% greater than the load:

$$13.5A \times 1.25 = 16.88A$$

The minimum size circuit breaker required is 20A since this is the next highest value greater than the 16.88A result.

Example 2:

What is the maximum continuous load that may be connected to a 30A, 120V fuse?

Solution:

Per *NEC Section 210.20(A)*, the overcurrent protective device must be 125% greater. To determine the maximum load, we can divide by 1.25 instead of multiplying:

$$30A \div 1.25 = 24A$$

The maximum continuous load cannot exceed 24A.

Example 3:

Receptacle outlet: A power outlet location where one or more receptacles are connected.

How many **receptacle outlet** devices in a nonresidential building can be connected to a 20A, two-wire, 120V circuit? Assume that the receptacles serve noncontinuous loads.

Solution:

Determine branch circuit capacity:

$$20A \times 120V = 2{,}400VA$$

Know the Code

Other Loads — All Occupancies
NEC Section 220.14

Per *NEC Section 220.14(I)*, each outlet is assigned a load of 180VA, so you can divide available load by the value of each individual outlet load:

$$2{,}400VA \div 180VA = 13.33 \text{ receptacle outlets}$$

You must round down to avoid exceeding the maximum number. Thus, 13 receptacle outlets can be connected to this circuit.

Example 4:

An office manager has purchased a new state-of-the-art copy machine. The nameplate rating on the copy machine is 17A at 120V. What is the minimum branch circuit rating required to serve this equipment?

Solution:

This is not a continuous load. However, per *NEC Sections 210.23(B)(1) and (C)*, the rating of any one cord-connected load shall not exceed 80% of the branch circuit current rating for 15A, 20A, and 30A circuits. Because the load is greater than 15A, determine the current-carrying capacity of a 20A circuit (the smallest logical size), to see if it can accommodate the circuit:

$$20A \div 1.25 = 16A$$

A 20A circuit is not sufficient either. The next step up is a 30A circuit:

$$30A \div 1.25 = 24A$$

Since the 30A circuit can manage the 17A load of the copier, the minimum branch circuit rating must be 30A. Note that this solution assumes that only multi-receptacle 20A and 30A branch circuits exist in the office. An alternative solution would be to install a 20A individual branch circuit exclusively for the copy machine.

***Example 5*:**

What is the maximum lighting load that may be connected to a 20A branch circuit supplying a piece of fixed equipment that has a rating of 8.5A at 120V?

Solution:

The equipment rating is less than 50% of the rating of the 20A branch circuit. Therefore, 11.5A of noncontinuous lighting may be added, as shown here:

$$20\text{A} - 8.5\text{A} = 11.5\text{A}$$

***Example 6*:**

A commercial dishwasher has a nameplate rating of 14.7A, 208V, 3ϕ. During busy times in the restaurant, it is anticipated that the dishwasher will be turned on and operated for more than three hours at a time, making it a continuous load. What is the minimum branch circuit rating required to power the dishwasher?

Solution:

Since this equipment is considered a continuous load, the load must be multiplied by 125% to determine the branch circuit size:

$$14.7\text{A} \times 1.25 = 18.38\text{A}$$

The minimum branch circuit rating required is 20A since this is the next highest value greater than the 18.38A result.

1.2.0 Derating Factors

The ampacity of branch circuit conductors must be derated when any of the following circumstances apply:

- The ambient temperature that the conductors will pass through exceeds the temperature rating for conductors listed in *NEC Table 310.15(B)(1)* or *NEC Table 310.15(B)(2)*.
- A voltage drop exists that exceeds the recommended 3% for branch circuits or 5% for the combination of the feeder and branch circuits, per the Informational Note in *NEC Section 210.19*. Refer to Informational Note No. 2 in *NEC Section 215.2(A)(2)* for additional information about voltage drop on feeder conductors.
- An installation where the number of conductors in a raceway or cable exceeds three; or where single conductors or multiconductor cables are installed without any separation for a distance of 24" (600 mm) or more, unless they are installed in raceways. In these cases, derating is required per *NEC Section 310.15(C)(1)*.

NEC Table 310.16 lists the acceptable ampacities for insulated conductors rated up to 2,000V. The listed ampacities apply when no more than three conductors are installed in a raceway, they are part of a cable assembly, or they are directly buried. The listed ampacities are based upon the ambient temperature being no more than 86°F (30°C).

If the number of current-carrying conductors in the conduit or cable does exceed three, the ampacity of the conductors must be derated using *NEC Section 310.15(C)(1)*. If the ambient temperature exceeds 86°F (30°C), the ampacity of the conductors must be derated using the temperature correction factors listed at the bottom of *NEC Section 310.15(B)(1)(1)*. Note that the multipliers listed in the table are called *correction factors*. You may also see them referred to as *adjustment factors* in a few places within the code.

Know the Code

Minimum Rating and Size
NEC Section 215.2

Know the Code

Ampacities of Insulated Conductors with Not More Than Three Current-Carrying Conductors in Raceway, Cable, or Earth (Directly Buried)
NEC Table 310.16

Where raceways are exposed to sunlight on or above rooftops, the conductors must be derated by applying the factors in *NEC Section 310.15(B)(2)* to the ambient temperature, in addition to the application of other correction factors.

Know the Code

Electrical Connections
NEC Section 110.14

Know the Code

Ampacities of Insulated Conductors in Raceway, Cable, or Earth (Directly Buried)
NEC Section 310.16

1.2.1 Temperature Derating

NEC Section 110.14(C) states that the lowest temperature rating of any component in a branch circuit must be used to determine the ampacity rating of the branch circuit conductors. The lowest temperature rating could be found on a circuit breaker, a fuse, a receptacle, conductors, or any other part of the circuit. When a component has a lower rating than the others, conductors with a higher temperature rating can be used to compensate, per *NEC Section 310.16.*

For example, if the termination rating of a circuit breaker is 140°F (60°C), the ampacity rating of the conductors cannot exceed the value given in the 140°F (60°C) column of *NEC Table 310.16* for overcurrent protection. However, if the branch circuit conductor has a higher temperature rating, such as THHN wire rated at 194°F (90°C), the higher ampacity rating can be used for ampacity adjustment, correction, or both.

Example 1:

It is determined that, by combining branch circuits on a project, two costly runs of conduit can be eliminated. The branch circuits are to be pulled in a single conduit. All the conductors in the conduit will be current-carrying conductors of the same size and insulation type. What is the derated ampacity for each of eight No. 12 THHN copper conductors when pulled in a single conduit?

Solution:

Per *NEC Table 310.16*, the ampacity of No. 12 THHN is 30A. Per *NEC Section 310.15(C)(1)*, the ampacity shown in *NEC Table 310.16* must be derated to 70% of its value because there are eight current-carrying conductors.

$$30A \times 0.7 = 21A$$

Thus, the new ampacity after derating is 21A.

Example 2:

What is the maximum load that may be connected on each of six No. 2 THWN copper conductors in a single conduit?

Solution:

Per *NEC Table 310.16*, the ampacity of No. 2 THWN is 115A. Per *NEC Table 310.15(C)(1)*, the ampacity shown in *NEC Table 310.16* must be derated to 80% of its value because there are six current-carrying conductors.

$$115A \times 0.8 = 92A$$

The maximum load after ampacity derating is 92A.

Example 3:

A branch circuit is required to supply a noncontinuous load with a nameplate rating of 55A. The equipment is in a plant, located in a room with plastic extrusion equipment that gets very warm. The ambient temperature in the room is 92°F (33.3°C). What is the minimum size of THHN copper conductors allowed to supply the new load?

Solution:

Per *NEC Table 310.16*, No. 8 THHN copper conductors have an ampacity rating of 55A (coincidently, the same as the load of the new equipment). Per *NEC Table 310.15(B)(1)(1)*, the ampacity of the conductors must be multiplied by a correction factor of 0.96.

$$55A \times 0.96 = 52.8A$$

The available ampacity is 52.8A. Thus, the ampacity of No. 8 THHN copper conductors operating in an area that is 92°F (33.3°C) does not meet the requirements for a load of 55A. The next size of wire, No. 6 THHN, has a rating of 75A. But before selecting it, the 0.96 correction factor must also be applied to it:

$$75A \times 0.96 = 72A$$

While the No. 8 THHN conductors may have looked just right at first, they aren't acceptable. No. 6 THHN conductors will be required to power the new load, due to the heat in the area.

Example 4:

A branch circuit will power 16A of continuous loads. The conductors to be used are copper THHN. They will be installed in a raceway with a total of 12 current-carrying conductors and routed through an area with an ambient temperature of 105°F (40.5°C). What is the minimum wire size and overcurrent protective device required for this circuit?

Solution:

Per *NEC Section 210.19(A)(1)* for conductors and *NEC Section 210.20(A)* for overcurrent protection, the minimum current rating for the conductor and the overcurrent protective device is calculated by multiplying the load by 125%:

$$16A \times 1.25 = 20A$$

The conductor and overcurrent protection rating required is 20A.

Per *NEC Table 310.15(C)(1)*, the adjustment factor for 12 conductors is 50%. Essentially, this means that the value above must be doubled:

$$20A \div 0.50 = 40A$$

Because the ambient temperature is 105°F (40.5°C), the conductor ampacity must be further derated per *NEC Section 310.15(B)(1)(1)*:

$$40A \div 0.87 = 45.98A$$

Thus, the conductor ampacity required is 45.98A.

Refer to *NEC Table 310.16* and find THHN under the 90°C (194°F) column. No. 8 THHN provides an ampacity of 55A, which is greater than the 45.98A required here. Therefore, the overcurrent protective device would be sized at 20A as noted in the first calculation, and the correct wire size would be No. 8 THHN.

Know the Code

Conductors — Minimum Ampacity and Size

NEC Section 210.19

Know the Code

Overcurrent Protection

NEC Section 210.20

1.2.2 Voltage Drop Derating for Single-Phase Circuits

The informational note in *NEC Section 210.19* states that reasonable efficiency is provided by branch circuits where its conductors do not experience more than a 3% voltage drop at the farthest outlet, or more than a 5% voltage drop for the combination of feeder and branch conductors at the farthest outlet. There are two formulas used to calculate voltage drop for single-phase circuits.

Formula 1:

$$VD = \frac{2 \times L \times R \times I}{1{,}000'}$$

Where:

VD = voltage drop

L = load center or total length in feet

R = conductor resistance per 1,000' (from ***NEC Chapter 9, Table 8****, Conductor Properties*)

I = current

Know the Code

Conductor Properties

NEC Chapter 9, Table 8

The numerator (upper part) of the fraction is divided by 1,000' to determine the total resistance of the conductor. For example, if a conductor has 10Ω of resistance per 1,000', but the conductor is only 100' long, its resistance would be 1Ω.

The second formula is slightly different.

Formula 2:

$$VD = \frac{2 \times L \times K \times I}{\text{cmil}}$$

Where:

VD = voltage drop

L = load center or total length in feet

K = constant of 12.9 for copper conductors or 21.2 for aluminum conductors

I = current

cmil = conductor cross-sectional area in circular mils (from ***NEC Chapter 9, Table 8**, Conductor Properties*)

The results of these formulas represent volts. The result is divided by the circuit voltage, which is then multiplied by 100 to determine the voltage drop as a percentage. If the calculated branch circuit voltage drop exceeds 3%, the conductor size should be increased to compensate. For a 120V circuit, the maximum voltage drop at 3% is 3.6V (120V × 0.03 = 3.6V). For a 240V single-phase circuit, the maximum voltage drop at 3% is 7.2V (240V × 0.03 = 7.2V).

In commercial or industrial fixed load applications, if the branch circuit load is not concentrated at the end of the branch circuit but is spread out along the circuit because of multiple outlets, the *load center length* of the circuit should be calculated and used in the two formulas provided. The load center length is the point in the circuit where, if the load were concentrated at that point, the voltage drop would be the same as the voltage drop to the farthest load in the actual circuit. This is because the total current does not flow the complete length of the circuit. If the full length were used in computing the voltage drop instead of the load center length, the result would be greater than the actual voltage drop.

To determine the load center length of a branch circuit with multiple outlets, multiply each outlet load by the actual routing distance from the supply end of the circuit (*Figure 1*). Add these products for all loads fed from the circuit and divide this sum by the sum of the individual loads. The resulting distance is the load center length (L) for the total load (I) of the branch circuit.

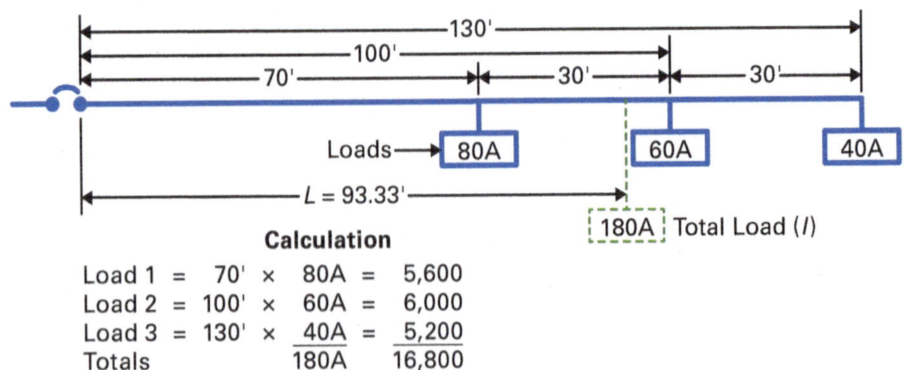

Figure 1 Calculating a load center length and total load for multiple fixed loads on a circuit.

***Example 1*:**

The length of a 120V, two-wire branch circuit is 95'. The noncontinuous load is 14.5A. If No. 12 THHN solid copper conductors are used, will the voltage drop for this branch circuit exceed 3%?

Solution:

Use the first voltage drop formula to determine voltage drop for this branch circuit. Look up the resistance for No. 12 solid copper in *NEC Chapter 9, Table 8, Conductor Properties.* You will find it to be 1.93Ω per 1,000'.

$$VD = \frac{2 \times L \times R \times I}{1{,}000'}$$

$$VD = \frac{2 \times 95' \times 1.93\Omega \times 14.5\text{A}}{1{,}000'}$$

$$VD = \frac{5{,}317.15}{1{,}000'}$$

$$VD = 5.31715\text{V, rounded to } 5.32\text{V}$$

Now divide by 120V to find the voltage drop as a percentage:

$$5.32\text{V} \div 120\text{V} = 0.0443\text{V}$$

$$0.0443 \times 100 = 4.43\%$$

Since 4.43% is greater than the 3% allowable voltage drop, the answer to the question is yes—the voltage drop is excessive. Larger conductors are required for this circuit to ensure the voltage remains consistent under load. Note that this solution uses the first formula for single-phase voltage drop calculations.

Example 2:

Which size THHN solid copper branch circuit conductors would be recommended for a single-phase, noncontinuous branch circuit load of 23A at 240V? The length of the circuit is 130'.

Solution:

Use the second voltage drop formula to determine the voltage drop for this branch circuit. Because No. 10 THHN copper would be the smallest size permitted for a branch circuit load of 23A, this size will be evaluated first. Look up the area, in circular mils, for a No. 10 solid copper conductor in *NEC Chapter 9, Table 8, Conductor Properties.* That value is 10,380 cmil. Once you have found the voltage drop using the formula, divide it by the total branch circuit voltage to find the percentage, as follows:

$$VD = \frac{2 \times L \times K \times I}{\text{cmil}}$$

$$VD = \frac{2 \times 130' \times 12.9\Omega \times 23\text{A}}{10{,}380\text{ cmil}}$$

$$VD = \frac{77{,}142}{10{,}380}$$

$$VD = 7.4317\text{V, rounded to } 7.43\text{V}$$

Now determine the voltage drop as a percentage:

$$7.43\text{V} \div 240\text{V} = 0.031$$

$$0.031 \times 100 = 3.10\%$$

Think About It

Voltage Drop Formula Factors

What do you think the number 2 in voltage drop Formula 1 represents?

Know the Code

Conductor Properties
NEC Chapter 9, Table 8

The result exceeds 3%, although it's close. Therefore, the No. 10 THHN conductors would not meet the *NEC*® requirements and the next larger size would be required. The next size up, No. 8 THHN, would be required.

Another way to find the minimum conductor size for the situation described in this example is to use the following formula to find the conductor size in circular mils:

$$\text{cmil} = \frac{2 \times L \times K \times I}{E \times VD}$$

In this formula, *E* represents the voltage of the circuit. Note that *VD* represents the allowable voltage drop as a decimal. Using this equation with the values from *Example 2* yields the following result:

$$\text{cmil} = \frac{2 \times 130' \times 12.9\Omega \times 23\text{A}}{240\text{V} \times 0.03}$$

$$\text{cmil} = \frac{77{,}142}{7.2}$$

$$\text{cmil} = 10{,}714.1667\text{, rounded to } 10{,}714$$

A conductor with this cross-sectional area is the minimum acceptable size. Using *NEC Chapter 9, Table 8, Conductor Properties*, the area of No. 10 copper is 10,380 cmil. Because that is less than 10,714 cmil, a No. 8 conductor with a much larger area of 16,510 cmil must be used. The conductor can always be larger, but never smaller than the calculated minimum size.

Example 3:

A 240V, single-phase, three-wire branch circuit has multiple power outlets consisting of three fixed 240V loads. The three loads are 30A at a length of 60', 30A at 80', and 20A at 100'. What size THHN copper conductors would be required?

Solution:

Determine the load center length of the circuit by first multiplying the outlet loads by their distance from the circuit source. Then, find the sum of the three results, and divide it by the sum of the three loads. Refer to *Figure 1* if needed:

$$\text{Outlet 1} = 60' \times 30\text{A} = 1{,}800$$

$$\text{Outlet 2} = 80' \times 30\text{A} = 2{,}400$$

$$\text{Outlet 3} = 100' \times 20\text{A} = 2{,}000$$

Now total the products above:

$$1{,}800 + 2{,}400 + 2{,}000 = 6{,}200$$

Divide this figure by the sum of the loads:

$$\text{Sum of loads} = 30\text{A} + 30\text{A} + 20\text{A} = 80\text{A}$$

$$6{,}200 \div 80\text{A} = 77.5'$$

The load center length is 77.5'. Thus, for a load of 80A, No. 4 copper THWN/THHN conductors at 167°F (75°C) will be selected, per *NEC Table 310.16*. The area of the conductor is 41,740 cmil. Use the second formula and substitute values to check the voltage drop. In this example, all terminals are rated for 167°F (75°C).

$$VD = \frac{2 \times L \times K \times I}{\text{cmil}}$$

$$VD = \frac{2 \times 77.5' \times 12.9\Omega \times 80A}{41{,}740 \text{ cmil}}$$

$$VD = \frac{159{,}960}{41{,}740 \text{ cmil}}$$

$$VD = 3.8322V\text{, rounded to } 3.83V$$

Because the permissible voltage drop is 240V × 3%, or 7.2V, the voltage drop using the No. 4 THWN/THHN conductors is well below the maximum.

1.2.3 Three-Phase Correction Factors for Voltage Drop

The voltage drop for balanced three-phase circuits can be calculated using the two formulas above by substituting the square root of 3, or 1.732, for the value of 2 in the formulas:

Formula 1:

$$VD = \frac{\sqrt{3} \times L \times R \times I}{1{,}000'}$$

Formula 2:

$$VD = \frac{\sqrt{3} \times L \times R \times I}{\text{cmil}}$$

Example 1:

What is the voltage drop in a 208V three-phase branch circuit with a load of 32A? It is 115' from the circuit breaker to the load, and No. 8 stranded copper conductors are in place.

Solution:

Use the first three-phase voltage drop formula to determine voltage drop for this three-phase circuit. Look up the resistance for No. 8 stranded copper conductor in *NEC Chapter 9, Table 8*, *Conductor Properties*. It is 0.778Ω per 1,000'.

Rather than using the calculator to determine the square root of 3 each time, we will use the value 1.732 in its place:

$$VD = \frac{1.732 \times L \times R \times I}{1{,}000'}$$

$$VD = \frac{1.732 \times 115' \times 0.778\Omega \times 32A}{1{,}000'}$$

$$VD = \frac{4{,}958.758}{1{,}000'}$$

$$VD = 4.9587V\text{, rounded to } 4.96V$$

The maximum voltage drop permitted for a 208V circuit is:

$$208V \times 0.03 = 6.24V$$

Since the 4.96V voltage drop is less than the maximum allowed, the conductor selection is acceptable.

Know the Code

Ampacities of Insulated Conductors with Not More Than Three Current-Carrying Conductors in Raceway, Cable, or Earth (Directly Buried) *NEC Table 310.16*

Example 2:

Which size THHN copper conductors would be recommended for a 208V, three-phase, four-wire feeder with a length of 150' from the source to a fixed continuous load of 160A?

Solution:

Since it is a continuous load, determine the minimum ampacity required by multiplying the 160A by 125% to obtain 200A. Using ***NEC Table 310.16***, you'll see that 3/0 THHN copper conductors at 167°F (75°C) are satisfactory for the load of 200A.

To determine whether the 3/0 conductors are adequate to prevent unacceptable voltage drop, use the second three-phase voltage drop formula. Note that the current used will be 160A and not the calculated value of 200A, which is used only to size branch circuit or feeder conductor sizes as well as the minimum overcurrent protective device rating.

$$VD = \frac{1.732 \times L \times K \times I}{\text{cmil}}$$

$$VD = \frac{1.732 \times 150' \times 12.9\Omega \times 160\text{A}}{167{,}800 \text{ cmil}}$$

$$VD = \frac{536{,}227.2}{167{,}800 \text{ cmil}}$$

$$VD = 3.1956\text{V, rounded to } 3.2\text{V}$$

Because the maximum recommended voltage drop is 208V × 3%, or 6.24V, the selected conductors are satisfactory.

1.3.0 Branch Circuit Ampacity

The branch circuit ampacity for single-phase circuits is calculated by dividing the volt-amps by the circuit voltage. For example, the ampacity of a 120V, single-phase load rated at 1,600VA is determined by dividing 1,600VA by 120V:

$$1{,}600\text{VA} \div 120\text{V} = 13.33\text{A}$$

Another example would be determining the ampacity of a single-phase 3,450VA load at 277V:

$$3{,}450\text{VA} \div 277\text{V} = 12.45\text{A}$$

The branch circuit ampacity for three-phase circuits is calculated by dividing the volt-amps by the circuit voltage, then multiplying by the square root of 3. As noted earlier, the square root of 3 is 1.732. For example, the ampacity (*I*) of a 208V three-phase load rated at 5,200VA is determined as follows:

$$I = \frac{\text{volt-amps}}{E \times \sqrt{3}}$$

$$I = \frac{5{,}200\text{VA}}{208\text{V} \times 1.732}$$

$$I = \frac{5{,}200\text{VA}}{360.256\text{V}}$$

$$I = 14.43\text{A}$$

Using the same equation, the ampacity of a 14,000VA three-phase load at 480V is determined as follows:

$$I = \frac{14{,}000\text{VA}}{480\text{V} \times 1.732}$$

$$I = \frac{14{,}000\text{VA}}{831.36\text{V}}$$

$$I = 16.84\text{A}$$

Example 1:

What is the ampacity of a single-phase load with a nameplate rating of 5.5 kW at 240V?

Solution:

Multiply 5.5 kW by 1,000 to determine the VA; remember that watts and volt-amps are basically the same value. Then divide the result by 240V as shown here:

$$5.5\text{ kW} \times 1{,}000 = 5{,}500\text{VA}$$

$$5{,}500\text{VA} \div 240\text{V} = 22.92\text{A}$$

The ampacity of this load is 22.92A.

Example 2:

What is the ampacity of a three-phase electric water heater with a nameplate rating of 20 kW at 208V?

Solution:

Multiply 20 kW by 1,000 to find the VA, then divide the result by the voltage multiplied by the square root of 3:

$$20\text{ kW} \times 1{,}000 = 20{,}000\text{VA}$$

$$20{,}000\text{VA} \div (208\text{V} \times 1.732) = 55.52\text{A}$$

The ampacity of this load is 55.52A.

1.0.0 Section Review

1. What is the *maximum* continuous load that may be connected to a 20A, 120V fuse?
   a. 10A
   b. 16A
   c. 20A
   d. 25A

2. The voltage drop for a single-phase 208V circuit with a load of 28A, served by No. 8 stranded copper conductors with a circuit length of 160', is _____.
   a. 5.87V
   b. 5.98V
   c. 6.78V
   d. 6.97V

3. The ampacity of a single-phase 120V load rated at 800VA is _____.
   a. 2.87A
   b. 4.38A
   c. 5.78A
   d. 6.67A

2.0.0 Residential Branch Circuit Requirements

Performance Tasks

There are no Performance Tasks in this section.

Objective

Identify residential branch circuit load requirements.

a. Calculate lighting loads.
b. Calculate receptacle loads.
c. Calculate small appliance loads.
d. Calculate laundry circuit loads.
e. Calculate cooking appliance loads.
f. Calculate water heater loads.
g. Calculate electric heating loads.
h. Calculate air conditioning loads.

Many branch circuits are often required for dwellings. The *NEC*® lists the requirements for these circuits, including the requirements for calculating branch circuit loads for specific residential loads. These loads, which are somewhat unique to residential construction, are discussed individually in this section.

2.1.0 Lighting Loads

Lighting load branch circuit calculations are based upon the type of lighting, the voltage, and whether the lighting is considered a continuous duty load. Like other loads, this means the lights would be on for three or more hours at a time.

Branch circuit loads for incandescent lighting are determined by calculating the total load and dividing the total by the circuit voltage. Incandescent lighting is a purely resistive load, so watts and VA are the same. For example, three 500W quartz lamps connected on a 120V circuit would be calculated by first adding the loads. Since, in this example, there are three equal loads of 500VA, multiply 500VA by 3:

$$3 \times 500\text{VA} = 1{,}500\text{VA}$$

Divide the result by the circuit voltage:

$$1{,}500\text{VA} \div 120\text{V} = 12.5\text{A}$$

Know the Code

Maximum Load
NEC Section 220.11

Branch circuit loads for electric discharge and LED lighting are determined by multiplying the number of fixtures times the ampacity rating of the ballast, assuming that all the ballasts are identical. Per *NEC Section 220.11(B)*, the calculated load shall be based upon the total ampere rating of the lighting units and not the total watts of the lamps. For example, if fifteen 150W, high-pressure sodium fixtures connected on a 277V circuit each have a ballast current of 0.79A, the total load would be 11.85A, as shown in the following equation:

$$15 \times 0.79\text{A} = 11.85\text{A}$$

Example 1:

Incandescent lighting utilizing 150W medium base lamps is required for temporary lighting on a construction site. This lighting will remain on all day and all night. How many 150W lamps can be connected on an available 20A, 120V circuit?

Solution:

Because the lighting will remain on for longer than three hours, this is a continuous load. For continuous duty, find 125% of 150W (shown in the equation as VA) by multiplying by 1.25:

$$1.25 \times 150\text{VA} = 187.5\text{VA}$$

Now, determine the total capacity, in VA, for the 20A circuit:

$$20\text{A} \times 120\text{V} = 2{,}400\text{VA}$$

Divide the circuit ampacity by the lamp load:

$$2{,}400\text{VA} \div 187.5\text{VA} = 12.8 \text{ lamps}$$

Twelve 150W lamps may be safely connected to a 20A, 120V circuit.

***Example 2*:**

Two 400W metal halide high-bay fixtures are required to provide additional lighting for a production machine. There is an existing 20A, 277V lighting circuit available near the equipment. After investigation, it is determined that the circuit has six of the same type of fixtures connected. The nameplate current rating for these fixtures is 2A each at 277V. During normal operation, the fixtures are on for 12 hours every day. Can two more 400W fixtures, bringing the total to eight fixtures, be safely added to the existing circuit?

Solution:

This is a continuous lighting load. Per *NEC*® requirements, find 125% of the of the circuit amperage rating to determine the maximum continuous load that may be connected:

$$20A \times 1.25 = 25A$$

Now, determine the load for all eight fixtures. Since all of these fixtures are on at the same time (and have an amperage rating of 2A at 277V), multiply their amperage by the number of fixtures to find the load:

$$8 \text{ fixtures} \times 2A = 16A$$

Since 25A is the maximum continuous load that may be connected and the total load for all eight fixtures is 16A, the two additional fixtures may be added to this circuit.

2.1.1 Recessed Lighting Loads

The load for recessed lighting fixtures, excluding the residential general lighting load of $3VA/ft^2$, is calculated per *NEC Section 220.41*. The load calculation can also be done using the maximum VA rating of the fixture and lamps. For example, a load of 250VA would be used to calculate the load for a recessed incandescent fixture rated at 250W. To calculate the load for a recessed fixture that uses compact fluorescent lamps, you need to know the voltage and ampacity rating of the fluorescent ballast.

Know the Code

Dwelling Units, Minimum Unit Load
NEC Section 220.41

***Example 1*:**

What is the needed ampacity for a circuit with seven incandescent recessed fixtures with a nameplate maximum lamp size of 150W that are to be connected to a 120V circuit?

Solution:

Multiply the number of fixtures by the VA rating of each fixture to obtain the total VA load:

$$7 \times 150VA = 1{,}050VA$$

To determine the ampacity, divide the total VA rating by the circuit voltage:

$$1{,}050VA \div 120V = 8.75A$$

***Example 2*:**

What is the total ampacity needed for seven fluorescent recessed fixtures operating continuously? Each fixture has a ballast rating of 0.20A at 277V, and each fixture takes two compact fluorescent lamps rated at 26W each.

Solution:

Because these fixtures are of the electric discharge type, multiply the current load of one ballast by the number of ballasts. Then, because they are a continuous load, find 125% of the result to calculate the total ampacity required:

$$7 \times 0.20A = 1.4A$$

$$1.4A \times 1.25 = 1.75A$$

2.2.0 Receptacle Loads

When the exact VA rating of a load that is cord- and plug-connected to a receptacle outlet is not known, a VA rating of not less than 180VA per outlet is used, per *NEC Section 220.14(I)*. Note that this *NEC*® reference does not apply to residential receptacles. Residential receptacles are included in the general illumination load (**general-purpose branch circuits**) or in specific residential loads such as receptacles required for small **appliance** and laundry loads.

For example, assume a 15A, 120V circuit is to be added to supply general-purpose receptacles. The receptacles are to be located above workbenches that are positioned against a wall. The owner states that the workers using the benches will plug small power tools into the outlets, but the tools will be occasionally used and only for short periods.

The question, then, is how many general-purpose duplex receptacles can be connected to a 15A, 120V circuit when the outlets are to be rated for noncontinuous duty?

The capacity of a 15A, 120V circuit is 1,800VA, as shown by the following equation:

$$15A \times 120V = 1{,}800VA$$

The total circuit capacity is then divided by the rating per receptacle to determine the total number of receptacles that may be connected. Per *NEC Section 220.14(I)*, each receptacle is assigned a value of 180VA since the exact VA ratings of the loads are unknown:

$$1{,}800VA \div 180VA = 10 \text{ receptacles}$$

So, 10 receptacles can be positioned along the benches on the 15A circuit. But what if the tools were going to be operated continuously?

If the loads must operate continuously, the 180VA value assigned to each receptacle must be increased by 125%:

$$180VA \times 1.25 = 225VA$$

As in the first example, the number of receptacles that can be installed is determined by dividing the circuit capacity by the new load information:

$$1{,}800VA \div 225VA = 8 \text{ receptacles}$$

Know the Code

Other Loads — All Occupancies
NEC Section 220.14

General-purpose branch circuits: Branch circuits that supply two or more outlets or receptacles for lighting and appliances.

Appliance: Equipment typically manufactured in standardized sizes and types that are installed as a unit to perform one or more specific functions, such as clothes washers and dryers, blenders, toasters, and similar appliances.

NOTE

When calculating the number of receptacles allowed on a circuit, always round down. For example, if the number of receptacles calculated resulted in 12.7, round down to 12.

2.3.0 Small Appliance Loads

NEC Sections 210.11(C)(1) and 210.52(B)(1) require at least two 20A small **appliance branch circuits** to be installed to supply receptacles in the kitchen, pantry, breakfast room, and dining room of homes. *NEC Section 220.52(A)* requires that the feeder load be computed at 1,500VA for each of these circuits. No small appliance branch circuit may serve more than one kitchen.

For example, what is the total feeder load for two small appliance branch circuits, each of which is rated for 20A at 120V?

To determine the answer, multiply the number of small appliance branch circuits by 1,500VA to determine the total load:

$$2 \text{ circuits} \times 1{,}500VA = 3{,}000VA$$

Think About It

Adding a Kitchen

You are adding a new kitchen to the basement of a residence. How does this affect the load calculation for the home?

Appliance branch circuits: Branch circuits supplying energy to outlets to which only appliances are connected.

Know the Code

Branch Circuits Required
NEC Section 210.11

Know the Code

Dwelling Unit Receptacle Outlets
NEC Section 210.52

Know the Code

Small-Appliance and Laundry Loads — Dwelling Unit
NEC Section 220.52

2.4.0 Laundry Equipment Loads

NEC Sections 210.11(C)(2) and 210.52(F) require the installation of at least one receptacle to supply laundry equipment. *NEC Section 220.52(B)* requires that the feeder load be computed at 1,500VA for each laundry circuit.

As specified in *NEC Section 220.54*, the load for household electric dryers must be calculated at either 5,000VA or the nameplate rating of the dryer—whichever is larger. *NEC Table 220.54* lists **demand factors** for dryers.

Example 1:

What is the demand load for one household electric dryer with a nameplate rating of 5.5 kW?

Solution:

Per *NEC Table 220.54*, the demand factor for one dryer is 100%. Therefore, the load would be calculated as 5.5 kW, or 5,500VA.

Example 2:

A new apartment building is planned. It consists of 10 residential units. Each unit has an electric dryer with a nameplate rating of 5.5 kW. There will be one utility transformer and one service entrance serving the building. What is the demand load, in VA, for these 10 dryers?

Solution:

Per *NEC Table 220.54*, the demand factor for 10 dryers is 50%. The reasoning is that it is highly unlikely that all 10 dryers will operate at the same time, and the load is not continuous in a residential setting. Find the total load for the 10 dryers, then reduce it by 50% to apply the demand factor:

Total load = 10 dryers × (5.5 kW × 1,000W/kW)

Total load = 10 dryers × 5,500W

Total load = 55,000VA

55,000VA × 0.5 = 27,500VA

NOTE

NEC Table 220.54 applies only to residential dryers used in single-family or multifamily residences. The demand factors listed do not apply to commercial dryers used in commercial facilities.

Demand factors: The ratio of the maximum anticipated demands of certain electrical loads compared to the actual connected load.

Think About It

Residential Laundry Room Circuit

Does the *NEC*® specifically require a dedicated branch circuit for a washing machine?

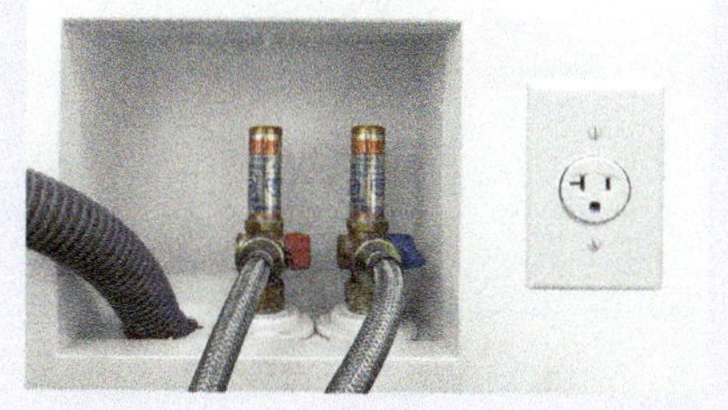

2.5.0 Cooking Appliance Loads

Loads for ranges, wall-mounted ovens, counter-mounted cooking units, and other household cooking appliances are calculated using the demand factors listed in *NEC Table 220.55* and the notes that accompany it. The most important piece of information needed to size the circuit for residential cooking equipment is the nameplate rating of the equipment. Although *NEC Table 220.55* lists loads in kW, *NEC Section 220.55* states that kVA shall be considered equivalent to kW for loads calculated using *NEC Table 220.55*.

Demand factors are calculated using *NEC Table 220.55* based upon the number of appliances, the maximum demand listed in Column C, and a demand factor percentage that is found using Columns A and B. Column C is used when the nameplate rating is over 8.75 kW but not over 12 kW. Column C is also used to calculate larger range loads per Notes 1 and 2.

Column A is used when the nameplate rating is less than 3.5 kW. Column B is used when the nameplate rating is from 3.5 kW to 8.75 kW. Note 3 provides a method to calculate the demand for multiple ranges, each of which has a nameplate rating of more than 1.75 kW but less than 8.75 kW. Note 4 provides a method to determine the size for a single branch circuit supplying one counter-mounted unit and no more than two wall-mounted ovens, provided the equipment is in the same room.

NOTE

There are seven notes associated with *NEC Table 220.55*. When using the table, it is important to review the notes, as one or more may apply to your application.

Know the Code

Electric Cooking Appliances in Dwelling Units and Household Cooking Appliances Used in Instructional Programs
NEC Section 220.55

Know the Code

Demand Factors and Loads for Household Electric Ranges, Wall-Mounted Ovens, Counter-Mounted Cooking Units, and Other Household Cooking Appliances over 1 3⁄4 kW Rating
NEC Table 220.55

Example 1:

What is the demand load for one range with a nameplate rating of 11.3 kW?

Solution:

Because the nameplate rating falls between 8.75 kW and 12 kW, Column C of *NEC Table 220.55* is used. The maximum demand shown for one range is shown as 8 kW.

Example 2:

What is the demand load, in amps, for one range with a nameplate rating of 15.7 kW at 240V?

Solution:

Because the nameplate rating exceeds 12 kW, Note 1 of *NEC Table 220.55* is applied. Subtract 12 kW from the nameplate rating of the range to determine the number of kW exceeding 12.

15.7 kW – 12 kW = 3.7 kW

Because 0.7 is a major fraction according to the *NEC*®, meaning that it is more than half, the result is rounded up to 4 kW. Per Note 1 of *NEC Table 220.55*, multiply 4 kW by 5 to determine the total percentage by which the demand must be increased:

4 kW × 5 = 20% demand increase required

The maximum demand listed in Column C for one range (8 kW) is then multiplied by 120% (or 1.2) to account for the 20% required demand increase:

8 kW × 1.2 = 9.6 kW

To determine the ampacity of this load, first convert the load to watts (VA) by multiplying by 1,000, since 1 kW is equal to 1,000W. Then, divide the result by the circuit voltage to find the demand load in amps:

9.6 kW × 1,000W/kW = 9,600VA

9,600VA ÷ 240V = 40A

Example 3:

One wall-mounted oven and one counter-mounted cooking unit are to be installed in the kitchen. To reduce the cost of the electrical installation, both units are to be connected to the same 240V circuit. The oven has a nameplate rating of 12.5 kW and the counter-mounted unit has a nameplate rating of 8 kW. What is the demand load, in amps, for a single circuit to supply this cooking equipment?

Solution:

Note 6 of *NEC Table 220.55* may be used for this calculation. First, obtain the total load by adding the nameplate ratings of the two units:

12.5 kW + 8 kW = 20.5 kW

The equivalent of one range is 20.5 kW. Using Note 1 of *NEC Table 220.55*, subtract 12 kW from 20.5 kW to determine the number of kW exceeding 12:

20.5 kW – 12 kW = 8.5 kW, rounded to 9 kW

Because 0.5 is a major fraction, the result is rounded up to 9 kW. Now multiply 9 kW by 5 to determine the total percentage by which the demand must be increased:

9 kW × 5 = 45%

This means that the demand must be increased by 45%. Find 145% of the maximum demand for one range in Column C to determine the demand load. The Column C value is 8 kW:

8 kW × 1.45 = 11.6 kW

Convert the kilowatts to VA by multiplying by 1,000W per kW:

11.6 kW × 1,000W/kW = 11,600VA

Divide the VA load by the circuit voltage to determine branch circuit ampacity:

11,600VA ÷ 240V = 48.33A

Example 4:

Using the resulting ampacity for the branch circuit in the previous example (48.33A), what is the minimum size conductor that may be used for this branch circuit, and what circuit breaker rating is required?

Solution:

According to *NEC Section 110.14(C)(1)(a)*, branch circuits of 100A or less must be terminated using the 60°C (140°F) column of *NEC Table 310.16*. Therefore, select a No. 6 copper conductor rated at 55A. A 50A circuit breaker would properly protect this circuit.

Think About It

Demand Loads

You are adding a hot tub with a nameplate rating of 4,000VA and have three other fixed appliances with a total rating of 13,000VA. Assuming this feeder does not also serve an electric range, clothes dryer, space heating units, or air conditioning equipment, how will this addition affect the total demand load?

Think About It

One Cooktop and One Oven

Which note under *NEC Table 220.55* refers to this installation?

Source: iStock@baloon111

Know the Code

Electrical Connections
NEC Section 110.14

2.6.0 Water Heater Loads

As specified in *NEC Section 422.13*, fixed storage water heaters having a storage capacity of 120 gallons or less shall have the branch circuit rated at no less than 125% of the nameplate rating of the water heater. This is required because it is considered a continuous load per *NEC Section 422.10(A)(3)*.

For example, what minimum circuit breaker rating and branch circuit conductor size is required for a water heater with a nameplate rating of 9,000VA at 240V? Assume that Type NM cable will be used for the circuit.

Begin by dividing the total VA by the voltage to determine the ampacity. Then find 125% of the result to account for the continuous load:

9,000VA ÷ 240V = 37.5A

37.5A × 1.25 = 46.88A

Per the 60°C (140°F) column of *NEC Table 310.16*, No. 6 NM copper cable is rated at 55A, so it will handle the load. The circuit breaker rating would be 50A.

Know the Code

Storage-Type Water Heaters
NEC Section 422.13

Know the Code

Branch Circuits
NEC Section 422.10

2.7.0 Electric Heating Loads

NEC Article 424 covers fixed electric space-heating equipment, including heating cables, unit heaters, boilers, central systems, and similar equipment. This article does not apply to process heating or room air conditioning.

Per *NEC Section 424.4(A)*, branch circuits can supply any size of fixed electric space-heating equipment for which the circuit is rated. If two or more outlets for heating equipment are supplied, the branch circuit size cannot exceed 30A. In nonresidential applications, fixed infrared heating equipment can be powered by branch circuits rated up to 50A.

When space-heating equipment uses electric resistance heating elements, protection for the elements cannot exceed 60A, per *NEC Section 424.22(B)*. If the unit is rated at more than 48A, the heating elements must be subdivided, and each subdivided load cannot exceed 48A.

Know the Code

Fixed Electric Space-Heating Equipment
NEC Article 424

Outdoor Loads

Parking lots may be equipped to supply the needed lighting circuits as well as branch circuits for other loads. This hotel parking lot in Alaska provides receptacles for engine block heaters to make starting the engine easier.

For space-heating equipment with resistance heating elements, regardless of the presence of an electric blower motor, the ampacity of the branch circuit shall be considered continuous per *NEC Section 424.4(B)* and therefore it must be calculated at no less than 125% of the total load.

Source: Tri-City Electrical Contractors Inc

Example 1:

What is the minimum conductor ampacity for the following single-phase baseboard electric heaters rated at 240V: one 1,500W unit, one 1,000W unit, and two 500W units?

Solution:

First, determine the total connected load by adding the load of each baseboard unit. Then divide the total load in VA by the circuit voltage. Remember that watts and VA are the same for this type of load:

$$1{,}500W + 1{,}000W + 500W + 500W = 3{,}500W\ (VA)$$

$$3{,}500VA \div 240V = 14.58A$$

Electric heaters like this must be considered continuous loads. Find 125% of the load to determine the minimum conductor ampacity:

$$14.58A \times 1.25 = 18.23A$$

Example 2:

Which size NM cable and circuit breaker rating is required for the load of 18.23A in the previous example?

Solution:

Per the 60°C (140°F) column of *NEC Table 310.16*, No. 12 Type NM cable is rated at 20A. The load of 18.23A requires a 20A circuit breaker.

Example 3:

What is the minimum conductor ampacity for an electric forced-air furnace with heating elements totaling 25 kW at 240V, and a blower motor with a nameplate full-load current of 4.9A?

Solution:

Begin by dividing the total of the resistance heat (in VA) by the circuit voltage to determine the ampacity for the heating elements:

$$25\ \text{kW} \times 1{,}000\text{W/kW} = 25{,}000\text{VA}$$

$$25{,}000\text{VA} \div 240\text{V} = 104.17\text{A}$$

Now add the motor current of 4.9A:

$$104.17\text{A} + 4.9\text{A} = 109.07\text{A}$$

Since it is considered a continuous load, calculate 125% of the ampacity to find the total circuit ampacity:

$$109.07\text{A} \times 1.25 = 136.34\text{A}$$

Remember that, because this equipment is rated at far more than 48A, the heating elements must be subdivided into groups of no more than 48A, per *NEC Section 424.22(B)*. In this example, the heating elements must be subdivided into three groups.

2.8.0 Air Conditioning Loads

Branch circuit conductors supplying a single air conditioning compressor must be rated for at least 125% of the compressor full-load current or the branch circuit selection current, whichever is greater (per *NEC Section 440.32*).

Branch circuit selection current is determined by the equipment manufacturer. The current is required to be greater than or equal to the compressor full-load current. If the branch circuit selection current is higher, its rating shall be used instead of the compressor full-load current.

Per *NEC Section 440.22(A)*, overcurrent protective devices cannot exceed 175% of the compressor full-load current or the branch circuit selection current. If the chosen breaker is not sufficient to manage the starting current of the compressor, the rating can be increased. However, it cannot exceed 225% of the compressor full-load current or the branch circuit selection current, whichever is greater.

Example 1:

Which size copper conductors are required for a three-phase branch circuit supplying an air conditioner with a nameplate rating of 45A at 480V?

Solution:

Find 125% of the nameplate ampacity to account for the continuous load and use the result to select the conductor size from *NEC Table 310.16*:

$$45\text{A} \times 1.25 = 56.25\text{A}$$

Per *NEC Section 110.14(C)*, use the 60°C (140°F) column of *NEC Table 310.16*. The table shows an ampacity of 70A for No. 4 copper conductors. Therefore, the conductor size would be No. 4 copper.

Example 2:

Which Type NM cable size and circuit breaker rating would be required to supply a single-phase air conditioner with a nameplate rating of 16A at 240V?

Solution:

First, multiply the nameplate ampacity by 1.25 to account for the continuous load:

$$16A \times 1.25 = 20A$$

Use the result to select the proper conductor size from the 60°C (140°F) column of *NEC Table 310.16*. The table shows a rating of 20A for No. 12 copper conductors, so this wire size is acceptable.

Now calculate 175% of the ampacity to determine the circuit breaker size:

$$16A \times 1.75 = 28A$$

Because *NEC Section 440.22(A)* states that the rating of the overcurrent device cannot exceed 175%, a 25A circuit breaker would be required. A 30A circuit breaker is not acceptable.

Know the Code

Application and Selection
NEC Section 440.22

Air Conditioning Unit Nameplates

The nameplate provides information about an air conditioning unit's operating voltage and current. An important piece of information given on the nameplate is the RLA or FLA value for each motor. RLA stands for *rated load amps*, and FLA stands for *full-load amps*. The two terms can generally be used interchangeably.

Nameplates for hermetic compressors built after 1972 are marked with the RLA, as shown in the image. The FLA of the outdoor fan motor is also shown. Note that RLA for the compressor motor has two values, although they are the same in this case. This is because the unit is designed to operate at either 208V or 230V.

RUUD AIR CONDITIONER
MODEL NO. UPMC-042JAZ MFD 04/99
SERIAL NO. 6288 M1799 22780 OUTDOOR USE
VOLTS 208-230 PHASE 1 HERTZ 60
COMPRESSOR R.L.A. 18.6/18.6 L.R.A. 109
OUTDOOR FAN MOTOR F.L.A. 2.0 HP(WATTS) 1/3 ()
MIN. SUPPLY CIRCUIT AMPACITY 26/26 AMP
MAX. FUSE OR CKT. BRK. SIZE* 40/40 AMP
MIN. FUSE OR CKT. BRK. SIZE* 30/30 AMP
DESIGN PRESSURE HIGH 300 PSIG
DESIGN PRESSURE LOW 150 PSIG
OUTDOOR UNITS FACTORY CHARGE 214 OZ. R22
TOTAL SYSTEM CHARGE OZ. R22
SEE INSTRUCTIONS INSIDE ACCESS PANEL
RUUD AIR CONDITIONING DIVISION
FORT SMITH, ARKANSAS
MADE IN THE USA
*HACR TYPE BREAKER FOR U.S.A.

2.0.0 Section Review

1. What is the calculated load, in amps, for a 120V circuit supplying eight recessed fluorescent fixtures with ballasts rated 0.5A each, all of which operate continuously?
   a. 0.5A
   b. 2A
   c. 4A
   d. 5A

2. What is the *maximum* number of continuous-duty duplex receptacles that can be connected to a 20A, 120V circuit in a commercial building?
   a. 10 receptacles
   b. 13 receptacles
   c. 15 receptacles
   d. 16 receptacles

3. What is the total load, in VA, for four small appliance branch circuits in a single-family dwelling?
   a. 1,500VA
   b. 3,000VA
   c. 4,000VA
   d. 6,000VA

4. What is the demand load, in VA, for one household dryer with a nameplate rating of 4.75 kW?
   a. 2,375VA
   b. 3,800VA
   c. 4,750VA
   d. 5,000VA

5. What is the demand load, in amps, for a single-phase, two-burner electric range with a nameplate rating of 7 kW at 240V?
   a. 23.3A
   b. 29.1A
   c. 45.6A
   d. 50.8A

6. What is the ampacity for wire sizing required for a water heater with a nameplate rating of 8,400VA at 240V?
   a. 35A
   b. 43.75A
   c. 52.55A
   d. 70A

7. What is the *minimum* conductor ampacity for four 1,200W, 240V, single-phase electric baseboard heaters?
   a. 20A
   b. 25A
   c. 45A
   d. 60A

8. What size copper branch circuit conductors are required for a three-phase branch circuit supplying an air conditioning unit with a nameplate rating of 18A at 240V?
   a. No. 10
   b. No. 8
   c. No. 6
   d. No. 4

3.0.0 Commercial Loads

Performance Tasks

There are no Performance Tasks in this section.

Objective

Calculate commercial electrical loads.

a. Calculate the loads on multioutlet assemblies.
b. Calculate show window loads.
c. Calculate sign loads.
d. Calculate loads for heavy-duty lamp holder outlets.
e. Calculate commercial kitchen equipment loads.
f. Calculate motor loads.
g. Calculate welding machine loads.

Commercial buildings have special branch circuit loads with different *NEC®* requirements than those for residential applications. These unique loads include multioutlet assemblies, such as those used in office buildings, as well as show window loads, sign loads, and commercial kitchen equipment loads.

3.1.0 Loads on Multioutlet Assemblies

Know the Code

Definitions
NEC Article 100

Know the Code

Other Loads — All Occupancies
NEC Section 220.14

Multioutlet assembly: A type of surface mounted, flush mounted, or freestanding raceway designed to hold conductors and receptacles.

NOTE

Plugmold® and Wiremold® are products offered by Legrand North America. If you are not familiar with these products, visit **https://www.legrand.us/**.

A **multioutlet assembly** is defined by *NEC Article 100* as a type of surface mounted, flush mounted, or freestanding raceway designed to hold conductors and receptacles. They may be factory assembled or fabricated in the field.

Multioutlet assemblies often consist of single outlets wired to one or more circuits. These outlets are typically spaced apart at equal distances of 6" (150 mm), 12" (300 mm), 18" (450 mm), etc. The *NEC®* rules for calculating loads for multioutlet assemblies do not apply to residential applications.

Per *NEC Section 220.14(H)(1)*, each 5' (1.5 m) of multioutlet assembly is considered as one outlet of at least 180VA capacity. In locations where many loads are likely to be connected at one time, each 1' (300 mm) of multioutlet assembly is considered as one outlet of at least 180VA capacity.

Example 1:

A total of 40' of Plugmold® includes one single receptacle per lineal foot. The multioutlet assembly is to be connected to a single 120V circuit. What is the minimum size 120V circuit that would safely supply this light-duty multioutlet assembly?

Solution:

Because this is a light-duty application, the load is to be calculated at 180VA per 5' of multioutlet assembly. Divide 40' of multioutlet assembly by 5' and multiply the result by 180VA to obtain the total anticipated VA load. Then divide the total VA load by 120V to determine the minimum circuit ampacity. This is illustrated in the following equations:

$$40' \div 5' = 8$$

$$8 \times 180\text{VA} = 1{,}440\text{VA}$$

$$1{,}440\text{VA} \div 120\text{V} = 12\text{A}$$

The minimum circuit ampacity would be 15A since it is the first available rating above 12A.

Example 2:

The conditions listed for the previous example have changed due to a change order associated with the project. It is now necessary to feed this Plugmold® with three circuits. The reason for the change is that more tools will be plugged in and operating to increase production. This means that the assembly must now be able to power several loads simultaneously. How many 20A, 120V circuits would be required to supply this heavy-duty multioutlet assembly?

Solution:

Because this is a heavy-duty application, the load is to be calculated at 180VA per foot of multioutlet assembly. First, determine the VA capacity for one 20A, 120V circuit using multiplication:

$$20\text{A} \times 120\text{V} = 2{,}400\text{VA}$$

Multiply 40' of multioutlet assembly by 180VA to obtain the total VA load:

$$40' \times 180\text{VA} = 7{,}200\text{VA}$$

Divide the total VA by 2,400VA—the 20A circuit capacity—to determine the number of circuits required:

$$7{,}200\text{VA} \div 2{,}400\text{VA} = 3 \text{ circuits}$$

Three 20A, 120V circuits would now be required to power this multioutlet arrangement.

3.2.0 Show Window Loads

The *NEC*® addresses three areas of requirements regarding show window lighting:

- Receptacle requirements
- Branch circuit load calculations
- Feeder or service load calculations

NEC Section 210.62 requires that at least one 120V, 15A or 20A receptacle outlet be installed within 18" (450 mm) from the top of a show window. They must be spaced so that no point along the top of the window is more than 6' (1.8 m) from an outlet.

NEC Section 220.14(G) provides two options for calculating branch circuit loads for show window lighting. One way is to use the calculated receptacle load of 180VA per receptacle, per *NEC Section 220.14(G)(1)*. Alternatively, you can calculate the load at 200VA per linear foot (300 mm) of show window, per *NEC Section 220.14(G)(2)*.

NEC Section 220.46 addresses the calculated load of show windows when determining feeder or service loads. It requires that the total load be calculated at not less than 200VA for each linear foot of show window.

Know the Code

Show Windows
NEC Section 210.62

NOTE

Because show window lighting is likely to be on for three hours or more, it must be considered a continuous load. See *NEC Section 210.19(A)(1)*.

Know the Code

Show-Window and Track Lighting
NEC Section 220.46

Example 1:

According to *NEC Section 210.62*, how many receptacles would be required for a show window area that measures 67' in length?

Solution:

Divide the length of show window by 12' to determine the number of receptacles required. Outlets placed no more than 12' apart ensure that no point along the windows is more than 6' from an outlet:

$$67' \div 12' = 5.58 \text{ receptacles}$$

Because *NEC Section 210.62* requires one receptacle for each 12' or fraction thereof, six receptacles would be required. You must always round up in this case, rather than down.

Example 2:

Using *NEC Section 220.14(I)*, what is the total load for the receptacles required in the previous example?

Solution:

NEC Section 220.14(I) states that you must multiply the number of receptacles by 180VA per receptacle, as follows:

6 receptacles × 180VA = 1,080VA

Because the load is continuous, find 125% of the result to determine the load:

1,080VA × 1.25 = 1,350VA

Thus, the total load for the show window receptacles is 1,350VA. Since the voltage is 120V, the VA load can be converted to 11.25A current load as follows:

1,350VA ÷ 120V = 11.25A

Example 3:

Applying *NEC Section 220.14(G)(2)*, what is the load for a 30' long show window?

Solution:

First, multiply the total length of show window area by 200VA, as follows:

30' × 200VA = 6,000VA

Calculate 125% of the result to account for the continuous load, and divide by the voltage to determine the ampacity of the circuit:

6,000VA × 1.25 = 7,500VA

7,500VA ÷ 120V = 62.5A

Think About It

Show Window Load Calculations

These individual show windows are each 5' long. According to the requirements given in *NEC Section 220.14(G)(2)*, what is the branch circuit load for each window?

3.3.0 Sign Loads

NEC Article 600 covers requirements for signs and outline lighting. *NEC Section 600.5(C)(1)* specifies that circuits supplying neon lighting must not exceed 30A. Those that supply all other sign lighting systems must not exceed 20A, per *NEC Section 600.5(C)(2)*.

NEC Section 600.5(A) requires at least one 20A sign circuit with no other loads on the circuit. This circuit must be provided for each commercial building and each commercial occupancy accessible to pedestrians. *NEC Section 220.14(F)* requires the load for the sign circuit to be calculated at a minimum of 1,200VA. Because commercial signs are expected to operate for more than three hours at a time, *NEC Section 600.5* indicates it must be considered a continuous load. Therefore, the branch circuit rating must be calculated at 125% of 1,200VA, or 1,500VA. The actual sign load cannot exceed 80% of the branch circuit rating.

Know the Code

Electric Signs and Outline Lighting
NEC Article 600

Know the Code

Branch Circuits
NEC Section 600.5

Example 1:

What is the maximum continuous load in VA that may be connected to a 20A, 120V sign circuit?

Solution:

To determine the total VA load for the circuit, multiply the amperage by the voltage:

20A × 120V = 2,400VA

Find 80% of the result (multiply by 0.8) to determine the maximum continuous VA load:

2,400VA × 0.8 = 1,920VA

3.4.0 Loads for Heavy-Duty Lamp Holder Outlets

Per *NEC Section 220.14(E)*, outlets for heavy-duty lamp holders are calculated at a minimum of 600VA each. For example, a noncontinuous load consisting of four 120V outlets for heavy-duty lamp holders could be connected to a 20A branch circuit. This is illustrated by the following calculations:

20A × 120V = 2,400VA

2,400VA ÷ 600VA per outlet = 4 outlets

3.5.0 Commercial Kitchen Equipment Loads

Loads for commercial kitchen equipment are calculated based upon the nameplate rating. *NEC Table 220.56* lists demand factors for commercial kitchen equipment. This section applies to a variety of commercial kitchen equipment, including but not limited to cooking equipment, dishwasher booster heaters, and water heaters.

Example 1:

What is the current load for one dishwasher booster heater? The booster heater has a nameplate rating of 15kW, 208V, 3ϕ.

Solution:

First, convert the nameplate power rating in kW to VA by multiplying by 1,000. Then, since this is a three-phase load, divide the result by the product of the voltage and the square root of 3 to determine the current load. Remember that 1.732 is the square root of 3:

15 kW × 1,000 = 15,000VA

Current load = 15,000VA ÷ (208V × 1.732)

Current load = 15,000VA ÷ 360.256V

Current load = 41.64A

Example 2:

What is the demand load for six units of commercial kitchen equipment with a total VA rating of 47,000VA?

Solution:

Per *NEC Table 220.56*, the demand factor for six units is 65%. Multiply the total VA rating by 0.65 to find the demand load:

47,000VA × 0.65 = 30,550VA

3.6.0 Motor Loads

Branch circuit conductors and overcurrent protective devices for motors are determined by using *NEC Table 430.248* for single-phase motors and *NEC Table 430.250* for three-phase motors. To use the *NEC*® motor tables, it is necessary to know the motor horsepower (hp), voltage, phase, and National Electrical Manufacturers Association (NEMA) design letter. If the motor nameplate includes a value for current but not the horsepower, *NEC Section 430.6(A)(1)* states that the horsepower rating shall be assumed to be the value given in the tables. The assumed horsepower would be the value that most closely corresponds to the current rating in the table.

The needed motor information can be obtained from its nameplate. *NEC Section 430.6(A)(1)* requires that the values given in *NEC Tables 430.247 through 430.250* be used to determine the ampacity of branch circuit conductors and the current rating of switches and the branch circuit short circuit and ground fault protection. The motor's nameplate current data is not permitted to be used in sizing these components.

To determine the branch circuit conductor size for motor circuits, the motor full-load current taken from the appropriate table is multiplied by 125%. The resulting ampacity is used to select branch circuit conductors.

Example 1:

According to the *NEC*®, what is the motor full-load current for a single-phase 2 hp, 230V motor?

Solution:

NEC Table 430.248 is used to determine the full-load current for single-phase AC motors. The left column in the table is used to find the correct motor horsepower and the full-load current is then determined based upon the motor voltage. Per *NEC Table 430.248*, a 2 hp, 230V motor has a full-load current of 12A.

Example 2:

What is the motor full-load current for a 10 hp, 460V, 3ϕ motor?

Solution:

NEC Table 430.250 is used to determine full-load current for three-phase AC motors. According to the table, a 10 hp, 460V, 3ϕ motor has a full-load current of 14A.

Example 3:

Which size copper THWN branch circuit conductors would be required to power a 25 hp, 460V, 3ϕ, motor?

Solution:

Using *NEC Table 430.250*, the full-load current for this motor is 34A. Increase the value to 125% to determine the minimum ampacity for sizing the branch circuit conductors:

34A × 1.25 = 42.5A

Know the Code

Full-Load Currents in Amperes, Single-Phase Alternating-Current Motors
NEC Table 430.248

Know the Code

Full-Load Current, Three-Phase Alternating-Current Motors
NEC Table 430.250

Think About It

Buck-and-Boost Transformers

This 240V commercial kitchen mixer is connected to a standard 120V/208V service using a buck-and-boost transformer. Will this affect the branch circuit load calculation? What would happen if a buck-and-boost transformer was not used?

Source: iStock@Kondor83

The branch circuit conductor size is then determined using *NEC Table 310.16*. No. 6 THWN copper is rated at 55A using the 60°C (140°F) column. This is the correct wire size for the motor.

Motor short circuit and ground fault protection is sized using *NEC Table 430.52(C)(1)*. To use this table, it is helpful to know the type of motor, horsepower, phase, full-load current, and the design letter, if any.

Know the Code

Rating or Setting for Individual Motor Circuit

NEC Table 430.52

Example 4:

What size dual-element time-delay fuse would be required for a 50 hp, 3ϕ squirrel-cage motor operating at 460V and stamped with NEMA design letter D?

Solution:

The motor full-load current is first determined using *NEC Table 430.250*. Then, using *NEC Table 430.52(C)(1)*, the full-load current is multiplied by the proper percentage listed for the characteristics of the motor.

From *NEC Table 430.250*, the full-load current for a 50 hp, 3ϕ, 460V motor is 65A. From *NEC Table 430.52(C)(1)*, 175% of the full-load current must be used to select a time-delay fuse for a three-phase motor if it has a NEMA design letter other than B (energy-efficient). This is calculated as follows:

$$65\text{A} \times 1.75 = 113.75\text{A}$$

Per *NEC Section 430.52(C)(1)(a)*, the next higher standard fuse size is 125A. If the 125A fuse proves to be inadequate to carry the starting load, the guidance found in *NEC Section 430.52(C)(1)(b)(2)* may be applied. In no case, however, can the rating exceed 225% of the full-load current.

Example 5:

Which size inverse-time circuit breaker would be required for a 10 hp, 3ϕ squirrel-cage motor operating at 208V and stamped with NEMA design letter B?

Solution:

The motor full-load current is 30.8A per *NEC Table 430.250*. Find 250% of full-load current (multiply by 2.5) to size an inverse-time circuit breaker for a three-phase motor with design letter B per *NEC Table 430.52(C)(1)*:

$$30.8\text{A} \times 2.5 = 77\text{A}$$

The next higher circuit breaker rating is 80A, and *NEC Section 430.52(C)(1)(b)(3)* would permit a 110A circuit breaker if the 80A breaker is inadequate for the starting load.

The power factor can be a significant issue with larger motors. In facilities with multiple large motors, the operating current of a specific motor will depend on the power factor of the facility. *NEC Table 430.250* provides adjustment factors for power factors of 80% and 90%. For an 80% power factor, the operating current must be increased to 125% to achieve 100% of the full-load current. With a 90% power factor, the operating current must be increased to 110%.

3.7.0 Welding Machine Loads

Branch circuit conductors and overcurrent protective devices for welding machines are sized using the machine's nameplate current and duty cycle. Using the relevant *NEC*® section, the demand factor is determined based on the *duty cycle* of the welding machine.

The duty cycle refers to the percentage of time that a machine can safely operate at a given current within a larger period. For example, a welder may have a duty cycle of 200A at 30%. This means that it will operate at 200A for 3 minutes within each 10-minute period. Operating the machine beyond that is likely to result in the equipment overheating. The primary current is multiplied by the demand factor to determine the minimum circuit ampacity for sizing the branch circuit conductors and overcurrent protection.

Know the Code

Electric Welders
NEC Article 630

NEC Article 630 covers electric welding machines. *NEC Article 630, Part II* covers arc welding machines (*Figure 2*), while *NEC Article 630, Part III* covers resistance welding machines (*Figure 3*). Arc welders are the familiar style used by many welders for pipe fabrication and many other tasks. Resistance welding is more commonly referred to as *spot welding*, where two contact points come together momentarily with the workpieces between them. An instant weld in one small spot is the result.

Figure 2 Arc welding machine.
Source: Miller Electric Mfg. Co.

Figure 3 Contacts of a resistance welding machine (spot welder).
Source: iStock@kckate16

Each *NEC®* section lists duty cycle multipliers for individual welding machines based upon the type of welding machine. *NEC Table 630.11(A)* provides multipliers for arc welding machines, and *NEC Table 630.31(A)* provides multipliers for resistance welding machines.

Per *NEC Section 630.12(A)*, the overcurrent protective device for an individual welding machine branch circuit must be rated (or set if it is adjustable) for no more than 200% of the welding machine current rating. For resistance welding machines, *NEC Section 630.32(A)* directs that the rating or setting for overcurrent protection cannot exceed 300% of the welding machine current rating.

NOTE

It is advisable to check local codes or project specifications regarding the setting or rating for welding machine branch circuit overcurrent protective devices. Many authorities will not permit the overcurrent device rating to exceed the rating of the branch circuit conductors.

Example 1:

What size THWN copper branch circuit conductors would be required to supply an arc welding machine with a non-motor-generator and a nameplate current of 70A with a duty cycle of 80%?

Solution:

Per *NEC Table 630.11(A)*, the nameplate current is multiplied by 0.89. This multiplier applies to an arc welding machine with a nonmotor generator and a duty cycle of 80%, reducing the required ampacity:

$$70A \times 0.89 = 62.3A$$

The copper THWN conductor size is selected from *NEC Table 310.16*. No. 4 THWN copper conductors, with a rating of 70A, would be required using the 60°C (140°F) column.

Know the Code

Duty Cycle Multiplication Factors for Arc Welders
NEC Table 630.11(A)

Example 2:

What size THWN copper branch circuit conductors and circuit breaker would be required to power a resistance welding machine with a nameplate current of 125A and a duty cycle of 40%?

Solution:

Per *NEC Table 630.31(A)*, the nameplate current is multiplied by 0.63. This multiplier applies to a resistance welding machine with a duty cycle of 40%:

$$125A \times 0.63 = 78.75A$$

Per *NEC Table 310.16* and using the 60°C (140°F) column, No. 3 THWN conductors with a capacity of 85A would be adequate. Per *NEC Section 630.32(A)*, the welding machine current must be multiplied by 300% to size the overcurrent protective device. The nameplate rating for the resistance welding machine in this example is 125A, so the calculation is as follows:

$$125A \times 3 = 375A$$

Because this value exceeds the rating of a standard 300A overcurrent protective device, *NEC Section 630.32* permits the next higher standard size listed in *NEC Section 240.6(A)* (400A) to be used. Therefore, a 400A protective device would be chosen.

Know the Code

Duty Cycle Multiplication Factors for Resistance Welders
NEC Table 630.31(A)

Know the Code

Overcurrent Protection
NEC Section 630.32

Know the Code

Standard Ampere Ratings
NEC Section 240.6

3.0.0 Section Review

1. The load for 8' of a multioutlet assembly with the capability of powering all the receptacles simultaneously is _____.
   a. 720VA
   b. 1,440VA
   c. 2,160VA
   d. 2,880VA

2. How many receptacles are required for 60' of show window length?
   a. 3 receptacles
   b. 5 receptacles
   c. 6 receptacles
   d. 8 receptacles

3. What is the *maximum* continuous load, in VA, that may be connected to a 20A, 120V sign circuit?
   a. 1,680VA
   b. 1,920VA
   c. 2,700VA
   d. 2,880VA

4. Per *NEC Section 220.14(E)*, outlets for heavy-duty lamp holders are each calculated at a minimum of _____.
   a. 180VA
   b. 240VA
   c. 480VA
   d. 600VA

5. What is the calculated demand load for four pieces of commercial kitchen equipment with a total VA rating of 36,000VA?
   a. 23,400VA
   b. 28,800VA
   c. 36,000VA
   d. 45,000VA

6. What is the motor full-load current for a single-phase 5 hp, 208V motor?
   a. 19.6A
   b. 30.8A
   c. 46.0A
   d. 57.5A

7. Which size copper branch circuit conductors would be required to supply a nonmotor generator arc welding machine with a nameplate current of 100A and a duty cycle of 50%?
   a. No. 3 AWG
   b. No. 4 AWG
   c. No. 6 AWG
   d. No. 8 AWG

Module 26301-23 Review Questions

1. The general lighting load for a store is _____.
   a. $1.9VA/ft^2$
   b. $1.6VA/ft^2$
   c. $1.5VA/ft^2$
   d. $1.3VA/ft^2$

2. The general lighting load for a warehouse is _____.
   a. $1.1VA/ft^2$
   b. $1.2VA/ft^2$
   c. $1.4VA/ft^2$
   d. $1.5VA/ft^2$

3. The *NEC*® defines a continuous load as one where the maximum current will be sustained for at *least* _____.
   a. 30 minutes
   b. 1 hour
   c. 2 hours
   d. 3 hours

4. What is the capacity, in amps, for a 20A circuit breaker supplying a continuous load?
   a. 10A
   b. 12A
   c. 16A
   d. 20A

5. Branch circuit conductors supplying continuous loads are calculated at _____.
   a. 100% of the rated load
   b. 125% of the rated load
   c. 150% of the rated load
   d. 175% of the rated load

6. What is the current-carrying ampacity for each of ten No. 10 THHN copper conductors installed in a single conduit?
   a. 20A
   b. 28A
   c. 32A
   d. 40A

7. The voltage drop between the point of service and the farthest outlet in a branch circuit should *not* exceed _____.
   a. 2%
   b. 3%
   c. 5%
   d. 10%

8. The voltage drop for a 208V single-phase circuit with a load of 26.5A, using No. 8 stranded copper conductors at a circuit length of 145', is _____.
   a. 5.87V
   b. 5.98V
   c. 6.08V
   d. 6.18V

9. The voltage drop for a 480V three-phase circuit with a load of 26.5A, using No. 8 stranded copper conductors at a circuit length of 145', is _____.
   a. 4.87V
   b. 4.98V
   c. 5.08V
   d. 5.18V

10. Using the resistance formula, the voltage drop for an 18A, 208V, 3ϕ load with a total circuit length of 105' using No. 10 solid copper conductors is _____.
    a. 3.96V
    b. 4.57V
    c. 5.06V
    d. 5.69V

11. Load calculations for circuits supplying lighting units with ballasts are based upon _____.
    a. the total wattage of all lamps in the fixtures
    b. the circuit voltage times lamp wattage
    c. the current ratings of the ballasts
    d. 180VA per ballast

12. What is the calculated load, in amps, for a 120V circuit supplying four 150W recessed incandescent fixtures and six recessed fluorescent fixtures operating continuously? Assume that each fluorescent ballast is rated at 0.65A.
    a. 3.965A
    b. 6.115A
    c. 10.525A
    d. 11.125A

13. What is the *maximum* number of general-purpose noncontinuous duty duplex receptacles that can be connected to a 20A, 120V circuit in a commercial building, and what would the *NEC*® required load rating be for each receptacle?
    a. 10 receptacles and 1.5A each
    b. 10 receptacles and 1.875A each
    c. 13 receptacles and 150VA each
    d. 13 receptacles and 180VA each

14. What is the total load, in VA, for two small appliance branch circuits in a single-family dwelling?
    a. 1,800VA
    b. 2,400VA
    c. 3,000VA
    d. 4,500VA

15. What is the demand load, in amps, for one single-phase household electric range with a nameplate rating of 16.75 kW at 240V?
    a. 39.79A
    b. 41.67A
    c. 48.25A
    d. 53.33A

16. What is the demand load, in amps, for a single circuit supplying one single-phase, wall-mounted oven rated 11.75 kW at 240V, and a countertop cooking unit rated 9.6 kW at 240V?
    a. 36.67A
    b. 48.33A
    c. 53.25A
    d. 88.96A

17. What is the *minimum* conductor ampacity for four single-phase 1,000W, 208V electric baseboard heaters?
    a. 19.23A
    b. 24.04A
    c. 33.33A
    d. 40.83A

18. Which size copper branch circuit conductors are required for a three-phase branch circuit supplying an air conditioning unit with a nameplate rating of 33.5A at 208V?
    a. No. 10
    b. No. 8
    c. No. 6
    d. No. 4

19. The load for 6' of a multioutlet assembly with the capability of powering all the receptacles simultaneously is _____.
    a. 180VA
    b. 225VA
    c. 1,080VA
    d. 2,160VA

20. How many receptacles are required for 90' of show window length?
    a. 3
    b. 5
    c. 6
    d. 8

Answers to odd-numbered Module Review Questions are found in *Appendix A*.

Module 26301-23 Supplemental Exercises

1. How are branch circuit conductors sized?

 ____

 ____

2. *NEC Section 210.20(A)* states that the rating of a branch circuit overcurrent protective device shall not be less than the ________ plus ________.

3. The maximum load that can be supported by a 20A, 240V, two-wire circuit is ________.

4. The maximum load that can be supported by a 15A,120V, two-wire circuit is ________.

5. How does *NEC Article 100* define a continuous load?

 ____

 ____

6. Describe situations in which the ampacity of conductors must be derated.

 ____

 ____

7. *NEC Section 110.14(C)* states that the ________ temperature rating of any component in a branch circuit must be used to determine the minimum circuit ampacity for the branch circuit conductors.

8. The minimum number of 20A small-appliance branch circuits required in a residential kitchen is ________.

9. The single most important piece of information needed to size the circuit for residential cooking equipment is the ________.

10. *NEC Section 220.55* states that kVA shall be considered equivalent to ________ for loads calculated using *NEC Table 220.55*.

11. The branch circuit supplying an 80-gallon water heater must be rated at no less than 125% of the nameplate rating of the water heater because it is considered a(n) ________.

12. According to *NEC Section 600.5(C)*, the maximum current rating for sign circuits that supply incandescent and fluorescent lighting shall not exceed ________.

13. Commercial applications usually require signs to operate for more than ________ hours at a time; therefore, the sign circuit is typically considered a(n) ________ load. Under this condition, the branch circuit calculation is multiplied by ________ and the actual sign load cannot exceed ________ of the branch circuit rating.

14. The number of heavy-duty outlets that can be connected to a 20A, 120V branch circuit is ________.

15. Branch circuit conductors and overcurrent protective devices for welding machines are sized using the ________ and ________ for the welding machine.

Answers to odd-numbered Supplemental Exercises are found in *Appendix B*.

Answers to Section Review Questions

Answer	Section Reference	Objective
Section 1.0.0		
1. b	1.1.0	1a
2. d	1.2.2	1b
3. d	1.3.0	1c
Section 2.0.0		
1. d	2.1.1	2a
2. a	2.2.0	2b
3. d	2.3.0	2c
4. d	2.4.0	2d
5. a	2.5.0	2e
6. b	2.6.0	2f
7. b	2.7.0	2g
8. a	2.8.0	2h
Section 3.0.0		
1. b	3.1.0	3a
2. b	3.2.0	3b
3. b	3.3.0	3c
4. d	3.4.0	3d
5. b	3.5.0	3e
6. b	3.6.0; ***NEC Table 430.248***	3f
7. a	3.7.0	3g

Section Review Calculations

Section 1.0.0

Question 1

Per *NEC Section 210.20*, the overcurrent protective device must be 125% greater. To determine the maximum load, divide by 1.25:

20A ÷ 1.25 = 16A

The maximum continuous load that may be connected is **16A**.

Question 2

Use the following formula:

$$VD = \frac{2 \times L \times R \times I}{1{,}000'}$$

Insert circuit values into formula, based on the resistance per 1,000' found in *NEC Chapter 9, Table 8*, *Conductor Properties*:

$$VD = \frac{2 \times 160' \times 0.778\Omega \times 28A}{1{,}000'}$$

$$VD = 6.97V$$

The voltage drop is **6.97V**.

Question 3

Divide the rating, in VA, by the circuit voltage to find the ampacity:

800VA ÷ 120V = 6.67A

The ampacity of the load is **6.67A**.

Section 2.0.0

Question 1

Multiply the number of fixtures by the current rating of each fixture to find the load. Then, since this is a continuous load, the result must be increased to 125% of the total current:

8 fixtures × 0.5A = 4A

4A × 1.25 = 5A

The calculated load is **5A**.

Question 2

Multiply the circuit amperage by the circuit voltage to determine the total load of the circuit. Next, since they are for continuous use, multiply the standard receptacle rating by 125%. Then divide the result by the rating per receptacle to find the maximum number of receptacles that may be connected:

20A × 120V = 2,400VA

180VA receptacle rating × 1.25 = 225VA

2,400VA available power ÷ 225VA per receptacle = 10.7 receptacles (round down to 10)

The maximum number of receptacles that can be connected is **10 receptacles**.

Question 3

Multiply the number of small appliance branch circuits by 1,500VA to determine the total load:

4 circuits × 1,500VA = 6,000VA

The total load is **6,000VA**.

Question 4

Since *NEC Section 220.54* requires that a household dryer load be calculated at either 5,000VA or the nameplate rating of the dryer, whichever is larger, you must first convert the nameplate rating of the dryer to VA. Then, compare it to 5,000VA to determine which is larger:

4.75 kW × 1,000 = 4,750VA

4,750VA < 5,000VA

Since 5,000VA is larger than the nameplate rating of the dryer, and since there is only one dryer, the demand load is **5,000VA.**

Question 5

Per *NEC Table 220.55*, calculate 80% of the nameplate power rating. Then, convert it to VA and divide the result by the voltage to determine the demand load:

7 kW × 0.8 = 5.6 kW

5.6 kW × 1,000 = 5,600VA

5,600VA ÷ 240V = 23.3A

The demand load in amps is **23.3A.**

Question 6

First, calculate the current load by dividing the power rating by the voltage. Then, since it is a continuous load, find 125% of the result to determine the ampacity for wire sizing, per *NEC Section 422.13*:

8,400VA ÷ 240V = 35A

35A × 1.25 = 43.75A

The ampacity is **43.75A.**

Question 7

Calculate the total connected load by combining the loads of the four heaters. Then, divide the total load in VA by the voltage, and increase the result to 125% of its value to determine the minimum conductor ampacity:

1,200W × 4 = 4,800VA

4,800VA ÷ 240V = 20A

20A × 1.25 = 25A

The minimum conductor ampacity is **25A.**

Question 8

Find 125% of the nameplate ampacity and use the result to select the proper conductor size from *NEC Table 310.16*:

18A × 1.25 = 22.5A

22.5A = No. 10 AWG

See the 60°C (140°F) column of *NEC Table 310.16*, which indicates a copper conductor size of **No. 10 AWG.**

Section 3.0.0

Question 1

Since each foot of the multioutlet assembly is considered as one outlet of at least 180VA capacity, multiply each foot by 180VA:

8' × 180VA per outlet = 1,440VA

The load is **1,440VA.**

Question 2

Divide the length of the show window by 12' to determine the number of receptacles required:

60' show window length ÷ 12' = 5 receptacles

The result is **5 receptacles**.

Question 3

First, determine the total VA load for the circuit by multiplying the current by the voltage. Then find 80% of the result to determine the maximum continuous VA load:

20A × 120V = 2,400VA total circuit VA load

2,400VA × 0.8 = 1,920VA

The maximum continuous load is **1,920VA**.

Question 5

Per ***NEC Table 220.56***, the demand factor for four units is 80%. Therefore, to determine the demand load, multiply the load by 0.8:

36,000VA × 0.8 = 28,800VA

The demand load is **28,800VA**.

Question 7

Per ***NEC Table 630.11(A)***:

0.71 × 100A = 71A

See the 60°C (140°F) column of ***NEC Table 310.16*** to find that an ampacity of 71A requires **No. 3 AWG**.

MODULE 26302-23

Conductor Selection and Calculations

Source: Southwire

Objectives

Successful completion of this module prepares you to do the following:

1. Select conductors for various applications.
   a. Identify overcurrent protection for branch circuits and feeders.
   b. Identify the properties of conductors.
2. Size conductors based on expected load and voltage drop.
   a. Calculate wire sizes based on resistance.
   b. Calculate conductor resistances.
   c. Calculate voltage drops for various applications.

Performance Tasks

This is a knowledge-based module. There are no Performance Tasks.

Overview

The ability to calculate conductor resistances and voltage drops is important when selecting conductors. This module explains how to make conductor calculations and covers other factors involved in conductor selection, including insulation types, current-carrying capacity, temperature ratings, and voltage drop.

NOTE

NFPA 70®, *National Electrical Code®*, and *NEC®* are registered trademarks of the National Fire Protection Association, Quincy, MA.

NCCER Industry-Recognized Credentials

If you are training through an NCCER-accredited sponsor, you may be eligible for credentials from NCCER. The ID number for this module is 26302-23. Note that this module may have been used in other NCCER curricula and may apply to other level completions. Contact NCCER at 1.888.622.3720 or go to **www.nccer.org** for more information.

You can also show off your industry-recognized credentials online with NCCER's digital badges. Transform your knowledge, skills, and achievements into badges that you can share across social media platforms, send to your network, and add to your resume. For more information, visit **www.nccer.org**.

Digital Resources for Electrical

Scan this code using the camera on your phone or mobile device to view the digital resources related to this craft.

1.0.0 Selecting Conductors for Various Applications

Performance Tasks

There are no Performance Tasks in this section.

Objective

Select conductors for various applications.

a. Identify overcurrent protection for branch circuits and feeders.

b. Identify the properties of conductors.

A variety of materials may be used to transmit electrical energy, but copper remains the most ideal conductor due to its excellent cost-to-conductivity ratio. Electrolytic copper, the type used in most electrical conductors, has the following general characteristics:

- *Method of stranding* — The number and arrangement of conductor strands affects the relative flexibility of the conductor. A conductor may consist of only one strand or many strands, depending on the rigidity or flexibility required for a specific need. *NEC Section 310.3(C)* requires all conductors size No. 8 AWG and larger to be constructed with more than one strand when installed in raceways, unless specifically permitted or required elsewhere in the *NEC*® to be solid. For example, a small-gauge wire that is to be used in a fixed installation is normally solid (one strand), whereas a wire that will be constantly flexed requires a higher degree of flexibility and must contain many strands. Solid wire is the least flexible form of a conductor. Stranded conductors may have a few strands or many, depending on the wire gauge. See *Figure 1* and *NEC Chapter 9, Table 8*.
- *Degree of hardness (temper)* — Temper refers to the hardness of the conductor. Common wire tempers include *soft drawn* (SD), *medium hard drawn* (MHD), and *hard drawn* (HD). Again, the application will determine the required temper. For example, where greater tensile strength is needed, MHD would be selected instead of SD.
- *Bare or coated* — Bare copper is plain, uninsulated copper that is available solid or stranded and in the various tempers described above. In this form, it is often referred to as *red copper.* Bare copper is also available with a coating of tin, silver, or nickel to facilitate soldering, impede corrosion, and prevent adhesion of the copper conductor to rubber or other types of conductor insulation. Various coatings can also affect the electrical characteristics of copper.

Know the Code

Conductors
NEC Section 310.3

Know the Code

Conductor Properties
NEC Chapter 9, Table 8

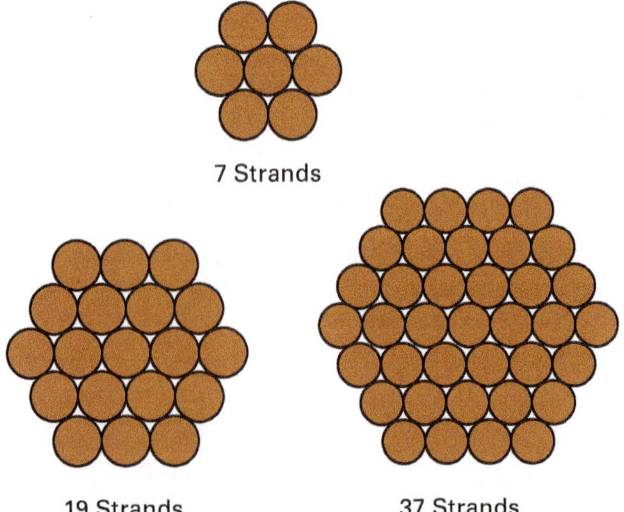

Figure 1 Common strand configurations.

The American Wire Gauge (AWG) is used in the United States to identify the sizes of wire and cable up to and including No. 4/0 (0000), which is commonly spoken in the electrical trade as *four aught* or *four naught*. These numbers run in reverse order as to size. In other words, No. 14 AWG is smaller than No. 12 AWG. Through No. 1 AWG, the larger the gauge number, the smaller the size of the conductor.

However, the next larger size after No. 1 AWG is No. 1/0 AWG, then 2/0 AWG, 3/0 AWG, and 4/0 AWG. Rather than using an AWG designation, these larger sizes of conductors are measured in circular mils (cmil). This represents the area of the end of the conductor. The larger the wire, the larger the number of circular mils. For example, 300,000 cmil is larger than 250,000 cmil.

When writing these sizes in circular mils, every 1,000 cmil is replaced by the letter *k*. For example, 500,000 cmil should be written as 500 kcmil, which is spoken in the craft as *five-hundred kay-cee-mil*. See *Figure 2* for a comparison of the different wire sizes.

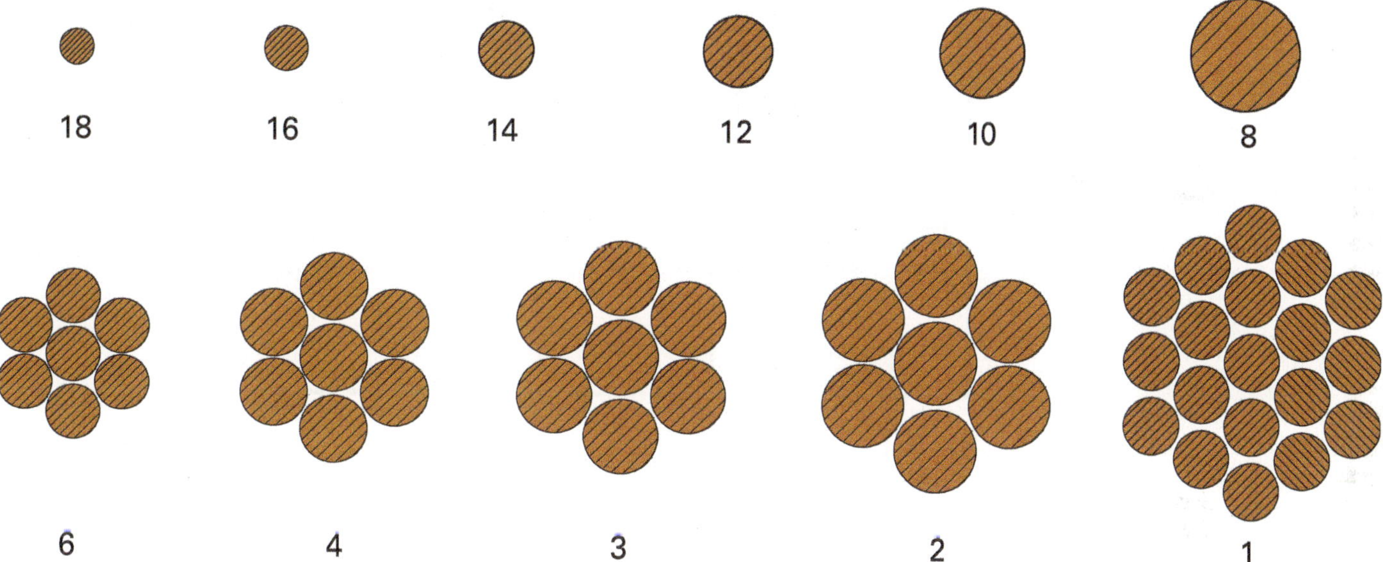

Figure 2 Comparison of various wire sizes.

Compacted conductors are conductors that have been compressed to reduce the air space between the strands. *Figure 3* shows a cross section of a 37-strand compact aluminum conductor.

Compacted conductors are smaller in diameter and rounder, allowing them to be installed in a smaller conduit size than the same noncompacted wire size. Compacted conductors are especially useful when increasing the ampacity of an existing service or feeder circuits. For example, assume an existing service is rated at 250A and is fed with four 350 kcmil THW conductors in 3" conduit (MD 78). To increase the ampacity of the service to 300A, 500 kcmil THW compacted conductors can replace the 350 kcmil standard conductors without increasing the size of the conduit. Compacted conductors are listed in NEC Chapter 9, Table 5A, *Compact Copper and Aluminum Building Wire Nominal Dimensions and Areas*.

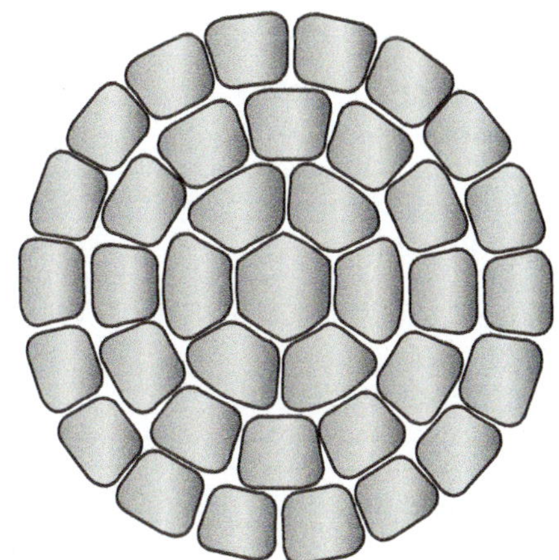

Figure 3 Cross section of a 37-strand compacted conductor.

1.1.0 Overcurrent Protection for Branch Circuits and Feeders

Know the Code

Definitions
NEC Article 100

NEC Article 100 defines *feeders* as the circuit conductors between the power source and the final branch circuit overcurrent protective device. A *branch circuit* is defined as the circuit conductors between the final overcurrent protective device and the outlet(s) it serves. The power riser diagram in *Figure 4* shows examples of both feeders and branch circuits.

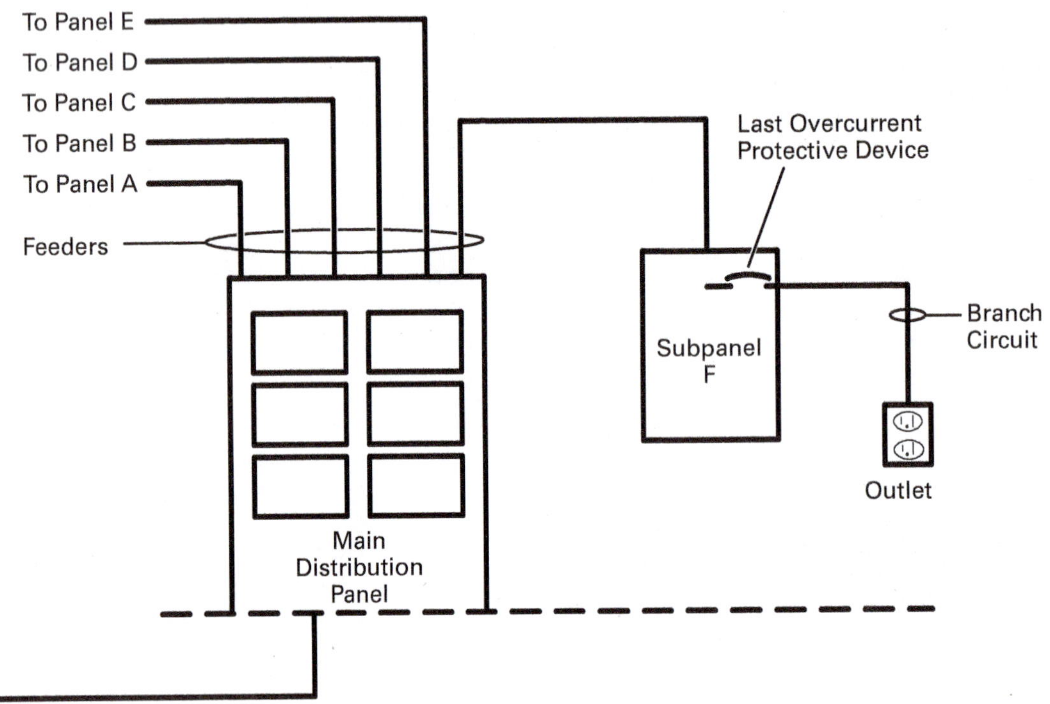

Figure 4 Power riser diagram showing feeders and branch circuits.

Know the Code

Protection of Conductors
NEC Section 240.4

Know the Code

Ampacities for Conductors Rated 0 Volts – 2000 Volts
NEC Section 310.14

Know the Code

Location in Circuit
NEC Section 240.21

Know the Code

Means of Identifying Grounded Conductors
NEC Section 200.6

The *NEC*® requires each ungrounded current-carrying conductor to be protected from damage by an overcurrent protective device such as a fuse or circuit breaker. The conductors must also be identified so that the ungrounded conductors can be distinguished from the grounded and grounding conductors. The minimum size of the conductors used in a circuit are selected based on *NEC*® rules and tables. The *NEC*® also provides tables that list the physical and electrical properties of conductors to support the selection of the proper conductor for any application.

NEC Section 240.4 covers the requirements for protecting conductors against overcurrent in accordance with their ampacities specified in *NEC Section 310.14*. In many cases, the overcurrent protective device setting must be larger than the ampacity rating of the conductors, since the ampacity of the conductors rarely falls on a listed size of overcurrent protective device. In that case, choose the next highest size of overcurrent protective device available, as long as it doesn't exceed 800A. For example, a branch circuit with a design load of 56A may be protected with a 60A overcurrent protective device, the next size available, because 56A is not a standard overcurrent device size.

In most cases, an overcurrent protective device must be connected at the point where the protected conductor receives its power (per *NEC Section 240.21*). The most common situations are shown in *Figure 5*, which illustrates the basic rule and several exceptions, including the 10' (3 m) tap rule. *Figure 6* illustrates the 25' (7.5 m) tap rule.

Wiring systems typically require a grounded conductor. A grounded conductor, such as a neutral, or a grounding conductor must be identified either by the color of its insulation, markings at the terminals, or other suitable means (per *NEC Section 200.6*).

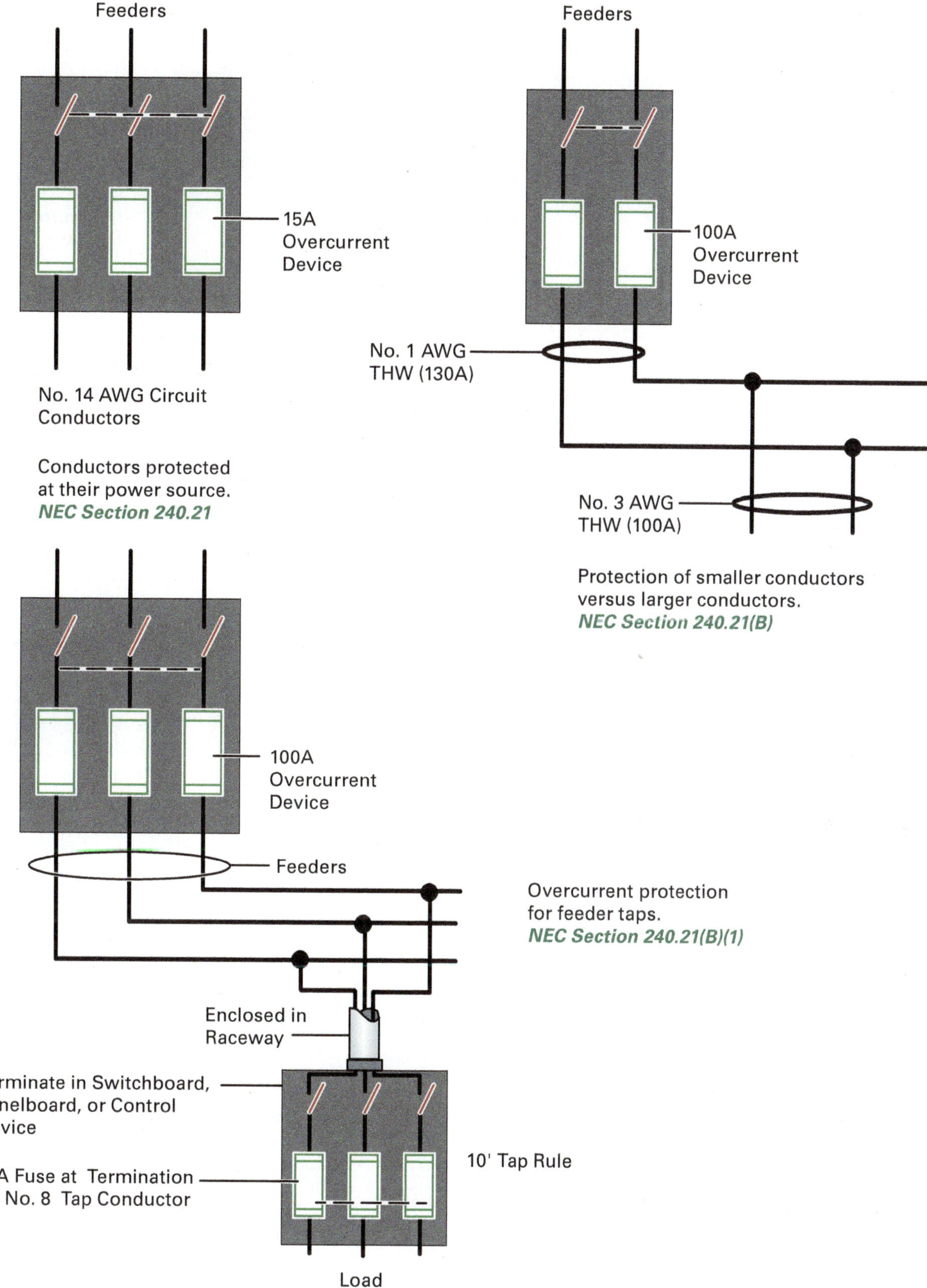

Figure 5 Location of overcurrent protection in circuits.

In general, a grounded conductor must have a white or gray finish. When this is not practical for conductors larger than No. 6 AWG, marking the terminations white is an acceptable method of identifying a grounded conductor. Tagging is also acceptable.

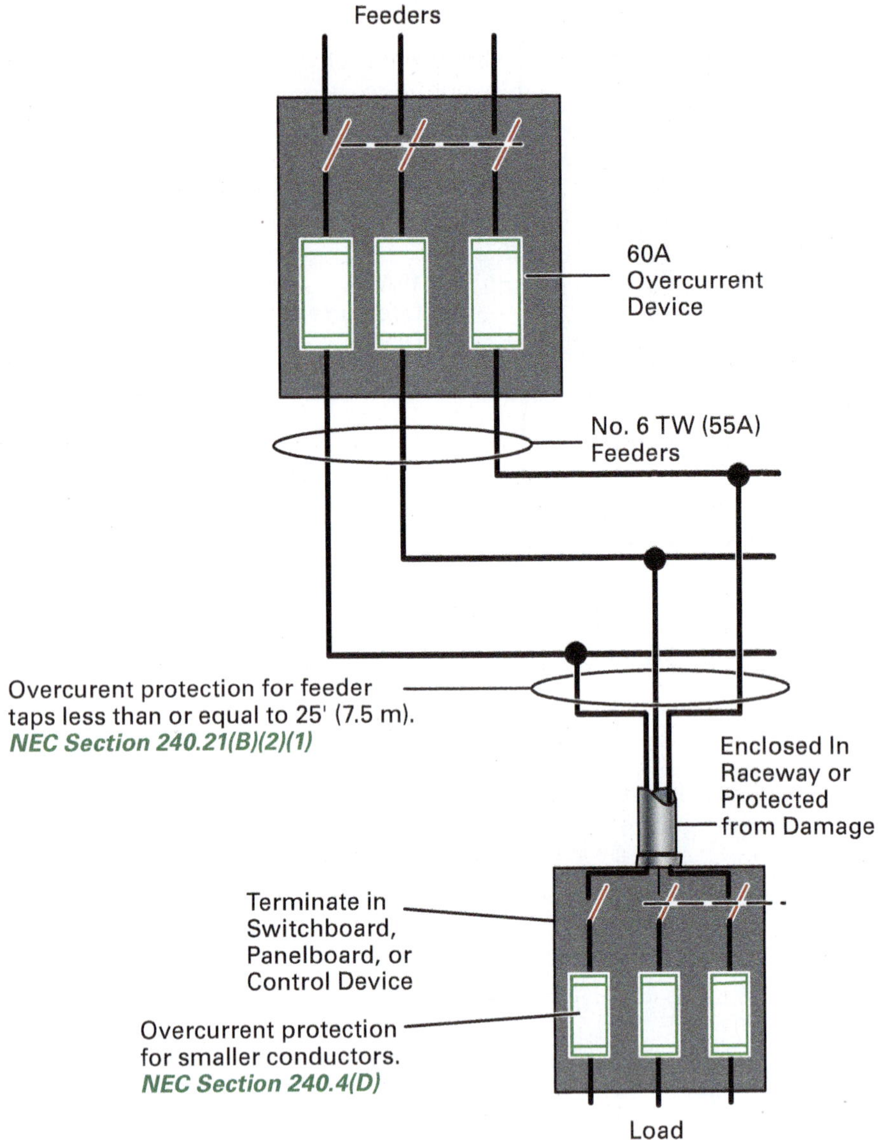

Figure 6 *NEC*® 25' (7.5 m) tap rule.

Roughed-In Residential Feeders

In the rough-in installation shown in the image, the large blue cable will feed a subpanel on the inside of the house, while the other three insulated cables will probably feed electrical heating units, an air conditioning unit, and other relatively large loads. The small wire is a bare grounding conductor that will connect the grounding electrode to the inside panel grounding bus.

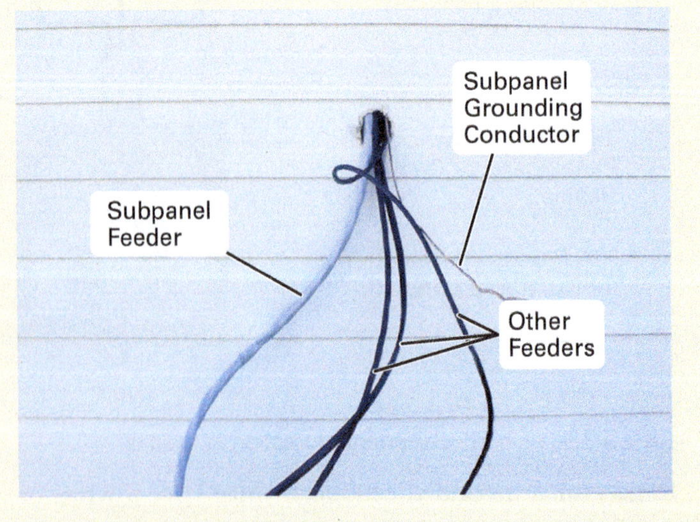

1.1.1 Branch Circuits

Now that an overview of *NEC*® overcurrent protection has been presented, we will examine the *NEC*® installation requirements for branch circuit conductors.

In general, the ampacity (current-carrying capacity) of a conductor must not be less than the maximum load served. However, there are exceptions to this rule, namely, a branch circuit supplying a motor. Motors and motor circuits are covered in *NEC Article 430*.

The rating of the branch circuit overcurrent device determines the rating of the branch circuit, per *NEC Section 210.18*. For example, if a No. 10 AWG 30A conductor is protected by a 20A circuit breaker, then the circuit is considered a 20A branch circuit, despite the ampacity of the conductor.

Furthermore, the current-carrying capacity of branch circuit conductors must not be less than the maximum load to be served. Where the branch circuit supplies receptacles for use with cord-and-plug-connected appliances and other utilization equipment, the conductor's ampacity must not be less than the rating of the branch circuit overcurrent protective device.

As mentioned previously, when the ampacity of the conductor does not match up with a standard rating of fuses or circuit breakers, the next higher standard overcurrent device rating may be used, provided that the overcurrent device does not exceed 800A per *NEC Section 240.4(B)*. This is not permitted, however, when the branch circuit supplies receptacles where cord-and-plug-connected appliances and similar electrical equipment could be used, because too many loads plugged into the circuit could result in an overload condition. The next standard fuse or circuit breaker rating may be used only when the circuit supplies a fixed load.

The permissible ampacity of conductors used on most electrical systems is found in *NEC Table 310.16*. However, the ampacities are subject to correction factors that must be applied where the ambient temperature around the conductor exceeds 86°F (30°C). This reduction is required even when the required ampacity reduction for more than three conductors in a raceway is also being applied.

For example, if six No. 10 AWG TW current-carrying conductors are installed in a single raceway where the ambient temperature is 104°F (40°C), the ampacity of 30A must be derated or reduced to 80% of its value because of the number of current-carrying conductors in the raceway. This is done in accordance with *NEC Table 310.15(C)(1)*. The ampacity must then be reduced further due to the ambient temperature by applying a correction factor of 0.82 per *NEC Table 310.15(B)(1)(1)*. The permissible ampacity for this condition is calculated as follows:

$$30\text{A} \times 0.80 \times 0.82 = 19.68\text{A}$$

The rating of the branch circuit overcurrent protective device serving continuous loads must be not less than the noncontinuous load plus 125% of the continuous load per *NEC Section 210.20(A)*.

1.1.2 Conductor Protection

According to the *NEC*®, conductors must be installed and protected from damage (both physically and electrically). Additional requirements specify the use of boxes or fittings for certain connections, specify how connections are made to terminals, and restrict the use of parallel conductors. When conductors are installed in enclosures or raceways, additional rules apply. Finally, if conductors are installed underground, the *NEC*® specifies the burial depth and other installation requirements.

All conductors must be protected against overcurrent in accordance with their ampacities as set forth in the *NEC*®. They must also be protected against overloads and ground fault / short circuit current damage.

Minimum Branch Circuit Conductor Size

The smallest size conductor permitted by ***NEC Section 310.3(A)*** for branch circuits in residential, commercial, and industrial locations is No. 14 copper. However, some local codes may require No. 12 or larger.

Know the Code

Branch Circuits Not Over 1000 Volts ac, 1500 Volts dc, Nominal

NEC Article 210

Know the Code

Overcurrent Devices Rated 800 Amperes or Less

NEC Section 240.4(B)

NOTE

NEC Table 310.15(B)(1)(1) or *NEC Table 310.15(B)(1)(2)* provides temperature correction factors that must be applied to the average ambient temperature for derating conductor ampacity when the conductors are installed in conduit exposed to direct sunlight on or above rooftops. *NEC Table 310.15(B)(2)* provides a temperature adder that is used when applying the correction factors.

NOTE

Remember that *NEC Article 100* defines a continuous load as one where the maximum current is expected to continue for three hours or more.

Think About It

Conductor Identification

Can you identify the ungrounded feeder conductors, the grounded feeder conductor, and the grounding feeder conductor in this picture?

Know the Code

Standard Ampere Ratings
NEC Section 240.6

Know the Code

Standard Ampere Ratings for Fuses and Inverse Time Circuit Breakers
NEC Table 240.6(A)

NOTE

Per *NEC Section 240.6(A)*, the small fuse ratings of 1A, 3A, 6A, and 10A were added to the *NEC*® to provide more effective overcurrent protection for small loads.

According to *NEC Section 240.6* and *NEC Table 240.6(A)*, standard overcurrent protective device ratings are as follows:

- 10A through 50A, in 5A increments
- 60A through 110A, in 10A increments
- 125A through 250A, in 25A increments
- 300A through 500A, in 50A increments
- 600A through 800A, in 100A increments
- 1,000A
- 1,200A
- 1,600A
- 2,000A
- 2,500A
- 3,000A through 6,000A, in 1,000A increments

Additional standard ratings for fuses include 1A, 3A, 6A, and 601A.

Protection of conductors under short circuit conditions is accomplished by determining the maximum short circuit current possible at the supply end of the conductor, the short circuit withstand rating of the conductor, and the short circuit let-through characteristics of the overcurrent protective device. *Let-through* refers to the current that can pass through the device before the circuit can be interrupted by the protective device.

When a noncurrent-limiting device is used for short circuit protection, the conductor's short circuit withstand rating must be properly selected based on the overcurrent protective device's ability to protect the circuit (*Figure 7*).

It is necessary to determine the energy let-through of the overcurrent protective device under short circuit conditions. Select a wire size that will be able to withstand the let-through current.

In contrast, the use of a current-limiting device (*Figure 8*) limits short circuit current to a level substantially less than possible in the same circuit if the overcurrent protective device were replaced with a solid conductor having comparable impedance (resistance). This eliminates the need for oversized ampacity conductors.

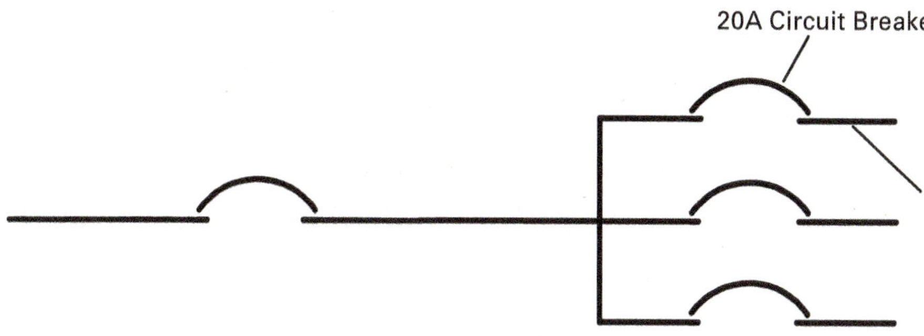

Figure 7 Noncurrent-limiting device.

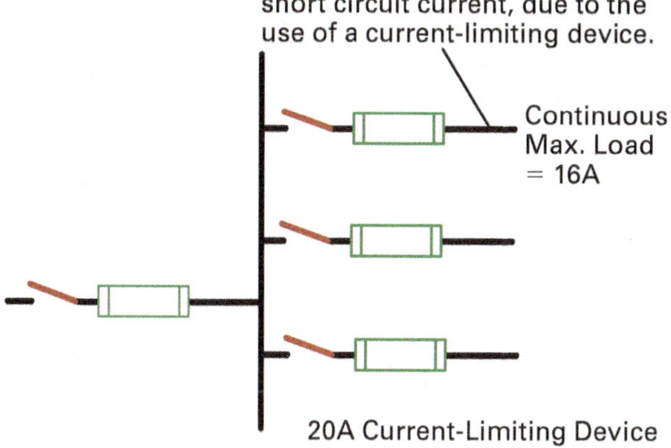

Figure 8 Current-limiting device used.

In many applications, it is desirable to have the convenience of a circuit breaker for a manual disconnect and general overcurrent protection, supplemented by current-limiting devices at strategic points in the circuits.

Per *NEC Section 240.5(A)*, flexible cords, including **tinsel cords** and extension cords, must be protected against overcurrent in accordance with their ampacities listed in *NEC Tables 400.5(A)(1) and (2)*. Supplementary overcurrent protection is acceptable. For example, for No. 18 AWG extension cords, a 10A overcurrent device would provide the necessary protection for a cord where only two conductors are carrying current. A 7A overcurrent device would provide the necessary protection for a cord where three conductors are carrying current.

In general, overcurrent protection must be installed at points where the conductors receive their supply, that is, at the beginning or line side of a branch circuit or feeder. See *NEC Section 240.21*. The exceptions to this rule are known as *tap rules*. The basic tap rules from the *NEC®* are presented in the sections that follow.

Circuit Breakers

Although a circuit breaker may appear to be a simple On/Off switch, it is actually a complex mechanical device. This picture shows the internal components of a typical circuit breaker.

Know the Code

Protection of Flexible Cords, Flexible Cables, and Fixture Wires
NEC Section 240.5

Tinsel cords: Cables typically used for telephones and similar applications that offer excellent flexibility but low current-carrying capacity.

Know the Code

Ampacity for Flexible Cords and Flexible Cables
NEC Tables 400.5(A)(1)

Know the Code

Ampacity of Cable Types SC, SCE, SCT, PPE, G, G-GC, and W
NEC Table 400.5(A)(2)

Know the Code

Location in Circuit
NEC Section 240.21

Know the Code

Feeder Taps
NEC Section 240.21(B)

Taps Up to 10' (3 m) Long

Per *NEC Section 240.21(B)(1)*, overcurrent protection is not required at the conductor supply if a tap conductor is not over 10' (3 m) long and meets all the following additional requirements:

- The tap conductors are enclosed in raceway.
- The tap conductor does not go beyond the switchboard, panelboard, disconnect, or a control device to which it provides power.
- The ampacity of the tap conductors is equal to or greater than the combined calculated loads and no less than the rating of the load they power.
- The ampacity of the tap conductors is not less than the rating of the overcurrent protective device at the tap conductor termination.

For field-installed taps, the ampacity of the tap conductors cannot be less than one-tenth of the rating of the overcurrent protective device protecting the feeder conductors.

Taps Up to 25' (7.5 m) Long

Per *NEC Section 240.21(B)(2)*, overcurrent protection is not required at the conductor supply if the tap conductor meets these criteria:

- The tap conductors are protected from physical damage.
- The ampacity of the tap conductors is not less than one-third that of the overcurrent protective device providing power to the feeder conductors.
- The tap conductors terminate at a single circuit breaker or a single set of fuses that limit the current to their ampacity.

Taps Associated with a Transformer (Primary and Secondary) Not Exceeding 25' (7.5 m) Long

Per *NEC Section 240.21(B)(3)*, additional overcurrent protection is not required at the conductor supply if all the conditions below are met:

- The total length of one primary conductor plus one secondary conductor does not exceed 25' (7.5 m), excluding any part of the primary conductor that is already protected at its ampacity.
- The primary and secondary conductors are protected from physical damage.
- The tap conductors terminate at a single circuit breaker or a single set of fuses that limit the current to the conductor's ampacity allowed by *NEC Section 310.14*.
- The tap conductors connected to the secondary winding of the transformer have an ampacity equal to or greater than the value of the transformer's primary-to-secondary voltage ratio multiplied by one-third of the feeder conductor's overcurrent protective device rating.
- The tap conductors powering the primary winding of the transformer have an ampacity at least one-third of the feeder conductor's overcurrent protective device rating.

Know the Code

Ampacities for Conductors Rated 0 Volts – 2000 Volts
NEC Section 310.14

Taps Exceeding 25' (7.5 m) Long

Per *NEC Section 240.21(B)(4)*, overcurrent protection is not required at the conductor supply in high-bay manufacturing buildings over 35' (11 m) high at its walls when all the following conditions are met:

- Facility policies restrict servicing of the system to qualified persons.
- The tap conductors are not over 25' (7.5 m) long horizontally and their total length does not exceed 100' (30 m).
- The ampacity of the tap conductors is equal to or greater than one-third of the rating of the feeder conductor's overcurrent protective device.

- The tap conductors terminate at a single circuit breaker or a single set of fuses that limit the current to their ampacity.
- The tap conductors are protected from physical damage.
- The tap conductors are at least No. 6 AWG copper or No. 4 AWG aluminum in size.
- The tap conductors have no splices (continuous from end to end).
- The tap conductors do not pass through any walls, floors, or ceilings.
- The taps are made at least 30' (9 m) above the floor.

WARNING!

Smaller conductors tapped to larger conductors can be a serious hazard. These unprotected conductors can vaporize or incur severe insulation damage when a short circuit occurs.

Outdoor Taps with Unlimited Length

Per *NEC Section 240.21(B)(5)*, overcurrent protection is not required at the conductor supply where the conductors are installed outside of a building or structure (except where they terminate), and when all the following conditions are met:

- The tap conductors are protected from physical damage.
- The tap conductors terminate at a single circuit breaker or a single set of fuses that limit the current to their ampacity.
- The disconnect serving the tap conductors is placed in an accessible location that is outside of the building, inside the building adjacent to the location where the tap conductors enter, or where they are installed as directed by *NEC Section 230.6*.
- The overcurrent protective device is integrated with the disconnect or installed immediately adjacent to it.

Know the Code

Conductors Considered Outside the Building
NEC Section 230.6

Transformer Secondary Conductors Not Exceeding 10' (3 m) Long

Per *NEC Section 240.21(C)(2)*, overcurrent protection is not required at a transformer secondary if the secondary conductor does not exceed 10' (3 m) in length and all the following conditions are met:

- The ampacity of the secondary winding conductors is equal to or greater than the combined calculated loads on the circuits powered by the transformer secondary.
- The ampacity of the secondary winding conductors is equal to or greater than the rating of the device they power, or equal to or greater than the rating of the overcurrent protective device located at their termination.
- The secondary winding conductors do not go beyond the switchboard, panelboard, disconnect, or a control device to which it provides power.
- The secondary winding conductors are installed in a raceway that travels from the transformer to the switchboard, switchgear, panelboard, or control device enclosure, or up to the conductor entrance to an open switchboard.
- When the secondary conductors leave the enclosure where the power connections are located, they must have an ampacity equal to or greater than the result of multiplying the primary-to-secondary voltage ratio by one-tenth of the rating of the overcurrent protective device serving the transformer primary winding.

Know the Code

Transformer Secondary Conductors
NEC Section 240.21(C)

Transformer Secondary Conductors Not Exceeding 25' (7.5 m) Long

Per *NEC Section 240.21(C)(6)*, overcurrent protection is not needed for a transformer secondary if the secondary conductors are less than 25' (7.5 m) long and all the following conditions are met:

- The ampacity of the secondary conductors must be equal to or greater than the result of multiplying the primary-to-secondary voltage ratio by one-third of the overcurrent protective device rating serving the primary winding of the transformer.
- The secondary conductors are protected from physical damage.
- The tap conductors terminate at a single circuit breaker or a single set of fuses that limit the current to their ampacity listed in *NEC Section 310.14*.

Other Tap Provisions for Transformers

Other tap provisions for single-phase, two-wire, and three-phase delta transformer secondaries are found in *NEC Section 240.21(C)(1)*. *NEC Section 240.21(C)(3)* covers secondary conductors in industrial installations.

NOTE

Switchboard and panelboard protection, along with transformer protection, must still be observed. See *NEC Sections 408.36 and 450.3*.

Think About It

Location of Overcurrent Protection

Is overcurrent protection required for the secondary conductors at the transformer terminals shown in the image?

1.2.0 Properties of Conductors

Various *NEC®* tables define the physical and electrical properties of conductors. Electricians use these tables to select the type of conductor and the size of conduit or raceway to enclose the conductors for specific applications. *NEC®* tables list the properties of conductors as follows:

- Name
- Operating temperature
- Application
- Insulation
- Physical properties
- Electrical resistance
- AC resistance and reactance

Know the Code

Conductor Applications and Insulations Rated 600 Volts

NEC Table 310.4(1)

NEC Table 310.4(1) lists the name, maximum operating temperature, applications, and insulation of various types of conductors. *NEC Chapter 9, Table 8* provides the physical properties and electrical resistance.

To better understand the *NEC®* tables and how they are used, the following steps will guide you through how to determine the properties of a 4/0 THHN copper conductor using the tables.

Step 1 Turn to *NEC Table 310.4(1)* and scan down the *Type Letter* column (second from the left) to find THHN. Scan to the left in this row to see that the trade name of this conductor is heat-resistant thermoplastic.

Step 2 Scanning to the right in this row, note that the maximum operating temperature for this wire type is 194°F (90°C). Continuing to the right in this row, under the *Application Provisions* column, you will see that this wire type is suitable for use in dry and damp locations. The next column reveals that the insulation is flame-retardant, heat-resistant thermoplastic.

Step 3 Continuing to the right in this row, the next column lists insulation thickness for various AWG or kcmil wire sizes. Note that the insulation thickness for 4/0 THHN copper wire is 50 mils.

Step 4 Look in the *Outer Covering* column, which states that this wire type has a nylon jacket or equivalent.

Think About It

Conductor Derating

If six current-carrying conductors are bundled tightly together with no spacing, but are installed in an open tray instead of an enclosed conduit, do they have to be derated?

Know the Code

Ampacities of Insulated Conductors with Not More Than Three Current-Carrying Conductors in Raceway, Cable, or Earth (Directly Buried)
NEC Table 310.16

If you want to find the maximum current-carrying capacity of this same conductor when it is used in a raceway, turn to *NEC Table 310.16* and proceed as follows:

Step 1 Scan down the left-hand column until the wire size is found. In this case, locate 4/0.

Step 2 Scan to the right in this row until the first *90°C (194°F)* column is found. This column covers the insulation types that are rated for this maximum operating temperature and includes Type THHN conductor insulation.

Step 3 Note that the current-carrying rating for this size and type of conductor is 260A. This rating, however, is for not more than three conductors in a raceway. If there are more than three conductors, this 260A value must be derated.

Step 4 Refer to *NEC Section 310.15(C)(1)*, which states that when there are more than three current-carrying conductors in a raceway or cable, the permissible ampacities shall be reduced as shown in *NEC Table 310.15(C)(1)*. The table indicates that, for four conductors, the previously determined current of 260A should be reduced to 80%.

$$260A \times 0.80 = 208A$$

These current tables are also based on the conductors being installed in areas where the ambient temperature is 86°F (30°C). If the conductors are installed in areas with different temperatures, a further reduction is required. For example, if more than three 4/0 THHN conductors were installed in an industrial area where the ambient temperature averaged 95°F (35°C), you would need to check the correction factors in *NEC Table 310.15(B)(1)(1)*. The correction factor is 0.96 for this particular conductor. Therefore, the previously calculated current capacity of 208A must be multiplied by 0.96 to obtain the capacity of the four conductors at the higher temperature:

$$208A \times 0.96 = 199.68A$$

Know the Code

Adjustment Factors for More Than Three Current-Carrying Conductors
NEC Table 310.15(C)(1)

Know the Code

Ambient Temperature Correction Factors Based on 30°C (86°F)
NEC Table 310.15(B)(1)(1)

Other conductors are handled in a similar manner. Find the appropriate table, determine the listed ampacity, and then multiply the ampacity by the appropriate correction factors from the tables.

It is sometimes necessary to know other properties of a conductor for some calculations, especially for voltage drop calculations. Voltage drop calculations are presented in *Section 2.3.0*.

There are many useful tables found in *NEC Chapter 9*. Examples of their practical use are presented throughout *Section 2.0.0*.

Know the Code

Tables
NEC Chapter 9

High Leg Connection

NEC Section 408.3(E)(1) requires that panelboards supplied by a three-phase, four-wire delta service have the high leg connected to the center phase, designated "B phase."

1.2.1 Color Coding

The *NEC®* specifies certain methods of identifying conductors used in wiring systems of all types. For example, the high leg of a 120/240V, grounded three-phase, four-wire delta system must be marked with an orange color for identification.

A grounded conductor must be identified either by the color of its insulation (white or gray finish), markings at the terminals, or other suitable means such as tagging. When this is not practical for conductors larger than No. 6 AWG, painting the terminals white is an acceptable method of identifying them.

Conductors bundled as cable are color coded for easy identification. When conductors are installed in raceway systems, any color insulation is permitted for the ungrounded phase conductors, except for the following:

- White or gray, which is reserved to identify the grounded conductor
- Green, which is reserved to identify a grounding conductor
- Orange, which is reserved to identify the high leg of a three-phase, four-wire delta system

1.2.2 Changing Colors

Should it become necessary to change the actual color of a conductor to meet *NEC®* requirements or to facilitate maintenance, the conductors may be reidentified with nonconductive colored tape or paint.

For example, assume that a two-wire cable containing a black-and-white conductor is used to feed a 240V, two-wire, single-phase motor. Since the white-colored conductor is supposed to be reserved for the grounded conductor, and none is required in this circuit, the white conductor may be marked with a piece of black tape at each end of the circuit so that qualified craftworkers will know that it is not a grounded conductor.

1.0.0 Section Review

1. The circuit between the final overcurrent device and the outlet is known as a _____.
   a. feeder
   b. branch circuit
   c. runner
   d. subfeeder

2. The insulation thickness of a 2/0 THHN conductor is _____.
   a. 20 mils
   b. 30 mils
   c. 40 mils
   d. 50 mils

2.0.0 Sizing Conductors

Performance Tasks

There are no Performance Tasks in this section.

Objective

Size conductors based on expected load and voltage drop.

a. Calculate wire sizes based on resistance.
b. Calculate conductor resistances.
c. Calculate voltage drops for various applications.

Conductors for power, heating, and lighting loads should be sized so that the voltage drop never exceeds 3%. In addition, the maximum total voltage drop for conductors, feeders, and branch circuits combined should not exceed 5%. These percentages are recommended by industry standards and by the informational note in *NEC Section 210.19*, but they are not requirements. However, it is considered good practice to respect these limits in all installations.

In some applications, such as for circuits feeding hospital imaging equipment, managing the voltage drop is even more critical. Equipment manufacturers typically require a maximum of 2% voltage drop in the power supply. It is extremely important to keep voltage drop in mind. Otherwise, serious problems are likely to occur. The resistance and voltage drop on long conductor runs can seriously interfere with the efficient operation of the connected equipment.

Electric motors are not adversely affected by small voltage variations, but they will not produce rated horsepower if the voltage is reduced below the nameplate value. When loaded motors operate at a reduced voltage, the current flow increases, as it requires more current to produce power at a lower voltage. This current increase is also caused by the opposition of the motor windings to current flow, which decreases as the motor speed decreases.

In long conductor runs, wire size must be determined by considering resistance and voltage drop along with the heating effect and *NEC Tables 310.16 through 310.21*.

Voltage drop problems can be avoided if you know a few simple facts about the areas and resistances of conductors and can apply some basic mathematical equations.

Know the Code

Conductors — Minimum Ampacity and Size

NEC Section 210.19

Know the Code

Conductors for General Wiring

NEC Article 310

Did You Know?

Voltage Drop Calculator

Southwire offers an online voltage drop calculator, which can be accessed by going to **Southwire.com** and using their Resources page. This calculator can also be downloaded to a mobile device for easy access in the field.

2.1.0 Wire Sizes Based on Resistance

You have learned that conductors are specified in kcmil or AWG sizes. The Brown & Sharpe Company developed the sizing system, which was originally called the *B&S gauge*. However, it was quickly adopted as the American Wire Gauge (AWG) and is now the standard in the United States.

AWG numbers are arranged according to the resistance of the wires, with the larger numbers representing the wires of greatest resistance and least cross-sectional area. A handy rule to remember is that decreasing the gauge by three numbers results in a wire of roughly twice the area and half the resistance. Conversely, increasing the gauge by three numbers results in a wire with roughly half the area and twice the resistance.

For example, a No. 3 AWG wire has a resistance of 0.245 ohm per 1,000' (Ω/1,000'), or 0.802 ohm per kilometer (Ω/km). Changing this wire to a No. 6 AWG results in a resistance of 0.491Ω/1,000' (1.608Ω/km). See *NEC Chapter 9, Table 8*.

Although *NEC Chapter 9, Table 8* only lists sizes down to No. 18 AWG, the American Wire Gauge numbers range from 0000 (4/0) to No. 40. The 4/0 conductor is more than ½" in diameter while the No. 40 wire is as fine as a human hair.

The most common sizes used for light and power installations range from 4/0 to No. 14 AWG. Light fixture wires are frequently size No. 16 or No. 18 AWG, since the load is relatively small. Low-voltage control wiring can be as small as No. 22 AWG as long as the run is reasonably short, since voltage drop is a significant factor.

2.1.1 The Mil

A unit called the *mil* is also used for measuring the diameter of conductors. One mil is equal to $^1/_{1,000}$", so it is a small enough unit to express wire sizes very accurately. For example, a wire with a diameter of 0.055" is more simply expressed as 55 mils. A wire that is 250 mils in diameter measures 0.250", or $^1/_4$".

Because the resistance and current-carrying capacity of conductors depend on a conductor's cross-sectional area, a unit is needed to express it as well. Electrical conductors are made in several different shapes as shown in *Figure 9*. You are already familiar with the circular mil. For square conductors, such as busbars, the square mil (mil^2) is used, which represents a square measuring $^1/_{1,000}$" on each side.

These units simplify conductor calculations. For example, to determine the area of a square conductor, multiply one side by the other, measuring one side in thousandths of an inch with a caliper and converting it to mils.

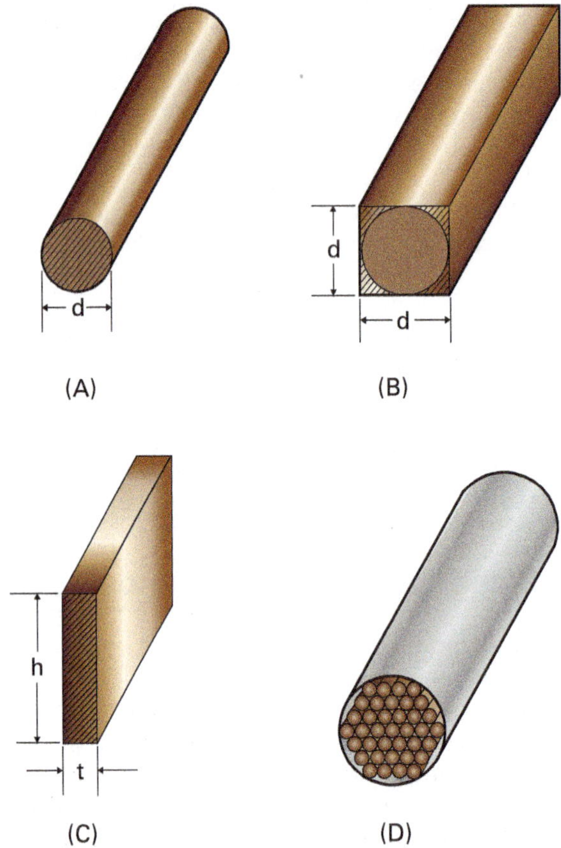

Figure 9 Various conductor shapes.

Busbars

Busbars are sized using square mils. They are often used in industrial installations, such as the one shown in the image.

Source: iStock@Burkay Dogan

2.1.2 Converting Square Mils to Circular Mils

When comparing round and square conductors, remember that the square mil and the circular mil are not equal units of area. For example, *Figure 9* (B) shows a circle within a square. Although the diameter of the circle is the same as the width of the square, the corners of the square provide more area. Therefore, the area in one circular mil is less than one square mil.

The actual ratio between the two is 0.7854. In other words, the circular mil has only 78.54% of the area of a square mil. To convert square mils to circular mils, divide the square mils by 0.7854. To convert circular mils to square mils, multiply the circular mils by 0.7854.

For example, the conductor in *Figure 9* (A) is a No. 4/0 conductor with an area of 211,600 circular mils (cmil). What is its area in square mils?

$$\text{Square mils} = 211{,}600 \text{ cmil} \times 0.7854$$

$$\text{Square mils} = 166{,}190.64 \text{ mil}^2\text{, rounded to } 166{,}191 \text{ mil}^2$$

If the busbar in *Figure 9* (C) is 1½" high and ¼" thick, what is its area in square mils, and what size of round conductor would be necessary to carry the same current as this busbar?

The dimensions of a ¼" × 1½" busbar, stated in mils, are 250 mils × 1,500 mils. Therefore, the area in square mils is determined by multiplying these two numbers like any other area calculation:

$$250 \text{ mils} \times 1{,}500 \text{ mils} = 375{,}000 \text{ mil}^2$$

To convert to circular mils, divide the square mil area by 0.7854:

$$375{,}000 \text{ mil}^2 \div 0.7854 = 477{,}463.71\text{, rounded to } 477{,}464 \text{ cmil}$$

The nearest standard wire size to this is a round conductor 500,000 cmil, or 500 kcmil, in size.

Busbars of the shape shown in *Figure 9* (C) are commonly used in panelboards and switchgear. Standard busbars range from 0.250" to 0.375" in thickness and from 1" to 12" in height.

Stranded conductors such as the one shown in *Figure 9* (D) are used on conductor sizes No. 6 AWG and larger. Because these conductors are not a solid single strand, the actual conductive area cannot be accurately determined. The diameter varies somewhat with the position of the strands, and there is often some air space between them.

To determine the cross-sectional area of these conductors, first determine the area of each strand using a table or by calculating it from the diameter. Then multiply this by the number of strands to get the total conductive area of the cable. *NEC Chapter 9, Table 8* can also be used to find the area of most conductors in circular mils. It also provides information necessary for other wiring calculations.

2.2.0 Conductor Resistances

You may need to determine the exact resistance of a length of conductor to calculate the voltage drop. Resistances can be calculated per foot or per mil foot.

2.2.1 Resistance of Copper per Foot

The resistance per 1,000' (Ω/km) of various conductors can be obtained from *NEC Chapter 9, Table 8*. These conductor specifications are needed to determine the voltage drop for various sizes of conductors.

For example, to find the total resistance of a 120V, two-wire circuit consisting of two No. 10 AWG solid copper conductors, each of which is 150' (45 m) long, proceed as follows:

2 × 150' = 300' total length

Step 1 The length of one conductor is first multiplied by 2 to obtain the total length of the conductors.

Step 2 Refer to *NEC Chapter 9, Table 8* to find the resistance of No. 10 solid copper wire. The table shows a resistance of 1.21Ω/1,000', or 3.984Ω/km, for uncoated copper wire at 167°F (75°C).

300' ÷ 1,000' = 0.3

Step 3 Since the circuit is less than 1,000', determine the resistance multiplier by dividing the actual footage by 1,000'.

90 m ÷ 300 m = 0.3

Stated in meters, the formula provides the same result for a multiplier.

1.21Ω × 0.3 = 0.363Ω

Step 4 Multiply 1.21Ω by 0.3 to determine the resistance in 300' of conductor.

Here is another example. Suppose you want to install an outside 120V, two-wire line between two buildings 1,650' apart using No. 1 AWG copper wire. What would be the total resistance of the conductors?

1,650' × 2 = 3,300' total length

Step 1 Determine the total length of both conductors by multiplying the length in one direction by 2.

Step 2 According to *NEC Chapter 9, Table 8*, No. 1 AWG uncoated copper wire has a resistance of 0.154Ω/1,000'.

3,300' ÷ 1,000' = 3.3

Step 3 The total length of 3,300' must be divided by 1,000' to determine the multiplier for the resistance.

0.154Ω × 3.3 = 0.5082Ω

Step 4 Multiply the resistance of 0.154Ω/1,000' by 3.3 to determine the total resistance of the 3,300' run.

Now we will see what happens if we apply the full current allowed by *NEC Table 310.16* to these conductors. Assuming that Type THHN conductors are used, *NEC Table 310.16* allows a maximum load on these conductors of 145A at 194°F (90°C). Note that you should always use this column in the table when derating conductors. If 145A of current flowed through this circuit for a distance of 3,300' (both ways), the voltage drop in the circuit is determined as shown here:

$$\text{Voltage drop} = I \times R$$

$$\text{Voltage drop} = 145\text{A} \times 0.5082\Omega = 73.7\text{V}$$

If the initial voltage is only 120V, this means that the voltage at the far end of the circuit will only be 46.3V. Devices designed to operate at 120V will not operate at this voltage. Consequently, the load will have to be reduced significantly to make this circuit functional.

Remember that the goal is to keep the voltage drop within 3% of the original voltage. Since the voltage is 120V, the permissible voltage drop can be found by using the following equation:

$$120\text{V} \times 0.03 = 3.6\text{V}$$

What level of current can be applied to this circuit to stay within this range? The maximum current for this circuit to avoid exceeding the permissible voltage drop is calculated using the following equation:

$$\text{Maximum current } (I) \times \text{total resistance} = \text{original voltage} \times \text{permissible voltage drop}$$

Substituting the known values in this equation results in the following:

$$I \times 0.5082\Omega = 120\text{V} \times 0.03$$

We can solve the right side of the equation easily, as was done above:

$$I \times 0.5082\Omega = 3.6V$$

To solve for I, divide both sides of the equation by 0.5082Ω, which results in the following:

$$I = 3.6V \div 0.5082\Omega$$

$$I = 7.084A$$

Therefore, to keep the voltage drop within 3% for this length of circuit with 120V applied, the amperage must not exceed 7.084A.

If a larger load must be powered by this circuit, either a larger conductor will have to be used or the voltage will have to be increased. For example, assume that this circuit feeds a 120/240V dual-voltage pump motor. If the motor connections were rewired to accept a 240V branch circuit, then the permissible 3% voltage drop would be calculated as follows:

$$0.03 \times 240V = 7.2V$$

Continuing as before, we can determine the maximum current available:

$$I = 7.2V \div 0.5082\Omega$$

$$I = 14.17A$$

Note that this is twice the current available at 120V. If 480V were applied to the circuit (even though it is not feasible with this two-wire arrangement), the available current would again double to 28.34A.

These examples show why power transmission lines must operate at extremely high voltages—to keep the available current high, the voltage drop low, and the conductor size reasonable.

2.2.2 Resistance of Copper per Mil Foot

In many cases, it may be necessary to calculate the resistance of a specific length of wire or a busbar of a given size.

This can be done very easily if the unit resistance of copper is known. A unit called the *mil foot* may be used. A mil foot represents a piece of round wire that is 1 mil in diameter and 1' in length. This is a small enough unit to be very accurate for calculations. Note that a round wire 1 mil in diameter has an area of 1 cmil.

The resistance of ordinary copper is 12.9Ω per mil foot (Ω / mil ft) at 167°F (75°C). This value is considered a *constant* and should be remembered. It is normally represented by the letter *K*.

For practice, determine the resistance of a piece of No. 12 copper wire that is 50' long. Remember that the resistance of any conductor increases as its length increases, while the resistance decreases as its area increases. For a wire that is 50' long, first multiply 50' by 12.9Ω, which results in 645Ω / mil ft. This is the resistance of a wire that is 1 cmil in area and 50' long.

NEC Chapter 9, Table 8 shows the area of a No. 12 wire is 6,530 cmil, which will reduce the resistance proportionally. Now divide the resistance per mil foot by the area in circular mils. The result is the resistance for 50' of No. 12 wire:

$$645\Omega \div 6{,}530 \text{ cmil} = 0.099\Omega \text{ / mil ft}$$

For additional practice, find the resistance of a coil of wire containing 3,000' of No. 18 stranded wire. First, multiply 3,000' by the constant 12.9Ω to find the resistance based on the length of 38,700Ω / mil ft:

$$3{,}000' \times 12.9\Omega = 38{,}700\Omega \text{ / mil ft}$$

Since the area of No. 18 wire is 1,620 cmil, the total resistance is found using the following equation:

$$38{,}700\Omega \text{ / mil ft} \div 1{,}620 \text{ cmil} = 23.89\Omega$$

To check the result, you can calculate it another way. *NEC Chapter 9, Table 8* shows a resistance of 7.95Ω/1,000' for No. 18 stranded AWG wire. For 3,000', use the following equation:

$$3 \times 7.95\Omega = 23.85\Omega$$

The small difference between this answer and the previous calculation is due to the use of approximate figures instead of numbers with many decimal places.

The mil foot unit and its resistance of 12.9Ω / mil ft for copper can also be used to calculate the resistance of square busbars. This can be accomplished by simply using the factor 0.7854 to convert from square mils to circular mils, as discussed in *Section 2.1.2*.

Suppose you need to find the resistance of a square busbar that measures ¼" × 2" and 100' long. The area in square mils is determined by multiplying the height by the width:

$$250 \text{ mils} \times 2{,}000 \text{ mils} = 500{,}000 \text{ mil}^2$$

To find the circular mil area, divide the area in square mils by the conversion factor:

$$500{,}000 \text{ mil}^2 \div 0.7854 = 636{,}618.28 \text{ cmil}$$

Since the busbar is 100' long, multiply 100' by the constant 12.9Ω / mil ft to determine the resistance based on length:

$$100' \times 12.9\Omega / \text{mil ft} = 1{,}290\Omega$$

Since the area of this bar was calculated to be 636,618.3 cmil, divide the resistance by the area in circular mils as follows:

$$1{,}290\Omega \div 636{,}618.28 \text{ cmil} = 0.002\Omega$$

total resistance of the busbar

NEC Tables 310.16 through 310.21 list the permissible ampacity of conductors with various types of insulation. These tables, however, do not take into consideration the length of the conductors or the voltage drop. Therefore, it is often necessary to use a larger size conductor than the tables indicate.

2.3.0 Calculating Voltage Drops

The size of the conductor required to connect an electrical load to the source of supply is determined by several factors, including:

- Load current
- Permissible voltage drop between source and load
- Total length of conductor
- Type of wire, such as copper, aluminum, copper-clad aluminum, etc., and its permissible ampacity based on *NEC Tables 310.16 through 310.21*

It is important to remember the following principles when performing voltage drop calculations.

The resistance (R) of a wire varies directly with its length (L):

$$R = \text{resistance per foot } (K) \times \text{length in feet } (L)$$

The resistance (R) varies inversely with its cross-sectional area (A):

$$R = 1 \div A$$

Combine these two statements to obtain the following equation:

$$R = (L \times K) \div A$$

Where:

A = wire size from *NEC Tables 310.16 through 310.21* for a given load, and then, if necessary, from *NEC Chapter 9, Table 8* for circular mils

L = length of wire in feet

R = total resistance of wire

K = resistance per mil foot

K is a constant whose value depends upon the units chosen and the type of wire (12.9Ω / mil ft for copper and 21.2Ω / mil ft for aluminum). Using the foot as the unit of length and the circular mil as the unit of area, the values for K represent the resistance in ohms per mil foot at 167°F (75°C).

The length in the equation is the total length of a single current-carrying conductor. For a two-conductor, 120V circuit or a 240V balanced, three-wire circuit (no current on the neutral), the length in the equation must be multiplied by 2. The equation for single-phase circuits would then look like this:

$$R = (2 \times L \times K) \div A$$

For a three-phase, four-wire balanced circuit with a power factor near the value of 1, the equation must be multiplied by √3, or 1.732. The equation for this application looks like this:

$$R = (\sqrt{3} \div 2) \times [(2 \times L \times K) \div A]$$

The 2s cancel each other out, so the equation can be simplified to the following:

$$R = (\sqrt{3} \times L \times K) \div A$$

Since $R = E \div I$, where E represents the voltage drop (VD), substitute $VD \div I$ for R in the previous equation for single-phase power supplies:

$$VD \div I = (2 \times L \times K) \div A$$

For three-phase balanced circuits, the equation now looks like this:

$$VD \div I = (\sqrt{3} \times L \times K) \div A$$

Multiply each side of both equations by I:

$$VD = (2 \times L \times K \times I) \div A$$

and:

$$VD = (\sqrt{3} \times L \times K \times I) \div A$$

Substitute circular mils for A where the area is initially determined from the ampacities listed in *NEC Tables 310.16 through 310.21* for the desired load, and where *NEC Chapter 9, Table 8* is used to convert the wire size to circular mils, if required. The equations for single-phase circuits and three-phase balanced circuits, respectively, become as follows:

$$VD = (2 \times L \times K \times I) \div \text{cmil}$$

and:

$$VD = (\sqrt{3} \times L \times K \times I) \div \text{cmil}$$

Using a similar exercise, the following equations can be derived for a given wire size resistance and load. Again, the equations for single-phase circuits and three-phase balanced circuits are shown in that order:

$$VD = (2 \times L \times R \times I) \div 1{,}000$$

and:

$$VD = (\sqrt{3} \times L \times R \times I) \div 1{,}000$$

Think About It

Voltage Drop

This parking lot fixture has three 1,000W lamps, is 300' from the panel, uses No. 12 wire, and is supplied by a 20A, 208V circuit. The fixture manufacturer specifies a maximum voltage drop of 3%. Can this installation adequately support the load?

Source: Tri-City Electrical Contractors Inc

2.3.1 Voltage Drop Example

To determine the voltage drop on an existing 120V, two-wire project, find the voltage drop of the circuit by using one of the single-phase formulas:

$$VD = (2 \times L \times K \times I) \div \text{cmil}$$

For example, assume that a 120V, copper two-wire circuit feeds a total load of 30A with a branch circuit length of 120'.

NEC Table 310.16 allows a No. 10 AWG conductor at 140°F (60°C) to carry this load. Note that this does not consider voltage drop or possible derating factors. *NEC Chapter 9, Table 8* shows that the area of a No. 10 AWG conductor is 10,380 cmil. Substitute these values and a value of 12.9Ω / mil ft for *K* in the equation:

$$VD = (2 \times 120' \times 12.9\Omega \text{ / mil ft} \times 30\text{A}) \div 10{,}380 \text{ cmil}$$

$$VD = 8.95\text{V}$$

The voltage drop can then be subtracted from the original voltage to determine the final voltage at the load:

$$120\text{V} - 8.95\text{V} = 111.05\text{V}$$

The resulting voltage at the load is 111.05V, which is not acceptable on most installations because it exceeds the 3% recommended maximum voltage drop of 3.6V for 120V circuits. Indeed, it is more than double the maximum voltage drop.

There are three ways to correct this situation:

- Reduce the load
- Increase the conductor size
- Increase the voltage

2.3.2 Miscellaneous Voltage Drop Equations

The final voltage drop equations presented earlier can be solved for length (*L*), yielding the following equations for single-phase circuits:

$$L = (VD \times \text{cmil}) \div (2 \times K \times I)$$

or:

$$L = (1{,}000 \times VD) \div (2 \times R \times I)$$

For three-phase balanced circuits, the equation is as follows:

$$L = (VD \times \text{cmil}) \div (\sqrt{3} \times K \times I)$$

or:

$$L = (1{,}000 \times VD) \div (\sqrt{3} \times R \times I)$$

These equations can be used to determine the maximum circuit length for a specified voltage drop at a given load using a wire size selected from *NEC Tables 310.16 through 310.21* for the ampacity of the load. *NEC Chapter 9, Table 8* may be used to convert the selected wire size to circular mils, if required.

For example, determine the maximum circuit length for a feeder serving a 208V balanced three-phase load of 400A at a voltage drop not to exceed 3%. From *NEC Table 310.16,* determine that a 750 kcmil conductor at 140°F (60°C) is rated to carry 400A. Using the three-phase length equation containing circular mils, substitute values and calculate length:

$$L = (VD \times \text{cmil}) \div (\sqrt{3} \times K \times I)$$

$$L = (208\text{V} \times 0.03 \times 750{,}000 \text{ cmil}) \div (1.732 \times 12.9 \times 400\text{A})$$

$$L = 523.6'$$

Therefore, the 400A load may be positioned up to 523.6' from the source and the voltage drop will be 3% or less using 750 kcmil conductors at 140°F (60°C).

The two final voltage drop equations provided earlier can also be solved for wire size in circular mils to directly determine, under certain conditions, the wire size for a given voltage drop, wire length, and load. Solving for circular mils yields the following for single-phase and three-phase balanced circuits, respectively:

$$\text{cmil} = (2 \times L \times K \times I) \div VD$$

and:

$$\text{cmil} = (\sqrt{3} \times L \times K \times I) \div VD$$

The results from these two circular mil equations must be used in accordance with the following two rules:

- *Rule 1* — If the wire size calculated using the circular mil equation results in a wire size with less ampacity than the given load determined from *NEC Tables 310.16 through 310.21*, discard the calculated result and use the wire size with the rated ampacity for the given load.
- *Rule 2* — If the wire size calculated using the circular mil equation results in a wire size equal to or greater than the wire size determined from *NEC Tables 310.16 through 310.21* for the given load, use the calculated wire size or the next larger wire size.

The reason for these rules is that the equation results are for non-derated bare wire. The wire sizes determined from the *NEC*® tables are derated for the number of wires, temperature, and other factors. Calculated sizes that are smaller than those sizes rated at the desired ampacity cannot be used. However, calculated wire sizes that are larger may be used when excessive voltage drop occurs solely due to the size of wire that has already been derated, and that is caused by additional length beyond the maximum length for a specified voltage drop and load.

To illustrate, solve for circular mils using the terms from the example for a 400A, three-phase load with a length of 523.6':

$$\text{cmil} = (\sqrt{3} \times L \times K \times I) \div VD$$

Substituting known values results in the following:

$$\text{cmil} = (1.732 \times 523.6' \times 12.9 \times 400\text{A}) \div (208\text{V} \times 0.03)$$

$$\text{cmil} = 749{,}916.03, \text{ or } 750 \text{ kcmil}$$

The length and load correlate exactly with the 3% voltage drop, and a 750 kcmil conductor was obtained from the calculation. In accordance with Rule 2, the size obtained equaled the size of a 140°F (60°C) conductor with a rated ampacity of 400A. Note that longer lengths for this same load will result in larger conductor sizes, each of which will be proven correct if cross-checked using the voltage drop equations.

Now take the same problem but change the 523.6' to 100' and solve for circular mils:

$$\text{cmil} = (\sqrt{3} \times L \times K \times I) \div VD$$

$$\text{cmil} = (\sqrt{3} \times L \times K \times I) \div VD$$

$$\text{cmil} = (1.732 \times 100' \times 12.9 \times 400\text{A}) \div (208\text{V} \times 0.03)$$

$$\text{cmil} = 143{,}223.08, \text{ rounded to } 143{,}223 \text{ cmil}$$

From *NEC Chapter 9, Table 8*, the next size standard conductor above this is 167,800 cmil or a No. 3/0 AWG conductor. However, when *NEC Table 310.16* is checked, the maximum permissible ampacity of a No. 3/0 AWG conductor is found to be only 165A—far below the load of 400A.

Therefore, in accordance with Rule 1, the 3/0 AWG conductor solution would be discarded and 750 kcmil conductors rated at 140°F (60°C), determined from *NEC Table 310.16* for the 400A load, would be used for the 100' run.

Even when sizing wire for low-voltage (24VAC) control circuits, the voltage drop should be limited to 3% because excessive voltage drop causes the following problems:

- Failure of relay and contactor solenoid coils to develop enough magnetism to operate the control
- Relay and contactor "chatter" (rapid opening and closing)
- Relay and contactor coil failure

The voltage drop calculations described previously may also be used for low-voltage wiring, but tables are quite common and can save on calculation time. One example is shown in *Table 1*.

TABLE 1 Capacity of 24VAC Wiring for Given Circuit Lengths at 3% Voltage Drop

Wire Size AWG	25' (7.5 m)	50' (15 m)	75' (22.5 m)	100' (30 m)	125' (37.5 m)	150' (45 m)	175' (52.5 m)	200' (60 m)	225' (67.5 m)	250' (75 m)	275' (82.5 m)	300' (90 m)
20	29VA	14VA	10VA	7.2VA	5.8VA	4.8VA	4.1VA	3.8VA	3.2VA	2.9VA	2.8VA	2.4VA
18	58VA	29VA	19VA	14VA	11VA	9.6VA	8.2VA	7.2VA	6.4VA	5.8VA	5.2VA	4.8VA
16	86VA	43VA	29VA	22VA	17VA	14VA	12VA	11VA	9.6VA	8.7VA	7.8VA	7.2VA
14	133VA	67VA	44VA	33VA	27VA	22VA	19VA	17VA	15VA	13VA	12VA	11VA

To use *Table 1*, assume there is a load of 35VA with a 50' (15 m) run for a 24V control circuit. Referring to the table, scan the 50' (15 m) column. Note that No. 18 AWG wire will carry 29VA, while No. 16 wire will carry 43VA. In this case, No. 16 wire would be the proper choice.

The 3% voltage drop limitation is imposed to assure proper operation when the power supply falls below the rated voltage. For example, if a 240V primary supply is decreased 10% (216V), a 240V/24V transformer secondary winding will only produce 21.6V. When the normal voltage drop expected is taken from this 21.6V, it approaches or can exceed the lower operating limit of many control devices.

In most installations, several lines connect the transformer to the control circuit. One lead, referred to as *Common*, usually carries the full load of the control circuit from the secondary side of the transformer. Therefore, the line from the secondary side of the transformer is the most critical regarding voltage drop and VA capacity and must be properly sized.

When low-voltage wiring is installed, it is suggested that one extra conductor be run for emergency/backup purposes. This wire can be substituted for any one of the existing lines that may be defective or damaged in the future.

Think About It

Voltage Drops in Low-Voltage Devices

You are adding a 12V/12W head 50' (15 m) away from an emergency pack. Use the voltage drop calculations to determine what size wire you should install.

Hollow Transmission Line

The copper cable shown in the image was part of the original 287.5 kV transmission line that supplied electricity to Los Angeles from the generators at the Hoover Dam. Its spiral-wound, interlocked design provided additional strength and flexibility while reducing the effects of thermal expansion.

The original cable had a tube within a tube (not shown in the image). This reduced the weight of the conductors and aided in heat dissipation. The hollow design enhanced the conductive quality of the cable due to a phenomenon known as the *skin effect.*

Because it travels in sine waves, alternating current tends to create internal interference within a conductor (eddy currents), causing most of the current to travel on the external surface or skin of the conductor rather than through the core. The skin effect of modern high-voltage transmission lines is often compensated for by using specially braided cable with a steel reinforcing core.

2.0.0 Section Review

1. The smallest wire size in the following list is _____.
   a. No. 40
   b. No. 18
   c. 1/0
   d. 3/0

2. The total resistance of a 120V, two-wire circuit consisting of two No. 12 AWG solid copper conductors, each 100' (30 m) long is _____.
   a. 0.386Ω
   b. 1.93Ω
   c. 2.47Ω
   d. 3.86Ω

3. If a 120V, two-wire circuit using No. 8 AWG copper wire feeds a total load of 40A with a branch circuit length of 100', the voltage drop is _____.
   a. 1.75V
   b. 3.65V
   c. 4.72V
   d. 6.25V

Module 26302-23 Review Questions

1. Which of the following is the *best* conductor?
   a. Bronze
   b. Brass
   c. Aluminum
   d. Copper

2. Which of the following is a standard configuration for stranded wire?
   a. 7 strand
   b. 12 strand
   c. 20 strand
   d. 26 strand

3. Of the following wire sizes, the *largest* standard AWG wire size is _____.
   a. No. 3 AWG
   b. No. 1 AWG
   c. No. 1/0 AWG
   d. No. 4/0 AWG

4. A branch circuit is *best* described as _____.
   a. the circuit conductors between the service equipment and the final branch circuit overcurrent device
   b. the circuit conductors between the final overcurrent device protecting the circuit and the outlet(s)
   c. a conductor tapped onto another conductor
   d. two conductors run in parallel

5. The rating of a branch circuit is determined by the _____.
   a. type and size of the branch circuit conductors
   b. rating of the branch circuit overcurrent device
   c. total connected load
   d. characteristics of the supply circuit

6. A continuous load is one in which the maximum current is expected to continue for _____.
   a. one hour or more
   b. two hours or more
   c. three hours or more
   d. four hours or more

7. Which of the following is a standard fuse rating?
   a. 201A
   b. 400A
   c. 620A
   d. 850A

8. You can find information on conductor applications and descriptions of insulation types in _____.
   a. *NEC Table 310.4(1)*
   b. *NEC Table 310.15(B)(1)(1)*
   c. *NEC Table 310.16*
   d. *NEC Chapter 9, Table 8*

9. Use *NEC Table 310.16* to find the maximum ampacity for a No. 2 AWG copper wire with THHN insulation at 194°F (90°C).
   a. 95A
   b. 110A
   c. 130A
   d. 150A

10. The constant (*K*) for the resistance of ordinary copper is _____.
    a. 12.9Ω per mil foot
    b. 21.2Ω per mil foot
    c. 1.732Ω per mil foot
    d. 3.141Ω per mil foot

Answers to odd-numbered Module Review Questions are found in *Appendix A*.

Module 26302-23 Supplemental Exercises

1. The electrical conductor with the best cost-to-conductivity ratio is __________.
2. True or False? Solid wire is the most flexible form of conductor.
3. Place the following wires in order from the smallest to the largest: No. 1/0 AWG, No. 14 AWG, No. 1 AWG, and No. 4/0 AWG.
   ____
   ____
4. The purpose of compact conductors is to __________.
5. The *NEC*® requires that each ungrounded conductor be protected from damage by a(n) __________.
6. The color of a grounded conductor must be __________.
7. The rating of the __________ determines the branch circuit rating.
8. If six No. 10 TW wires are installed in a single raceway where the temperature is 104°F (40°C), the ampacity of 30A would be derated to 19.68A. Why?
   ____
   ____
9. The fuse size that should be used with a No. 18 AWG extension cord where only two conductors are carrying current is __________.
10. For field-installed taps, the ampacity of tap conductors cannot be less than __________ of the rating of the overcurrent device protecting the feeder conductors per *NEC Section 240.21(B)(1)*.
11. Eight 4/0 THHN conductors are placed in a raceway. What is the maximum current allowed for each wire?
    ____
    ____
12. State the color-coding requirements for conductor insulation.
    ____
    ____
13. If it becomes necessary to change the actual color of a conductor to meet *NEC*® requirements or to facilitate maintenance on circuits and equipment, the conductor may be re-identified with __________ or __________.

14. Determine the resistance of 1,500' of two No. 1 AWG copper wires in a raceway, and then calculate the voltage drop using Type THHN conductors.

____

____

15. Name two ways to increase the current to a load.

____

____

Answers to odd-numbered Supplemental Exercises are found in *Appendix B*.

Answers to Section Review Questions

Answer	Section Reference	Objective
Section 1.0.0		
1. b	1.1.0	1a
2. d	1.2.0; *NEC Table 310.4(1)*	1b
Section 2.0.0		
1. a	2.1.0	2a
2. a	2.2.1	2b
3. d	2.3.1	2c

Section Review Calculations

Section 2.0.0

Question 2

2 × 100' = 200' total length

200' ÷ 1,000' = 0.20 multiplier

0.20 × 1.93Ω (from *NEC Chapter 9, Table 8*) = 0.386Ω

The total resistance of the conductors is **0.386Ω**.

Question 3

$VD = (2 \times L \times K \times I) \div$ cmil (from *NEC Chapter 9, Table 8*)

$VD = (2 \times 100' \times 12.9\Omega \times 40A) \div 16{,}510$ cmil

$VD = 6.25V$

The voltage drop is **6.25V**.

User Update

Did you find an error? Submit a correction by visiting **https://www.nccer.org/olf** or by scanning the QR code using your mobile device.

Source: iStock@Franck-Boston

MODULE 26303-23

Practical Applications of Lighting

Objectives

Successful completion of this module prepares you to do the following:

1. Identify characteristics of luminaires.
   a. Identify benefits and difficulties with different light distribution patterns.
   b. Categorize luminaires by mounting location.
   c. Explain how to choose luminaires for adverse locations.
   d. Describe how to determine special lighting requirements.
2. Select lighting systems for various applications.
   a. Identify types of lighting for indoor applications.
   b. Identify types of lighting for outdoor applications.
3. Select lighting controls for various applications.
   a. Describe how to select dimmer systems.
   b. Explain how to select occupancy and photo sensors.
   c. Explain how to apply lighting timers.
   d. State how various components relate to energy management systems.

Performance Tasks

Under supervision, you should be able to do the following:

1. Using manufacturers' catalogs, select the appropriate luminaires for specific lighting situations.
2. Tour an instructor-selected facility to identify various lighting systems and describe the specific purpose of each type encountered.

Overview

There are many examples of incandescent, fluorescent, LED, induction, and HID luminaires, each with their own unique characteristics. Lighting systems are needed in many different environments, both indoors and outdoors. This module describes various luminaires and their suitability for various applications. It also presents lighting controls, dimming, and energy management system applications.

Digital Resources for Electrical

Scan this code using the camera on your phone or mobile device to view the digital resources related to this craft.

NCCER **Industry-Recognized Credentials**

If you are training through an NCCER-accredited sponsor, you may be eligible for credentials from NCCER. The ID number for this module is 26303-23. Note that this module may have been used in other NCCER curricula and may apply to other level completions. Contact NCCER at 1.888.622.3720 or go to **www.nccer.org** for more information.

You can also show off your industry-recognized credentials online with NCCER's digital badges. Transform your knowledge, skills, and achievements into badges that you can share across social media platforms, send to your network, and add to your resume. For more information, visit **www.nccer.org**.

1.0.0 Characteristics of Luminaires

Objective

Identify characteristics of luminaires.

a. Identify benefits and difficulties with different light distribution patterns.
b. Categorize luminaires by mounting location.
c. Explain how to choose luminaires for adverse locations.
d. Describe how to determine special lighting requirements.

Performance Tasks

1. Using manufacturers' catalogs, select the appropriate luminaires for specific lighting situations.
2. Tour an instructor-selected facility to identify various lighting systems and describe the specific purpose of each type encountered.

Lighting plays an important role in how people feel about their surroundings and the buildings they occupy. Different lighting techniques can be used to highlight the special architectural features of a building while also providing functional light. Lighting can be used to enhance the decor of a building and its furnishings, to help people perform tasks more easily, and to make people feel safer and more comfortable indoors and out.

Luminaires can be classified in different ways, such as the following examples:

- Lamp characteristics
- Light distribution characteristics
- Installation location
- Mounting method
- Specific purpose

NCCER Module 26203-23, *Electric Lighting*, looked primarily at lamp characteristics and mounting methods. This module will discuss light distribution, installation location, and specific purposes.

The choice of light source depends on many factors, such as cost, efficacy, appearance, controllability, size, use, and other factors. *Table 1* below compares the performance of the most common light sources used. Most light sources need additional components to make them work properly. Ballasts, igniters, and drivers are examples of components required to operate various lights.

TABLE 1 Characteristics of Various Light Sources

Type of Lamp	Life Span (Hours)	Efficacy (LPW)	CRI	Color Temperature (Kelvin)
Incandescent	700–2,000	5–17	100	2,400–3,000
Halogen	1,000–3,000	16–30	100	3,000–3,400
Fluorescent	7,000–15,000	45–105	60–98	2,800–6,000
LED	10,000–100,000	75–200	60–99	2,700–6,000
Metal halide (HID)	24,000	65–115	65–95	3,600–4,000
High-pressure sodium (HID)	30,000–80,000	85–150	22	2,700
Induction	40,000–100,000	30–50	95–100	3,000–5,000

The fourth column of *Table 1* shows the color rendering index (CRI). The CRI is a measure of how well a light source accurately reproduces the colors of the surfaces it strikes, indexed to a natural source of light. As you can see, high-pressure sodium lamps do a very poor job of showing an object's color, with a CRI of 22, but they have other characteristics that are desirable in other applications.

1.1.0 Control of Light Distribution

The light intensity and distribution pattern from a luminaire are determined by the design and spacing between them. Luminaire design aims to illuminate a desired location while mitigating or eliminating unwanted effects such as glare and shadows. Components used to execute a design and manage light include baffles, reflectors, lenses, diffusers, and louvers. Each of these is described below:

- *Baffles* — Baffles are a series of light absorbing ridges that ring the lower portion of the luminaire interior. They restrict the illumination to a small range of angles, effectively shielding the light source from view. However, they also significantly reduce the luminaire's light output, spread, and efficacy.
- *Reflectors* — Reflectors have polished surfaces, usually in a parabolic or ellipsoidal shape for precise reflection to focus the light into a controlled beam. Theatrical spotlights are a good example. They are also used to control light distribution across a surface such as a roadway. Reflectors often use narrow flat segments instead of a continuous curve, as can be seen in many luminaires. Reflectors decrease the loss of light inside the luminaire and thus increase its efficacy.
- *Lenses* — Lenses are located at the surface of a luminaire. To focus light, an adjustable Fresnel lens is used to control the beam angle in theater, television, and photographic lighting, as well as some recessed and track lighting. Prismatic lenses are a common choice where brightness and a wide light distribution are desired along with a reduction in glare and eye strain. A prismatic lens will reduce the apparent brightness of luminaires when viewed from a distance.
- *Diffusers* — Diffusers are also located at the surface of a luminaire. Their translucent material causes the transmitted light to leave the luminaire in all directions, which greatly reduces glare and shadowing. They are sold in many patterns, textures, and even murals, and are often used with fluorescent and LED luminaires. Diffusers are not recommended for use in open office areas where personal computers are widely used because they can cause unacceptable reflections on the monitors.
- *Louvers* — Louvers are usually white or silver and made of metal or plastic strips in a vertical orientation. If made in a small honeycomb or square pattern, they can often replace diffusers or prismatic lenses to give better glare control and reduce eyestrain. In deeper fluorescent and LED luminaires, the louvers are spaced apart with their openings centered beneath the lamps. This creates the desired cutoff angle. With parabolic instead of flat-sided louvers, even more light is directed downward, which increases the luminaire's overall efficacy while reducing glare and eyestrain. This is especially desirable when viewing computer monitors.

Choosing the right accessory for an application is as important as choosing the luminaire itself. For this reason, the selection for large projects is usually made by a lighting engineer and/or a lighting manufacturer's representative.

Then and Now

Over one hundred years ago, the safety, efficiency, and ease of use of incandescent lighting led to the obsolescence of gas mantles, lime lights, and other sources of lighting. The demand today for higher efficiency, longer life span, and flexibility in use have led us to using LEDs instead of incandescent, fluorescent, and similar light sources. LEDs are also referred to as *solid-state lighting (SSL)* sources.

1.1.1 Indoor Light Distribution

The light distribution of indoor luminaires for general lighting is classified by the Commission Internationale de l'Eclairage (CIE). The distribution is classified by the percentages of total light output emitted above and below the horizontal plane. Refer to *Figure 1*. These classifications include the following:

- *Direct lighting* — 90% to 100% of the luminaire's emitted light is downward, below the horizontal plane. Any remaining light not directed downward will be directed upward.
- *Semi-direct lighting* — 60% to 90% of the luminaire's emitted light is downward, with the remainder directed upward to illuminate the ceiling and upper walls.
- *General diffuse lighting* — 40% to 60% of the luminaire's emitted light is downward and the balance upward, sometimes with a strong component at the horizontal plane.
- *Direct-indirect lighting* — Similar to general diffuse lighting, but very little light is emitted at horizontal angles.
- *Semi-indirect lighting* — 60% to 90% of the luminaire's emitted light is upward. The downward component usually produces a luminance closely matching that of the ceiling.
- *Indirect lighting* — 90% to 100% of the luminaire's emitted light is upward.

1.1.2 Outdoor Light Distribution

For streets, parking lots, and similar areas, the Illuminating Engineering Society of North America (IESNA) describes six patterns for horizontal light distribution. They are designated as Type I through Type V (circular and square), as shown in *Figure 2*.

The narrow Type I has a front-back distribution of only 15°, while the much wider Type IV has a front-back distribution of 60°. The symmetrical Type V circular and Type V square (sometimes referred to as *Type VS*) have round or square distribution patterns. In addition, the following set of three letters are used to describe the vertical light distribution:

- A Type C (cutoff) luminaire directs almost no light above 90°.
- A Type S (semi-cutoff) luminaire directs some light above 90°.
- A Type N (non-cutoff) luminaire is not restricted to how much light is emitted in any vertical direction.

For flood lighting and sport lighting luminaires, the National Electrical Manufacturers Association (NEMA) developed a second light distribution classification. The NEMA types range from 1 (narrowest) to 7 (widest) as shown in *Table 2*. Luminaires have separate designations for horizontal and vertical beam spreads, such as Type 2 × 3 to describe a narrow flood.

TABLE 2 NEMA Classifications for Sports Lighting Distribution

NEMA Type	Beam Spread	Beam Description	Beam Projection Distance
1	10° to 18°	Very Narrow	240' and up
2	18° to 29°	Narrow	200' to 240'
3	29° to 46°	Medium Narrow	175' to 200'
4	46° to 70°	Medium	145' to 175'
5	70° to 100°	Medium Wide	105' to 145'
6	100° to 130°	Wide	80' to 105'
7	130° and up	Very Wide	Less than 80'

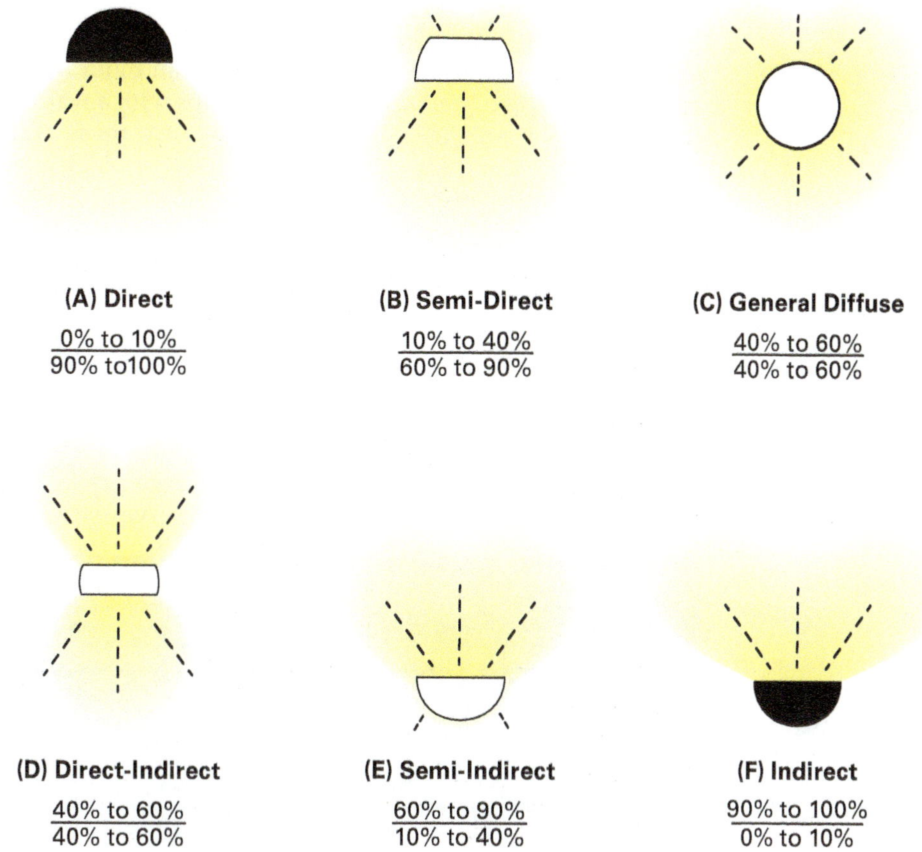

Figure 1 CIE classifications of indoor light distribution.

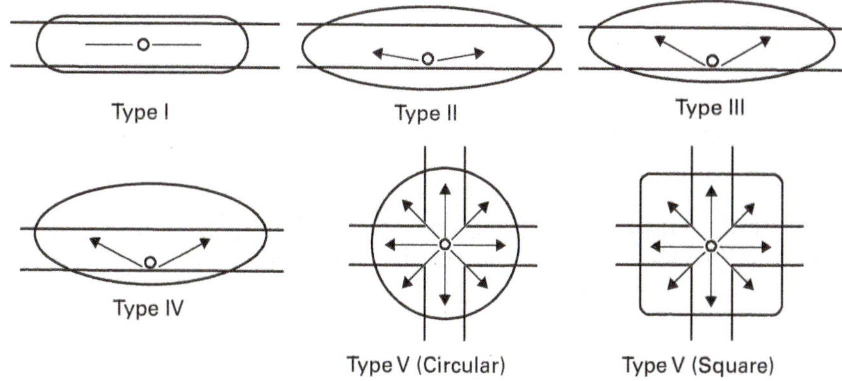

Figure 2 IESNA classification of roadway light distribution.

1.1.3 Light Pollution

Light pollution is caused by outdoor luminaires. For example, bright floodlighting from a business or residence can blind drivers approaching the area or using adjacent roadways. Using lamps with a warmer color temperature reduces the amount of light pollution because these wavelengths are more readily absorbed by the atmosphere. Excessive light at night can have the following harmful effects:

- Interference with astronomy
- Disrupting the travel patterns and endangering birds, sea turtles, and many insects
- Changes to human circadian rhythms and decreased levels of melatonin needed for sleep
- Difficulty for airplane pilots landing or taking off

The images in *Figure 3* were taken near a large city in Eastern Canada before and during a major blackout. They show the pervasive nature of light pollution.

Figure 3 Light pollution over a city.

1.2.0 Luminaire Locations

Luminaires are designed to suit a specific application. The type of construction, aesthetics, finish, lighting distribution, glare, heat control, size, and cost are only a few of the characteristics that result in the wide variety of luminaires.

This section will look at the following classifications based on mounting location:

- *Concealed* — Used indoors
- *Recessed* — Used in offices, stores, and public buildings
- *Surface mounted* — Used indoors and outdoors
- *Suspended* — Used indoors
- *Pole mounted* — Lighting of streets, roads, highways, amphitheaters, and stadiums
- *Security and architectural lighting* — Lighting of outdoor areas such as recreational grounds, parking lots, sidewalks, building exteriors, and other spaces

1.2.1 Concealed Luminaires

Concealed luminaires provide indirect lighting to an area by directing it upward from a concealed location, commonly called a *cove* or *cornice*. Concealed luminaires are used in hallways, above sections of suspended ceilings in larger rooms, for general lighting from above cabinets or draperies, in crown and cove molding, and any other location where diffuse, glare-free indirect illumination is desired.

Concealed luminaires are mounted in end-to-end strips. Staggered, or *offset*, strips are used to avoid dark spots where the ends of two luminaires meet. In the past, cove lighting typically used fluorescent or cold cathode lamps. Cathode lamps are a special type of fluorescent lamp with a very long lifespan. Today, however, most of these luminaires use LEDs (*Figure 4*).

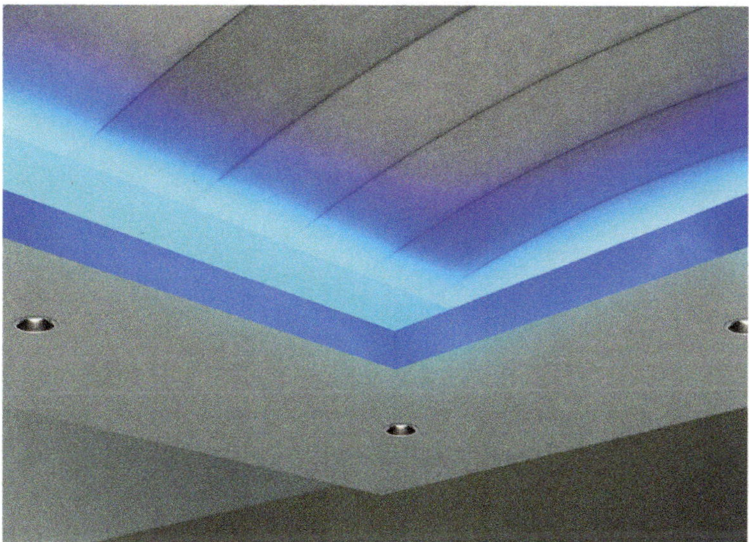

Figure 4 LED cove lighting.
Source: iStock@angsabiru

1.2.2 Recessed Luminaires

Recessed luminaires (downlights) offer light with little or no intrusion of a luminaire. They can provide a variety of lighting effects including general lighting, accent or task lighting, and wall washing. They are made in a variety of shapes, sizes, and uses, such as traditional cans, shallow puck-type luminaires, troffers, ceiling panels, and strips.

A recessed can luminaire is basically a domed reflector with a lamp set in the top. It is matched with a variety of trims to produce the desired lighting effect (*Figure 5*). If used in damp or wet areas, they must be rated for this use. If used in insulated ceilings, they must have an insulated ceiling approval. The wiring can be accessed by removing the can from the luminaire frame (*Figure 6*). A common and versatile type is the adjustable eyeball type with a partially recessed sphere that can be rotated and tilted to direct light for wall-washing and accent lighting purposes.

LEDs, besides their use as efficient replacements of other lamp types in recessed can luminaires, have allowed the creation of a new shallow puck-type recessed luminaire. These can be mounted directly to the ceiling in very tight locations and don't need a separate housing. However, the driver and its housing must be mounted in a separately accessible location and must be connected to the luminaire with the specified cable.

When used over sinks, countertops, desks, and workstations, a downlight spreads a strong task light over the work surfaces. Downlight luminaires are also used for lighting stairways and entries. A series of wall-wash luminaires can be used to produce consistent, balanced lighting on a wall of artwork or bookcases. Low-voltage downlights, especially those that use MR-16 lamps and black baffles, are widely used for accent lighting.

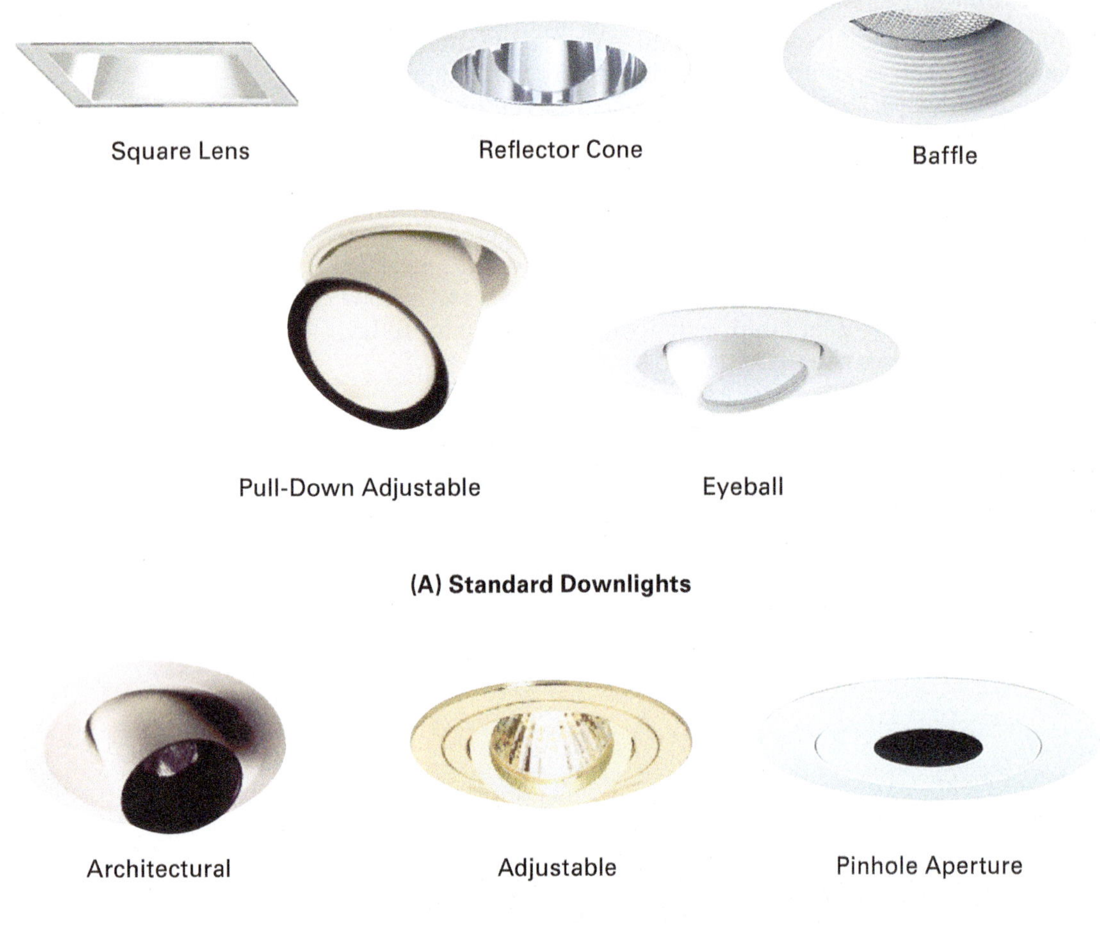

Figure 5 Recessed downlights.
Source: Hubbell Incorporated

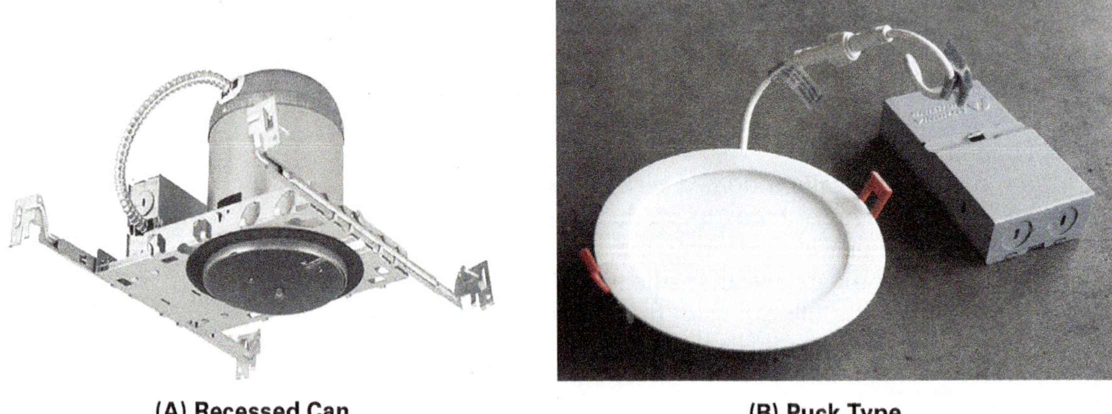

Figure 6 Recessed luminaires.
Source: Hubbell Incorporated (6A)

Recessed troffer-type luminaires are widely used to provide uniform general lighting and task lighting in areas such as offices, corridors, conference rooms, laboratories, and schools. These square and rectangular luminaires are sized to lay into T-bar type grid ceilings of all standard sizes and can also be installed in the recesses of plaster or drywall ceilings. They are made with diffusers, lenses, and louvers for the desired light control.

Backlit decorative ceiling panels are used in areas where a gentle and soothing level of illumination is needed. They are available with pictures printed on the diffusers, as seen in *Figure 7* in a room used for magnetic resonance imaging (MRI).

Recessed strip lighting was introduced as a narrow (4" or 100 mm) variation of recessed troffers mounted end to end. These have been supplanted by even narrower LED strips, such as the ones designed for the ceiling T-bar support system (*Figure 8*).

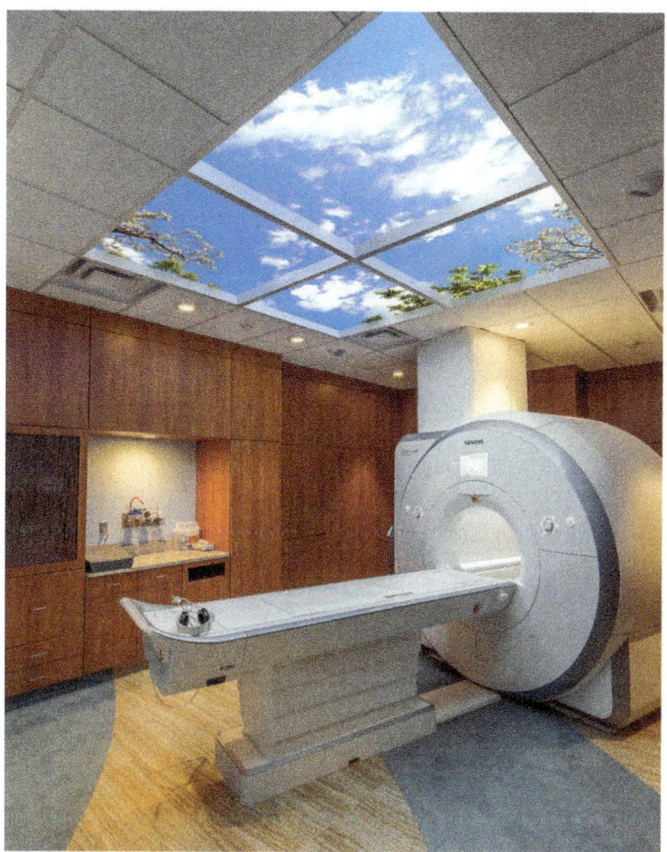

Figure 7 Backlit ceiling panel luminaire.
Source: The Sky Factory

Figure 8 Recessed T-bar LED strip luminaire.
Source: JLC-Tech LLC

1.2.3 Individual Surface-Mounted Luminaires

Individual ceiling-mounted luminaires equipped with diffusers (*Figure 9*) direct their light in a wide pattern, providing what is referred to as **diffused lighting**. The diffuser completely or partially hides the lamp behind it. The position of these luminaires allows them to produce more consistent light distribution than similar luminaires hung lower into the line of sight. Ceiling-mounted directional luminaires, called *downlights,* are typically used to provide accent lighting or task illumination.

Diffused lighting: Light that has been filtered or scattered, without the normal reflective characteristic of direct light.

Many large and attractive ceiling-mounted luminaires are designed for use in offices, kitchens, and other commercial or institutional applications (*Figure 10*). These luminaires are typically available with smooth, white acrylic diffusers. Some have solid or simulated wood frames. Others have acrylic diffusers with tapered round edges that offer a floating cloud look. Small, shallow surface-mounted luminaires with lenses are made for use under cabinets and similar locations.

Wall luminaires (*Figure 11*) can provide either diffused or directional lighting. Some have exposed clear lamps for decorative purposes. Wall luminaires designed as downlights are typically used to provide accent and/or display lighting. Those designed as uplights are used to provide indirect light for the room. Wall sconces are commonly used in hallways and stairways or as accent lights in a room to supplement larger luminaires. By design, they cast their light toward the wall, diffuse light into a room, or direct light toward the ceiling.

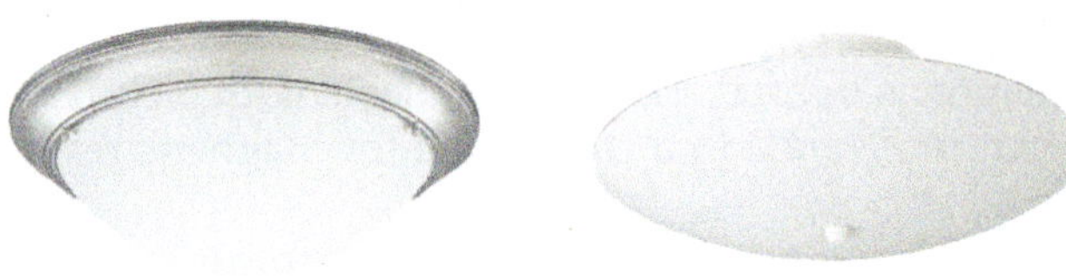

Diffused Fixtures

Downlights

Figure 9 Small ceiling-mounted luminaires.
Source: Hubbell Incorporated

Framed

Wraparound

Soft Cloud

Dome

Under Cabinet

Shallow Mushroom

Figure 10 Larger ceiling-mounted luminaires.
Source: Hubbell Incorporated

Going Green

LED Luminaires for Commercial Use

These images show the result of converting the luminaires in a parking garage from sodium vapor to long-life white LEDs with multiple LED arrays.

Old Sodium Vapor Lighting

New LED Lighting

Source: Lighting Science Group Corporation

Figure 11 Wall-mounted luminaires.
Source: Hubbell Incorporated

Chandeliers and Pendant Luminaires

When chandeliers and pendant luminaires are used in foyers and entranceways, the clearance for door swings must be considered. Also, the luminaires should be hung well above people's heads (generally at least 7' or 2.1 m high) unless installed over a table.

1.2.4 Individual Suspended Luminaires

Suspended luminaires are hung from a ceiling by a chain, aircraft wire, cord, conduit, or stem (*Figure 12*). Suspended luminaires can have single or multiple arms or a single lamp holder and diffuser. Those that diffuse the light are used for both functional and decorative purposes. Ideally, the light should be managed to prevent glare and to obtain the proper light ratios, particularly if the luminaires are hung low enough to be in the line of sight.

Suspended luminaires that provide indirect and direct-indirect lighting are widely used in commercial applications, especially where personal computers are in use. Direct-indirect luminaires with opaque or diffused shades or shields may be placed in the direct line of sight. Glass or plastic diffusers should not be visible below the shade or shield.

The main function of suspended downlights is to provide accent lighting. Over work surfaces, downlights are normally suspended at or below eye level. When mounted above eye level, such as in stairways, halls, and entryways, they are normally equipped with a louver, diffuser, or other bottom shielding to keep the lamp out of the line of sight.

In commercial and residential applications, chandeliers and pendant luminaires are widely used for decorative purposes as well as lighting. These luminaires can produce direct light, diffused light, or both.

The size and mounting height of a pendant or chandelier in relation to its surroundings is important. If the luminaire is hung low over a table, the width of the luminaire should be at least 12" (300 mm) less than the width of the table to minimize the possibility of someone hitting their head on it as they rise from a chair. Suspended luminaires equipped with enamel steel reflectors are made in several shapes for commercial applications.

High-bay luminaires (*Figure 13*) are used extensively in industrial applications and in many commercial applications. They are available with beam spreads from 45° to 120°, as well as asymmetric spreads for aisles or similar spaces. High-bay applications are those with a mounting height of 20' (6 m) or more above the **work plane**, while low-bay luminaires (*Figure 14*) are mounted less than 20' high. The major difference between the two is a higher level of light output from high-bay lighting where glare is reduced because of the height. These luminaires are available with reflectors, diffusers, housings, and mounting methods according to the specific application's needs and restrictions.

Work plane: The distance above the floor where a task is to be performed. In offices and schools, for example, the distance is usually 30" (750 mm).

Figure 12 Typical suspended luminaires.
Source: Hubbell Incorporated

Figure 13 High-bay suspended luminaire.
Source: Hubbell Incorporated

Figure 14 Low-bay luminaires.
Source: iStock@Andrii-Shablovskyi

High-bay and low-bay luminaires are typically surface mounted or pendant mounted. Some are suspended from and connected to the building's lighting distribution system by *power hook assemblies*. A power hook assembly consists of a hook with an attached power cord and plug that is connected between the luminaire and related power receptacle above. The hook, via its attachment to a loop on the receptacle outlet box cover, supports the suspended luminaire. For added protection, safety chains are often installed between the building structure and the luminaire.

1.2.5 Strip Luminaires

Strip luminaires are used for general lighting applications. They are designed for use in either surface-mounted or suspended commercial or industrial applications and can be mounted end to end, using *channel connectors*, as needed. Depending on the specific design and mounting orientation, they can provide direct lighting or direct, indirect, or direct-indirect lighting.

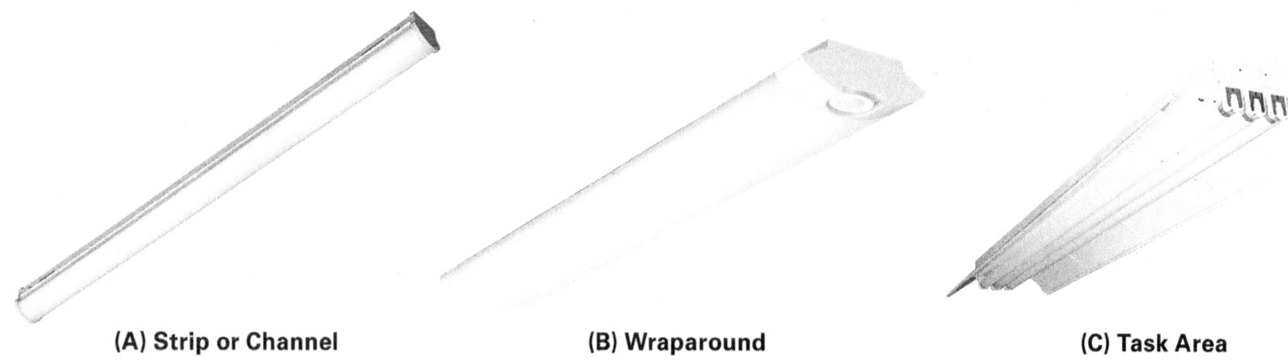

(A) Strip or Channel (B) Wraparound (C) Task Area

Figure 15 Examples of strip luminaires.
Source: Hubbell Incorporated

Traditionally, strip luminaires have been made with fluorescent tubes. The need for lower maintenance costs and higher efficiency have driven the market toward retrofit LED lamps and new LED luminaires. Strip-type luminaires include the following:

- *Open strip or channel luminaires* — Open strip or channel luminaires are used in areas in which a lot of light is needed, and aesthetics are not important, such as in storage rooms and warehouses. They are made in single-lamp and multiple-lamp assemblies of various lengths, shown in *Figure 15* (A).
- *Staggered strips* — Staggered strips are similar to strip or channel luminaires. However, the lamps are staggered so that the light from at least one lamp is always available at each joint to provide more uniform illumination.
- *Wraparound luminaires* — Wraparound luminaires have a lens or diffuser enclosing the lamps. They are commonly used when horizontal and vertical light from aesthetically pleasing, moderately bright, area-type sources is required, shown in *Figure 15* (B). Their narrow profile is well-suited to areas such as corridors, stairwells, hallways, and between aisles.
- *Task area luminaires* — Task area luminaires are generally used where low-mounted task lighting is needed in workstations, storage areas, and utility rooms, shown in *Figure 15* (C). They have reflectors for more efficient illumination as well as to provide protection for the lamps. The reflectors can be of the nonapertured or apertured type. If used with apertured reflectors (slots cut into their tops), about 10% to 20% of the light is directed upward toward the ceiling. This uplighting helps improve the overall lighting in the work area, especially when used with a reflective ceiling. The slots also allow air to circulate through the reflector, keeping the luminaire cooler than with a nonapertured type.
- *Commercial/industrial luminaires* — Commercial/industrial luminaires are typically used as general downlighting in low-bay and high-bay applications that include factories, schools, warehouses, and retail stores.

Think About It

Strip Lamps

Why would some applications require that the lamps of strip luminaires be fitted with shields, covers, or guards?

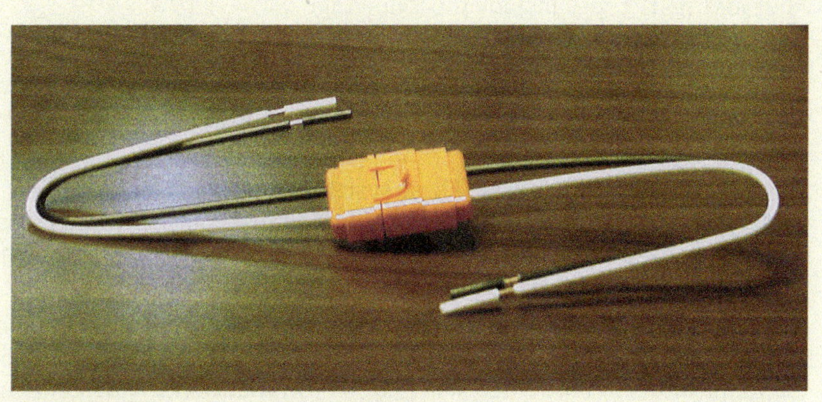

Fluorescent Luminaire Disconnects

Fluorescent luminaires in commercial/industrial locations that utilize double-ended lamps and contain ballasts that can be serviced in place require an accessible disconnect. See *NEC Section 410.71* for details and exceptions. The disconnect shown in the image is wired between the ballast and supply leads to provide a safe, convenient method of physically disconnecting power during luminaire maintenance.

1.2.6 Track and Other Luminaires

Track lighting allows light to be directed where needed to achieve a variety of effects, such as accent lighting, task lighting, and wall-wash lighting. A variety of track-mounted luminaires are available. Luminaires and the required track must be from the same manufacturer or be listed for use together.

By design, most have a modern look (*Figure 16*). Adjustable track luminaires typically rotate and can be adjusted horizontally along the track to provide precise, effective lighting control. Low-voltage (12V) luminaires designed for use with MR-16 and PAR-36 lamps are also made. Some of these luminaires snap into a low-voltage transformer, which in turn locks into the track section. Others have an integral low-voltage transformer.

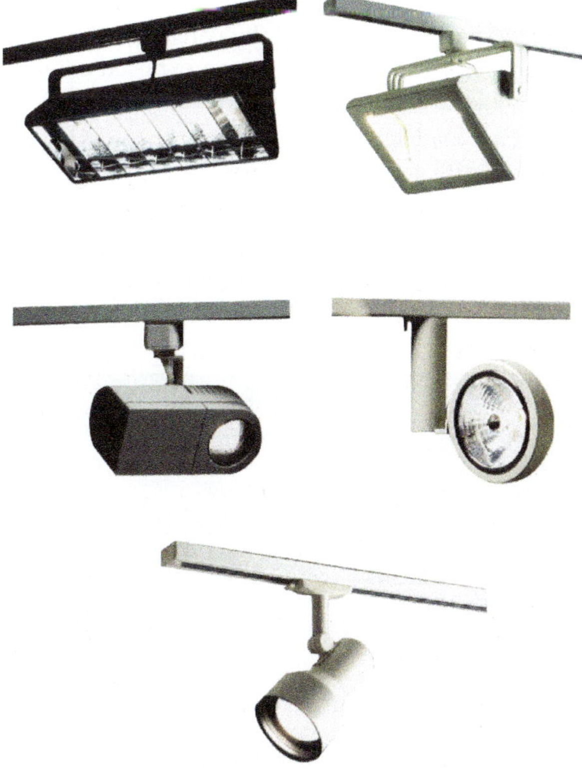

Figure 16 Typical track luminaires.
Source: Hubbell Incorporated

Track Lighting

Track lighting is used for accent or task lighting intended to light a specific area or object with a pool of light. Ideally, the track luminaire is positioned so its light is at a 60° angle from the horizontal surface of the illuminated object. For this use, track lighting should be installed 24" to 36" (600 mm to 900 mm) from the wall.

Other luminaire types include the following:

- *Monotrack* — A variation of track lighting operating at 12V or 24V. It is made of two bare tubes separated by an insulating strip in a vertical configuration so it can be field bent. A variety of connectors, luminaires, and lamp types are made for use with it.
- *Bare or covered cable systems* — These systems also operate at 12V or 24V. The cables are stretched tightly in straight lines between anchor points, and the luminaires have individual clamps that attach between the cables. Again, a variety of lamp types can be used.
- *Rope and tape systems* — These systems generally operate at 12V or 24V, but a few are made for 120V operation. They come in different lengths up to 100' (30 m) or more and are fitted with LEDs or miniature incandescent lamps. The LED versions come in a variety of colors including color-changing RGB types. Many rope and tape systems can be used outdoors, often for decorative or seasonal lighting.

1.3.0 Adverse Locations

Adverse locations refer to those where the luminaire must be protected from damage by environmental conditions and where the environment must be protected from fire or explosion caused by the luminaire.

1.3.1 Protection from Environmental Conditions

A variety of specialized luminaires are made for operation in adverse conditions. They include the following:

- *Dustproof luminaires* — Constructed so that dust will not interfere with the luminaire operation.
- *Dust-tight luminaires* — Constructed so that dust is prevented from entering the enclosed case.
- *Vapor-tight luminaires* — Gasketed and fully enclosed luminaires made for use in wet or very humid environments.
- *Corrosion-resistant luminaires* — Typically constructed with stainless steel or nonmetallic housings so that they have no exposed exterior parts that can be attacked by moisture or other corrosive agents.
- *Wash-down resistant luminaires* — Used in food processing and similar areas subjected to hose-directed water. These are typically given a rating with the pressure of the hose stream that the luminaire can withstand. See *NEC Sections 300.6(D) and 410.10(A)* for additional guidance.
- *Submersible luminaires* — Used in the walls of swimming pools and in waterfalls and fountains (*Figure 17*). Special rules for them can be found in *NEC Section 680.23*.

Know the Code

Indoor Wet Locations
NEC Section 300.6(D)

Know the Code

Wet and Damp Locations
NEC Section 410.10(A)

Know the Code

Underwater Luminaires
NEC Section 680.23

Figure 17 Fountain luminaires can be used to produce special effects.
Source: iStock@TTshutter

Guards and grilles are available for use on luminaires in gymnasiums and similar areas where they are subject to breakage by balls and similar objects.

Many processes used in electronics and pharmaceuticals, as well as in industries such as medicine and aerospace, require the contamination-free environment provided by a clean room. Luminaires used in clean rooms are specially designed for that purpose. Typically, they are designed without crevices where contaminants can accumulate. The housings are free of holes, slots, and other openings that would allow passage of air or particles into or out of the luminaire. Lenses and other components are sealed to the housing with special gasketing. Some are made to be used in conjunction with high-efficiency particulate arrestance (HEPA) filters that allow air flow of well-filtered air through the luminaire and into the clean room or operating room space.

Luminaires subject to physical abuse include vandal-resistant and prison-grade types, although these terms are not well-defined. Vandal resistance usually is provided by lenses or diffusers that are difficult to break and fasteners that require special tools to remove. Prison-grade luminaires use housings and lenses or diffusers that are much thicker, have concealed hinges, no places to conceal contraband, and heavier fasteners.

1.3.2 Hazardous (Classified) Environments

An explosion-proof luminaire is a completely enclosed luminaire capable of preventing the ignition of a gas or vapor surrounding the enclosure by containing any sparks, flashes, or explosion within the luminaire. Also, it must operate at an external temperature that will not cause a surrounding flammable atmosphere to be ignited.

To maintain the explosion-proof integrity of such a luminaire, all fasteners must be undamaged and fully tightened, all mating surfaces must be free from scratches or other damage, all gaskets and seals must be intact, and the globe must be fully threaded into the luminaire housing. *Figure 18* shows examples of hazard warning lights.

NEC Article 500 defines and categorizes luminaires in terms of class, division, and group. The definitions for each class are as follows:

- *Class I locations* — Locations in which flammable gases or vapors are or may be present in the air in quantities sufficient to produce explosive or ignitable mixtures (*NEC Article 501*).
- *Class II locations* — Locations that are hazardous because of the presence of combustible dust (*NEC Article 502*).
- *Class III locations* — Locations that are hazardous because of the presence of easily ignitable fibers or other airborne materials, but where such materials are not likely to be in suspension in the air in quantities sufficient to produce ignitable mixtures (*NEC Article 503*).

Figure 18 Hazard warning lights.
Source: iStock@Berkut_34

Vandal-Resistant (Tamper-Resistant) Bollard Luminaires

Because bollard luminaires are used mainly outdoors and are installed at ground level, they can be subjected to accidental damage or vandalism. For this reason, manufacturers offer heavy-duty bollard luminaires like the one shown in the image.

Source: Hubbell Incorporated

Know the Code

Hazardous (Classified) Locations, Classes I, II, and III, Divisions 1 and 2
NEC Article 500

Know the Code

Class I Locations
NEC Article 501

Know the Code

Class II Locations
NEC Article 502

Know the Code

Class III Locations
NEC Article 503

Know the Code

Luminaires
NEC Section 501.130

Know the Code

Luminaires
NEC Section 502.130

Know the Code

Luminaires – Class III, Divisions 1 and 2
NEC Section 503.130

Each class is further defined as being either Division 1 or Division 2:

- *Division 1* — An environment that is normally hazardous.
- *Division 2* — An environment that is not normally hazardous but could become hazardous.

Each of these divisions can be further classified using the letters A through G, according to the ignition characteristics of the gas, vapor, or dust in the atmosphere.

It is important to be aware that the classification of a hazardous area is not the responsibility of the electrician. The classification of a given area as to class, division, and group is the responsibility of the owner, insurance company, or Authority Having Jurisdiction (AHJ). The proper selection of luminaires for use in hazardous, nonhazardous, or adverse locations should always be made by someone with expertise in this area. Refer to *NEC Sections 501.130, 502.130, and 503.130* for more specific information.

Any luminaire used in hazardous locations should meet the UL 844 standard and be labeled "LISTED ELECTRIC LUMINAIRE FOR HAZARDOUS LOCATIONS."

1.4.0 Special Lighting Requirements

Emergency and exit luminaires (*Figure 19*) provide minimum levels of lighting as required for personnel exit routes and other areas as specified in local and national codes (primarily NFPA 101®, *Life Safety Code®*).

This type of lighting must be maintained for at least 90 minutes whenever the normal source of power to illuminate the area is lost. Local switches and lighting controls must not disable the emergency circuit in the luminaires. Exit luminaire locations and directional arrows tell people where to go, while emergency luminaires provide enough illumination to safely get there.

Because exit luminaires mounted above door height may be obscured by smoke or fire, many codes also require duplicate exit luminaires mounted no more than 24" (600 mm) above the floor. In addition to a battery, charger, and automatic controls, these luminaires normally have an accessible test switch and pilot light for testing at intervals as specified by the manufacturer and applicable codes.

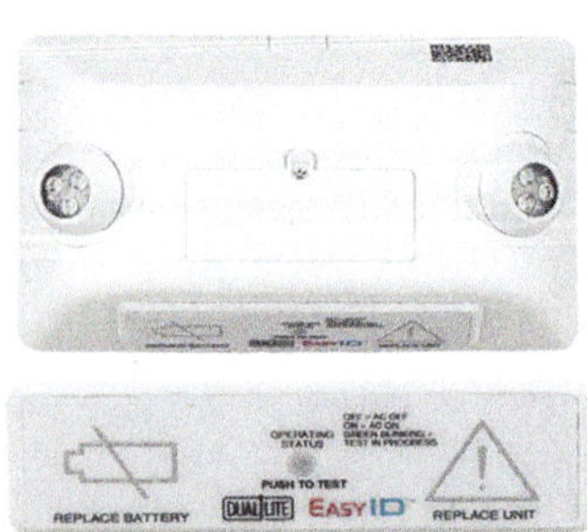

Figure 19 Emergency and exit luminaires.
Source: Hubbell Incorporated

Going Green

LED Lighting

The Paris Hotel in Las Vegas, Nevada features an exact half-scale replica of the Eiffel Tower in Paris, France. Like the original, it is lit with energy-efficient LED lighting modules.

Energy-efficient LED lamps provide a mile-long beacon of color on the Vincent Thomas Bridge in Los Angeles Harbor, the busiest port in the United States.

Source: LEDtronics

Source: iStock@VitalyEdush

1.0.0 Section Review

1. The type of lighting that distributes 90% to 100% of the light downward is referred to as _____.
   a. direct-indirect
   b. general diffuse
   c. indirect
   d. direct

2. Recessed luminaires are likely to be found in a _____.
   a. cove
   b. ceiling
   c. wall
   d. track

3. Wash-down resistant luminaires are *most* likely to be used in _____.
   a. gymnasiums
   b. garages
   c. food processing plants
   d. clean rooms or operating rooms

4. Following the loss of normal power, an exit luminaire *must* remain lit for _____.
   a. 30 minutes
   b. 60 minutes
   c. 90 minutes
   d. 120 minutes

2.0.0 Selecting Luminaires for Various Applications

Performance Tasks

1. Using manufacturers' catalogs, select the appropriate luminaires for specific lighting situations.
2. Tour an instructor-selected facility to identify various lighting systems and describe the specific purpose of each type encountered.

Objective

Select lighting systems for various applications.

a. Identify types of lighting for indoor applications.

b. Identify types of lighting for outdoor applications.

This section provides an overview of common commercial/industrial lighting applications and the types of luminaires used in them. A discussion of all the factors related to the design of lighting systems, the criteria for selection of luminaire types, and the exact methods for determining the quantity and placement of luminaires in a room or area are beyond the scope of this module. These design decisions are normally made by an architect, lighting engineer, or other person with lighting system design expertise. Information specific to the design of lighting systems can be found in the technical section of many luminaire manufacturers' product catalogs. Manufacturers also want their products to be properly applied.

It is important that electricians be aware of the basic units of measure used to describe lighting systems. The amount of light energy from a light source is called a *lumen*. The amount of illumination is a **footcandle (fc)**. Designers use tables published by IESNA and specify the footcandle levels for different activities and areas.

Footcandle (fc): Unit of measure for light reaching a surface. One lumen falling on one square foot of surface produces an illumination level of one footcandle. The international unit for illumination levels is the *lux*, which equates to one lumen per square meter. A footcandle is equal to approximately 10 lux.

The terms *footcandle* and *candlepower* are often confused. The footcandle is the unit used to measure the illuminance or amount of total light that strikes a horizontal or vertical surface, such as a table or wall. *Candlepower* measures the intensity of a light source in any given direction, regardless of anything being lit.

Recommended Light Levels

The recommended light levels, expressed in footcandles, for various lighting applications can be found by consulting the appropriate IESNA publication. Some typical examples include the following:

- 1 — Parking lots, streetlights, and egress lighting
- 5 — Parking garages
- 20 — Depots, terminals, and/or station waiting rooms
- 30 — Auditorium/exhibition areas, gymnasiums, and library reading areas
- 50 — Offices
- 70 — Food service kitchens
- 75 — Bank teller stations
- 100 — School laboratories
- 200 — Clothing manufacturer cutting and pressing areas and store merchandising self-service areas

Light levels can be measured using a light meter, such as the one shown in the image.

Source: Ridge Tool Company

Portable Lamps

The focus of this module is on the various types of permanently installed luminaires used to provide indoor and outdoor lighting. However, lighting designers and interior decorators also make extensive use of decorative portable lamps (plug-in lamps) as part of a building's lighting scheme to meet both lighting and decorating requirements. This is especially true in residential and professional office applications. Plug-in lamps include floor lamps, table lamps, desk lamps, task lamps, spotlights, picture lamps, and curio and kitchen cabinet lamps.

2.1.0 Indoor Lighting

The following sections will discuss applications of indoor luminaires according to their purposes. The following purposes are commonly used:

- *General lighting* — Lighting designed to provide a uniform level of illumination for large rooms, entrances, corridors, lobbies, etc., as shown in *Figure 20* (A).
- *Localized general lighting* — Lighting that uses luminaires above the visual task area that also contribute to the lighting in the local area.
- *Local lighting* — Lighting that provides illuminance over a small area or confined space without providing any significant general lighting in the local area.
- *Task lighting* — Lighting that directs well-focused illumination to a specific surface or area such as a work surface, desk, or counter, as shown in *Figure 20* (B).
- *Accent (directional) lighting* — Lighting used to point out the best-selling features of merchandise, pinpoint high-key decor, emphasize artwork, or make a room look larger by adding contrast and shadows, as shown in *Figure 20* (C).
- *Wall wash* — Lighting that can illuminate vertical surfaces from the ceiling to the floor. It can be used to expand space visually, define a space, light a wall of pictures, or emphasize color or texture, as shown in *Figure 20* (D).
- *Egress lighting* — Lighting that provides the minimum light levels needed to assist people in exiting a building.

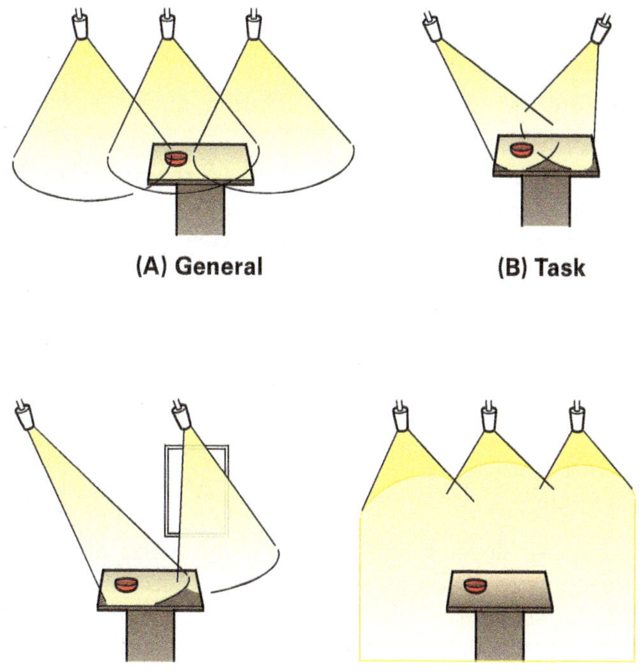

Figure 20 Classification of indoor lighting by purpose.

2.1.1 Office Buildings

Office lighting affects the appearance of the space and its occupants, the mood or effect, and the productivity level. Most modern office lighting uses surface-mounted or recessed LED luminaires as the main light source. However, fluorescent luminaires remain popular.

One approach to office lighting is to use general lighting for all illumination. This is commonly done in private offices or other areas where task lighting is not needed. A second method is to have a lower level of general illumination supplemented with task lighting in work areas. This method is commonly used in open office plan areas with partial-height partitions between individual or common work areas. Each separate office area should have its own means to control the illumination.

Spacing criterion (SC): The ratio of spacing to mounting height that will produce uniform lighting when the luminaire is used. Exceeding the recommended spacing criterion results in uneven lighting on the work plane.

Regardless of the illumination method used, care must be taken to mount luminaires within the **spacing criterion (SC)** given by the manufacturers. SC is the maximum ratio of spacing to mounting height that will produce uniform lighting when the luminaire is used. For example, if a luminaire is mounted at a height of 12' and the SC is 1.5, the maximum spacing between adjacent luminaires is calculated as follows:

$$12' \times 1.5 = 18'$$

Therefore, the maximum spacing between adjacent luminaires should not exceed 18'. Exceeding the recommended spacing results in uneven lighting on the work plane.

General lighting can be from luminaires that provide direct (downward) lighting, indirect (upward) lighting, or some combination of the two, depending on ceiling height and luminaire mounting height.

The most efficient task lighting with the lowest glare comes from luminaires with deep parabolic louvers mounted over the work area. Luminaires with diffusers may provide slightly greater light output but their wider light distribution causes unacceptable levels of glare on computer monitors.

In addition to the work areas, office buildings normally contain other areas with different lighting requirements, including conference rooms, public areas, restrooms, and exits.

- *Conference rooms* — Because of the variety of activities that occur in conference rooms, the rooms may have one or more lighting systems. Rooms with one general lighting system typically control illumination by switching and by using luminaires with dimming capability. Some rooms have separately controlled task lighting to allow use of overhead projectors, or wall-washing luminaires for presentation materials mounted on the room walls.
- *Reception and waiting areas* — These areas require glare-free lighting provided by ceiling luminaires, wall luminaires, or both, sometimes with accent lighting on pictures and other objects. Lighting levels must be sufficient to allow reading and similar activities.
- *Public areas* — Public areas in a building include entrances, corridors, and stairways. These areas usually must remain lit either continuously or for long periods of time. Codes also require emergency lighting for corridors, stairways, lobbies, and entries when they are part of the egress (exit) path.
- *Building entrances* — Lighting is typically provided by ceiling and/or wall luminaires. The lighting must provide for a safe transition from a brightly lit exterior to a darker interior so eyes can adapt to a change in light levels.
- *Elevator lobbies* — Elevator and similar lobby areas are typically casual viewing areas where differences in lighting are acceptable. The important thing is to have a relatively high level of light to illuminate the elevator threshold to see possible height differences between the elevator cab and the floor.

- *Corridors* — Corridor lighting levels should provide at least 20% of the illuminance of adjacent areas. Linear luminaires oriented crosswise to the corridor can be used to make the corridor appear wider. Continuous linear luminaires located adjacent to the walls generally give a feeling of spaciousness.
- *Stairways* — The stair treads must be well illuminated, and the luminaires should be located to avoid both glare and shadows cast by occupants using the stairs.
- *Restrooms* — Like elevator lobbies, lighting differences in restrooms are acceptable. Luminaires should be placed so that they provide light in the vicinity of mirrors to illuminate the face adequately. Other luminaires should be positioned so that they provide adequate lighting for the plumbing fixtures and stalls.
- *Emergency and exit luminaires* — These luminaires provide minimum levels of lighting as required for egress routes and other areas as specified in local and national codes.

Think About It

Exit Lighting

In some localities, exit lighting is required to be placed within 24" (600 mm) of the floor as well as over the door. Why?

2.1.2 Schools

The lighting used in schools should produce a visual environment that supports the learning and instructional process. The lighting must be visually comfortable and amenable to the psychological and emotional needs of the students. Lighting can make schools more pleasant and attractive, reinforce a feeling of spaciousness, and be used to separate areas involving different activities or functions.

With some exceptions, offices, meeting rooms, and public areas located within schools use lighting designs and a control system similar to those used for office buildings. Emergency and exit lighting are also required as they are in other public buildings.

In new construction, luminaires with a high CRI are preferred in schools. In high-ceiling spaces, suspended direct-indirect luminaires provide reflected light from the ceiling and downlighting. For classrooms with low ceilings, surface-mounted or recessed luminaires are used. Recessed luminaires are less obtrusive than surface-mounted luminaires of the same size.

The lighting design and luminaires used in classrooms for computer-based training should be the same as those used for offices and conference rooms. Such designs reduce glare and reflections on the monitors. Similarly, the lighting design and luminaires used in task-oriented laboratories or shop classrooms should be the same as those used in the related industrial or commercial environments where the same tasks are being performed. For example, a machine shop classroom should have the same type of lighting as found in an industrial machine shop.

A school gymnasium is a multipurpose area. Besides physical education and athletics, it is commonly used for graduations, dances, and community meetings. Some existing gymnasiums have HID luminaires, but the lack of good control and higher operating cost are leading to the use of LED luminaires for remodeling and new construction. Mercury vapor and metal halide luminaires in these locations must be specially designed or use guards to prevent breakage. See *NEC Section 410.10(E)*.

Know the Code

Luminaires in Indoor Sports, Mixed-Use, and All-Purpose Facilities

NEC Section 410.10(E)

Schools may also use vandal-resistant luminaires in various applications. Vandal-resistant luminaires are heavy-duty luminaires typically equipped with virtually unbreakable lenses that protect the lamp and socket assemblies (*Figure 21*). Many have lenses with radius corners that increase the ability of the lens to deflect blows and absorb impact. Some are also equipped with guards to protect the lenses.

Vandal-resistant luminaires typically have four-point mounting bases and tamperproof hardware, making them more difficult to detach from their mounting surfaces and/or to disassemble. In addition to schools, vandal-resistant luminaires are used to light areas with high degrees of exposure to unsupervised public traffic, such as public restrooms, hallways, stairwells, loading areas, transit stations, and sidewalks.

Figure 21 Vandal-resistant luminaire.
Source: Hubbell Incorporated

2.1.3 Retail Store Merchandise Areas

The function of the lighting systems in store merchandise areas is to attract and guide the customer to the merchandise, allow the customer to visually evaluate the merchandise, and provide illumination to complete the sale and perform other administrative tasks. To a degree, proper illumination here is part of a marketing strategy.

The current trend for merchandising is to provide variations in illumination for different kinds of stores and for different departments within a store. LED, fluorescent, and HID luminaires are all used in retail lighting applications. The luminaire types used depend on the application.

Spotlights, floodlights, and wall-wash luminaires are all used for merchandise highlighting. The form of objects and texture of surfaces can be revealed by directional lighting such as that produced by small, highly focused downlights. Three-dimensional merchandise is generally displayed using a mix of directional lighting to accent the form and diffuse lighting to relieve the harshness produced by significant shadows.

Retail stores can use a layout of luminaires designed to provide general light throughout the sales area, with or without regard for the location of the merchandise. However, such lighting schemes produce a store environment that lacks emphasis or focal points. Other stores have the luminaires permanently located, such as in supermarkets or similar stores in which the locations of the display units and the types of merchandise being displayed do not change.

Flexible lighting systems are widely used in department and similar stores in which the locations of the displays and the types of merchandise being displayed are continuously changing. These systems typically use adjustable and pull-down luminaires in fixed locations as well as track and similar luminaires. This flexibility allows for a general pattern of lighting or for specific lighting. It also has the added feature of allowing a change of luminaire types to suit the application.

Merchandise within store display cases is normally lit with strip-type luminaires having a high CRI. Flush, surface-mounted, or suspended adjustable luminaires are commonly used for accent lighting of merchandise displayed in free-standing vertical display cases or open-front wall cases. Clothing racks located in cased walls may be lit from above by concealed, baffled luminaires. Those located in open areas are typically lit by flush or surface-mounted adjustable ceiling downlights.

Retail Store Lighting

Track lighting is often used to highlight merchandise and furnishings in retail stores, as shown in the image.

Source: iStock@fiphoto

2.1.4 Health Care Facilities

Lighting systems used in support of health care vary widely in purpose and application. For example, the lighting requirements for hospitals are more varied and complex than those for a doctor's office.

In hospitals, the lighting in the patient rooms must be suitable both for the patient's comfort and for the observation and examination of the patient. Lights for reading are positioned so that the light falls directly on the patient's reading material with no reflected light. Switch-controlled lighting used for the observation and/or examination of the patient typically requires luminaires with excellent color rendition. Good color rendition is important so that the color of the patient's skin and eyes can be properly evaluated.

In critical or intensive care rooms, several lighting arrangements are used. A typical arrangement might include indirect general lighting luminaires, ceiling-mounted or recessed luminaires as downlights for reading, and luminaires mounted near the floor to provide low-level nighttime illumination for nursing activities while the patient is resting.

General lighting at a nurses' station is normally multilevel via dimmer and/or switch control. This allows for higher illumination during the day and a lower level at night. Task lighting is designed to facilitate the reading of patient information from charts and computer monitors. Often, under-counter task lights also serve as night lights. Operating rooms, computerized tomography (CT) areas, MRI scanning rooms, and similar task areas all require very specialized luminaires to provide the appropriate general and task lighting. Emergency lighting in case of a power outage is critical in operating rooms.

Lighting requirements for patient rooms in extended care facilities are similar to those for hospital patient rooms. However, extended care facilities are often populated by elderly patients with reduced vision. Lighting systems here must take this into account, in addition to the lighting needed to support patient care tasks. To compensate for the poorer vision of older patients, these systems generally incorporate the following:

- Indirect lighting whenever possible
- Constant floor illumination
- Higher levels of task-appropriate light
- Adjustable task-oriented luminaires whenever possible
- High color rendering sources of LED or fluorescent lighting
- Luminaires located to provide lighting patterns that aid in environment orientation and minimize confusion
- Lighting controls placed with the patient (corded switches)

2.1.5 Theaters and Auditoriums

Theaters and auditoriums can be divided into two general areas—performance space and audience space. Theater designs and shapes vary widely, from the traditional form with seating in the front to forms with audience seating on three or even all sides of the stage. Theater and auditorium spaces can be used for motion pictures, lectures, teaching, recitals, concerts, worship, live entertainment, and more.

The audience space has comfortable lighting under dimmer control that can be operated from two or more locations. This lighting may include general downlight luminaires, coves, sidewall sconces, curtain and mural lights, and step or aisle lights. Selected lighting circuits may also be used to provide work lights for cleaning and rehearsals. In addition to general lighting requirements, lighting systems used in live performance theaters must provide for visibility, motivation, composition, and mood relative to an activity occurring onstage. For lectures, teaching, worship, and similar activities, the light levels must be sufficient for ease of reading.

In the performance space, the location, aim, intensity, color, and luminaire type are all chosen to create the desired program mood and dimensionality, focus attention onto the action as well as the setting, and produce specific visual effects. For plays, dances, operas, and similar types of entertainment, lighting must come from the front, back, sides, and top of the stage to produce the effects needed. For concerts, lighting needs to come from the front, above, and behind the stage area so the performers can read their music.

Professional designers generally handle theatrical lighting design, due to the complex interaction of the various luminaires. The following types of luminaires are commonly found around theaters and stages (*Figure 22*):

Fresnel spotlights: Focusable spotlights that can be adjusted using a control knob on the back of the light from a narrow, focused beam to a wide beam.

- Parabolic (reflector) spotlights
- **Fresnel spotlights** (*Figure 23*)
- Ellipsoidal spotlights
- Follow spotlights
- Scoop floodlights
- Soft lights
- Strip lights
- Foot lights
- Lasers

Control of theater lighting systems can be very complicated due to the constantly changing requirements during a performance. The control systems must make extensive use of memory, use accurately timed faders, and coordinate simultaneous operations. Most systems incorporate both manual and computer-controlled devices to accomplish the task.

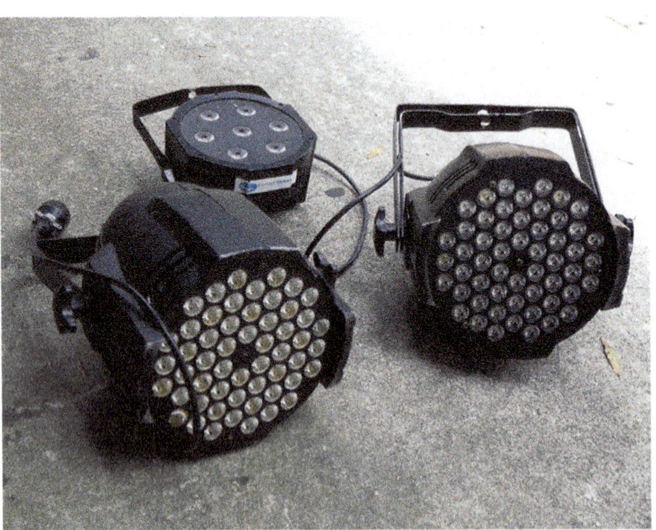

Figure 22 Stage lighting luminaires.

Figure 23 Fresnel spotlight.
Source: iStock@Sergey-Spritnyuk

Case History

Lighting Installations

A ceiling contractor finished the installation on an 86' (26 m) theater ceiling but failed to test the luminaires before the scaffolding was removed. Shortly after, it was discovered that one of the luminaires was wired incorrectly. To replace the single luminaire cost the contractor $6,000 for the scaffolding and rework hours.

The Bottom Line: Because of the difficulty in accessing luminaires in some environments, it is essential to test and verify the operation of new luminaires before the scaffolding or lifts are removed.

2.1.6 Industrial Locations

The lighting used in industrial settings must provide adequate visibility for transforming raw materials into finished products. Because physical hazards exist in manufacturing processes, lighting is also a major factor in the prevention of accidents. The lighting scheme used in an industrial application depends on the specific industry and/or the tasks involved.

Most industrial applications use either direct or semi-direct lighting. However, semi-direct luminaires with an upward light component are preferred because they contribute to visual comfort by balancing the light between the luminaires and their backgrounds.

The interiors of many industrial buildings contain work areas commonly called *bays*. These are typically classified by height above the work plane as *low bay* (up to 20' or 6 m) or *high bay* (over 20' or 6 m). Low-bay luminaires can be individual or continuous strip designs. High-bay luminaires are individual in nature. Both low-bay and high-bay luminaires use reflectors to direct light downward. Low-bay luminaires generally also use a refractor to spread out the light for better light distribution and reduced brightness, since they are mounted closer to the floor.

The luminaires are typically surface mounted or suspended from bar joists, beams, or on lighting busways where they can be relocated. This arrangement forms a uniform lighting array called a *ceiling plane* and provides for general lighting in the bay. Localized general lighting using a greater number of luminaires in a specific area is often used where workers perform tasks that require higher light levels.

Where more difficult visual tasks are performed, additional task lighting is provided to supplement the general lighting. This task lighting is chosen to provide a certain illuminance or color, or to permit special aiming or positioning of light sources to produce or avoid highlights and shadows and best illuminate the details of the visual task. The types of luminaires used to provide task lighting in industrial situations are as varied as the individual tasks themselves.

Industrial lighting applications frequently require specialized luminaires designed for use in areas that are environmentally harsh or classified by the *NEC®* and/or local codes as hazardous. As appropriate to the environment, the luminaires used would be explosion proof, dust tight, dustproof, vapor tight, or have other special features.

2.2.0 Outdoor Lighting Systems

Walks, driveways, streets, and parking lots should be illuminated at night using appropriate luminaires. Entrance and exit areas of the building should be more brightly lit with wall-mounted luminaires or overhead lights. Building outside surfaces, approaches, and outdoor activity areas should be illuminated both for the activity itself and for general safety, as well as for protection against vandalism and theft. Facade lighting typically relies on carefully located, vandal-resistant luminaires mounted on poles, trees, adjacent walls, or special pedestals.

The following sections will discuss these applications of outdoor luminaires:

- Architectural lighting
- Landscape lighting
- Roadway lighting
- Sports lighting
- Security and safety lighting

Going Green

Low-Voltage and Solar Lighting Systems

Low-voltage and solar lighting systems are widely used for energy-efficient residential outdoor landscape and/or aesthetic lighting. Low-voltage lighting systems operate at 30VAC or less. They can have one or more secondary circuits, each limited to 25A maximum. A basic low-voltage outdoor lighting system consists of an isolating power supply (step-down transformer), usually with a built-in timer, luminaires, and associated equipment and wiring identified for such use.

The low-voltage wires are buried at least 6" (150 mm) under the ground. The length of cable runs in a low-voltage system are limited by the allowable voltage drop. ***NEC Article 411*** governs low-voltage lighting systems.

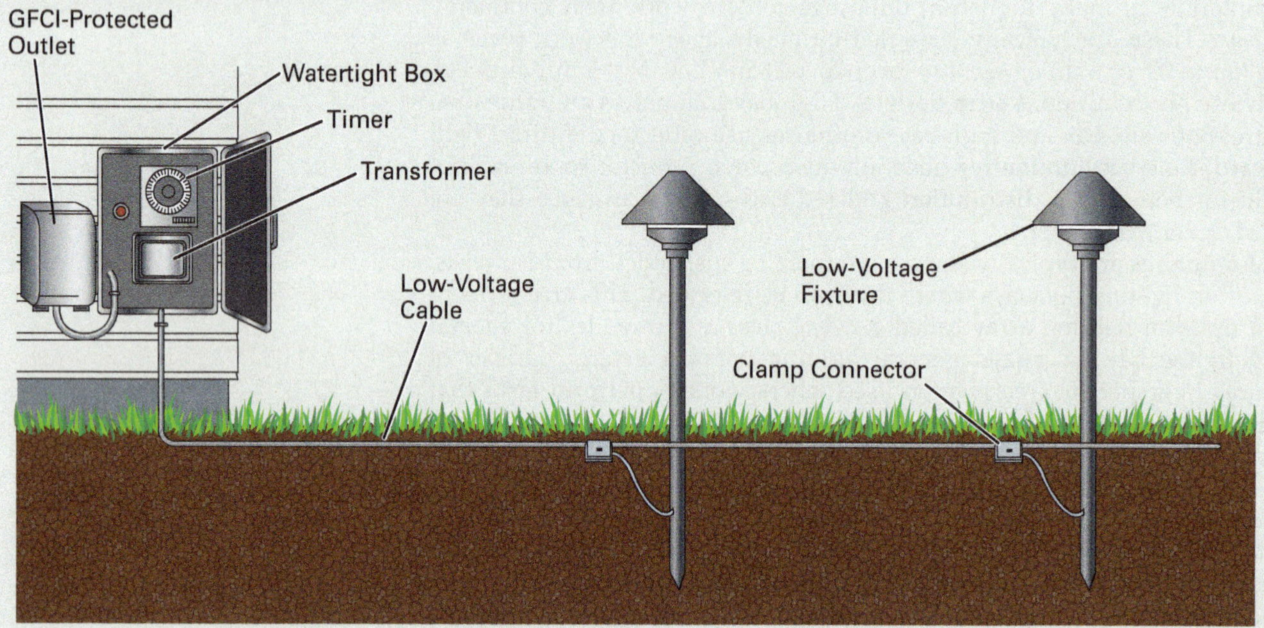

A solar outdoor luminaire is powered by the sun's energy using a built-in solar panel. During daylight hours, the panel gathers and stores energy in an internal battery to power the luminaire at night. This luminaire is controlled by a built-in ambient light sensor that turns it on at dusk and off at dawn.

Source: Shutterstock.com/Lamyai

2.2.1 Architectural and Landscape Lighting

Outdoor lighting for buildings should facilitate nighttime approach and entry (whether on foot or by vehicle), provide security for the building and its contents, and enhance the architectural features of the building. Some artistic or aesthetic uses of outdoor area lighting include the following:

- Downlighting mounted close to the ground to emphasize flower beds, steps, and paths and to visually extend the property and spaciousness.
- Downlighting mounted high in trees or on poles to floodlight a broad area.
- Uplighting to highlight interesting trees, statuary, or textured walls.
- Accent or spotlighting to emphasize a special object.
- Diffuse light for patios, decks, driveways, and walkways.
- Shadow lighting to provide a sense of depth and project intriguing shadows on vertical surfaces.
- Silhouetting objects from behind to obscure details and emphasize their shape (the reverse of shadow lighting).
- Grazing an object with light to highlight an interesting surface and emphasize its texture.
- Using multiple luminaires to provide cross lighting to create a softer, deeper, and more pleasing effect.

Because of their efficiency and long life, LED luminaires are now common in outdoor lighting applications.

Outdoor ceiling-mounted and wall-mounted luminaires (*Figure 24*) are widely used to illuminate building entranceways, parking areas next to buildings, and parking garages. A wall-mounted luminaire, commonly called a *wall pack*, has an uneven (asymmetrical) distribution to light a wider area. Luminaires installed on the walls or ceilings inside parking garages or similar structures are designed to produce high levels of reflected light in the structure.

Walkway and landscape lighting is frequently done using localized lighting supplied by *bollard luminaires*. They are short, thick posts with a lamp installed on top (*Figure 25*).

Figure 24 Typical outdoor surface-mounted luminaires.
Source: Hubbell Incorporated

Figure 25 Examples of outdoor luminaires.
Source: Hubbell Incorporated

Floodlight and spotlight luminaires are commonly used to light the outsides of buildings. A spotlight is usually equipped with lenses and reflectors to provide a fixed or adjustable narrow beam.

Building facade lighting uses luminaires with either narrow or wide distribution patterns. The pattern used depends on the portion of the building to be lit and how far away the luminaire is located from the building. Lighting large areas with a luminaire mounted near the building requires the use of a very wide distribution pattern. On the other hand, column lighting or accent lighting using a luminaire located at a distance from the building requires the use of luminaires with narrow distribution patterns.

Another type of outside luminaire frequently used is a step light. They provide safety lighting and to accent obstacles around steps, corridors, walkways, and entrances. Step luminaires are normally mounted in brick, concrete, or other building materials.

Illuminated Walking Aids

Because of their small size, LEDs are being integrated into architectural components, such as illuminated handrails, step edges, and more.

Source: iStock@necati bahadir bermek

2.2.2 Roadway Lighting

Good roadway lighting is designed to promote safer conditions for drivers and pedestrians. There are several classes of roadways, including freeways, expressways, major streets, and local streets. An example of a pole-mounted roadway luminaire is shown in *Figure 25*.

Roadway luminaires have wide light distribution patterns to permit greater spacing between poles. Luminaires with oval-shaped reflectors mounted on long arms are commonly used for roadway applications. Because of their appearance, they are often referred to as *cobra head luminaires*.

Parking lot lighting is often accomplished using luminaires with flat-bottomed lenses that are mounted on short arms. These can be mounted in single, twin, triple, or quad arrangements. Small luminaires mounted on short poles can also be used to provide walkway and grounds lighting.

Going Green

Protecting Endangered Loggerhead Turtles

In some areas of the country, loggerhead turtle hatchlings are protected from excessive light levels to help ensure more hatchlings make it from the nesting area to the ocean. Young hatchlings are confused by man-made light and move toward it instead of the natural light reflected off the ocean. To prevent this, all exterior lights, including nearby streetlights, are required to be turned off during the hatching season.

2.2.3 Sports Lighting

The goal of sports lighting is to provide appropriate lighting by controlling the brightness of an object and its background so that the object will appear clear and sharp to those watching. The level of illumination required depends on several factors, such as the environment, the speed of the game, the skill of the players, and the position of spectators relative to the field of play. It also depends on the direction players must look—toward the ground or in the air.

Because many sports involve a fast-moving object, such as a football or ice hockey puck, sports facilities require very high levels of light directed and/or focused in a manner appropriate for the sport.

Some sports have more than one illuminance requirement, such as the difference between infield and outfield lighting in a baseball stadium. The lighting layout used in a sports facility depends on all the factors mentioned here. Information about lighting relative to specific sports can be found in the technical section of lighting product catalogs.

For outdoor sports requiring direct lighting, LED luminaires (*Figure 26*) provide better control, better lumen maintenance, and lower operating costs than the previously popular HID luminaires. Multiple gangs of pole-mounted or tower-mounted enclosed and/or shielded floodlight luminaires are used for this purpose.

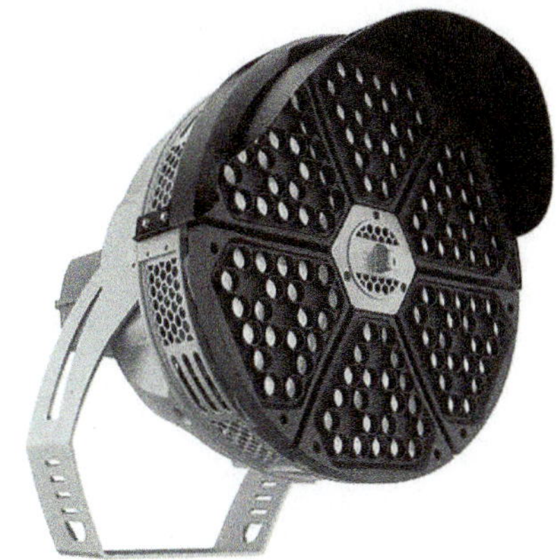

Figure 26 Sports lighting luminaire.
Source: Hubbell Incorporated

For indoor sports and where luminaires are less than 20' (6 m) from the lit area, smaller luminaires are needed. Depending on use, the lighting for indoor facilities may be totally indirect lighting or it may be direct-indirect lighting. To accommodate television cameras, the CRI for sports lighting must be above 90, preferably 97 or better.

Because outdoor sports lighting can be visible at distances beyond the sports facility, it must be adequately controlled to limit light pollution. *Spill light* is light that shines beyond the facility that can annoy occupants of adjacent properties. As a result, many municipalities have enacted ordinances to control it. It can be controlled using luminaires with an intensity distribution pattern that does not light areas outside the facility. It can also be controlled by using cutoff luminaires or high mounting poles with the luminaires set to a low aiming angle. *Sky glow,* typically seen in urban areas, is caused by stray light shining through the atmosphere and being reflected and scattered by dust and water vapor. This can be controlled somewhat by using luminaires that have no direct upward light emission.

2.0.0 Section Review

1. In a health care facility, luminaires mounted near the floor would be found in _____.
   a. examining rooms
   b. patient rooms
   c. operating rooms
   d. nurses' stations

2. A cobra head luminaire may be used for _____.
   a. landscape lighting
   b. step lighting
   c. roadway lighting
   d. sports lighting

3.0.0 Lighting Controls

Performance Tasks

There are no Performance Tasks in this section.

Objective

Select lighting controls for various applications.

a. Describe how to select dimmer systems.
b. Explain how to select occupancy and photo sensors.
c. Explain how to apply lighting timers.
d. State how various components relate to energy management systems.

According to the US Energy Information Administration, electric lighting consumes over 30% of the electrical energy used in commercial buildings. The increasing cost of energy and growing concerns for environmental protection have led to the development of many codes, such as California Title 24, ASHRAE Standard 90.1, and the International Energy Conservation Code (IECC). Voluntary standards, such as those from Leadership in Energy and Environmental Design (LEED), by the Green Building Council have also been introduced. Many utilities in North America are working together in the Design Lights Consortium, with standards and specifications for energy-efficient lighting and rebate listings made available to consumers for lighting retrofits and upgrades.

The methods discussed in this section are all used individually or in combination to conserve energy and meet these codes and standards. These methods include dimming, occupancy and photo sensors, timers, control panels, and similar devices.

Traditionally, these systems had hardwired connections between all the components. Modern systems integrate sensors, luminaires, controllers, and users via communication methods such as **Wi-Fi®** and Bluetooth® wireless technology that can provide control of individual luminaires throughout a facility.

Wi-Fi®: Acronym for *wireless fidelity*; the technology allowing communication via radio signals over a local or wide-area network equipped with a wireless access point or the internet.

3.1.0 Dimming

Dimmer systems control lighting levels to meet varying illumination requirements and conserve energy. Many different dimming technologies are in use. The choice depends significantly on the type of lamp being dimmed. Types of dimming technologies include the following:

- *Autotransformer control* — The oldest dimming technology, with high power output and no harmonic distortion, used almost exclusively with incandescent and halogen lamps.
- *Two-wire leading edge control (forward phase)* — The most common dimming technology used today, with silicon-controlled rectifiers (SCRs) reducing the voltage. Although it is used with incandescent or halogen lamps, it does not work well with LEDs.
- *Two-wire trailing edge control (reverse phase)* — Similar to the leading edge dimming technology using SCRs but applied in a different way. Requires a neutral wire to the control device and is more expensive. Also used with incandescent and halogen lamps, and it does work well with some LED drivers.
- *Three-wire leading edge control* — Originally made for fluorescent dimming but works seamlessly with LED drivers designed with a three-wire (three-phase) control input.
- *0–10V control* — A system with two power and two analog signal wires. Used with compatible fluorescent ballasts and LED drivers.

- *Digitally addressable lighting interface (DALI)* — A system with two power and two low-voltage data wires. Originally for use with fluorescent ballasts, but now widely used with LED drivers.
- *DMX512* — A system with power wires to each device as well as a daisy-chained RS-485 cable connecting them with a terminator at the end. Originally for theatrical applications but now common with white or red-green-blue (RGB) LEDs where color-changing effects are desired. DMX512 also controls luminaires that can pivot and/or swing.
- *Radio frequency (RF)* — A growing group of systems used to control LED drivers individually. Different technologies are used, such as Wi-Fi® and Bluetooth®. RF interference as well as security vulnerability may be a concern, so RF isn't used when there is a risk of personal injury or property damage.

Incandescent dimming reduces the lamp voltage and prolongs lamp life, but it isn't very energy efficient. The controls usually mount into a switch box with touch, switch, slide, and rotary control interfaces. They are made in single-pole or three-way styles and can be integrated into larger dimmer control panels.

Fluorescent dimming differs from incandescent dimming because the light does not dim to zero, the color temperature stays constant, and the ballast and control must be matched. When dimming fluorescent lamps, the cathode voltage must be maintained while the arc current is reduced—usually by a high-frequency pulse width modulated (PWM) method. Switching between fixed light levels can be done with magnetic and electronic ballast designs. Dimming can be done with wall-switch controls but is more often done at larger control panels commercially.

HID lamps are designed to operate at full power. Depending on lamp construction, HPS and MH lamps can be dimmed to between 50% and 70% of their full output, but only after they have been operating at full power for 15 minutes. Switching between fixed light levels is applied in warehouses, parking garages, and similar applications. Dimming and light level switching require the ballast and control to be matched. Slow rates of light level changes are required, and unintended arc extinguishment can occur with related restrike delay times.

You'll recall that the low-voltage power supplies required for LEDs are called *drivers*. They can support a single LED or an array of LEDs connected in series. They function much like ballasts for a fluorescent or HID lamp, but protection of their components requires internal or external surge protection.

The voltage and current rating of the driver must be matched to the requirements of the LEDs. Dimming is usually done by PWM of the low-voltage DC at high frequencies and does not affect efficiency or color temperature. Unlike fluorescent lamps, dimming LEDs increases LED life by reducing the operating temperature of the LED substrate.

LED dimming requires very careful matching of the dimmer and driver. Because of the very large number of luminaire and driver manufacturers, this can be confusing and difficult, so consultation with the manufacturer is often required. Poorly matched controls can cause early driver failure, flashing of the LEDs, other optical aberrations, and limited dimming range. With properly matched controls, dimming can be done down to 1% of full light output with no flicker or other problems, longer life for the LED and driver, and integration into different energy management systems.

Dimming can be combined with occupancy sensors, photo sensors, and timers to provide substantial energy savings in many applications.

Lighting Control

This conference room lighting control panel provides both incandescent and fluorescent control.

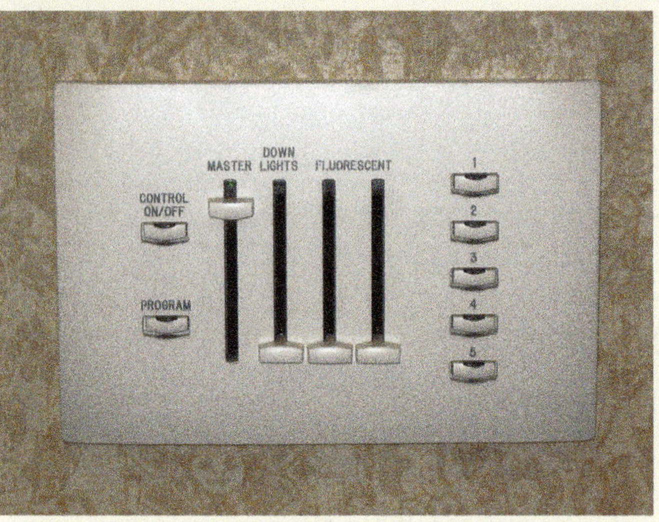

3.1.1 Dimmer Control Racks

Modern theater lighting control systems are a good example of dimmers and other automatic controls used to meet constantly changing lighting requirements. Power to all the luminaires flows through a *dimmer rack*. Typically, each lighting circuit in the rack has its own overcurrent protection with digital control provided through a remote console.

Luminaires without any local control are powered through dimmers in the rack. Luminaires with addressable control are also connected to the digital control link, which is usually a DMX512 system or its successors. The console is located to provide the operator an unobstructed view of the performance and the related lighting. A typical theater lighting control console is shown in *Figure 27*.

Older installations have manually operated dimmers located in a dimmer rack, usually with resistance or autotransformer-type dimmers.

Figure 27 Theater lighting control console.
Source: iStock@Carlos-Pascual

3.1.2 Special Uses

Color-changing RGB LEDs and luminaires can be used for architectural and special-effects lighting, such as wall-wash and accent lighting in homes and retail establishments, or for special effects and theatrical lighting at concerts and in theaters.

Control of the color and brightness is accomplished using many methods, from a simple preset sequencer to keypads, smartphone applications, and DMX and DALI protocols. LED billboards, signs, and large video displays use arrays of LEDs in separate modules assembled into the final product (*Figure 28*). These modules are made in standard and custom sizes and are available in rigid and flexible forms. *Figure 29* shows the use of flexible organic LED, also known as *OLED*, modules to form a curved billboard display. They also are made in transparent forms for mounting against the inside of a window.

Figure 28 Large RGB video display.
Source: Graham Hack

Figure 29 Curved billboard display.
Source: iStock@naumoid

3.2.0 Occupancy and Photo Sensors

Occupancy sensors (*Figure 30*) can control the operation of luminaires automatically. Occupancy sensors can be motion detecting (ultrasonic), heat sensing (infrared), or sound sensing (auditory). They can be recessed or surface mounted on a wall or ceiling, replace wall switches, or plug into receptacles. The sensor turns the luminaires on when it senses personnel entry and turns the luminaires off after a preset time when the area is vacated. The time delay is often adjustable, as is their sensitivity.

Ultrasonic sensors transmit sound at a frequency between 25 kHz and 40 kHz and receive a reflected signal to sense the movement of occupants. Passive infrared sensors detect the changes in heat across their segmented detection regions. Sound sensors passively listen for changes in background sounds made by occupants as they move about.

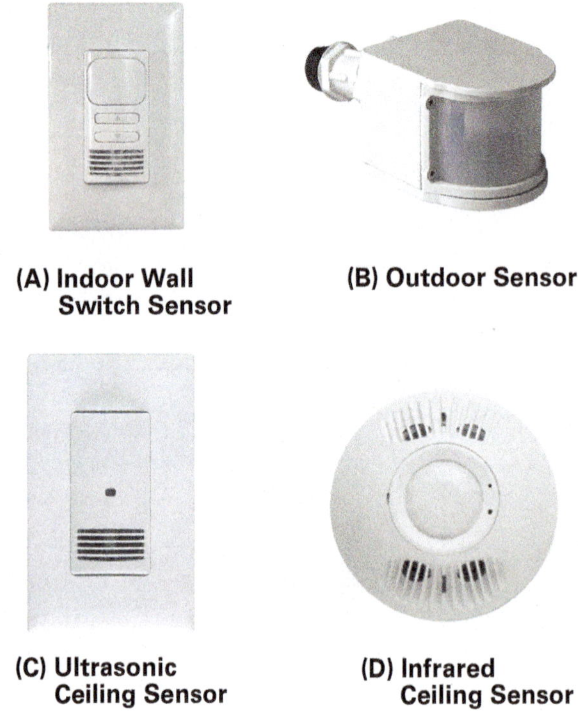

Figure 30 Examples of occupancy sensors.
Source: Hubbell Incorporated

The type of sensor used must be compatible with the application. For example, a motion or sound detector may not be the right choice where occupants sit quietly at desks. Some complain that they must deliberately move or make noise from time to time to prevent the sensor from turning off the luminaires unexpectedly. On the other hand, infrared sensors must be placed so that their sensing field isn't blocked.

Electronic photo sensors sense the level of visible light in the surrounding area and convert it into an electrical signal. This signal can be used in one of two ways, depending on the type of system. In the first type, the signal activates a simple On/Off switch or relay that controls the power to the related luminaires. They are typically used for dusk-to-dawn control of outdoor luminaires. In the second type, a variable output (analog or digital) signal is sent to a controller that continuously adjusts the output of electric lighting in the area to maintain a constant level of illumination. This approach is called **daylight harvesting** and can achieve substantial lighting energy savings because sunlight through windows or skylights often exceeds the needed illumination level. The devices shown in *Figure 31* include an infrared sensor to turn on the luminaires when motion is detected plus a photoelectric sensor to dim them as needed so they can supplement the available sunlight.

Daylight harvesting: Making use of available sunlight in an area while maintaining required illumination levels using lighting control systems to dim or switch electric lighting as needed to supplement the natural light; also referred to as *sunlight harvesting*.

Photo sensors can be an integral part of a luminaire, remote from the luminaire, or may control a circuit relay that operates several luminaires. It is important that the area controlled by one photo sensor have the same amount and direction of daylight illumination, and that the area be contiguous with no high walls or partitions to divide it. When photo sensors are used outdoors, they should be aimed due north.

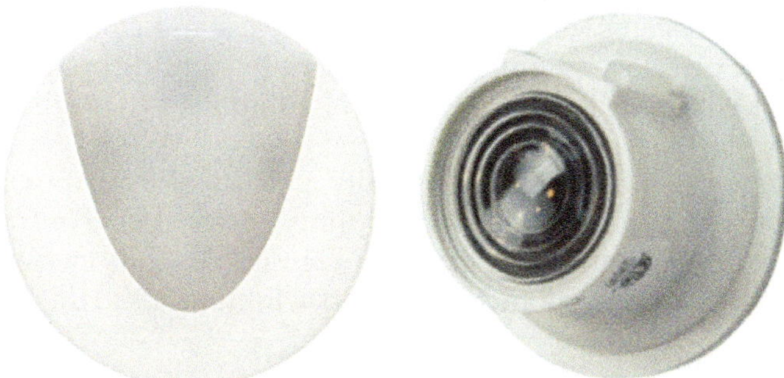

Figure 31 Daylight harvesting sensors.
Source: Hubbell Incorporated

3.3.0 Lighting Timers

Timers (*Figure 32*) are used to turn luminaires on or off in response to known or scheduled sequences of events. Timers can be very simple mechanisms with a mechanical clock, or they can be microprocessor-based with programming capability.

An electromechanical time clock / timer is driven by an electric motor, with contacts actuated by mechanical stops or arms attached to the clock face. Time clocks are either 24-hour or 7-day models. Some models offer a hold position for continuous service.

Simple time clocks can initiate several operations. Some can actuate a momentary contact switch to provide on and off pulses to actuate compatible relays or contactors. Electronic time clocks / timers provide more flexibility since they are programmable. They can typically control an output to the nearest minute over a 7-day period. Most can control multiple channels and offer time-of-day scheduling, holiday programming, automatic daylight savings time adjustment, leap year correction, annual override, and battery carryover for protection against power outages.

Time clocks are often used in conjunction with photo sensors to turn off luminaires when they are no longer needed. For example, an industrial building may use a photo sensor to activate outdoor lighting at dusk, then use a time clock to turn off that lighting after the last worker has departed. Another example is using a photo sensor to turn the luminaires on in the evening while a timer is used to dim the luminaires at midnight.

(A) Basic Electromechanical

(B) Simple Electronic

(C) Multifunction Programmable

Figure 32 Typical timer controls.
Source: Intermatic Incorporated

3.4.0 Energy Management Systems

In energy-efficient lighting designs, sensors, time clocks, and/or dimmer circuits are used to control the lighting based on occupancy and/or the required levels of lighting. The following are some of the energy conservation measures that should be incorporated in lighting:

- Zone the work areas in large open spaces to allow the lighting to be controlled in groups.
- Use controllers and matched ballasts or drivers to provide independent control when desired.

- Control luminaires near windows separately from those in the interior of a room to apply daylight harvesting.
- Ensure energy conservation measures such as dimming do not disrupt workers.

Modern buildings use some form of an energy management system (EMS) to economically control the amount of energy consumed by the building's lighting circuits and other systems. An EMS can be a simple stand-alone unitary control connected to one piece of equipment, such as a room lighting or a rooftop HVAC unit, or it can be more extensive and control equipment throughout an entire large facility.

Whether large or small, an EMS typically consists of a computer or control processor, energy management and scheduling software, sensors and controls located where needed, and one or more communications networks. When programmed, an EMS can automatically control luminaires to do the following:

- Turn off luminaires in unoccupied areas.
- Maintain partial lighting before and after working hours or public-use hours.
- Schedule lighting operation by hour of day and time of year.

The EMS receives signal inputs from sensors and processes them to perform tasks reliably and precisely. Lighting control can be implemented in a building by a local approach, a central system, or both. A local lighting system divides the building into small, independently controllable zones based on size, environment, or according to functional need. The inputs from the sensors located in the zone are wired directly to a control that is also located in the zone (*Figure 33*). In central systems, the sensor inputs from the individual zone sensors are all connected back to a central processing unit.

Hardwired systems use their own communications cable and must be installed when the structure is built. They are the most reliable option, but they are also the most expensive, cannot be easily installed in retrofit installations, and are difficult to modify.

RF systems use radio signals to control selected devices and luminaires, either from a central location or remote units such as a key fob or a mobile phone. The luminaires are plugged into RF modules or controlled by special dimmer switches.

Power line carrier communication (PLCC) systems operate through the existing wiring. In homes, these range from simple lamp timers to whole-house lighting, security, motorized window treatments, and home entertainment systems. Like RF systems, PLCC systems use plug-in modules and/or special dimmer switches and a programmable controller. Sometimes a signal booster module may be needed. At the utility or industrial level, additional components (filters, couplers, line traps, arresters, and more) are needed when PLCC systems must communicate past transformers onto feeders and transmission lines. Because PLCC systems rely on the power wiring as a network, they may be subject to signal problems in certain areas of the home. Booster modules are available for signal strengthening in problem areas.

Advanced features are available with X10 and similar systems (for example, Insteon®, Lutron, and others) which can be configured to use both RF (or infrared) control and PLCC signaling, along with computer software. Many other communication standards have been used or are being developed, to use existing power line wiring for smart metering, distributed peak shaving, substation breaker control, and more.

Other energy-saving options are also available with all these systems, including daylight harvesting and demand response control. Demand response control allows building managers to reduce lighting in entire facilities during periods of high use. Electricity billing is often structured to encourage the use of demand response control.

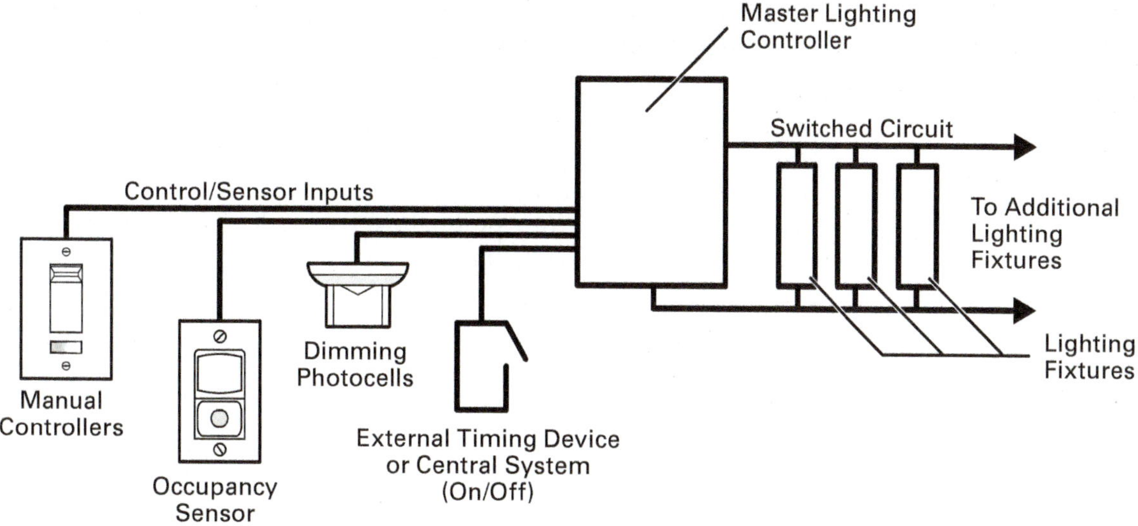

Figure 33 Simplified zone lighting control.

Think About It

Putting It All Together

Examine the lighting system in your kitchen. What changes, if any, would maximize the lighting in the key work areas? What types of lighting could you use to make the other rooms in your home appear warmer and more inviting?

3.0.0 Section Review

1. The type of dimming system that provides color-changing ability and operates through the full range of illumination without flickering is a(n) _____.
   a. LED dimming system
   b. fluorescent dimming system
   c. HID dimming system
   d. incandescent dimming system

2. An office area in which the occupants sit quietly at desks is *best* suited for a lighting system controlled by a(n) _____.
   a. motion detector
   b. photo sensor
   c. ultrasonic sensor
   d. infrared sensor

3. Time clocks are often used with _____.
   a. wall switches
   b. photo sensors
   c. three-way switches
   d. four-way switches

4. HVAC equipment as well as lighting and other systems can be controlled by a(n) _____.
   a. photo sensor
   b. infrared sensor
   c. EMS
   d. ultrasonic sensor

Module 26303-23 Review Questions

1. Luminaires that distribute 40% to 60% of their emitted light downward and the balance upward, sometimes with a strong horizontal component, are classified as _____.
   a. direct luminaires
   b. general diffuse luminaires
   c. indirect luminaires
   d. semi-direct luminaires

2. A luminaire with an IESNA Type I light distribution would be used for a _____.
   a. parking lot
   b. parking garage
   c. wide roadway
   d. narrow roadway

3. Light pollution does *not* interfere with _____.
   a. astronomy
   b. animal and insect travel patterns
   c. human sleep patterns
   d. crime

4. Ceiling luminaires equipped with diffusers are typically used to provide _____.
   a. general diffused lighting
   b. accent lighting
   c. display lighting
   d. exit lighting

5. Industrial fluorescent luminaires with apertured reflectors typically direct which percentage of light downward toward the work surface?
   a. 10% to 20%
   b. 20% to 30%
   c. 80% to 90%
   d. 100%

6. *NEC Article 501* governs the application and installation of Class I _____.
   a. emergency lighting
   b. hazardous location lighting
   c. corrosive location lighting
   d. vandal-resistant lighting

7. One lumen falling on one square foot of surface produces an illumination of _____.
   a. one footcandle
   b. one lux
   c. one candlepower
   d. one watt

8. Lighting that directs well-focused illumination to a specific surface such as a kitchen counter is classified as _____.
   a. general lighting
   b. accent lighting
   c. task lighting
   d. wall-wash lighting

9. Two identical low-bay HID luminaires are to be mounted at a height of 18'. If these luminaires have a SC of 1.5, what is the *maximum* distance that should exist between them to provide proper lighting?
   a. $15\frac{1}{2}$'
   b. 20'
   c. 25'
   d. 27'

10. In a school, task-oriented lighting is *most* likely found in a(n) _____.
    a. gymnasium
    b. restroom
    c. office
    d. cafeteria

11. An industrial application that allows for frequent lighting relocation is *most* likely to use _____.
    a. track-mounted luminaires
    b. fixed lighting outlet systems
    c. lighting trolley busways
    d. manufactured system wiring

12. Bollard luminaires are commonly used to light _____.
    a. parking lots
    b. roadways
    c. walkways and landscapes
    d. building facades

13. Two codes or standards for lighting efficiency are found in _____.
    a. California Title 24 and *NEC Article 410*
    b. ASHRAE Standard 90.1 and California Title 20
    c. IECC and ASHRAE Standard 90.1
    d. *NEC Article 410* and ASHRAE Standard 90.1

14. All the following sensors are normally used as occupancy sensors, *except* a(n) _____.
    a. motion detector
    b. infrared sensor
    c. photo sensor
    d. sound sensor

15. Daylight harvesting uses an occupancy sensor along with a _____.
    a. photo sensor
    b. timing mechanism
    c. sound sensor
    d. dimmer control panel

Answers to odd-numbered Module Review Questions are found in *Appendix A*.

Module 26303-23 Supplemental Exercises

1. Lighting that provides illumination over a small area or confined space without providing any significant general lighting in the local area is classified as _________ lighting.
2. DALI, DMX512, RF, and _________ can all be used for dimming luminaires.
3. True or False? Ceiling-mounted luminaires equipped with diffusers direct their light in a narrow pattern.
4. Track lighting is normally used to provide accent lighting, task lighting, and _________.
5. _________ luminaires are used in areas where a high amount of light is needed, and aesthetics are not important.
6. Low-bay luminaires are generally designed for installation at mounting heights of _________ and below.
7. _________ luminaires are the preferred luminaire when constructing schools.
8. Merchandise contained in store display cases is frequently illuminated by _________ luminaires with a high CRI.
9. To compensate for the poorer vision of older patients in extended care facilities, use adjustable task-oriented luminaires and _________ whenever possible.
10. A backlit ceiling panel is a type of _________ luminaire.
11. Two identical high-bay luminaires are to be mounted at a height of 25'. If these luminaires have a SC of 1.5, the maximum distance that should exist between them is _________.
12. True or False? Dustproof luminaires are constructed so that dust is prevented from entering the enclosed case.
13. An outdoor roadway luminaire is designed so that almost no light is directed above 90°. By IESNA classification, this luminaire is classified as a(n) _________ luminaire.
14. Light that shines beyond an outdoor sports facility, possibly annoying the occupants of adjacent properties, is called _________.
15. True or False? Fluorescent lamps can be dimmed to zero light output.

Answers to odd-numbered Supplemental Exercises are found in *Appendix B*.

Answers to Section Review Questions

Answer	Section Reference	Objective
Section 1.0.0		
1. d	1.1.1	1a
2. b	1.2.2	1b
3. c	1.3.1	1c
4. c	1.4.0	1d
Section 2.0.0		
1. b	2.1.4	2a
2. c	2.2.2	2b
Section 3.0.0		
1. a	3.1.0	3a
2. d	3.2.0	3b
3. b	3.3.0	3c
4. c	3.4.0	3d

User Update

Did you find an error? Submit a correction by visiting **https://www.nccer.org/olf** or by scanning the QR code using your mobile device.

MODULE 26304-23

Hazardous Locations

Source: iStock@Sergey Ryzhov

Objectives

Successful completion of this module prepares you to do the following:

1. Identify hazardous locations and their classifications.
   a. Identify Class I locations.
   b. Identify Class II locations.
   c. Identify Class III locations.
   d. Locate *NEC*® requirements for hazardous locations.
2. Explain how to prevent ignitions and explosions in hazardous locations.
   a. Identify sources of ignition.
   b. Explain how explosion-proof equipment and seals are selected and installed.

Performance Task

Under supervision, you should be able to do the following:

1. Using two rigid metal conduit nipples, a sealing fitting, three pieces of No. 12 THHN wire, and a packing fiber / sealing kit, perform the following operations:
   - Secure one conduit nipple in each end of the seal with the required number of threads engaged.
   - Pull the three conductors through the nipples and seal so that about 6" (150 mm) is protruding from each nipple.
   - Pack the fiber per the instructions furnished with the sealing kit.
   - Mix the sealing compound, position the assembly, and apply the compound.

Overview

The *NEC*® outlines the requirements for electrical equipment and wiring in locations where fire or explosion hazards may exist due to flammable vapors and liquids, combustible dust, or similar materials. This module describes the classifications and *NEC*® requirements for power supplies in hazardous locations.

NOTE

NFPA 70®, *National Electrical Code*®, and *NEC*® are registered trademarks of the National Fire Protection Association, Quincy, MA.

Digital Resources for Electrical

Scan this code using the camera on your phone or mobile device to view the digital resources related to this craft.

NCCER Industry-Recognized Credentials

If you are training through an NCCER-accredited sponsor, you may be eligible for credentials from NCCER. The ID number for this module is 26304-23. Note that this module may have been used in other NCCER curricula and may apply to other level completions. Contact NCCER at 1.888.622.3720 or go to **www.nccer.org** for more information.

You can also show off your industry-recognized credentials online with NCCER's digital badges. Transform your knowledge, skills, and achievements into badges that you can share across social media platforms, send to your network, and add to your resume. For more information, visit **www.nccer.org**.

1.0.0 Hazardous Locations

Performance Tasks

There are no Performance Tasks in this section.

Objective

Identify hazardous locations and their classifications.

a. Identify Class I locations.
b. Identify Class II locations.
c. Identify Class III locations.
d. Locate *NEC*® requirements for hazardous locations.

Hazardous locations: Locations where ignitable vapors, dust, or fibers can cause a fire or explosion when exposed to a source of ignition.

Explosion proof: Designed and constructed to withstand an internal explosion without allowing an external explosion or fire.

Know the Code

Classifications of Locations
NEC Section 500.5

The *NEC*® outlines the requirements for electrical equipment and wiring in locations where fire or explosion hazards exist due to flammable vapors or liquids, combustible dust, or ignitable airborne materials. These locations, referred to as **hazardous locations**, are classified according to the hazards present and the extent of the hazard.

Any area where the atmosphere or materials are subject to ignition from the arcs of operating electrical devices is considered a hazardous location. In a hazardous location, even the smallest of arcs has the potential to initiate a fire or explosion under the right conditions. To minimize the hazard, a wiring system that is **explosion proof** must be created using specific components approved for the application.

The *NEC*® divides hazardous materials into three classes—Class I, Class II, and Class III—with two additional divisions in each class. Class I, Division 1 hazards represent the most risk. The classes have been established based on the explosive character of the atmosphere. The testing and approval of equipment that can be used for each class are also based on this information. Considerable skill and judgment are applied to the classifications. Many factors such as the volume of explosive or ignitable material, temperature, barometric pressure, humidity, ventilation rates, and distance from the vapor source are all considered. When these factors are properly evaluated, a consistent classification for the selection and location of electrical equipment results.

NEC Section 500.5 identifies and defines the various classes and divisions discussed in the following sections.

1.1.0 Class I Locations

Know the Code

Materials
NEC Section 500.6

Places where flammable or combustible gases or vapors may be present in a volume sufficient to produce explosive or ignitable mixtures are identified as Class I locations. Class I atmospheric hazards are divided into two divisions—Division 1 and Division 2. In addition, there are four groups identified with the alpha characters A through D. Class I locations are defined in *NEC Sections 500.5(B)(1) and 500.5(B)(2). NEC Sections 500.6(A)(1) through 500.6(A)(4)* identify the materials included in the four Class I alpha groups.

Class I, Division 1 locations are those with one or more of the following characteristics:

- Flammable or combustible gases or vapors are present during normal operations.
- The gases or vapors are frequently released during maintenance and repairs.
- Frequent leaks of the gases or vapors occur.
- Breakdowns or operating problems with the equipment can release gases or vapors and may also cause a simultaneous failure of the electrical equipment.

Examples of these locations include, but are not limited to, the following (*Figure 1*):

- Paint spray booths and nearby areas.
- Areas with open tanks of **volatile** liquids that produce vapors.
- Fat and oil processing areas where these materials are extracted using volatile solvents.
- Anesthetizing locations in health care facilities where hazardous gases are likely to be present.
- Specific areas in gas manufacturing plants where hazardous gases or vapors often escape.
- Industrial refrigerators and freezers where volatile substances are stored in open, poorly sealed, or easily damaged vessels.
- Drying rooms where materials wet with volatile substances are held while they dry.

Volatile: Describes a substance that evaporates (vaporizes) readily at room temperatures.

Class I, Division 2 includes the following locations:

- Areas where volatile liquids or flammable/combustible vapors are handled or processed, but they are confined in a manner that prevents their escape unless there is an accident or equipment failure. If their escape during handling is a common or expected occurrence, the area falls into Division 1.
- Areas where mechanical ventilation usually controls the concentration unless an equipment failure occurs.
- Areas near Class I, Division 1 areas that can be contaminated by the gases or vapors but the migration of them is generally controlled by proper ventilation systems.

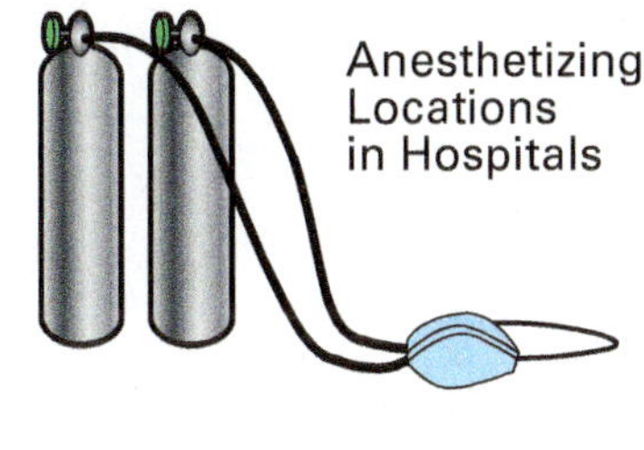

Figure 1 Typical *NEC*® Class I locations.

1.2.0 Class II Locations

Know the Code

Class II Locations
NEC Section 500.5(C)

Class II locations are defined in *NEC Sections 500.5(C)(1) and 500.5(C)(2).* Class II locations are hazardous due to the presence of combustible dust. Examples of these locations are portions of grain handling and storage plants as well as rooms where grains are ground or pulverized (*Figure 2*).

Class II, Division 1 areas have one or more of the following characteristics:

- Areas where combustible dust in sufficient volume to create an explosive or combustible mixture with air is present during normal operations.
- Areas where equipment failures or abnormal operations can lead to explosive or combustible dust mixtures and lead to the simultaneous failure of electrical systems.
- Areas where Group E dusts can reach hazardous concentrations.

Class II, Division 2 locations have one or more of the following characteristics:

- Areas where there is enough dust to produce explosive or combustible mixtures may be present due to abnormal operating conditions.
- Areas where combustible dust accumulates but usually not in sufficient volume to interfere with electrical equipment operation, but where a sufficient volume could develop due to occasional handling or processing equipment problems.
- Areas where combustible dust could accumulate on the electrical equipment in a volume that interferes with heat dissipation, or the accumulated dust could be ignited by the failure or abnormal operation of the electrical equipment.

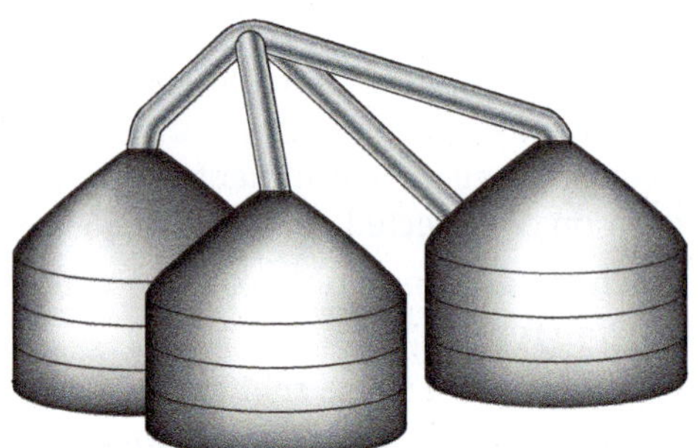

Grain Handling and Storage Plants

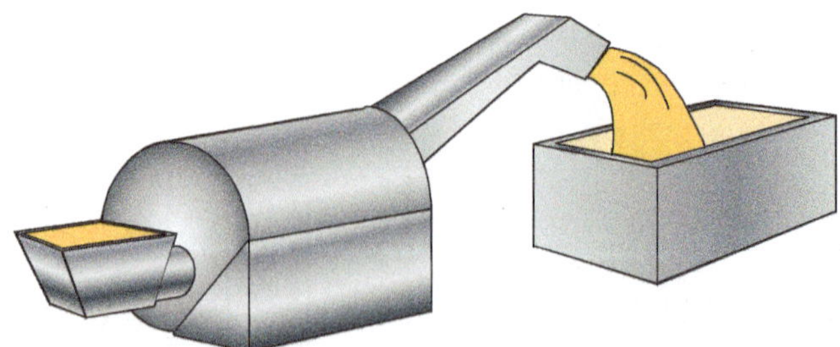

Rooms Containing Grain Grinders and Pulverizers

Figure 2 Typical *NEC*® Class II locations.

Because the *NEC*® is considered a definitive classification tool and contains explanatory data about hazardous atmospheres, refer to *NEC Section 500.5(C)* for exact definitions of Class II, Divisions 1 and 2.

Note that *NEC Sections 500.6(B)(1) through 500.6(B)(3)* identify the combustible dust classifications for Class II locations. There are three alpha groups in Class II used to classify dust—Group E, Group F, and Group G. Group E dusts include dust from metals such as magnesium and its alloys, aluminum, and other materials with similar properties. Group F includes carbonaceous materials such coal, charcoal, and coke dust. Group G includes materials that are not found in Group E or F, as well as flour, wood, and plastics. The groupings are based on the electrical resistivity and ignition temperature of the material. Group E materials are typically related to Division 1 atmospheres, while Group F and Group G materials may be related to either division.

Know the Code

Class II Combustible Dust Group Classifications
NEC Section 500.6(B)

Zone Classifications

The zone system covered in *NEC Articles 505 and 506* is an international alternative to the class/division system in *NEC Articles 500 through 504*. Because the zone system and components for it are infrequently seen, this module will not cover *NEC Articles 505 or 506*. In the plant or on the jobsite, the two systems can each be used on-site but generally must be in separate areas. See *NEC Sections 505.7(A) through (C)* and *NEC Sections 506.7(A) through (C)* for details.

1.3.0 Class III Locations

Class III locations are those with easily ignitable fibers or similar materials that are airborne. They are not broken down into alpha groups like Class I and Class II materials. *NEC Sections 500.5(D)(1) and 500.6(D)(2)* identify locations that are considered Class III applications.

Class III, Division 1 locations have one or more of the following characteristics:

- Areas where nonmetal combustible fibers are airborne in sufficient volume to create an explosive mixture during normal operations.
- Areas where equipment problems might cause airborne fibers to be produced or might lead to the simultaneous failure of electrical equipment or the actuation of an overcurrent protective device that could ignite the material.
- Areas where combustible fibers are airborne due to being handled, manufactured, or used in other processes.

Class III, Division 2 locations have one or more of the following characteristics:

- Areas where nonmetal combustible fibers might become airborne in a sufficient volume to produce an explosive mixture due to abnormal operations.
- Areas where nonmetal combustible fibers accumulate but not normally enough to interfere with the operation of the electrical equipment but could, due to problems with handling or processing equipment, become airborne.
- Areas where nonmetal combustible fibers are stored or handled, other than in the manufacturing process.

Class III locations for both divisions include certain areas of rayon, cotton, and textile mills; clothing manufacturing plants; and woodworking plants (*Figure 3*).

Know the Code

Class III Locations
NEC Section 500.5(D)

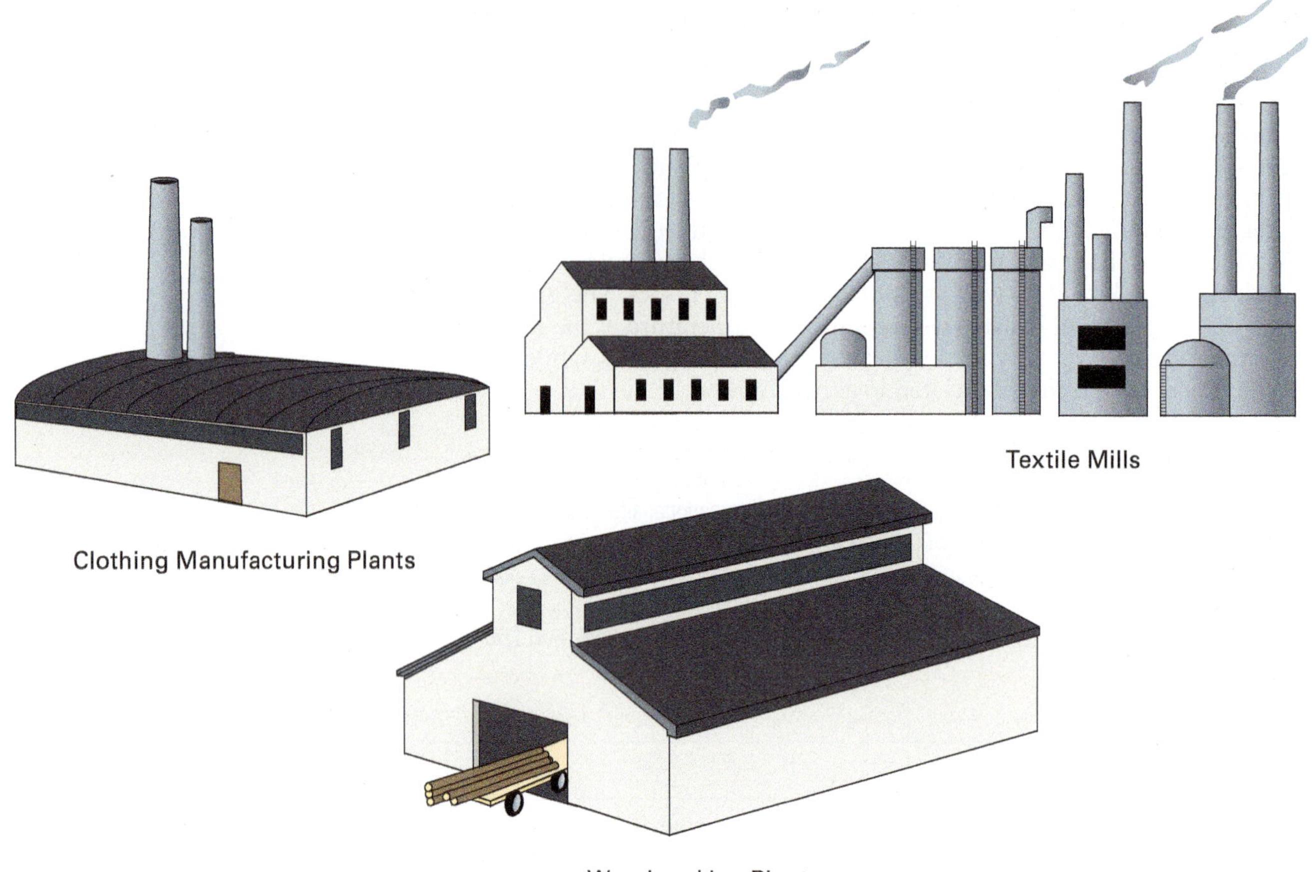

Figure 3 Typical *NEC*® Class III locations.

1.4.0 *NEC*® Requirements for Hazardous Locations

Hazardous atmospheres are summarized in *Table 1*. For a more complete listing of flammable liquids, gases, and solids, see NFPA 497, *Recommended Practice for the Classification of Flammable Liquids, Gases, or Vapors and of Hazardous (Classified) Locations for Electrical Installations in Chemical Process Areas*.

Once the class of an area is determined, the conditions under which the hazardous material is present determine the division. In Class I and Class II, Division 1 locations, the hazardous gas or dust is generally present in dangerous concentrations during normal operations. In Division 2 locations, the hazardous material is not normally in the air, but it might be released if there is an accident or if there is faulty operation of equipment.

It is recommended that electrical equipment serving hazardous locations be installed in less hazardous areas when possible. It also suggested that the hazards be reduced or eliminated by positive-pressure ventilation using a clean source of outside air. In many cases, the installation of a dust collection system can greatly reduce the hazards in a Class II area.

The following tables summarize the application rules for the various classes and divisions:

- *Table 2* — Class I, Division 1 locations
- *Table 3* — Class I, Division 2 locations
- *Table 4* — Class II, Division 1 locations
- *Table 5* — Class II, Division 2 locations
- *Table 6* — Class III locations, both divisions

TABLE 1 Summary of Hazardous Atmospheres

Hazardous Area	Class Subdivisions	Groups
	Class I Divisions	**Class I, Division Groups**
Class I: Material present is a flammable gas or vapor.	*Division 1*: Areas where hazardous concentrations of flammable gases or vapors are expected frequently.	*Group A*: Atmospheres containing acetylene
	Division 2: Areas where hazardous concentrations of flammable gases or vapors are present due to rare equipment or container failures.	*Group B*: Atmospheres containing hydrogen, manufactured gases containing more than 30% hydrogen by volume, or equal gases or vapors
		Group C: Atmospheres containing ethylene, cyclopropane, or equal gases or vapors
		Group D: Atmospheres containing propane, gasoline, or equal gases or vapors
	Class II Divisions	**Class II, Division Groups**
Class II: Material present is a combustible dust.	*Division 1*: Areas where hazardous concentrations of combustible dust are usually present or can develop because of equipment problems or where conductive dusts are present in hazardous quantities.	*Group E*: Atmospheres containing combustible metal dusts such as aluminum, magnesium, and some magnesium alloys
	Division 2: Areas where hazardous concentrations of combustible dust are not normally in the air but may be due to rare equipment problems or where dust accumulation might slow heat dissipation or be ignited by an electrical equipment malfunction.	*Group F*: Atmospheres containing combustible carbon-bearing dusts such as carbon black, charcoal, and coal
		Group G: Atmospheres containing combustible nonconductive dusts not included in Group E or F, such as flour and plastic
	Class III Divisions	**No Division Groups**
Class III: Material present is an ignitable fiber or flying.	*Division 1*: Areas where fibers or materials that can generate combustible concentrations are handled, manufactured, or applied.	
	Division 2: Areas where combustible fibers are stored or handled, but not those engaged in the manufacturing process.	

TABLE 2 Application Rules for Class I, Division 1

Components	Characteristics	Code References
Boxes and fittings	Explosion proof and threaded for connection to conduit	*NEC Section 501.10(A)*
Wiring methods	Rigid metal conduit, steel intermediate metal conduit, Type MI cable, and Types ITC and MC cables (under certain conditions)	*NEC Section 501.10(A)(1)*
Flexible connections	Class I, explosion proof	*NEC Section 501.10(A)(2)*
Sealoffs	Approved for purpose	*NEC Sections 501.15(A), (B), (C), and (D)*
Liquid-filled transformers	Installed in an approved vault outside the area	*NEC Section 501.100(A)(1)*
Circuit breakers, fuses, switches, and motor controls	Class I enclosure*	*NEC Section 501.115*
Dry-type transformers	Class I, Division 1 enclosure or outside the area	*NEC Sections 501.100(A)(2) and 501.120(A)*
Motors and generators	Class I, Division 1 and totally enclosed or submerged	*NEC Section 501.125(A)*
Lighting fixtures (luminaires)	Approved for Class I, Division 1	*NEC Section 501.130(A)*
Portable lamps	Approved as a portable assembly for Class I, Division 1	*NEC Section 501.130(A)(1)*
Utilization equipment	Class I, Division 1	*NEC Section 501.135(A)*
Receptacles	Listed for the location	*NEC Section 501.145*
Alarm systems	Class I, Division 1	*NEC Section 501.150(A)*

*Class 1, Division 1 enclosures include explosion-proof or purged and pressurized enclosures per *NEC Section 501.105(A)*.

TABLE 3 Application Rules for Class I, Division 2

Components	Characteristics	Code References
Flexible connections	Class I, explosion proof	*NEC Section 501.10(B)(2)*
Wiring methods	Rigid metal conduit; steel intermediate metal conduit; Type MI, MC, MV, TC, ITC, or PLTC cable; or enclosed gasketed busways or wireways	*NEC Section 501.10(B)(1)*
Sealoffs	Approved for purpose	*NEC Sections 501.15(B), (C), and (E)*
Boxes, fittings	Do not have to be explosion proof unless current interrupting contacts are exposed	*NEC Section 501.10(B)(4)*
Liquid-filled transformers	Approved vault outside the area	*NEC Sections 501.100(B), 450.26, and 450.27*
Switches, motor controls, and circuit breakers	Class I enclosure*	*NEC Section 501.115(B)(1)*
Fuses	Location-identified enclosure marked factory sealed or seal not required or equivalent	*NEC Section 501.115(B)(3)*
Dry-type transformers	Class I, general purpose except switching mechanism Division 1 enclosures	*NEC Sections 501.120(B) and 501.100(B)*
Motors and generators	General purpose unless motor has sliding contacts, switching contacts, or integral resistance devices; if so, use Class I, Division 1	*NEC Section 501.125(B)*
Lighting fixtures (luminaires)	Protected from physical damage	*NEC Section 501.130(B)(1)*
Portable lamps	Explosion proof	*NEC Section 501.130(B)(4)*
Utilization equipment	Depends on equipment type	*NEC Section 501.135(B)*
Receptacles	Approved for the location	*NEC Section 501.145*
Alarm systems	Class I, Division 2	*NEC Section 501.150(B)*

*Class 1, Division 2 enclosures include explosion-proof or purged and pressurized enclosures per *NEC Section 501.105(B)*.

TABLE 4 Application Rules for Class II, Division 1

Components	Characteristics	Code References
Wiring methods	Rigid metal conduit, steel intermediate metal conduit, or Type MI cable and Type MC cable (under certain circumstances) listed for use in Class II, Division 1 locations	*NEC Section 502.10(A)(1)*
Flexible connections	Extra-hard usage cord, liquidtight, and others	*NEC Section 502.10(A)(2)*
Boxes and fittings	Class II boxes required when using taps, joints, or other connections; otherwise, use dust-tight boxes with threaded openings	*NEC Section 502.10(A)(3)*
Liquid-filled transformers	Install in an approved vault	*NEC Section 502.100(A)(1)*
Dry-type transformers	Class II, vault	*NEC Section 502.100(A)(2)*
Circuit breakers, fuses, switches, and motor controls	Enclosure listed for the location	*NEC Section 502.115(A)*
Motors and generators	Class II, Division 1 or totally enclosed pipe-ventilated	*NEC Section 502.125(A)*
Lighting fixtures (luminaires) and portable lamps	Class II, identified for location	*NEC Section 502.130(A)*
Utilization equipment	Class II, identified for location	*NEC Section 502.135(A)*
Receptacles	Class II, identified for location, permanent wiring	*NEC Section 502.145(A)*

TABLE 5 Application Rules for Class II, Division 2

Components	Characteristics	Code References
Wiring methods	Rigid metal conduit; steel intermediate metal conduit; electrical metallic tubing (EMT); Type MI, MC, TC, ITC, or PLTC cable; or enclosed dust-tight busways or wireways	*NEC Section 502.10(B)(1)*
Flexible connections	Extra-hard usage cord, liquidtight, and others	*NEC Section 502.10(B)(2)*
Boxes and fittings	Dust tight	*NEC Section 502.10(B)(4)*
Liquid-filled transformers	Install in vault	*NEC Section 502.100(B)(1)*
Dry-type transformers	Class II or vault	*NEC Section 502.100(B)(3)*
Circuit breakers, fuses, switches, and motor controls	Dust-tight enclosure	*NEC Section 502.115(B)*
Motors and generators	Class II, Division 1 or totally enclosed types	*NEC Section 502.125(B)*
Lighting fixtures (luminaires)	Identified for location	*NEC Section 502.130(B)*
Portable lamps	Identified for location	*NEC Section 502.130(B)(1)*
Utilization equipment	Dust-tight enclosure	*NEC Section 502.135(B)*
Receptacles	Identified for location, permanent wiring	*NEC Section 502.145(B)*

TABLE 6 Application Rules for Class III, Divisions 1 and 2

Components	Characteristics	Code References
Wiring methods	RMC, IMC, EMT, PVC, RTRC, Types MI and MC cables, enclosed dust-tight busways or wireways, and tray cables	*NEC Section 503.10(A)(1)*
Boxes and fittings	Dust tight	*NEC Section 503.10(A)(2)*
Flexible connections	Extra-hard usage cord and other flexible conduit/fittings	*NEC Section 503.10(A)(3)*
Liquid-filled transformers	Install in an approved vault	*NEC Sections 503.100 and 502.100(B)*
Dry-type transformers	Dust-tight enclosure	*NEC Sections 503.100 and 502.100(B)*
Circuit breakers, fuses, switches, and motor controls	Dust-tight enclosure	*NEC Section 503.115*
Motors and generators	Totally enclosed, with exceptions	*NEC Section 503.125*
Lighting fixtures (luminaires)	Tight enclosure with no openings	*NEC Section 503.130*
Portable lamps	Unswitched, guarded with tight enclosure for lamp	*NEC Section 503.130(D)*
Utilization equipment	Class III	*NEC Section 503.135*
Receptacles	Minimize accumulation of fibers or flyings	*NEC Section 503.145*

Know the Code

Class III Locations
NEC Article 503

The *NEC®* also provides guidelines for other potentially hazardous locations. These locations include garages, airport hangars, health care facilities, and petrochemical facilities, among others. Always reference the latest edition of the *NEC®*, along with state and local codes and standards, when installing systems in these or similar locations.

Think About It

Identifying Hazardous Locations

This fixture is installed in a grain conveyor tunnel. What class, division, and group do you think would apply to this installation?

Source: Tim Dean

Delayed Action Receptacles

The receptacle shown in the image is rated for Class I, Division 1 and 2 locations and features a delayed-action rotating sleeve that prevents complete withdrawal of the plug in one continuous movement. This delay allows any arc-generated heat to be dissipated before the plug is released.

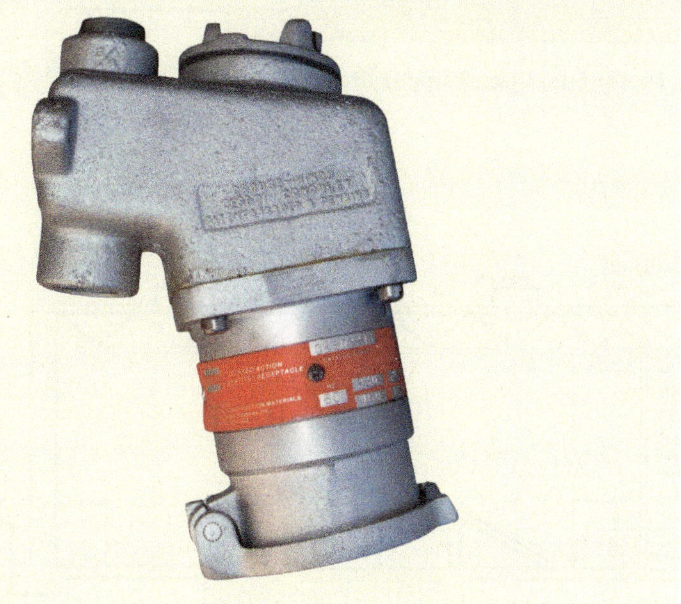

1.4.1 Garages and Similar Locations

Garages and similar locations where volatile and flammable liquids are used in vehicles such as automobiles, buses, tractors, and boats are not generally considered to be excessively hazardous. However, the entire area from the floor to 18" (450 mm) above the floor is considered a Class I, Division 2 location. This is because the vapors from fuels are heavier than air and tend to settle at the lowest level. Likewise, any pit or depression below floor level carries the same classification. The depression may even be judged as a Class I, Division 1 location if it is not ventilated, per *NEC Table 511.3(C)*. These areas require certain precautionary measures.

Normal raceways and conductors may be used above the hazardous level of 18" (450 mm), except where there is evidence the area is more hazardous than usual for a garage. In this case, the applicable type of explosion-proof wiring is often required.

Approved fittings are used on all conduit passing from Class I hazardous areas into nonhazardous areas, per *NEC Sections 501.15(A)(4) and 501.15(B)(2)*. The requirements set forth throughout *NEC Section 501.15* apply to horizontal as well as vertical boundaries of defined hazardous areas. Raceways embedded in a masonry floor or buried beneath a floor are considered within the floor-level hazardous area if any connections lead into or through the classified area. However, conduit systems terminating at an open raceway in an outdoor unclassified area do not require sealing between the point where the conduit leaves the classified area and enters the raceway.

Figure 4 shows a typical automotive service facility with the applicable *NEC*® requirements. The space in the immediate vicinity of the motor fuel dispensing island is denoted as Class I, Division 1. The surrounding area within a radius of 20' (6 m) falls under Class I, Division 2 to a height of 18" (450 mm) above grade. Bulk storage plants for motor fuel are subject to comparable restrictions.

Note that *NEC Article 514* covers motor fuel dispensing facilities. *NEC Article 511* covers commercial garages.

A summary of *NEC*® rules governing the installation of electrical wiring at and near gasoline dispensing pumps is shown in *Table 7*.

NOTE

In major repair garages that service vehicles using lighter-than-air fuels such as hydrogen and compressed natural gas, the *NEC*® provides classifications in ***NEC Table 511.3(D)***.

Know the Code

Commercial Garages, Repair and Storage
NEC Article 511

Know the Code

Sealing and Drainage
NEC Section 501.15

Know the Code

Motor Fuel Dispensing Facilities
NEC Article 514

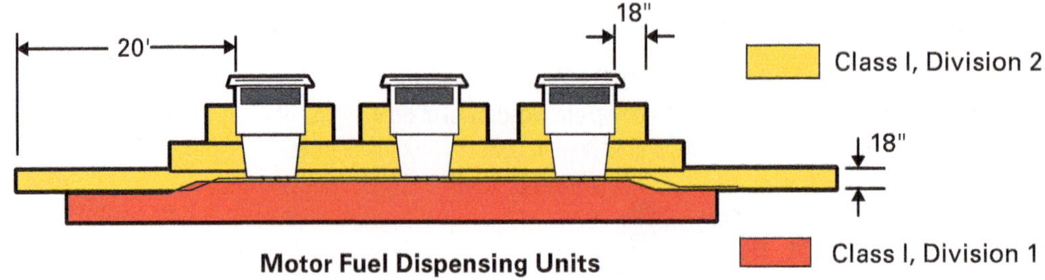

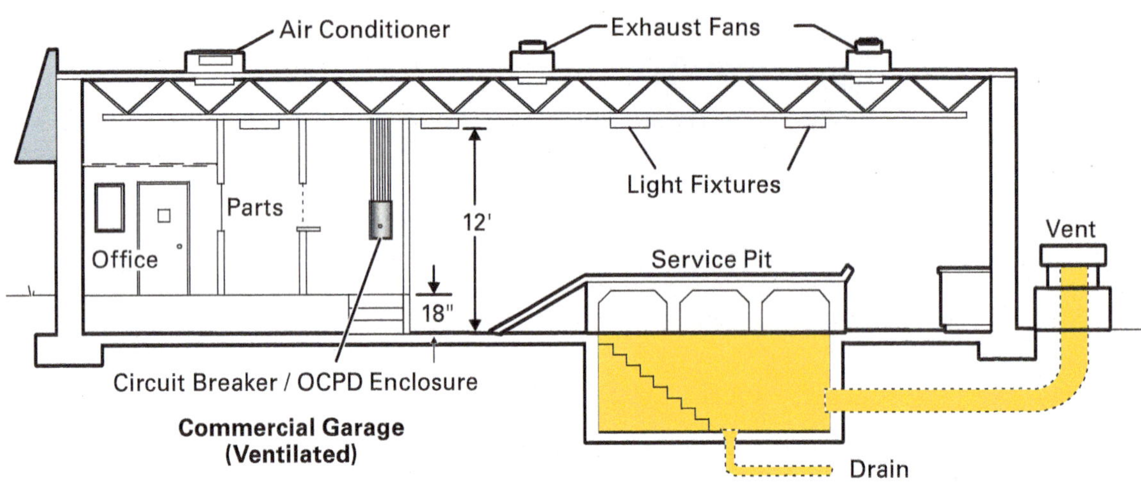

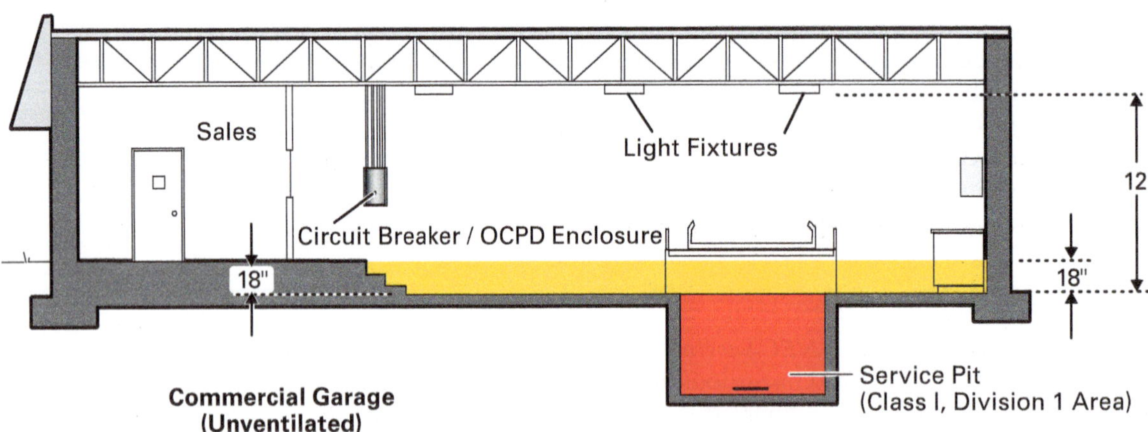

Figure 4 Commercial service station and garage classifications.

TABLE 7 *NEC*® Application Rules for Motor Fuel Dispensing Facilities

Application	Code Requirement	Code Reference
Equipment in hazardous locations	All wiring and components must conform to the rules for Class I locations.	*NEC Section 514.4*
Equipment above hazardous locations	All wiring must conform to the rules for such equipment in commercial garages.	*NEC Section 514.7*
Gasoline dispenser	A means of disconnect must be provided for each circuit leading to or through a dispensing pump, including feedback, during service and maintenance. An approved sealoff is required for each conduit entering or leaving a fuel dispenser.	*NEC Section 514.13*
Grounding	Metal parts of all noncurrent-carrying dispenser parts must be grounded and bonded.	*NEC Section 514.16*
Underground wiring	Underground wiring installed within 2' (600 mm) of the ground must be installed in threaded RMC or IMC. If underground wiring is buried 2' (600 mm) or more, rigid nonmetallic conduit may be used in addition to the types listed above. Type MI cable may also be used in some cases.	*NEC Section 514.8*

1.4.2 Airport Hangars

Hangars for aircraft where fuels and other volatile or flammable materials are used fall under *NEC Article 513*. Like garages, any depression below the level of the hangar floor is a Class I, Division 1 location. The entire floor area of the hangar, including any adjacent and communicating area not suitably cut off from the hangar, is a Class I, Division 2 location up to 18" (450 mm) above the floor.

The area within 5' (1.5 m) horizontally from aircraft power plants, fuel tanks, or anything else containing fuel is considered a Class I, Division 2 hazardous location. The area extends upward from the floor to a level 5' (1.5 m) above the upper surface of engine or fuel vessel.

Adjacent areas where hazardous vapors are not likely to be released, such as stockrooms and electrical control rooms, are not classified as hazardous if they are adequately ventilated and effectively separated from the hangar area by walls or partitions. Per *NEC Section 513.7(A)*, all fixed wiring in a hangar not within a hazardous area as defined in *NEC Section 513.3* must be installed in metallic raceways or shall be Type MI, TC, or MC cable.

The only exception to this is wiring in nonhazardous locations as defined in *NEC Section 513.3(D)*. It may be of any type recognized in *NEC Chapter 3*. *Figure 5* summarizes the *NEC*® requirements for airport hangars.

Know the Code

Aircraft Hangars
NEC Article 513

Know the Code

Wiring Methods and Materials
NEC Chapter 3

1.4.3 Health Care Facilities

Hospitals and other health care facilities fall under *NEC Article 517*. *NEC Article 517, Part II* covers the general wiring in patient areas of health care facilities. *NEC Article 517, Part III* covers essential electrical systems for health care facilities. *NEC Article 517, Part IV* provides equipment performance criteria and wiring methods used in inhalation anesthetizing locations. *NEC Article 517, Part V* covers the requirements for electrical wiring and equipment where diagnostic imaging and treatment equipment is used. *NEC Article 517, Part VI* covers communications, signaling systems, and fire alarm systems. *NEC Article 517, Part VII* covers isolated power systems.

Anesthetizing locations of health care facilities are considered Class I, Division 1 to a height of 5' (1.5 m) above the floor. Anesthetizing gases are often lighter than fuel vapors, resulting in a higher measurement. Gas storage rooms are designated as Class I, Division 1 throughout a health care facility. Most of the wiring in these areas, however, can be limited to light fixtures only by locating their switches and other devices outside of the hazardous area.

Know the Code

Health Care Facilities
NEC Article 517

NOTE

The rounding conventions for metric conversions in the *NEC*® vary depending on the section and are often more specific when detailing critical requirements, such as anesthetizing locations. Always refer to the latest edition of the *NEC*® for specific requirements.

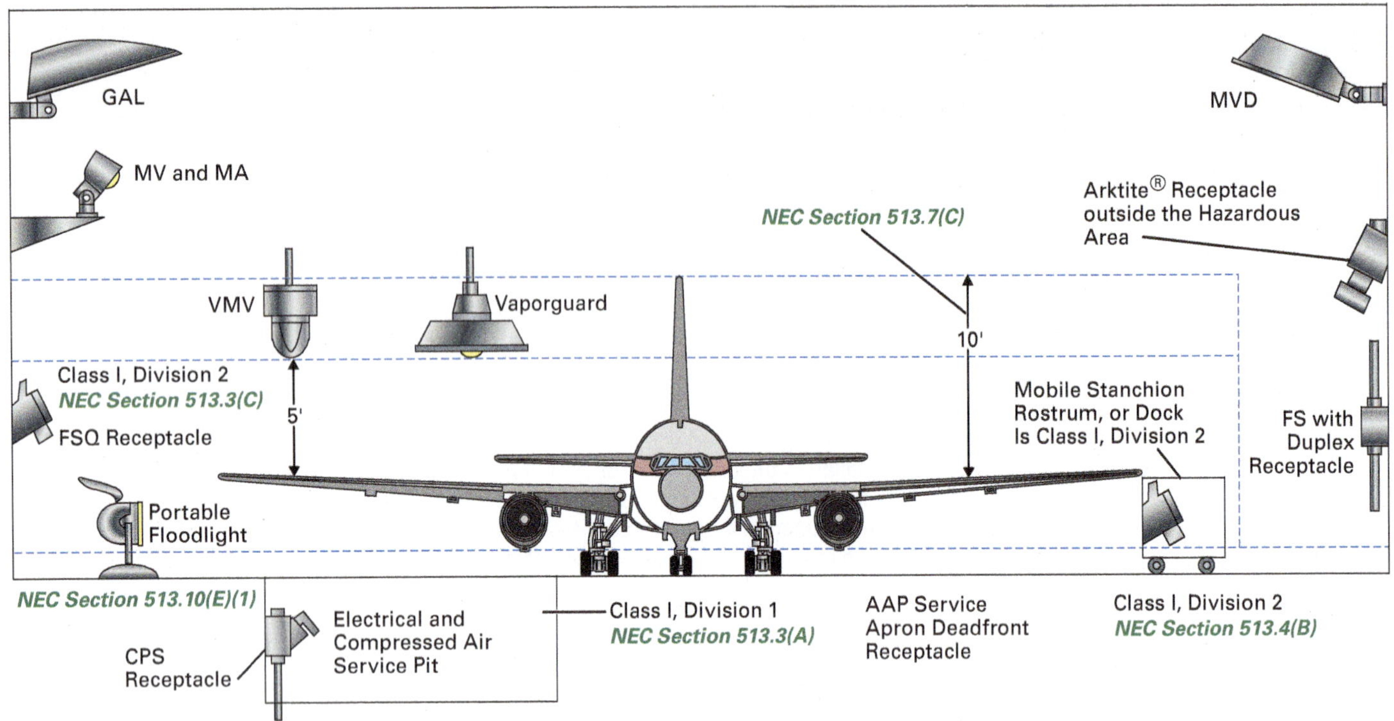

Figure 5 Sections of an airport hangar showing hazardous locations.

1.4.4 Petrochemical Facilities

Most manufacturing facilities related to flammable liquids, vapors, or fibers must conform strictly to the *NEC*® as well as governmental, state, and local codes. Therefore, like health care facilities, experts in the field are responsible for designing their electrical systems.

Industrial installations dealing with petroleum and other chemicals are subject to restrictions from many governmental agencies. Electrical installations for petrochemical plants have many pages of drawings and specifications, all of which require approval from various agencies. Even the smallest change must also flow through those same agencies for approval.

Explosion-Proof Equipment and Fittings

Manufacturers of explosion-proof equipment and fittings expend considerable time, energy, and money in developing guidelines and brochures to ensure their products are used correctly and in accordance with the latest *NEC*® requirements. The many helpful charts, tables, and application guidelines available from manufacturers are invaluable to anyone working on projects involving hazardous locations.

Obtain as much of this data as possible and study it thoroughly. Doing so will enhance your qualifications for working in and around hazardous locations. Data related to specific equipment is usually available and can be obtained from local distributors or directly from the manufacturer's website.

1.0.0 Section Review

1. An area near a Division 1 location that can be contaminated by gases or vapors should be designated _____.
   a. Division 1
   b. Division 2
   c. Division 3
   d. unclassified

2. A grain grinding and pulverizing area would *most* likely be designated as _____.
   a. Class I, Division 1
   b. Class I, Division 2
   c. Class II, Division 1
   d. Class II, Division 2

3. A cotton or textile mill would *most* likely be designated as _____.
   a. Class I, Division 1
   b. Class II, Division 1
   c. Class II, Division 2
   d. Class III

4. A Class II atmosphere containing charcoal or coal dust would *most* likely be designated as _____.
   a. Group D
   b. Group E
   c. Group F
   d. Group G

2.0.0 Preventing Ignitions and Explosions

Objective

Explain how to prevent ignitions and explosions in hazardous locations.

a. Identify sources of ignition.
b. Explain how explosion-proof equipment and seals are selected and installed.

Performance Task

1. Using two rigid metal conduit nipples, a sealing fitting, three pieces of No. 12 THHN wire, and a packing fiber / sealing kit, perform the following operations:
   - Secure one conduit nipple in each end of the seal with the required number of threads engaged.
   - Pull the three conductors through the nipples and seal so that about 6" (150 mm) is protruding from each nipple.
   - Pack the fiber per the instructions furnished with the sealing kit.
   - Mix the sealing compound, position the assembly, and apply the compound.

The main purpose of using explosion-proof fittings and wiring methods in hazardous areas is to prevent the ignition of flammable and combustible substances. However, when hazardous concentrations of air and substances are present, a source of ignition is still required before a disaster occurs.

2.1.0 Sources of Ignition

When flammable gases or combustible dusts are mixed in the proper proportion with air, any source of ignition can cause an explosion.

One prime source of ignition is an electric device, especially those with contacts that open and close. An arc across opening and closing contacts is common. Equipment such as switches, circuit breakers, motor starters, push-button stations, or plugs and receptacles normally produce arcs. Each arc represents an ignition source.

Other ignition sources are devices that produce heat, such as lighting fixtures and motors. Elevated temperatures may exceed the safe limit for some combustible mixtures.

Finally, other parts of an electrical system become potential ignition sources if a short circuit occurs. This group includes any portion of the electrical devices where a short circuit can develop for a variety of reasons.

Nonelectrical hazards such as small sparks from a hammer can also cause ignition. A steel tool that is dropped on masonry or on a ferrous surface can cause a spark unless the tool is made from nonsparking materials. When working in an area where such hazards are present, the choice of tool and how it is used can be the difference between a standard day on the job and a disaster.

It is crucial to prevent the accidental ignition of flammable liquids, vapors, and dusts released to the atmosphere. In addition, because electrical equipment is also used outdoors and in corrosive atmospheres, the material and finish must be protected from the elements.

The following three basic conditions must be satisfied for a fire or explosion to occur:

- A flammable fuel must be present in sufficient volume.
- The fuel must be mixed with oxygen (in the air) in a proportion that forms an explosive mixture.
- A source of ignition must ignite the mixture.

Volatile liquids vaporize and disperse in the atmosphere and can become explosive, but they can be rapidly diluted to concentrations below their lower explosive limit with proper ventilation.

WARNING!

Gas concentrations above the upper explosive limit are also less likely to ignite, but that does not provide any degree of safety. The concentration must first pass through the explosive range to reach the upper explosion limit, and a change in the concentration can quickly place it in the explosive range.

Explosion-Proof Equipment

Explosion-proof equipment must be marked to show class and group. See ***NEC Section 500.8(C)***. Always make sure that the device selected is rated for the hazardous location in which it will be installed.

2.2.0 Explosion-Proof Equipment and Seals

Areas with hazardous atmospheres must be carefully evaluated to make certain the correct electrical equipment is selected. Conformity with the *NEC*® requires the use of fittings and enclosures approved for the specific hazard. The equipment is rated by both classification and temperature. See *NEC Table 500.8(C)(4)*.

A wide assortment of explosion-proof equipment is available for all hazardous environments. To select the correct components, you must be thoroughly familiar with all *NEC*® requirements and know what fittings are available, how they are installed, and where to use them.

Some are under the false impression that a fitting rated for Class I, Division 1 can be used for any other class or division. This is not the case. For example, a fitting rated for Class I, Division 1, Group C cannot be used in areas classified as Groups A or B. On the other hand, fittings rated for use in Group A may be used in areas rated as Group A or any group beneath A; fittings rated for use in Class I, Division 1, Group B can be used in areas rated as Group B or below.

Know the Code

Classification of Maximum Surface Temperature
NEC Table 500.8(C)(4)

WARNING!

Never interchange fittings or covers between one hazardous area and another. The items must be rated for the appropriate class, division, and group.

Explosion-proof fittings are rated by both classification and group. All parts of the fittings must carry the same rating. Therefore, if a Class I, Division 1, Group A fitting is required, a Group B (or below) fitting cover must not be used. The cover itself must be rated for Group A locations. Consequently, when working on electrical systems in hazardous locations, always make certain that fittings and their related components have matching specs.

2.2.1 Intrinsically Safe Equipment

Intrinsically safe equipment is incapable of releasing sufficient electrical energy to cause ignition of a hazardous atmosphere in its most easily ignited concentration. The use of intrinsically safe equipment is primarily limited to process control instrumentation because these electrical systems have low energy requirements.

Installation rules for intrinsically safe equipment are covered in *NEC Article 504*. In general, intrinsically safe equipment and its wiring must be installed so that it is positively separated from the non-intrinsically safe circuits. Otherwise, induced voltages could defeat the purpose of intrinsically safe circuits. UL Solutions, formerly known as Underwriters Laboratories, Inc., and FM Global list multiple intrinsically safe devices.

Know the Code

Intrinsically Safe Systems
NEC Article 504

What's Wrong with This Picture?

Source: Dan Lamphear

2.2.2 Explosion-Proof Conduit and Fittings

In locations where threaded metal conduit is required, the conduit must be threaded with national pipe taper (NPT) or metric threads. A typical floor plan for a hazardous area is shown in *Figure 6*. Conduit and fittings shall be made wrench-tight to prevent arcing when fault current flows through the system, per *NEC Section 500.8(E)*.

NPT threads shall be made using National (American) pipe taper threads, as shown in *Figure 7*. *NEC Section 344.28* requires the taper to be $\frac{3}{4}$" taper per foot (1 in 16 taper). *NEC Section 500.8(E)(1)* requires conduit and fittings to be made up with five threads fully engaged.

For metric threads, conduit fittings and cable fittings shall be listed in *NEC Section 500.8(E)(2)*. All boxes, fittings, and joints must be threaded for connection and must be an approved type (*Figure 8*). When it becomes necessary to employ flexible connectors at motor or fixture terminals, flexible fittings approved for the class location are used (*Figure 9*). Unions are provided to facilitate the future installation and removal of equipment.

Know the Code

Threading
NEC Section 500.8(E)

Know the Code

Reaming and Threading
NEC Section 344.28

2.2.3 Seals and Drains

Sealoff fittings, also known as *sealing fittings* or *conduit seals* (*Figure 10*), are required in conduit systems so vapors and flames are not able to pass from one portion of the installation to another. Sealoffs also help limit explosions to the enclosure.

Sealoff fittings: Fittings required in conduit systems to prevent the passage of gases, vapors, or fire from one portion of an electrical installation to another through the conduit. Also referred to as *sealing fittings* or *conduit seals*.

For Class I, Division 1 locations, *NEC Section 501.15(A)(1)* states that, in each conduit run entering an enclosure for components that may produce arcs, sparks, or high temperatures, seals must be placed in the conduit within 18" (450 mm) from the enclosure. Explosion-proof fittings and bodies are the only components permitted between the sealing fitting and the enclosure. The conduit bodies cannot be larger than the largest conduit trade size.

Know the Code

Conduit Seals, Class I, Division 1
NEC Section 501.15(A)

However, there is one exception to this rule. The connecting conduits are not required to be sealed if the current interrupting contacts are enclosed within a hermetically sealed chamber to prevent the entrance of explosive mixtures. They are immersed in a dielectric oil in accordance with the exception in *NEC Section 501.15(A)(1)* or enclosed within a factory-sealed, explosion-proof chamber. The chamber must be marked "Factory Sealed" or in a similar fashion.

Note that the allowable fill for a sealoff can't exceed 25% of the cross-sectional area. This is less than what is allowed for most conduit. If the planned fill exceeds 25% of the cross-sectional area of the sealoff, a larger seal size might be necessary.

Seals are also required in Class II locations where a raceway provides a connection between an enclosure that is required to be dust tight / ignition proof and one that is not (per *NEC Section 502.15*). However, the seal does not need to be explosion proof. Electrical sealing putty is an approved solution.

Know the Code

Sealing, Class II, Divisions 1 and 2
NEC Section 502.15

A permanent and effective seal is one method of preventing the entrance of dust. A horizontal raceway, not less than 10' (3.05 m) long, is another approved method, as is a vertical raceway not less than 5' (1.5 m) long and extending downward from the dust-tight/ignition-proof enclosure.

Where a raceway provides a connection between an enclosure that is required to be dust tight / ignition proof and an enclosure in an unclassified location, seals are not required. If sealing fittings are installed, they must all be accessible.

Pressure Piling

Pressure piling occurs when an explosion in one section of a raceway creates expanding gases that cause compression of the air all along the raceway. This can cause a secondary explosion of even greater magnitude. The second explosion causes additional compression that may create additional explosions, initiating a chain reaction.

While it is not an *NEC*® requirement, many electrical designers sectionalize long conduit runs by inserting seals not more than 50' to 100' (15 m to 30 m) apart, depending on the conduit size. This is done to minimize the effects of pressure piling.

In general, seals are installed at the same time as the conduit system. However, the conductors are installed after the raceway system is complete and prior to packing and sealing the fittings.

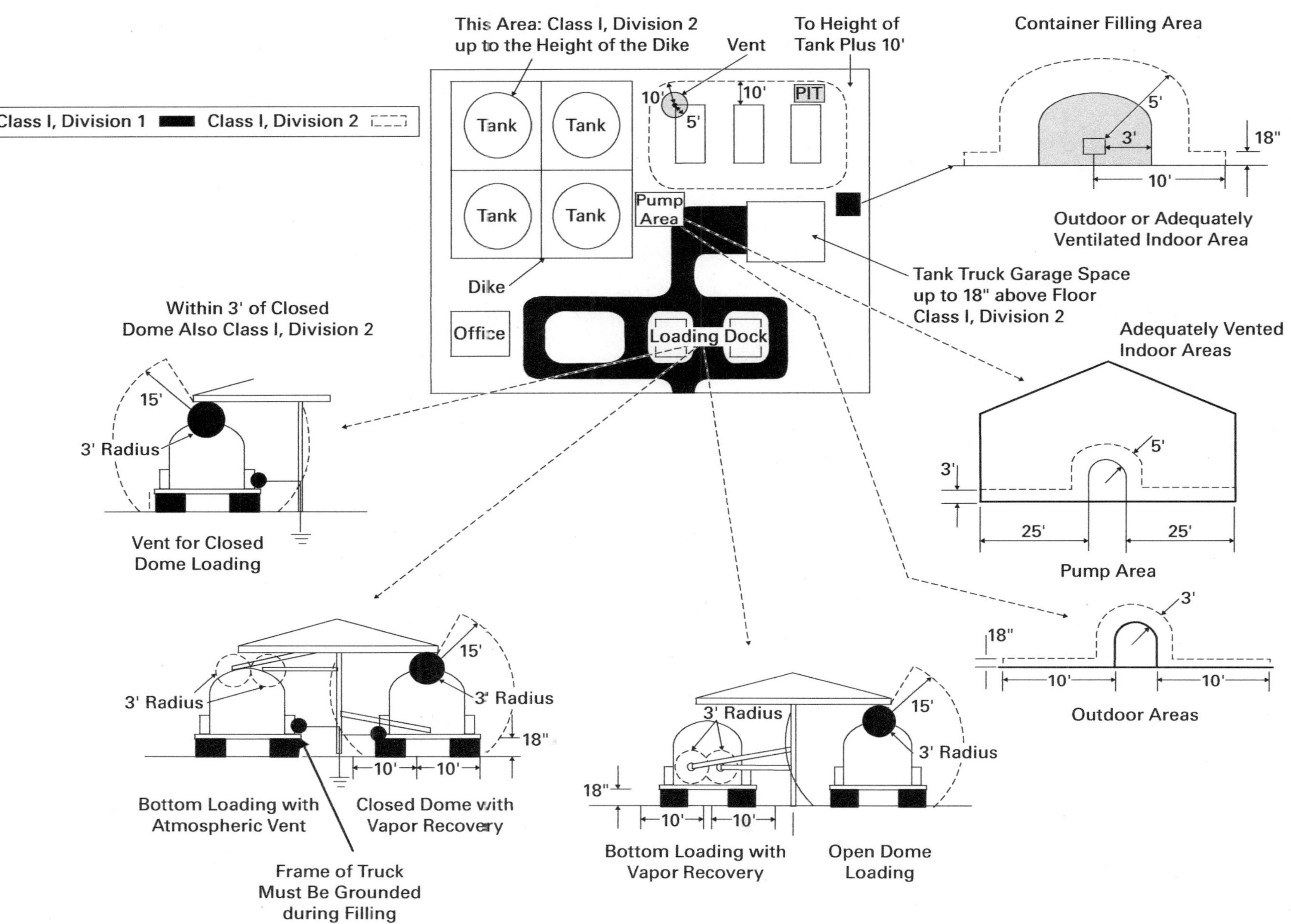

Figure 6 Floor plan of a hazardous location.

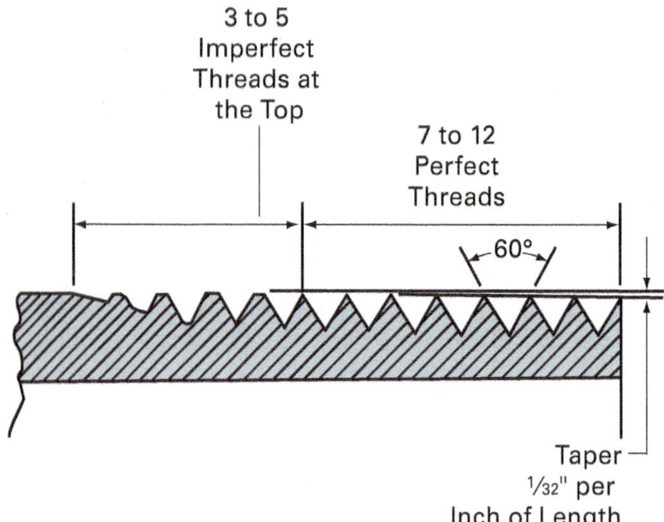

Figure 7 NPT threads.

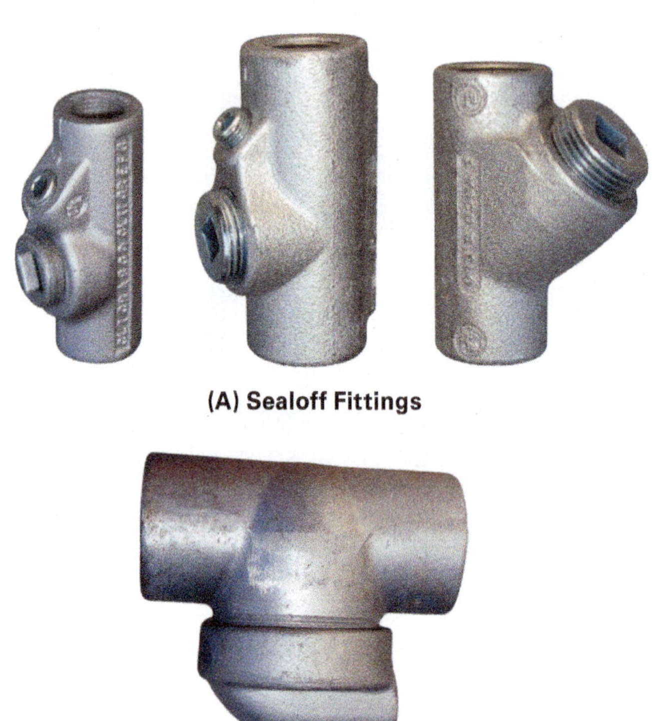

(A) Sealoff Fittings

(B) Explosion-Proof Seal

(C) Explosion-Proof Conduit Body

Figure 8 Typical fittings approved for hazardous areas.

Figure 9 Explosion-proof flexible connector.

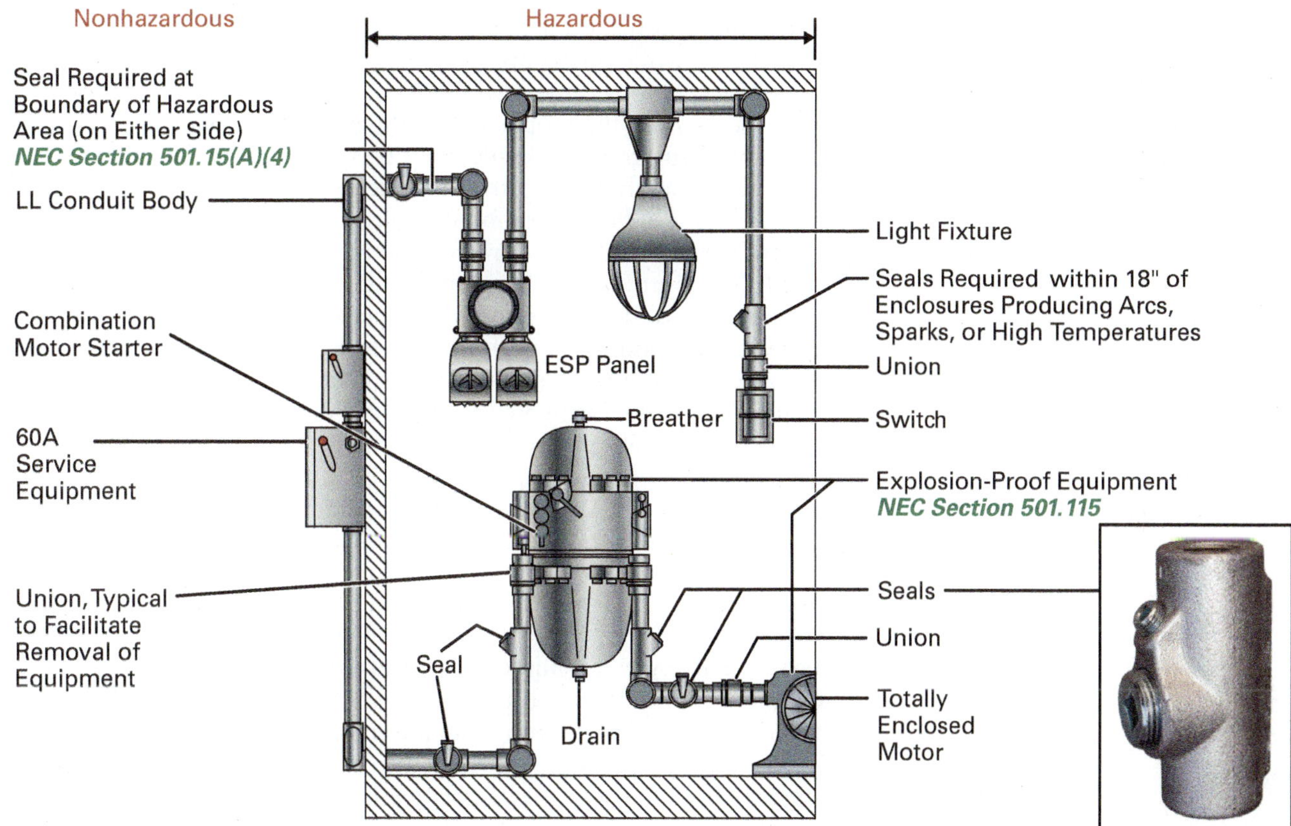

Figure 10 Installation of seals in Class I, Division 1 locations.

Think About It

Fill Requirements

Why do you think the standards for listing sealoff fittings allow only a 25% fill for these fittings, while conduits are permitted to have a 40% fill?

Figure 11 Typical drain seal.

In humid atmospheres or wet locations where water could enter the system, the raceways should be inclined so that water will not collect in enclosures or on seals but will drain to low points where it can be eliminated.

If the arrangement of the raceway makes this impractical, special drain/seal fittings should be used, such as the type shown in *Figure 11*. These fittings prevent water from accumulating above the seal and meet the requirements of *NEC Section 501.15(F)*.

Even if the location is not typically humid or wet, surprising amounts of water can still collect in conduit systems. Most conduit systems aren't airtight. Alternating weather changes cause air to move in and out. Since water vapor travels with it, any cooling of the air in the conduit results in condensation. The water collects and remains there as time passes.

To avoid this, install drain/seal fittings with drain covers or fittings that have inspection covers. This is recommended even if prevailing conditions don't indicate a that condensation inside the raceway system will be an issue.

Always ensure the selected sealoff fitting is appropriate for the mounting position. This is particularly critical when the conduit run crosses between hazardous and nonhazardous areas. Some seals are designed to be mounted in any position, while others are restricted to horizontal or vertical mounting. *Figure 12* shows various types of seals.

Install the seals on the proper side of the partition or wall, as recommended by the manufacturer. Seals should only be installed by trained personnel in strict compliance with the furnished instructions.

NEC Section 501.15(C)(4) prohibits splices or taps in sealoff fittings. Sealoff fittings are listed by UL Solutions for use in Class I hazardous locations, but only with an approved **sealing compound**. The compound, when properly mixed and poured, hardens into a dense mass that is insoluble in water and resists heat and chemicals. It is also designed to withstand the pressure caused by explosions. Conductors sealed in the compound may be any approved thermoplastic or rubber insulated type. A lead covering is optional.

Sealing compound: A material poured into an electrical fitting as a liquid that cures and hardens to positively prevent the passage of air, gases, and vapors.

Improperly poured seals serve no purpose; they must be applied carefully. Sealing compounds must be approved for the purpose and not have a melting point less than 200°F (~93°C). The sealing compound and dams must also be approved for the fitting. For example, Crouse-Hinds CHICO® A is the only sealing compound approved for use with Crouse-Hinds ECM fittings.

To pack the sealoff, remove the threaded plug or plugs from the fitting and insert the fiber supplied with the kit. Tamp the fiber between the wires and the hub before pouring the sealing compound into the fitting. Then pour in the sealant and reset the threaded plug tightly. The fiber packing will prevent the sealing compound from draining into the conduit lines.

Most sealing compound kits contain a powder in a polyethylene bag within an outer container. Remove the bag of powder, fill the outside container, pour in the powder, and mix to prepare the compound.

CAUTION

Always make certain that the sealing compound is compatible with the packing material, brand, fitting, and conductors used in the system.

Dozens of seals may be required for a significant project. To prevent the possibility of overlooking a seal, a certain color of paint is often sprayed on the seal hub to indicate that the seal has been packed. When the sealing compound is poured, a different color paint is sprayed on the seal hub to indicate a finished job. This method permits the job supervisor to visually inspect the conduit run, looking for seals without the appropriate color. A missed seal can then be addressed.

The sealoff fittings in *Figure 13* are typical. The type in *Figure 13* (A) is for vertical mounting and is provided with a threaded, plugged opening into which the sealing cement is poured.

The sealoff in *Figure 13* (B) has an additional plugged opening in the lower hub to facilitate the installation of packing fiber around the conductors to form a dam for the sealing compound. *Figure 13* (C) shows a sealoff fitting along with fiber material and sealing compound.

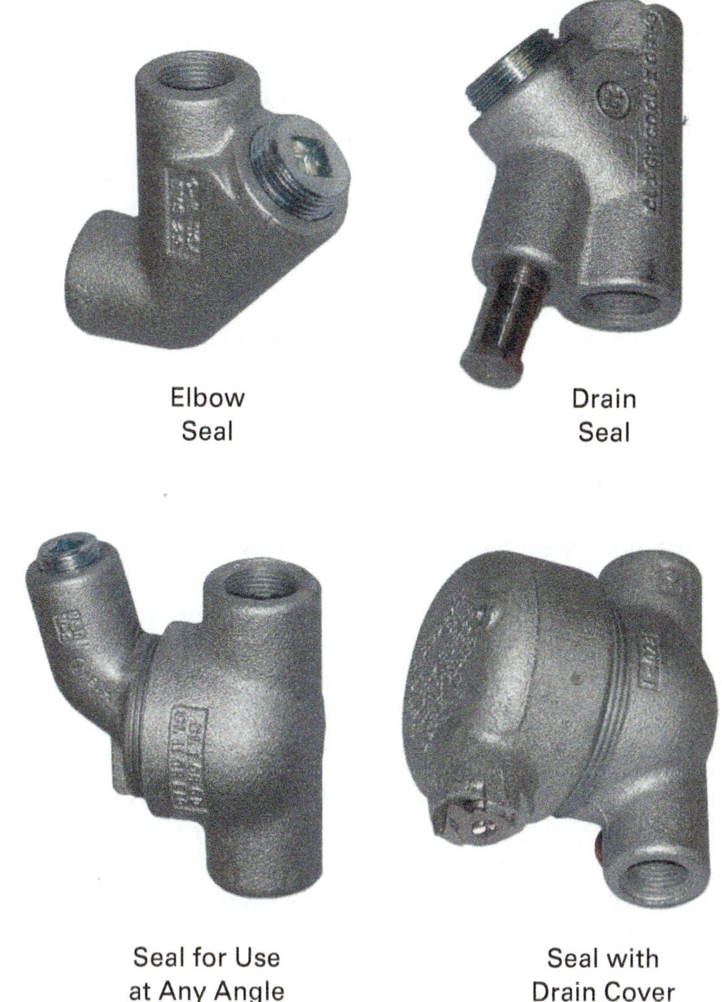

Figure 12 Various types of seals.

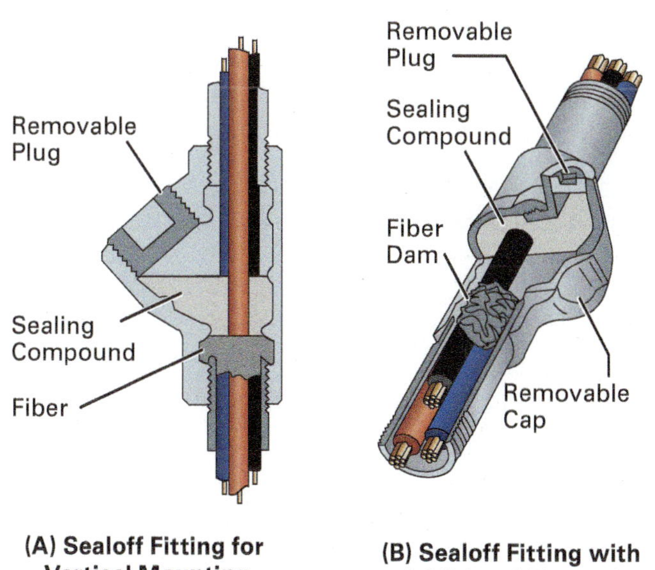

(A) Sealoff Fitting for Vertical Mounting

(B) Sealoff Fitting with Additional Plugged Opening

(C) Sealoff Fitting, Fiber Material, and Sealing Compound

Figure 13 Seals made with fiber dams and sealing compound.

Sealing Conductors

On a multiconductor cable traveling between areas where sealoff fittings are required, strip the insulation from the cable in the middle of the fitting and slightly unravel and separate (birdcage) the conductors. This ensures that the sealing compound fully surrounds each conductor to prevent the passage of gas through the seal. See ***NEC Section 501.15(D)(2)***. This may require special consideration when selecting a sealoff fitting.

Pouring Vertical Seals

When pouring a vertical seal that penetrates the top of an enclosure, open the door while you are pouring the seal to observe whether the sealant is leaking into the enclosure.

The following guidelines should be observed when preparing sealing compound:

- Use a clean mix vessel for every batch. Particles of previous batches or dirt will spoil the seal.
- Recommended proportions are by volume—usually two parts powder to one part clean water. Slight deviations in these proportions will not usually affect the results, but do not assume this is the correct recipe for all products.
- Do not mix more than can be poured in 15 minutes. Use cold water; warm water increases the setting speed. Stir immediately and thoroughly.
- If the batch starts to set, do not attempt to thin it by adding more water or by stirring. This will spoil the sealant. Discard the partially set material and make a fresh batch. After pouring, close the opening immediately.
- Do not pour sealing compound in subfreezing temperatures or when these temperatures are likely to occur as it cures.
- Ensure the compound level or depth is in accordance with the instruction sheet for the specific fitting.

Most other explosion-proof fittings are provided with threaded hubs for securing the conduit. Typical fittings include switch and junction boxes, conduit bodies, unions and connectors, flexible couplings, explosion-proof lighting fixtures, receptacles, and motor starter enclosures. *Figure 14* and *Figure 15* show typical Class 1 installations, while *Figure 16* shows Class II installations.

Anti-Seize Compound

When preparing sealing fittings, apply an approved anti-seize compound to the screw threads and interior surfaces of the caps and plugs. This will ease removal of the plugs for inspection.

Sealing Compound

SpeedSeal™ compound eliminates the need for packing material required in a typical seal filling. To use this compound, you simply pump the special applicator to mix the compound, and then inject it into the fitting. This greatly reduces the time required to complete a seal.

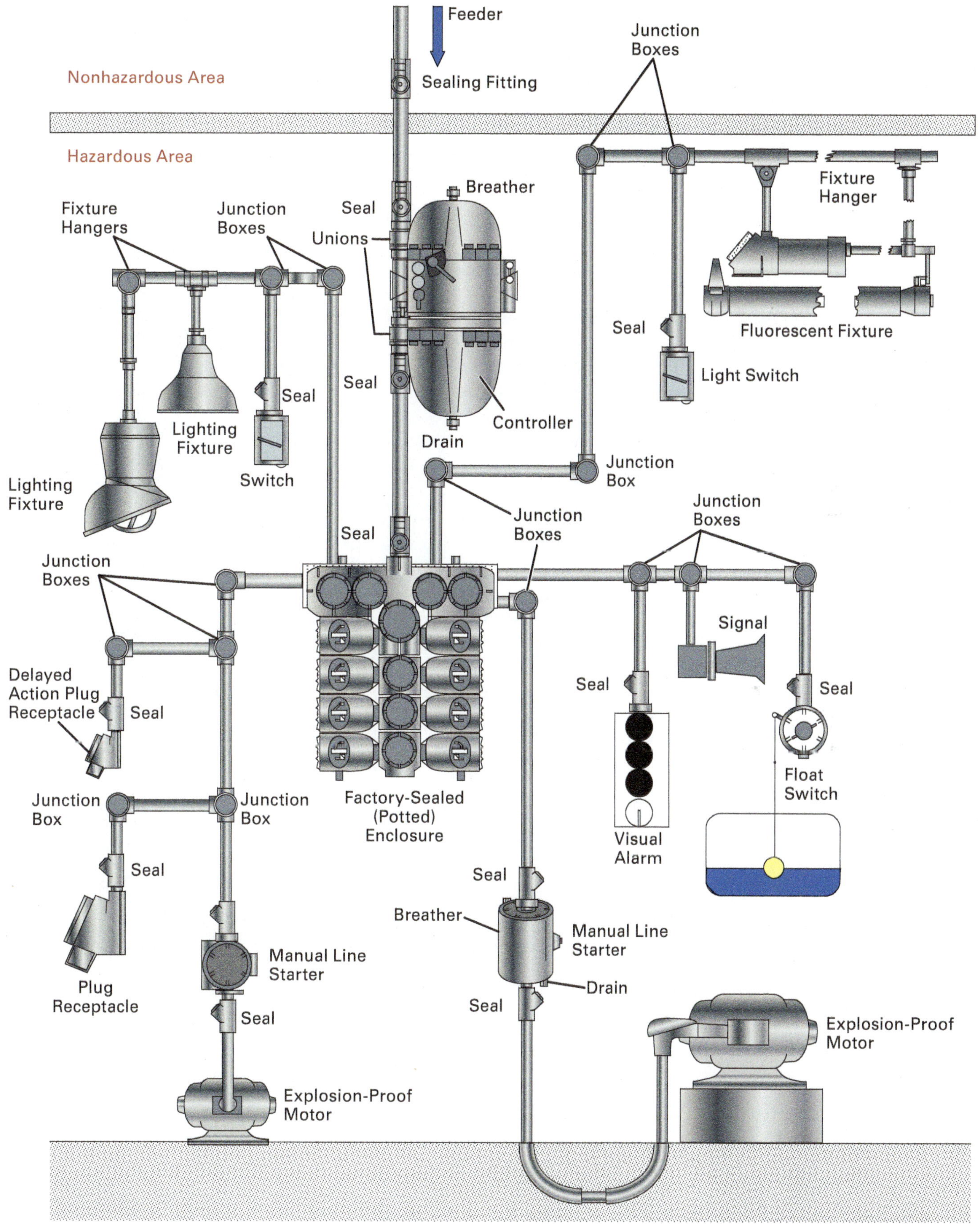

Figure 14 Class I, Division 1 electrical installation.

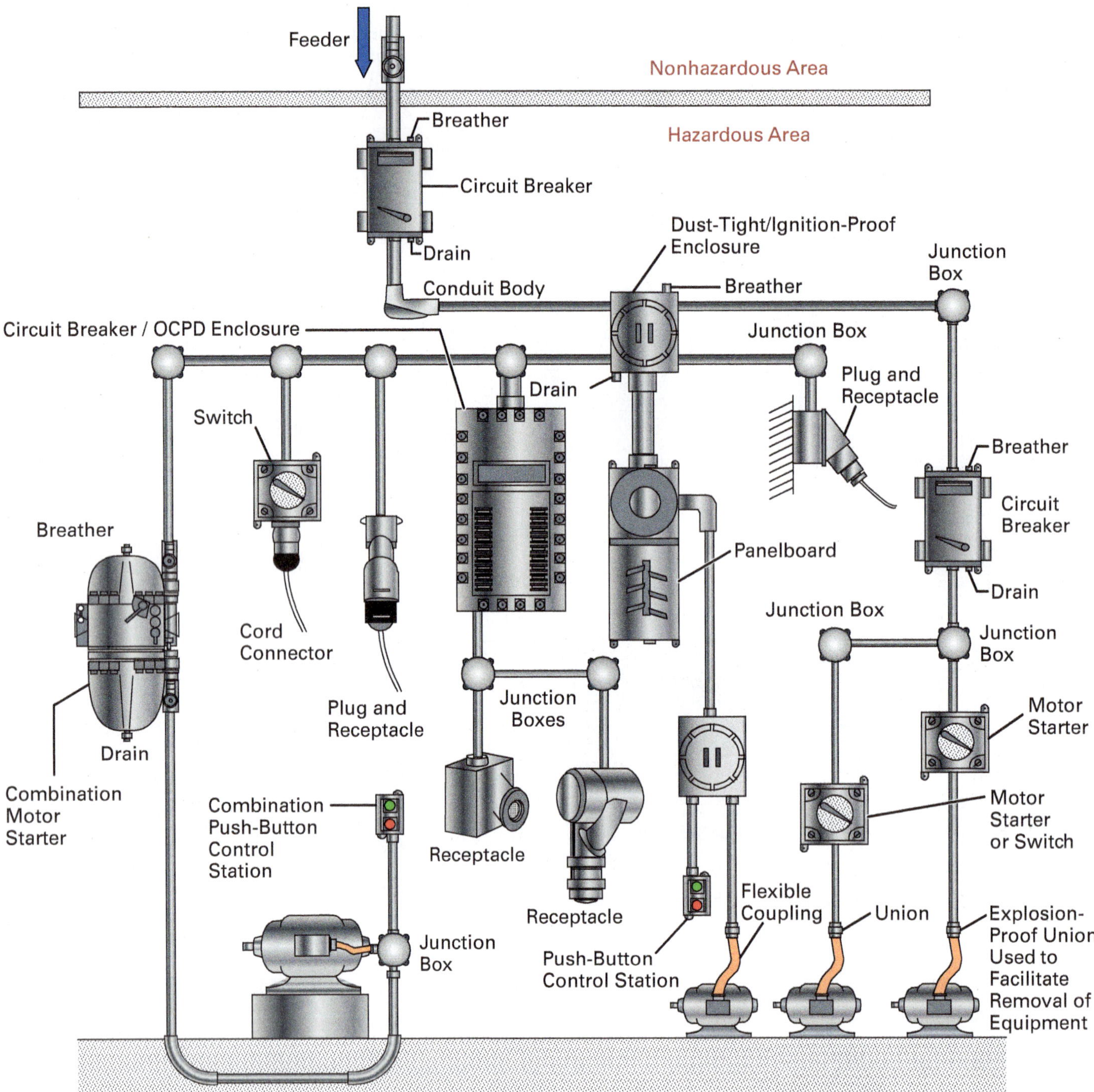

Figure 15 Class I, Division 2 electrical installation.

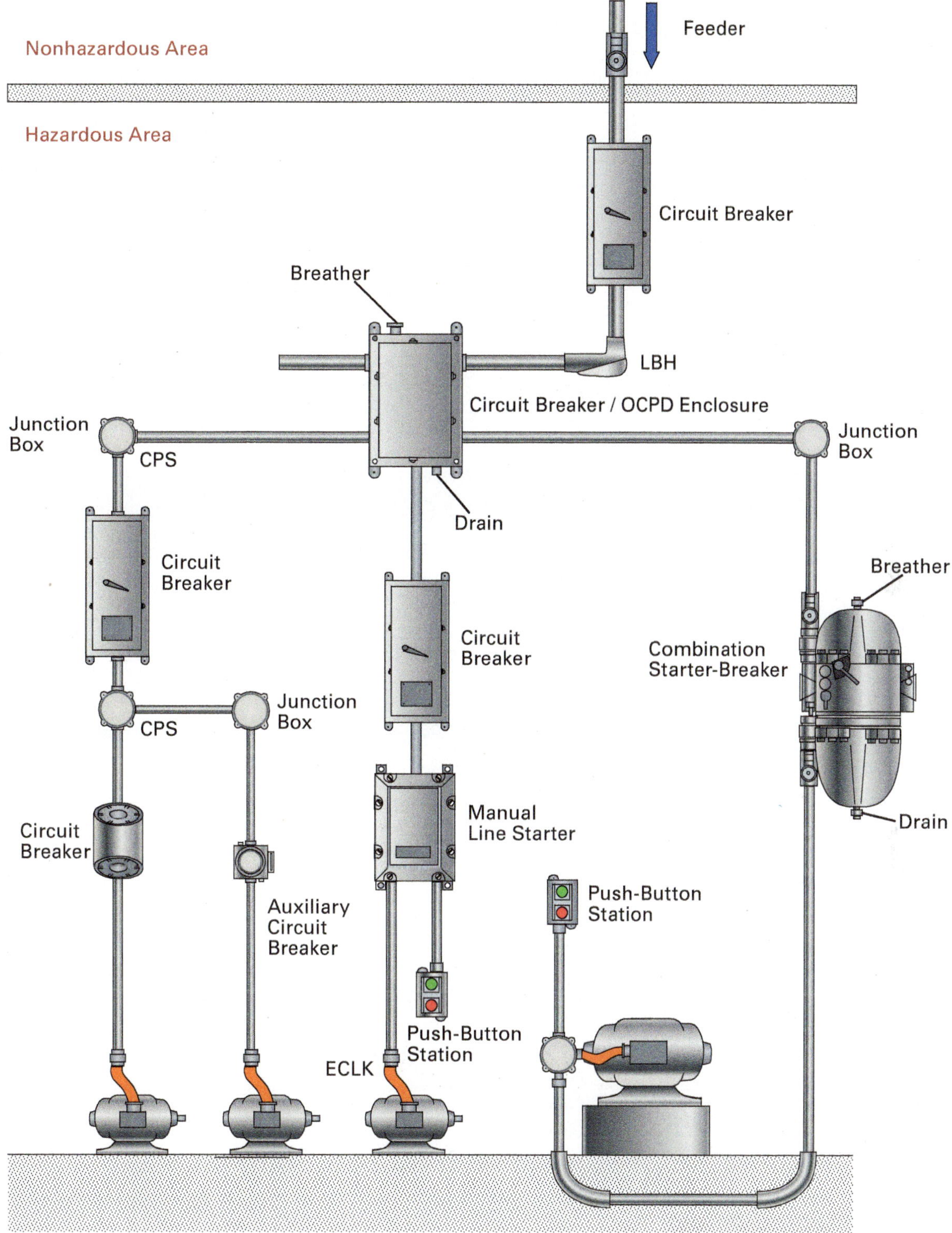

Figure 16 Class II power installation.

2.0.0 Section Review

1. Which of the following statements related to the ignition of combustible mixtures is *true*?
   a. Heat by itself will not produce an ignition without a spark.
   b. Volatile vapors will not cause ignition if diluted below the lower explosion limit.
   c. Electrical hazards are unlikely to cause ignition.
   d. Sparking metal tools are unlikely to ignite a mixture because it is so short-lived.

2. Explosion-proof equipment is rated by both classification and _____.
   a. weight
   b. volume
   c. pressure
   d. temperature

Module 26304-23 Review Questions

1. How many major classifications of hazardous atmospheres are listed in the *NEC*®?
   a. One
   b. Two
   c. Three
   d. Four

2. How many divisions are there for each hazardous atmosphere classification?
   a. One
   b. Two
   c. Three
   d. Four

3. How many substance groups are listed under Class I, Division 1?
   a. One
   b. Two
   c. Three
   d. Four

4. Which of the following are groups listed under Class II, Division 1?
   a. Groups A, B, C, D
   b. Groups E, F, G
   c. Groups H, I, J
   d. Groups L, M, N

5. When installing circuit breakers in Class II, Division 1 hazardous locations, a(n) _____.
   a. enclosure identified for the location is required
   b. dust-tight enclosure is required
   c. tight metal enclosure with no openings is required
   d. standard enclosure with several openings is required

6. When rigid metal conduit is required in hazardous locations, the threads *must* be cut at _____.
   a. $\frac{1}{2}$" taper per foot
   b. $\frac{3}{4}$" taper per foot
   c. 1" taper per foot
   d. $1\frac{1}{4}$" taper per foot

7. The main purpose of a union in conduit runs is to _____.
   a. facilitate the installation and removal of equipment
   b. provide vibration control
   c. seal the system from flammable gases or vapors
   d. form tighter joints in the system

8. When installing switches or other arc-producing apparatus in Class I, Division 1 locations, sealing fittings shall be installed within _____.
   a. 6" of the switch or other apparatus
   b. 12" of the switch or other apparatus
   c. 18" of the switch or other apparatus
   d. 24" of the switch or other apparatus

9. The purpose of packing fiber in a sealing fitting is to _____.
   a. identify the type of seal and the sealing compound to use
   b. prevent any liquids, gases, or vapors from passing through the fitting
   c. prevent flammable vapors from mixing with the sealing compound
   d. provide a dam to contain the sealing compound until it hardens

10. Which of the following *best* describes the correct time to pour sealing compound into a sealoff fitting?
    a. Within five minutes after it is installed in a raceway system
    b. After the conductors and packing fiber have been installed
    c. Before installing the packing fiber
    d. After the threaded plug is set tight

Answers to odd-numbered Module Review Questions are found in *Appendix A*.

Module 26304-23 Supplemental Exercises

1. What is considered a hazardous location?

   ____

   ____

2. Describe a Class I hazardous location.

   ____

   ____

3. Describe a Class II hazardous location.

   ____

   ____

4. Describe a Class III hazardous location.

   ____

   ____

5. True or False? One way to reduce the hazards in a Class II area is by installing dust-collecting systems.

6. The *NEC*® article that covers motor fuel dispensing stations is __________.

7. The classification and division of an unvented service station pit in a commercial garage is __________.

8. The entire area of an airport hangar, including any adjacent area not suitably cut off from the hangar, is a Class I, Division 2 location up to a level of ________ above the floor.
9. The *NEC*® article that covers the general wiring of health care facilities is ________.
10. The use of intrinsically safe equipment is primarily limited to ________.
11. Sealoff fittings prevent the passage of ________, ________, and________.
12. True or False? Where sealing fittings are used, they must be accessible.
13. In humid atmospheres where it is not possible to incline the raceway for the purpose of draining, special ________ fittings should be used.
14. True or False? Conductor splices or taps must *not* be made in enclosures where sealing compound is to be used.
15. Sealing compound must *not* have a melting point of less than ________.

Answers to odd-numbered Supplemental Exercises are found in *Appendix B*.

Answers to Section Review Questions

Answer	Section Reference	Objective
Section 1.0.0		
1. b	1.1.0	1a
2. c	1.2.0	1b
3. d	1.3.0	1c
4. c	1.4.0; *Table 1*	1d
Section 2.0.0		
1. b	2.1.0	2a
2. d	2.2.0	2b

MODULE 26305-23

Overcurrent Protection

Source: iStock@Pongchart

NOTE

NFPA 70®, *National Electrical Code*®, and *NEC*® are registered trademarks of the National Fire Protection Association, Quincy, MA.

Objectives

Successful completion of this module prepares you to do the following:

1. Identify and describe various overcurrent events.
   a. Identify and describe overload conditions.
   b. Identify and describe short circuit conditions.
   c. Identify and describe ground faults.
   d. Identify and describe arc faults.
2. Identify fuses and their characteristics.
   a. Describe fuse ratings.
   b. Identify fuse operating characteristics.
   c. Describe fuse classes and their relevance.
3. Identify circuit breakers by classification and interrupt capacity.
   a. Describe circuit breaker classifications.
   b. Describe circuit breaker interrupting capacity ratings.
4. Explain how to size and select overcurrent protective devices.
   a. Select overcurrent protective devices for various applications.
   b. Apply short circuit calculations.
5. Describe how to test and troubleshoot overcurrent protective devices.
   a. Describe how to test and troubleshoot circuit breakers.
   b. Describe how to test and troubleshoot fuses.

Performance Tasks

This is a knowledge-based module. There are no Performance Tasks.

Overview

Electrical distribution systems are often complicated and cannot be made entirely fail-safe. Circuits may experience destructive overcurrent events due to harsh environments, deterioration, damage, thermal expansion, or system overload. Overcurrent protective devices are essential to prevent costly damage to electrical system components, fire, and personal injury. This module explains how overcurrent protective devices are sized and selected for various applications and presents common testing and troubleshooting practices.

Digital Resources for Electrical

Scan this code using the camera on your phone or mobile device to view the digital resources related to this craft.

NCCER Industry-Recognized Credentials

If you are training through an NCCER-accredited sponsor, you may be eligible for credentials from NCCER. The ID number for this module is 26305-23. Note that this module may have been used in other NCCER curricula and may apply to other level completions. Contact NCCER at 1.888.622.3720 or go to **www.nccer.org** for more information.

You can also show off your industry-recognized credentials online with NCCER's digital badges. Transform your knowledge, skills, and achievements into badges that you can share across social media platforms, send to your network, and add to your resume. For more information, visit **www.nccer.org**.

1.0.0 Overcurrent Conditions

Performance Tasks

There are no Performance Tasks in this section.

Objective

Identify and describe various overcurrent events.

a. Identify and describe overload conditions.
b. Identify and describe short circuit conditions.
c. Identify and describe ground faults.
d. Identify and describe arc faults.

Electrical systems can be complicated. It is difficult, if not impossible in some cases, to make them entirely fail-safe. Circuits can experience destructive overcurrent events at any moment. Harsh environments, expected deterioration, accidental damage, and thermal expansion are just some of the factors that can lead to a system **overload**.

Overload: Any overcurrent condition that exceeds the normal full-load current of a circuit.

Reliable protective devices prevent or minimize costly damage to transformers, conductors, motors, and many other components. Reliable circuit protection is essential to avoid costly losses and personal injury. Overcurrent protective devices are a necessity in all electrical systems, both large and small.

Know the Code

Overcurrent Protection
NEC Article 240

Know the Code

Overcurrent Protection for Systems Rated Over 1000 Volts ac, 1500 Volts dc
NEC Article 245

Overcurrent protection of electrical circuits is so important that the *NEC®* devotes an entire article to this topic. *NEC Article 240* provides the general requirements for overcurrent protection and overcurrent protective devices. *NEC Article 240, Parts I through VIII* cover systems not more than 1,000V (nominal). *NEC Article 245* covers overcurrent protection over 1,000V. *NEC Article 240* will be covered in this module, along with practical examples.

All circuits must be protected against overcurrent events in accordance with their ampacities as set forth in *NEC Section 240.4*. They must also be protected against **short circuit current** per *NEC Sections 110.10 and 240.1*, Informational Notes No. 1 and No. 2. The two basic types of overcurrent protective devices that are in common use are fuses and circuit breakers.

Short circuit current: An overcurrent condition that greatly exceeds the normal current-carrying capacity of a circuit. This occurs when the current leaves the normal current-carrying path of the circuit, bypassing the load and returning to the source.

The function of overcurrent protective is to protect electrical conductors and equipment from damage due to fault currents. This is done by opening the circuit very quickly, but not so fast as to cause nuisance trips from normal surges when **inductive loads** like motors are energized. Types of fault current include the following:

Inductive loads: Electrical loads that consume significantly more current, referred to as *inrush current*, when first energized. The current rapidly declines to the normal, anticipated load current.

- Overloads
- Short circuits
- Ground faults
- Arc faults

Know the Code

Circuit Impedance, Short-Circuit Current Ratings, and Other Characteristics
NEC Section 110.10

1.1.0 Overload Conditions

Momentary overloads are most often between one and six times the normal current level. Usually, they are caused by temporary surges that occur when motors start or transformers are energized. Such overload currents or transients are expected. Because they are brief, any temperature rise is momentary and has no harmful effect on the circuit components when the system is designed to accommodate them.

Continuous overloads can result from defective motors (motors with worn bearings, for example) or too many loads on one circuit. Sustained overloads are destructive and must be terminated by protective devices before they damage the system or its loads. Sustained overload currents overheat conductors and other components and can cause insulation deterioration if not interrupted. However, because they are of relatively low magnitude compared to short circuit currents, terminating a sustained overload within a few seconds will generally prevent any significant damage.

1.2.0 Short Circuit Conditions

A short circuit is a conducting connection, whether intentional or accidental, between any of the conductors of an electrical system, either line-to-line or line-to-ground. It is a circuit that allows the current to bypass the load.

The **amperes interrupting capacity (AIC)** rating of a circuit breaker or fuse is the maximum fault current that the breaker will safely interrupt. This AIC rating is determined at the device's rated voltage and frequency.

Amperes interrupting capacity (AIC): The maximum short circuit current that a circuit breaker or fuse can safely interrupt.

Whereas sustained overloads occur at rather modest levels, a short circuit or fault current can be hundreds of times greater than the normal operating current. A high-level fault may reach 50,000A or more. If not terminated within a few thousandths of a second, damage and destruction can be extensive. Severe insulation damage, melted conductors, vaporized metal, and fires are possible, among other problems. High-level short circuit currents can also cause huge magnetic field stresses. The magnetic forces created between busbars and other conductors can reach many hundreds of pounds per lineal foot. Even heavy bracing may not be adequate to keep them from being warped or distorted beyond repair.

NEC Section 110.9 states that equipment intended to interrupt current at fault levels (fuses and circuit breakers) shall have an interrupting rating at their nominal circuit voltage at least equal to the current that is available at the line terminals of the equipment. Equipment intended to interrupt current at other than fault levels must have an interrupting rating at least equal to the current that must be interrupted at the circuit's nominal voltage.

Know the Code

Interrupting Rating
NEC Section 110.9

These *NEC*® statements mean that fuses and circuit breakers designed to interrupt fault or operating currents must be rated to withstand such currents. This section emphasizes the difference between clearing fault level currents and clearing excessive operating currents. Protective devices such as fuses and circuit breakers are designed to clear fault currents and therefore must have short circuit interrupting ratings sufficient for fault levels. Contactors, safety switches, and other equipment have interrupting ratings for currents at other than fault levels. Thus, the interrupting rating of electrical equipment is now divided into two parts:

- Current at fault (short circuit) levels
- Current at operating levels

Most people are familiar with the normal ampere (current) rating of fuses and circuit breakers. If an overcurrent protective device is designed to open a circuit when the circuit load exceeds 20A, the overcurrent protective device begins to overheat as the current approaches 20A. If the current barely exceeds 20A, the circuit breaker will open normally or a fuse link will melt after a given time with little, if any, arcing. For example, if 40A of current were instantaneously applied to the circuit, the overcurrent protective device would open, but again with little arcing. However, if a ground fault occurs on the circuit and increases the amperage to 5,000A, an explosion will occur within the protective device. One simple indication of this is the blackened windows of plug fuses.

If this fault current exceeds the interrupting rating of a fuse or circuit breaker, the protective device can be damaged or destroyed. Such a current can also cause severe damage to equipment and personal injury. Therefore, selecting overcurrent protective devices with the proper interrupting capacity is extremely important in all electrical systems.

There are several factors that must be considered when calculating the required interrupting capacity of an overcurrent protective device. *NEC Section 110.10* states that the overcurrent protective devices, total impedance, equipment short circuit current ratings, and other characteristics of the circuit to be protected shall be selected and coordinated to ensure the circuit protective devices can clear a fault without extensive damage to the electrical equipment of the circuit. This fault shall be assumed to be either between two or more of the circuit conductors, or between any circuit conductor and the equipment grounding conductor as permitted by *NEC Section 250.118*.

Know the Code

Types of Equipment Grounding Conductors
NEC Section 250.118

The component short circuit rating is a current rating given to conductors, switches, circuit breakers, and other electrical components, which, if exceeded by fault currents, will result in extensive damage to the component. The rating is expressed in terms of time intervals and/or current values. Short circuit damage can be the result of heat generated or the electromechanical force of a high-intensity magnetic field.

The intent of the *NEC®* is to ensure that short circuit currents cannot exceed the short circuit current ratings of the selected components. Given the level of available short circuit currents that could occur, overcurrent protective devices must be used that will limit fault currents to levels within the withstand ratings of the system components.

1.3.0 Ground Faults

A ground fault is an unintentional conducting connection between any of the conductors of an electrical system and any conducting material, such as a metal enclosure, that is grounded. A low-resistance ground fault behaves in the same way as a short circuit, and the protective device operates to clear the fault. A ground fault of one or more conductors gets the same response. Ground faults often progress and can be very destructive.

In many electrical systems, it is common to use a dedicated ground fault protective relay to detect and respond to levels of ground fault below full-load current values. There are electronic trip units available for circuit breakers that add ground fault protection to the breaker.

Know the Code

Ground-Fault Protection of Equipment
NEC Section 230.95

A typical response to a ground fault in a distribution system is to trip immediately when low-level ground faults occur and after a time delay for high-level ground faults. *NEC Section 230.95* discusses the requirements for ground fault protection of solidly grounded, wye-connected systems not over 1,000V for service disconnects rated at 1,000A and above.

1.4.0 Arc Faults

Although faults of any type that cause trips are often called *shorts,* an *arc fault* is different from a short circuit as it is the result of an intermittent or high-resistance fault rather than a direct short circuit. Most arc faults begin as ground faults and then grow to line-to-line and line-to-ground faults.

Arc faults can be catastrophic and expose a worker to arc blast hazards. Injuries can include thermal burns, flash burns to the eyes, inhalation of molten metal vapor, injuries from shrapnel and molten metal splatter, and hearing damage.

In systems 1,000V and above, arc fault currents are near the magnitude of short circuit currents and are usually detected and cleared in the same way. For 480V circuits, however, the arc current may be as low as 38% of the available fault current and seldom exceeds 60%. The arc fault current in systems at or below 600V is usually detected as an overload instead of a fault and will not initiate an instantaneous trip of the protective device.

1.0.0 Section Review

1. Worn motor bearings are *most* likely to result in a(n) _____.
   a. continuous overload
   b. short circuit
   c. ground fault
   d. arc fault

2. A high-level short circuit current can exceed _____.
   a. 50A
   b. 500A
   c. 5,000A
   d. 50,000A

3. Accidental contact between an electrical conductor and a grounded enclosure is called a(n) _____.
   a. overload
   b. short circuit
   c. ground fault
   d. arc fault

4. Most arcing faults begin as _____.
   a. ground faults
   b. temporary overloads
   c. sustained overloads
   d. short circuits

2.0.0 Fuses and Their Applications

Performance Tasks

There are no Performance Tasks in this section.

Objective

Identify fuses and their characteristics.

a. Describe fuse ratings.
b. Identify fuse operating characteristics.
c. Describe fuse classes and their relevance.

The fuse is a reliable overcurrent protective device. It consists of a fusible link or links captured in a tube and connected to conductive terminals. The electrical resistance of the link is so low that it simply acts as a conductor. However, when destructive currents occur, the link quickly melts and opens the circuit to protect conductors and other circuit components. Fuse characteristics are stable over time and do not require periodic maintenance or testing.

2.1.0 Fuse Ratings

Know the Code

Plug Fuses, Fuseholders, and Adapters
NEC Article 240, Part V

Know the Code

Cartridge Fuses and Fuseholders
NEC Article 240, Part VI

Know the Code

General
NEC Section 240.50

Know the Code

Edison-Base Fuseholders
NEC Section 240.52

NEC Article 240, Part V contains the requirements associated with plug fuses, their fuse holders, and compatible adapters. *NEC Article 240, Part VI* lists the requirements that apply to cartridge fuses and their fuse holders. Plug fuses are normally found in general-purpose branch circuits. Cartridge fuses may be found in branch circuits, feeder circuits, motor circuits, and many other common and special applications.

Per *NEC Section 240.50(A)*, plug fuses are only permitted in circuits not exceeding 125V between conductors or in circuits supplied by a system having a grounded neutral where the line-to-neutral voltage does not exceed 150V.

Plug fuses rated at 15A or less have hexagonal windows or caps so that they may easily be distinguished from fuses of higher current ratings. This is a requirement of *NEC Section 240.50(C)*. Plug fuses with a current rating above 15A have a round window.

There are two screw-in configurations associated with plug fuses. The Edison-base socket is a standard screw-in base, which allows any Edison fuse to be used. The Type S socket is matched to a specific fuse rating and will not accept any other fuse. An adapter is available to convert an Edison-base socket to a Type S socket.

NEC Section 240.52 prohibits the use of Edison-base fuse holders in new installations except where they are adapted to accept Type S fuses.

2.1.1 Voltage Rating

Voltage rating: The maximum voltage a fuse can be applied to and safely interrupt an overcurrent event.

Know the Code

Maximum Voltage — 300-Volt Type
NEC Section 240.60(A)

Most low-voltage power distribution fuses have 250V or 600V ratings. The **voltage rating** of a fuse must be at least equal to or greater than the circuit voltage. It can be higher, but never lower. For example, a fuse rated for 600V can be applied in a 240V circuit.

Note that there are some cartridge fuses rated for 300V. Per *NEC Section 240.60(A)*, cartridge fuses and their fuse holders rated at 300V are only permitted to be used in circuits not exceeding 300V between conductors, or in single-phase, line-to-neutral circuits powered by a three-phase, four-wire system where the voltage does not exceed 300V.

Arcing time: The amount of time passing from the instant a fuse link has melted until the overcurrent event is interrupted.

The voltage rating of a fuse affects its capability to open a circuit under an overcurrent condition. Specifically, the voltage rating determines the ability of the fuse to suppress the internal arcing that occurs when a fuse link melts and an arc is produced. The time that passes during an arc is called the **arcing time**. If a fuse with a voltage rating below that of the circuit voltage is used, arc suppression will be impaired and may not safely clear the overcurrent.

2.1.2 Current Rating

Every fuse has a specific current rating. When selecting the current rating, consideration must be given to the size and type of load as well as the *NEC®* requirements. The current rating of a fuse should not normally exceed the current-carrying capacity of the circuit.

For instance, if a conductor is rated to carry 20A, a 20A fuse is the largest that should be used. However, there are specific circumstances in which the current rating is permitted to be greater than the current-carrying capacity of the circuit.

A typical example is a motor circuit. In general, dual-element fuses can be sized up to 175% and nontime-delay fuses up to 300% of the motor full-load current. Typically, the current rating of a fuse and switch combination should be selected at 125% of the continuous load current. This usually corresponds to the circuit capacity, which is also selected at 125% of the load current. There are exceptions, though, such as when the fuse-switch combination is approved for continuous operation at 100% of its rating.

Fuse Position

Always make a point of installing fuses right side up, with the label facing out and up. When a fuse requires replacement and the label is upside down or backwards, the typical response would be to reach in and twist it around to read the label. This can result in serious shocks or even death.

2.1.3 Interrupting Rating

A protective device must be able to withstand the destructive energy of short circuit currents. If a fault current exceeds the capability of the protective device, the device can fail catastrophically. Thus, it is essential to choose a protective device that can manage the largest potential short circuit current. The rating that defines the capacity of a protective device to maintain its integrity when reacting to fault currents is called its *interrupting rating*.

NEC Section 110.9 requires devices intended to interrupt fault level current to have an interrupting rating at least equal to the current available at the line terminals of the equipment. Interrupting rating and interrupting capacity are covered in detail in NCCER Module 26210-23, *Circuit Breakers and Fuses*.

Fault Current Protection

In a *series-rated* system, the main overcurrent protective device has an interrupting capacity equal to or greater than the maximum available fault current, but the downstream protective devices can be rated at lower values. This reduces their cost. Series-rated systems have their place in the industry in areas where the tripping of a main breaker would cause a nuisance but would not cause an unsafe condition or interrupt a critical process. For this reason, series rated systems are not approved for use in all locations. See ***NEC Sections 110.22(B) and (C)*** and ***NEC Section 240.86***.

In a *fully rated* system, each overcurrent protective device has an interrupting rating that is equal to or greater than the maximum available fault current. Fully rated systems are approved for use in all locations.

2.1.4 Selective Coordination

The coordination of protective devices prevents system power outages or blackouts caused by overcurrent events. When only the protective device nearest a faulted circuit opens and larger upstream fuses remain closed, the protective devices are said to be *selectively coordinated*. The word *selective* is used to denote the isolation of a faulted circuit by the opening of only the local protective device.

Figure 1 shows the minimum ratios of current rating of low peak fuses that are required to provide selective coordination of upstream and downstream fuses.

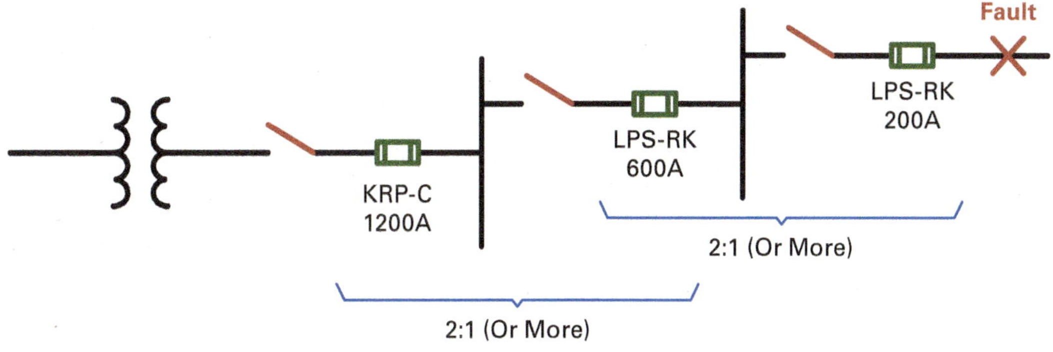

Figure 1 Selective coordination.

2.1.5 Current Limitation

If a protective device cuts off a short circuit current in less than one half cycle, before it reaches its peak value, the device is a **current-limiting device**. Most modern fuses are current-limiting devices. They restrict fault currents to such low values that a high degree of protection is given to circuit components. They permit breakers with lower interrupting ratings to be used and can reduce bracing of bus assemblies. They also minimize the need for other components to have high short-circuit current ratings.

Current-limiting device: A device designed to clear a short circuit in less than one half cycle. This limits the instantaneous peak let-through current to a value substantially less than would pass if the device were replaced with a solid conductor of equal impedance.

If not limited, short circuit currents can reach levels of 30,000A or 40,000A or higher in the first half cycle (about 0.008 seconds at 60 Hz). The heat that can be produced by the immense energy of short circuit currents can cause severe insulation damage or even an explosion. At the same time, the huge magnetic forces developed between conductors can crack insulators and distort and destroy bus bracing structures. It is extremely important that a protective device limit fault currents before they reach their full potential.

A noncurrent-limiting protective device, by permitting a short circuit current to build to its peak value, can let an immense amount of destructive short circuit heat energy through before opening the circuit, as shown in *Figure 2*. On the other hand, a current-limiting device, such as a current-limiting fuse, has such a rapid response that it cuts off a short circuit before it reaches its peak value, as shown in *Figure 3*.

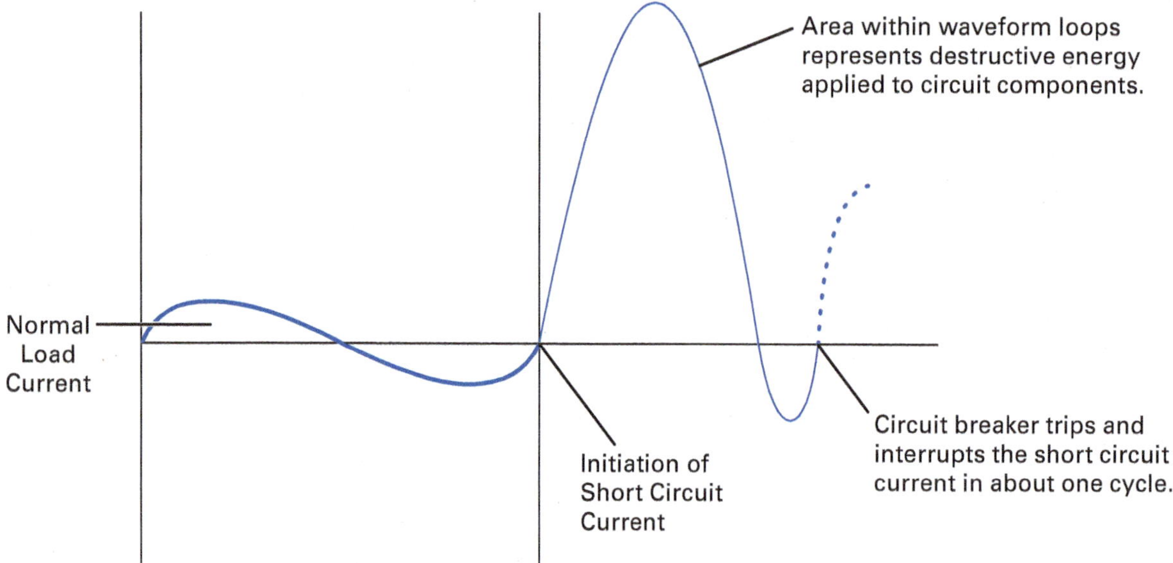

Figure 2 Characteristics of a noncurrent-limiting protective device.

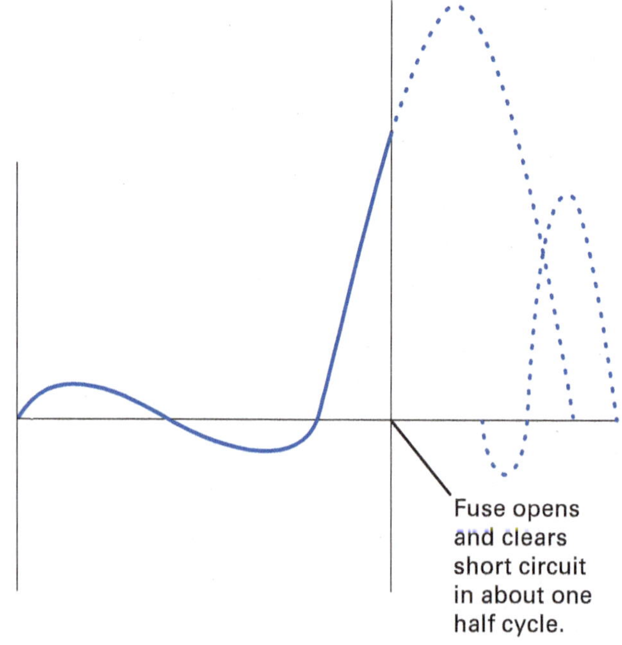

Figure 3 Characteristics of a current-limiting fuse.

2.2.0 Types of Fuses and Operating Characteristics

The following sections describes the operating characteristics of various types of cartridge fuses. These fuses include the nontime-delay and the dual-element, time-delay fuses.

2.2.1 Nontime-Delay Fuses

The basic component of a fuse is the link inside. Depending upon the current rating of the fuse, the single-element, nontime-delay fuse may have one or more links. They are electrically connected to the end blades or *ferrules* and enclosed in a tube surrounded by an arc-quenching filler.

Under normal operation, when the fuse is operating at or near its current rating, it simply functions as a conductor. However, as illustrated in *Figure 4*, if an overload current occurs and persists for more than a short interval of time, the temperature of the link eventually reaches a level that causes a restricted segment of the link to melt. A gap is formed, and an electric arc is established. As the arc causes the link metal to burn back, the gap becomes progressively larger. The electrical resistance of the arc eventually reaches such a high level that the arc cannot be sustained and is extinguished. The current flow is then terminated completely. Suppression or quenching of the arc is accelerated by the filler material.

Current-limiting fuses respond much faster to such currents. The individual sections of the fuse link will melt simultaneously within a matter of two-thousandths or three-thousandths of a second.

The high resistance of the multiple arcs, together with the quenching effects of the filler, results in rapid arc suppression. The short circuit current is cut off in less than one half cycle, long before the short circuit current can reach its peak value.

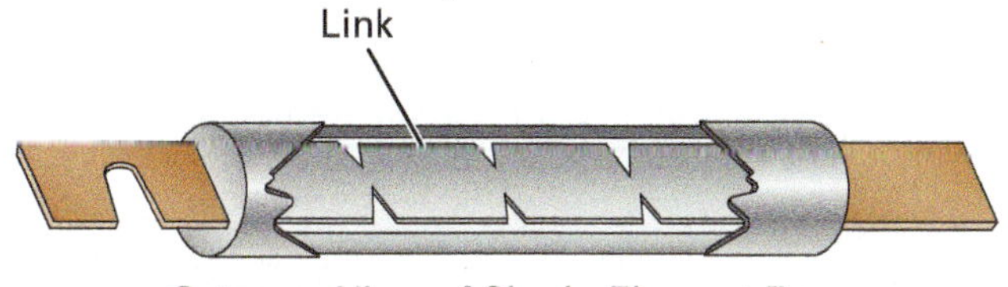

Cutaway View of Single-Element Fuse

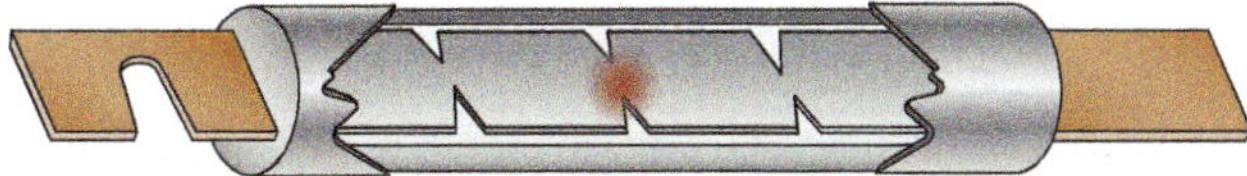

Under sustained overload, a section of the link melts and an arc is established.

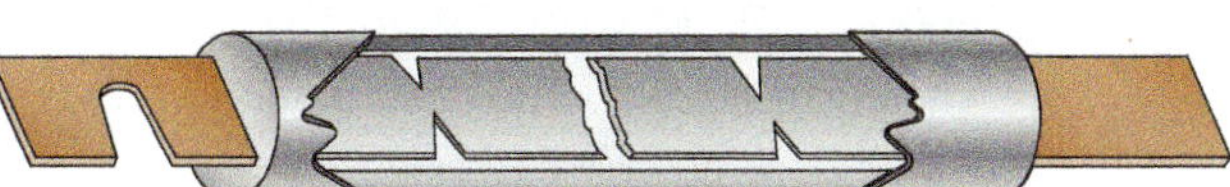

The open single-element fuse after opening a circuit overload.

(A) Response to Sustained Overload

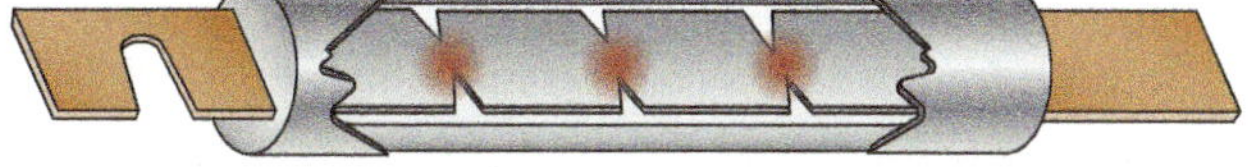

When subjected to a short circuit, several sections of the fuse link melt almost instantly.

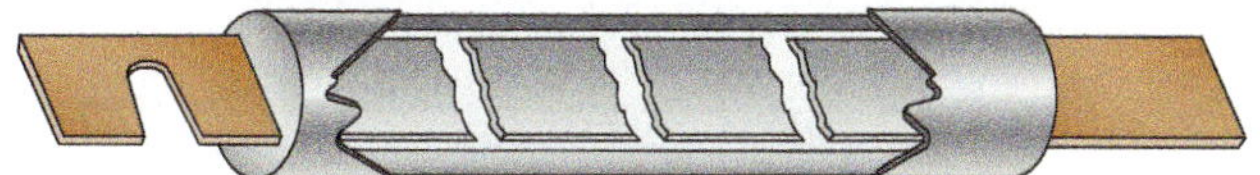

The appearance of an open single-element fuse after opening a short circuit.

(B) Response to a Short Circuit

Figure 4 Characteristics of a single-element fuse.

2.2.2 Dual-Element, Time-Delay Fuses

Dual-element, time-delay fuses, unlike single-element fuses, are often applied in circuits subject to temporary motor overloads and surge currents to provide both high-performance short circuit and overload protection. Oversizing the fuse to prevent nuisance openings is not necessary.

The dual-element, time-delay fuse contains two distinctly separate types of elements. Electrically, the two elements are connected in series. The fuse links perform the short circuit protection function; the overload element provides protection against low-level overcurrent events or overloads and will hold an overload that is five times greater than the current rating of the fuse for a minimum of 10 seconds.

As shown in *Figure 5*, the overload section consists of a copper heat absorber and a spring-operated trigger assembly. The heat absorber bar is permanently connected to the heat absorber extension and the short circuit link on the opposite end of the fuse by the S-shaped connector of the trigger assembly. The connector electrically joins the short circuit link to the heat absorber in the overload section of the fuse. These elements are joined by a calibrated fusing alloy.

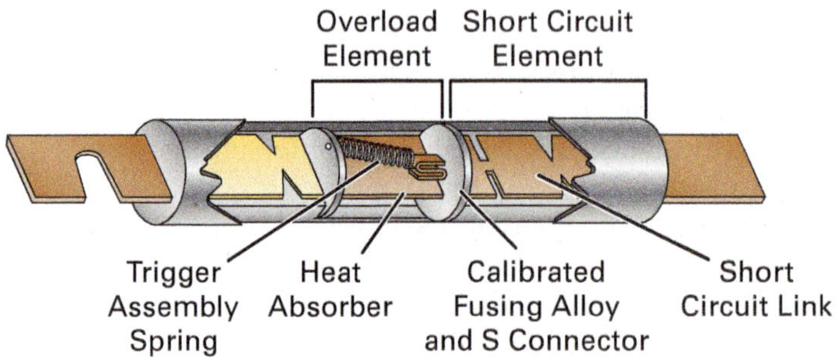

The true dual-element fuse has distinct and separate overload and short circuit elements.

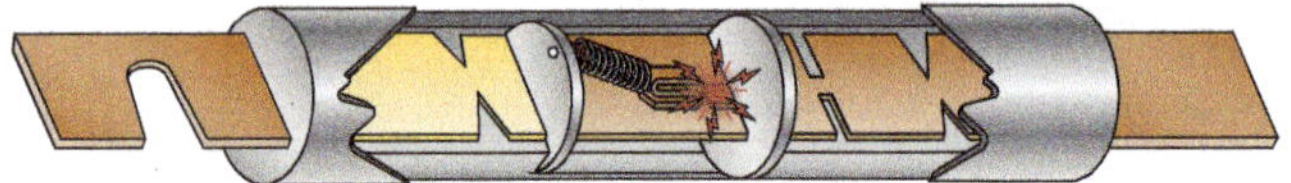

Under sustained overload conditions, the trigger spring fractures the calibrated fusing alloy and releases the connector.

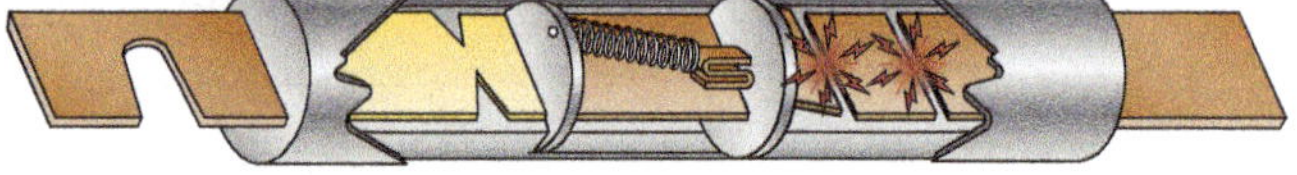

Like the single-element fuse, a short circuit current causes the restricted portions of the short circuit elements to melt and arcing to burn back the resulting gaps until the arcs are suppressed by the arc-quenching material and increased arc resistance.

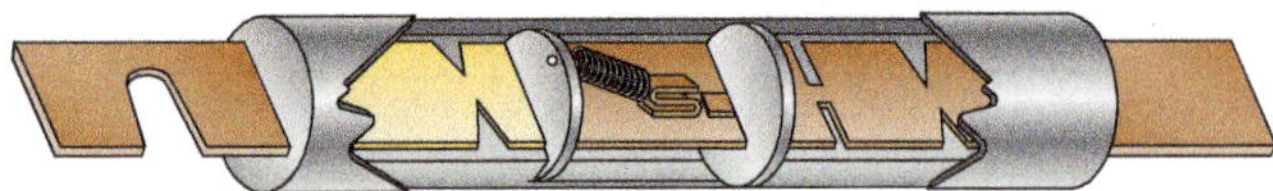

The dual-element fuse opens under an overload.

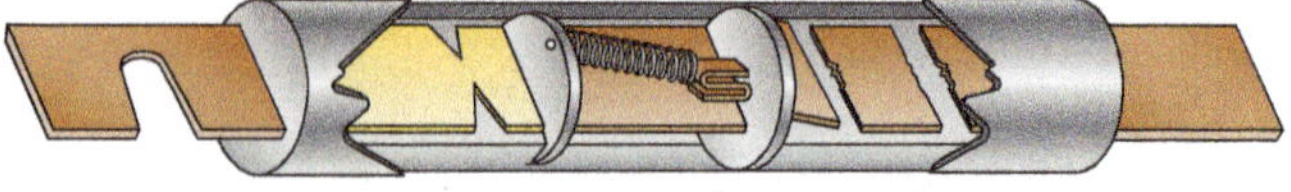

The dual-element fuse opens under a short circuit.

(A) Response to Sustained Overload

(B) Response to Short Circuit

Figure 5 Characteristics of a dual-element, time-delay fuse.

Overload current causes heating of the short circuit link connected to the trigger assembly. The transfer of heat from the short circuit link to the heat absorber begins to raise the temperature of the heat absorber. If the overload is sustained, the temperature of the heat absorber eventually reaches a level that permits the trigger spring to fracture the calibrated fusing alloy and pull the connector free of the short circuit link and the heat absorber. As a result, the short circuit link is electrically disconnected from the heat absorber, the conducting path through the fuse is opened, and the overload current is interrupted. A critical aspect of the fusing alloy is that it retains its original characteristics after repeated temporary overloads without degradation.

The advantages of dual-element fuses are as follows:

- Provide motor overload, ground fault, and short circuit protection
- Permit the use of smaller and less costly switches
- Give a higher degree of short circuit protection in circuits where surge currents or temporary overloads occur
- Simplify and improve blackout prevention through selective coordination

2.3.0 Fuse Classes and Applications

Safety is the primary consideration of UL Solutions, formerly known as Underwriters Laboratories, Inc. The proper selection, overall performance, and reliability of a product are factors that are not within the basic scope of UL Solutions' activities. However, to develop its safety test procedures, UL Solutions does develop the basic performance and physical specification standards of a product. In the case of fuses, these standards have culminated in the establishment of distinct classes of low-voltage (600V or less) fuses. Various UL Solutions fuse classes are described in the following paragraphs.

Class R (rejection) fuses are high-performance 0.1A to 600A units rated at either 250V or 600V. They have a high degree of current limitation and a **root-mean-square (rms)** symmetrical short circuit interrupting rating of up to 200,000A. This type of fuse is designed to be mounted in rejection-type fuse clips to prevent older Class H fuses from being installed. Because Class H fuses are not current-limiting devices and are noted by UL Solutions as having only a 10,000A interrupting rating, serious damage could result if a Class H fuse were inserted in a system designed for Class R fuses. Consequently, *NEC Section 240.60(B)* requires fuse holders for current-limiting fuses to reject noncurrent-limiting fuses.

Figure 6 shows standard Class H and Class R cartridge fuses. A grooved ring in one ferrule of the Class R fuse provides the rejection feature of the Class R fuse in contrast to the lower interrupting capacity, nonrejection type. *Figure 7* shows fuse rejection clips for Class R fuses.

Root-mean-square (rms): The multiplication of the square root of the mean (average) value by the squared function of the instantaneous values throughout the period. The rms value of an AC voltage or current is the effective value of the voltage or current.

Know the Code

Noninterchangeable – 0–600-Ampere Cartridge Fuseholders
NEC Section 240.60(B)

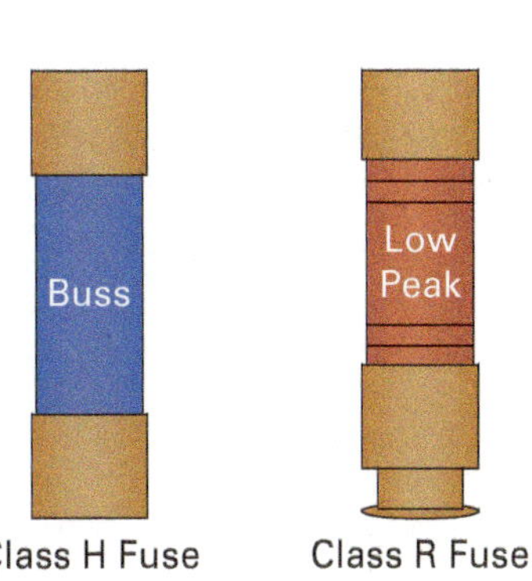

Figure 6 Comparison of Class H and Class R fuses.

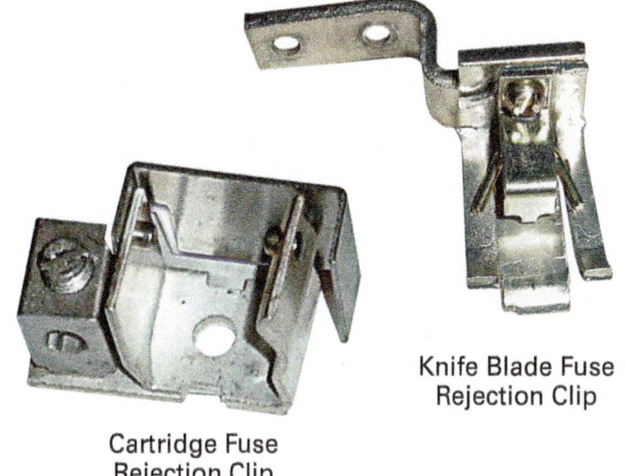

Figure 7 Fuse rejection clips.
Source: Ed Cockrell

NOTE

Metric fuse sizes may vary; always consult the manufacturer's application data when selecting and installing any overcurrent protective device.

Ampere squared seconds: A unit of measure for heat energy developed within a circuit during a fuse's clearing of a fault. It can be expressed as *melting* I^2t, *arcing* I^2t, or the sum of them as *clearing* I^2t. *I* stands for the effective let-through current and *t* stands for time to open in seconds.

Peak let-through: The instantaneous value of peak current allowed to pass through a current-limiting fuse when operating within its current-limiting range. Represented by the variable I_p.

Fast-acting fuse: A fuse that opens very quickly in response to overloads and short circuits and is not designed to withstand temporary overload events associated with common inductive loads.

Class CC fuses are 600V branch circuit fuses rated at 200,000A with overall dimensions of $^{15}/_{32}$" × $1^{1}/_{2}$". Their design incorporates rejection features that allow them to be inserted into rejection fuse holders and fuse blocks that refuse all lower voltage, lower interrupting rating $^{15}/_{32}$" × $1^{1}/_{2}$" fuses. They are available from 0.1A through 30A.

Class G fuses are 300V branch circuit fuses with a 100,000A interrupting rating that are size rejecting to eliminate the installation of a larger fuse. The fuse diameter is $^{13}/_{32}$", while the length varies from $^{15}/_{16}$" to $2^{1}/_{4}$". They are available in ratings from 1A through 60A.

Class H fuses are 250V and 600V branch circuit fuses with a 10,000A interrupting rating that may be renewable or nonrenewable. They are available in current ratings of 1A through 600A.

Class J fuses are rated to interrupt 200,000A. They are labeled as current limiting, are rated for 600VAC, and are not interchangeable with other classes.

Class K fuses are listed by UL Solutions as K-1, K-5, or K-9. Each subclass has designated maximums for **ampere squared seconds**, represented by I^2t, and **peak let-through**, represented by I_p. These are dimensionally the same as Class H fuses, and they can have interrupting ratings of 50,000A, 100,000A, or 200,000A. These fuses are current-limiting devices; however, they are not marked current limiting because they do not have a rejection feature. These are normally made to current-limiting standards as Class RK-1, Class RK-5, and other classes.

Class L fuses are available in current ratings of 601A through 6,000A and are rated to interrupt 200,000A. They are labeled *current limiting* and are rated for 600VAC. They are intended to be bolted into their mountings and are not normally used in clips. Some Class L fuses have time-delay features for all-purpose use.

Class T fuses are rated at 300V and 600V, with current ratings from 1A through 1,200A. They are physically very small and can be applied where space is at a premium. They are a **fast-acting fuse** type with an interrupting rating of 200,000A.

Three basic factors that must be considered when applying any fuse include voltage, continuous current-carrying capacity, and interrupting rating:

- *Voltage* — The fuse must have a voltage rating, equal to or greater than the normal frequency recovery voltage, that will be seen across the fuse under all conditions. On three-phase systems, it is a good rule of thumb that the voltage rating of the fuse be greater than or equal to the line-to-line voltage of the system.
- *Continuous current-carrying capacity* — Continuous current values that are shown on the fuse represent the level of current the fuse can carry continuously without exceeding the temperature rises as specified in *ANSI C37.46*. An application that exposes the fuse to a current slightly above its continuous rating but below its minimum interrupting rating may damage the fuse due to excessive heat. This is the main reason overload relays are used in series with backup current-limiting fuses for motor protection.
- *Interrupting rating* — All fuses are given a maximum interrupting rating. This rating is the maximum level of fault current that the fuse can safely interrupt. Backup current-limiting fuses are also given a minimum interrupting rating. When using backup current-limiting fuses, it is important that other protective devices are used to interrupt currents below this level.

When choosing a fuse, it is important that the fuse be properly coordinated with other protective devices located upstream and downstream. To accomplish this, one must consider the **melting time** and **clearing time** of the devices. Two curves, the minimum melting curve and the total clearing curve, provide this information. To ensure proper coordination, the following rules should be observed:

- The total clearing curve of any downstream protective device must be below a curve representing 75% of the minimum melting curve of the fuse being applied.
- The total clearing curve of the fuse being applied must lie below a curve representing 75% of the minimum melting curve for any upstream protective device.

Melting time: The amount of time required for a fuse link to melt during an overcurrent event.

Clearing time: The total time between the beginning of an overcurrent event and the complete opening of the circuit by an overcurrent protective device operating within its rated voltage. Clearing time is the sum of the melting time and arcing time.

2.3.1 Branch Circuit Listed Fuses

Branch circuit listed fuses and their holders are designed to prevent the installation of fuses that cannot provide a comparable level of protection to equipment. The characteristics of branch circuit fuses are as follows:

- They must have a minimum interrupt rating of 10,000A.
- They must have a minimum voltage rating of 125V.
- They must be size-rejecting so that a fuse of a lower voltage rating cannot be installed in the circuit.
- They must be size-rejecting so that a fuse with a current rating higher than the fuse holder rating cannot be installed in the circuit.

2.3.2 Medium-Voltage Fuses

As defined in *ANSI/IEEE STD C37.40-1981*, fuses above 600V are classified as *general-purpose current-limiting fuses, backup current-limiting fuses,* or *expulsion fuses.*

- *General-purpose current-limiting fuses* — Fuses that can interrupt all currents from the rated interrupting current down to the current that causes melting of the fusible element in one hour.
- *Backup current-limiting fuses* — Fuses that can interrupt all currents from the maximum rated interrupting current down to the minimum rated interrupting current.
- *Expulsion fuses* — Vented fuses in which the expulsion effect of gases produced by the arc and lining of the fuse holder, either alone or aided by a spring, extinguishes the arc.

A current-limiting fuse is a sealed, non-venting fuse that, when melted by a current within its interrupting rating, produces arc voltages exceeding the system voltage, which in turn forces the current to zero. The arc voltages are produced by introducing a series of high-resistance arcs within the fuse. The result is a fuse that typically interrupts high fault currents within the first half cycle.

In contrast, an expulsion fuse depends on one arc to initiate the interruption process. The arc acts as a catalyst, causing the generation of deionizing gas from its housing. The arc is then elongated either by the force of the gases created or by a spring. At some point, the arc elongates far enough to prevent a restrike after passing through a current zero. Therefore, it is not unusual for an expulsion fuse to take many cycles to clear.

Many of the rules for applying expulsion fuses and current-limiting fuses are the same. However, some additional rules must be applied to current-limiting fuses because they respond much faster to high fault currents.

Knife Blade Fuses

In a Class R knife blade fuse, a notch in one blade provides the rejection feature. This picture shows both Class H and Class R knife blade fuses.

Class H Fuse

Class R Fuse

2.3.3 Applying Current-Limiting Fuses

To ensure proper application of a current-limiting fuse, it is important that the following additional rules be applied:

- Current-limiting fuses produce arc voltages that exceed the system voltage. Care must be taken to ensure that the peak voltages do not exceed the insulation level of the system. If the fuse voltage rating is not permitted to exceed 140% of the system voltage, there should not be a problem. This does not mean that a higher rated fuse cannot be used, but one must be assured that the **basic impulse level (BIL)** of a system can handle the peak arc voltage produced.
- As with the expulsion fuse, current-limiting fuses must be properly coordinated with other protective devices on the system. For this to happen, the rules for applying an expulsion fuse must be used at all currents that cause the fuse to interrupt in 0.01 seconds or greater.

Basic impulse level (BIL): A means of expressing the maximum impulse voltage surge that electrical insulation can withstand without damage; also referred to as the *basic insulation level*.

When other current-limiting protective devices are on the system, it becomes necessary to use I^2t values for coordination at currents causing the fuse to interrupt in less than 0.01 seconds. These values may be supplied as minimum and maximum values or minimum melting and total clearing I^2t curves. In either case, the following rules should be followed:

- The minimum melting I^2t of the fuse should be greater than the total clearing I^2t of the downstream current-limiting device.
- The total clearing I^2t of the fuse should be less than the minimum melting I^2t of the upstream current-limiting device.

The fuse selection chart in *Figure 8* should serve as a guide for selecting fuses on circuits of 600V or less. Other valuable information may be found in catalogs furnished by manufacturers of overcurrent protective devices. These are usually obtainable from electrical supply houses or from manufacturers' representatives.

2.3.4 Fuses for Selective Coordination

The larger the upstream fuse is relative to a downstream fuse, the less likely it is for an overcurrent in the downstream circuit to cause both fuses to open. Fast-acting, nontime-delay fuses require at least a 3:1 ratio between the current rating of a large upstream, line-side, time-delay fuse to that of the downstream, load-side fuse to be selectively coordinated. In contrast, the minimum selective coordination ratio necessary for dual-element fuses is only 2:1 when used with low-peak load-side fuses (*Figure 9*).

Circuit	Load	Ampere Rating	Fuse Type	Symbol	Voltage Rating (AC)	UL Class	Interrupting Rating (K)	Remarks
Main, Feeder, and Branch (Conventional Dimensions)	All load types (optimum overcurrent protection)	0A to 600A	LOW PEAK® (dual-element time-delay)	LPN-RK	250V	RK1	200	All-purpose fuses. Unequaled for combined short circuit and overload protection.
				LPS-RK	600V			
		601A to 6,000A	LOW PEAK® (time-delay)	KRP-C	600V	L	200	
	Motors, welders, transformers, capacitor banks (circuits with heavy inrush currents)	0A to 600A	FUSETRON® (dual-element time-delay)	FRN-R	250V	RK5	200	Moderate degree of current limitation. Time-delay passes surge currents.
				FRS-R	600V			
		601A to 4,000A	LIMITRON® (time-delay)	KLU	600V	L	200	All-purpose fuse. Time-delay passes surge currents.
	Nonmotor loads (circuits with no heavy inrush currents)	0A to 600A	LIMITRON® (fast-acting)	KTN-R	250V	RK1	200	Same short circuit protection as LOW PEAK® fuses but must be sized larger for circuits with surge currents, i.e., up to 300%.
				KTS-R	600V			
	LIMITRON® fuses particularly suited for circuit breaker protection	601A to 6,000A	LIMITRON® (fast-acting)	KTU	600V	L	200	A fast-acting, high-performance fuse.
	All load types (optimum overcurrent protection)	0A to 600A	LOW PEAK® (dual-element, time-delay)	LPJ	600V	J	200	All-purpose fuses. Unequaled for combined short circuit and overload protection.
	Nonmotor loads (circuits with no heavy inrush currents)	0A to 600A	LIMITRON® (quick-acting)	JKS	600V	J	200	Very similar to KTS-R LIMITRON®, but smaller.
		0A to 1,200A	T-TRON®	JJN	300V	T	200	The space saver (one-third the size of KTN-R/KTS-R).
				JJS	600V			

Figure 8 Fuse selection chart (600V or less).

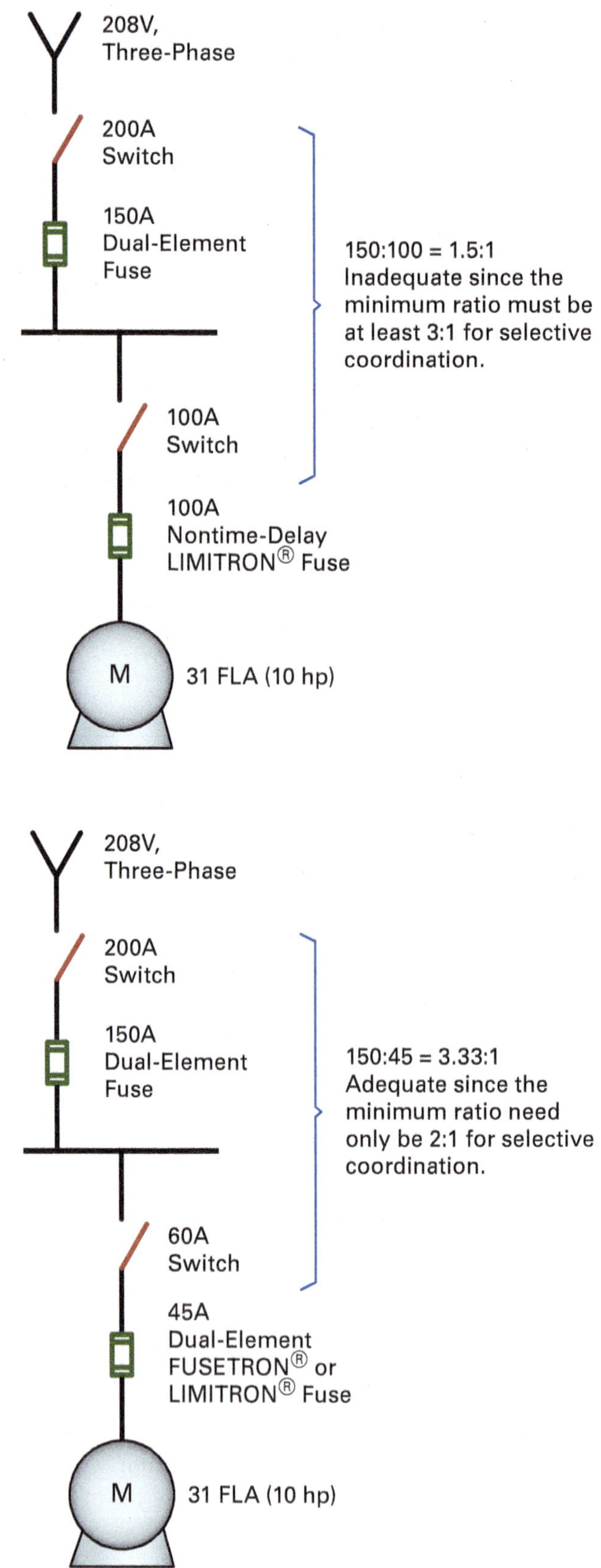

Figure 9 Fuses used for selective coordination.

The use of dual-element, time-delay fuses affords easy selective coordination, which requires little more than a routine check of a tabulation of required selectivity ratios. As shown in *Figure 9*, close sizing of dual-element fuses in the branch circuit for motor overload protection provides a large difference (ratio) in the current ratings between the feeder fuse and the branch fuse compared to the single-element, nontime-delay fuse.

Selective coordination requires the study of time-current curves and peak let-through charts. When a low-level overcurrent occurs, a long interval of time will be required for a fuse to open (melt) and clear the fault. On the other hand, if the overcurrent is large, the fuse will open very quickly. The opening time is a function of the magnitude of overcurrent level.

Overcurrent levels and the corresponding intervals of opening times are logarithmically plotted in graph form, as shown in *Figure 10*. Levels of overcurrent are scaled on the horizontal axis, with time intervals on the vertical axis. The curve is therefore called a *time-current curve*.

The plot in *Figure 10* reflects the characteristics of a 200A, 600V, dual-element fuse. Note that at the 1,000A overload level, the time required for the fuse to open is 10 seconds. Yet, at the 2,200A overcurrent level, the opening (melt) time of the fuse is only 0.01 seconds. It is apparent that the time intervals become shorter as the overcurrent levels become larger.

This relationship is called an *inverse time-to-current characteristic*. Time-current curves are available for most fuses showing their minimum melt, average melt, and/or total clear characteristics. Although upstream and downstream fuses are easily coordinated by adhering to simple ampere ratios, these time-current curves permit close or critical analysis of coordination.

Peak let-through charts enable you to determine both the peak let-through current and the apparent prospective rms symmetrical let-through current. Such charts are commonly referred to as *current limitation curves*.

Figure 11 shows a simplified chart with explanations of the various functions. Point 1 shows the system available rms short circuit current of 40,000A. Upward from Point 1 is the intersection with the fuse size of 100A (Point 2). Follow this left to Point 3, where you can see the peak let-through current of 10,400A. This line represents the point at which the fuse becomes current limiting. From Point 3, go down to the X axis and read the apparent symmetrical rms let-through current of 4,000A (Point 4).

Three factors that affect the let-through performance of a fuse are the short circuit power factor, the point in the sine wave where fault interruption occurs, and the available voltage. The published charts typically represent worst-case scenarios.

Think About It

Arc Fault

In the case of an arc fault, why might the overcurrent or short circuit protection not clear the fault?

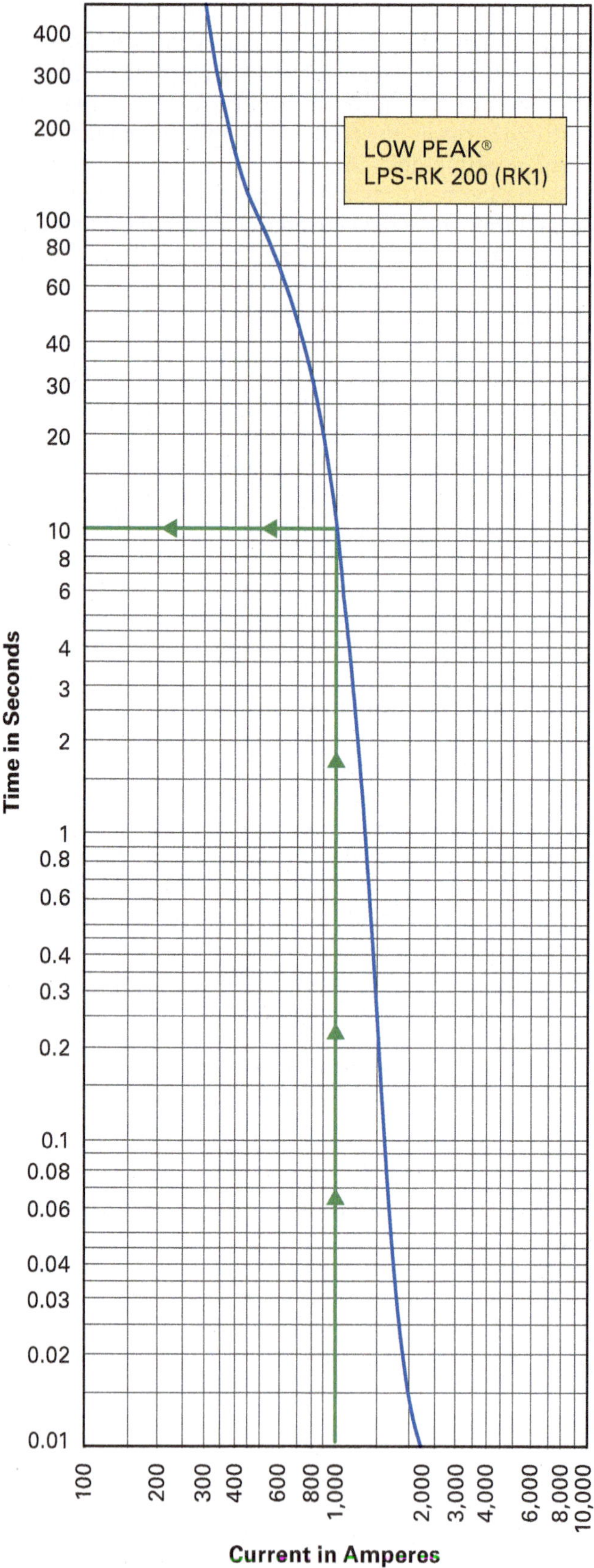

Figure 10 Typical time-current curve of a fuse.

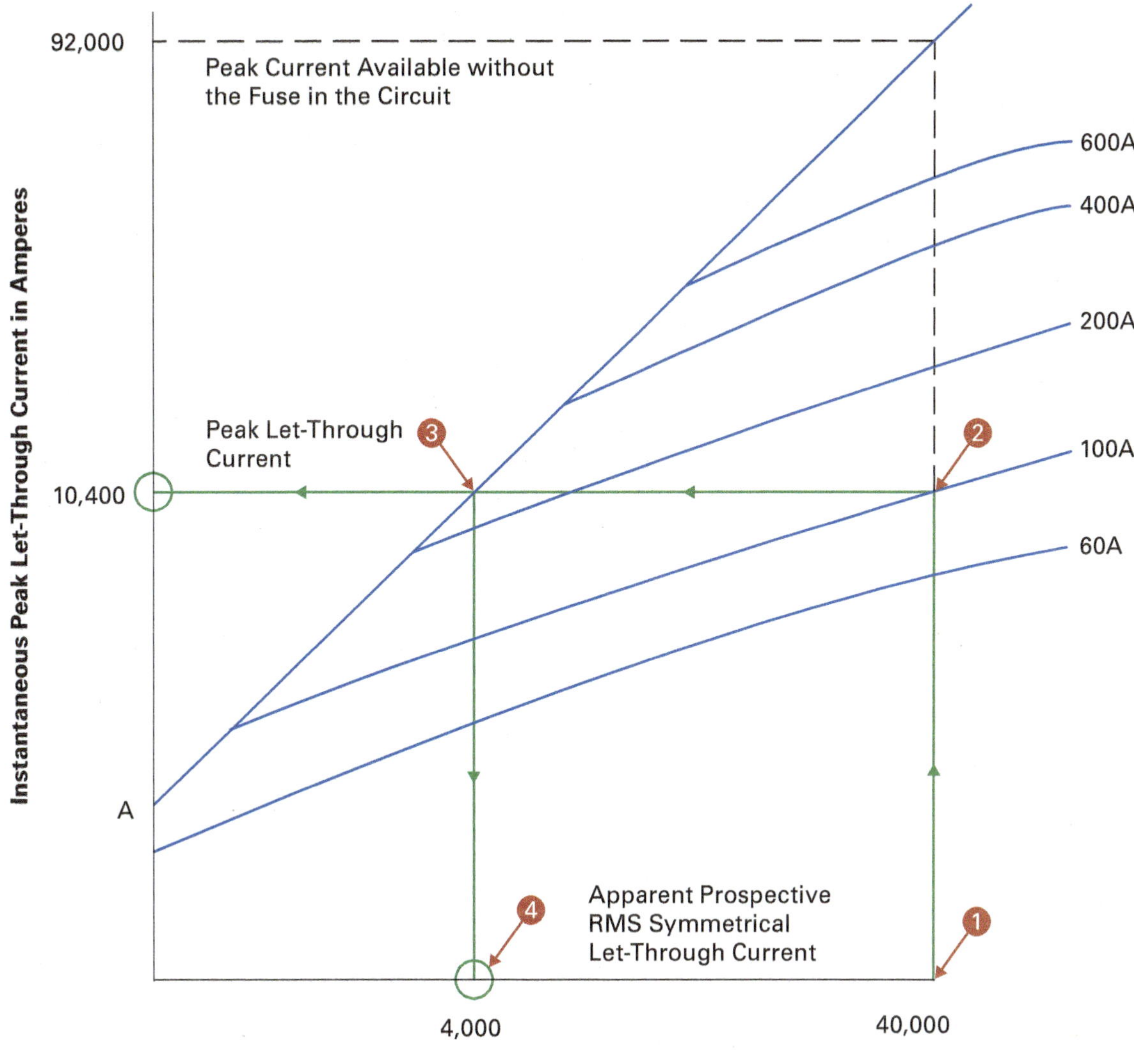

Figure 11 Principles of forming current limitation curves.

2.3.5 Fuses for Motor Overload and Short Circuit Protection

When used in circuits with surge currents such as those caused by inductive loads, dual-element, time-delay fuses can be sized close to full-load amperes (FLA) to give maximum overcurrent protection. For example, assume that a 10 hp, 208V three-phase motor with integral thermal protection has a full-load current of 31A. *Table 1* shows the fuse type, size, and switch size required by *NEC Sections 430.52 and 430.110* for a 10 hp, three-phase motor.

Table 1 shows that a ≤50A, dual-element fuse will protect the 31A motor compared to the much larger 100A, single-element fuse necessary. It is apparent that if a sustained, harmful overload of 300% occurred in the motor circuit, the 100A, single-element fuse would never open and the motor would likely be damaged. The nontime-delay fuse provides only ground fault and short circuit protection, meaning separate overload protection is required as per the *NEC®*.

Know the Code

Rating or Setting for Individual Motor Circuit
NEC Section 430.52

Know the Code

Current Rating and Interrupting Capacity
NEC Section 430.110

TABLE 1 Fuse and Switch Size for a 10 hp Motor (208V, 3ϕ, 31 FLA)

Fuse Type	Maximum Fuse Size	Required Switch Size
Dual-element, time-delay	≤50A	60A
Single-element, nontime-delay	100A	100A

Single phasing: A condition that occurs when one phase of a three-phase system is lost, leaving loads energized by only the two remaining phases. Single phasing results in unbalanced loads in polyphase motors. Unless protective devices respond, it will cause overheating and failure.

In contrast, the ≤50A, dual-element fuse provides ground fault and short circuit protection plus a higher degree of backup protection against motor burnout from overload or **single phasing** should other overload protective devices fail. If thermal overloads, relays, or contacts fail to operate, the dual-element fuses will act independently to protect the motor.

Aside from providing only short circuit protection, the single-element fuse also makes it necessary to use switches with higher ratings because the rating must be equal to or larger than the current rating of the fuse. As a result, the larger switch may cost two or three times more than would be necessary if a dual-element fuse were used (*Figure 12*).

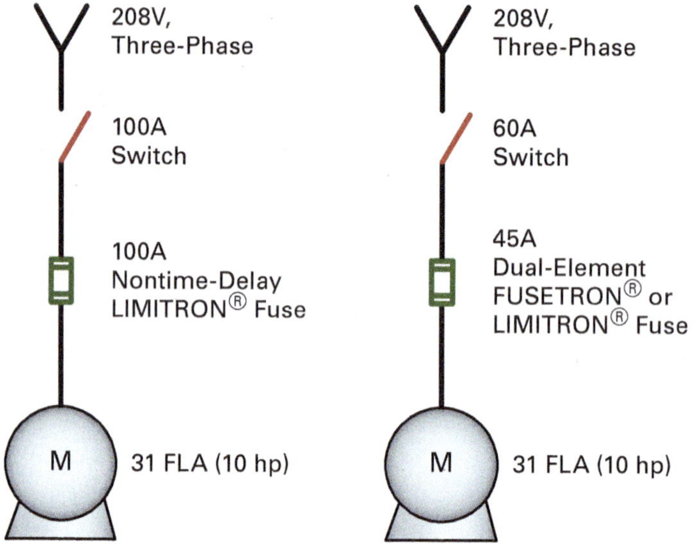

Figure 12 Dual-element fuses permit the use of smaller and less costly switches.

When single phasing occurs in a transformer secondary, the current in the remaining phases increases to 170% to 200% of the rated full-load current. When single phasing occurs on the primary side of a transformer, unbalanced voltages that occur in the motor circuit cause excessive current. Dual-element fuses sized for motor overload protection can protect motors against the overload damage caused by single phasing.

The nontime-delay, fast-acting fuse must be oversized in circuits serving inductive loads. The response of the oversized fuse to short circuit currents is slower. Current builds up to a high level before the fuse opens, causing the current-limiting action of the oversized fuse to be less than a fuse whose current rating is closer to the normal full-load current of the circuit. Consequently, oversizing sacrifices some component protection and, although the *NEC®* permits it, the practice is not recommended.

In actual practice, dual-element fuses used to protect motors keep short circuit currents to approximately half the value of the nontime-delay fuses, because the nontime-delay fuses must be oversized to carry the temporary starting current of a motor per *NEC Table 430.52(C)(1)*.

While dual-element, time-delay fuses can provide motor overload protection, they are nonrenewable and must be replaced after they have blown. In most motor branch circuits, a motor starter typically fitted with a resettable overload relay is used for overload protection. However, a branch circuit supplying a motor starter requires short circuit protection in the form of a backup overload/overcurrent device or instantaneous-trip overcurrent device. From this device, the motor branch circuit is routed to the motor starter.

In some cases, the motor starter and overcurrent protective devices are contained in a combination motor starter. Branch circuit power, switched by the motor starter, is applied to the motor through the overload relay, which is selected specifically for the motor. Like dual-element, time-delay fuses, the overload relay provides protection for the motor for overload conditions due to low line voltage, single phasing, and excessive loads.

Know the Code

Maximum Rating or Setting of Motor Branch-Circuit Short-Circuit and Ground-Fault Protective Devices
NEC Table 430.52(C)(1)

Overload relays can be thermal, magnetic, or solid-state. Thermal overload relays are commonly used because they can be matched to the heating curve of the motor for maximum motor protection. Overload relays are designed to withstand numerous trip and reset cycles without replacement.

When the overload relay in a manual motor starter is actuated by an overload, a trip lever opens the motor starter contacts. In a magnetic motor starter, power to the coil for the contactor is interrupted. This de-energizes the contactor and opens the contacts. When the problem has been resolved, the overload relay is manually reset, and the motor can be restarted. In certain applications, overload relays with an automatic reset function are used, but they are relatively rare.

NOTE

Overload relays and their characteristics are presented in greater detail in NCCER Module 26211-23, *Control Systems and Fundamental Concepts*.

2.0.0 Section Review

1. A plug fuse with a hexagonal cap has a *maximum* amperage rating of _____.
   a. 15A
   b. 20A
   c. 25A
   d. 30A

2. Under normal operation and within its current rating, a fuse operates as a _____.
   a. diode
   b. resistor
   c. conductor
   d. heat sink

3. UL Solutions classifies low-voltage fuses as those operating at _____.
   a. 24V or less
   b. 60V or less
   c. 120V or less
   d. 600V or less

3.0.0 Circuit Breakers and Applications

Objective

Identify circuit breakers by classification and interrupt capacity.

a. Describe circuit breaker classifications.
b. Describe circuit breaker interrupting capacity ratings.

Performance Tasks

There are no Performance Tasks in this section.

A circuit breaker is a device used to close or interrupt a circuit under both normal and abnormal conditions. A circuit is controlled manually by using the handle to switch it to the On or Off position. However, a circuit breaker is also designed to open a circuit automatically during an overload or ground fault without damaging itself or its associated equipment. As long as a circuit breaker is applied within its rating, it will automatically interrupt a fault and is therefore classified as an *inherently safe overcurrent protective device.*

The internal arrangement of a circuit breaker is shown in *Figure 13*, while its external operating characteristics are shown in *Figure 14*. Note that the handle on a circuit breaker resembles an ordinary toggle switch. On an overload, the circuit breaker "trips." In the tripped position, the handle jumps to the middle position as shown in *Figure 14*. To reset it, move the handle to the Off position and beyond it as far as it will go. Then move it to the On position.

WARNING!

When a breaker trips, always determine why it tripped before you reset it and ensure all personnel are clear of any equipment connected to the breaker.

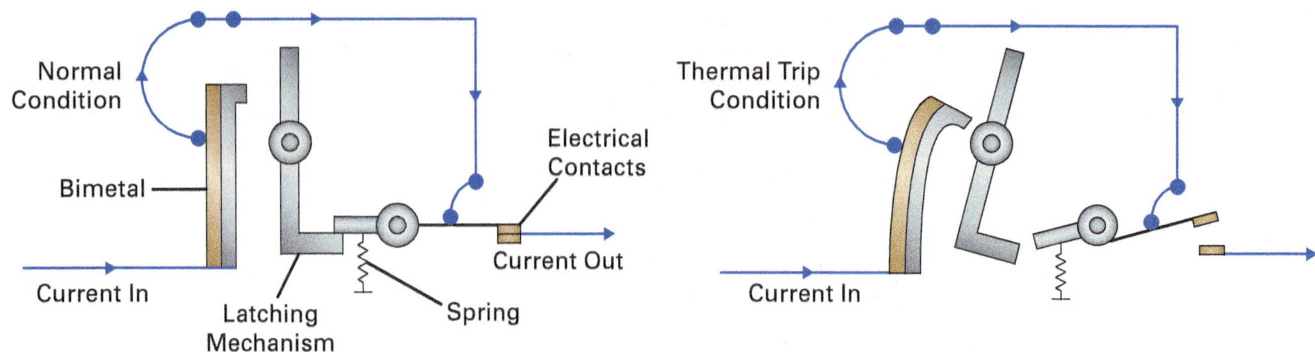

Figure 13 Internal arrangement of a circuit breaker.

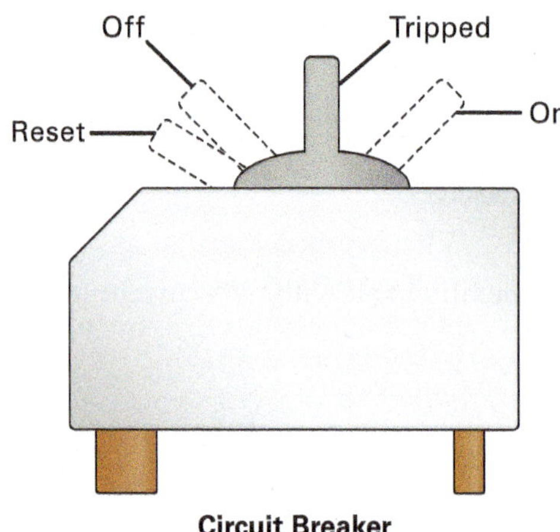

Figure 14 Circuit breaker handle positions.

A standard molded case circuit breaker usually contains:

- A set of contacts
- A magnetic trip element
- A thermal trip element
- Line and load terminals
- Busing used to connect these individual parts
- An enclosing housing of insulating material

The circuit breaker handle manually opens and closes the contacts and resets the automatic trip units after an interruption. Some circuit breakers also contain a manually operated push-to-trip testing mechanism.

Molded Case Circuit Breakers

UL Solutions requires that the handles of molded case circuit breakers be trip-free. This means that the circuit breaker must open under trip conditions even if the handle is physically held in the On position.

3.1.0 Circuit Breaker Classifications

Circuit breakers are grouped for identification according to given current ranges. These groups are as follows:

- 10A–100A
- 125A–225A
- 250A–400A
- 500A–1,000A
- 1,200A–2,000A

Each group is classified by the highest current rating in it. Therefore, the groups are identified as 100A, 225A, 400A, 1,000A, and 2,000A frames. The numbers are referred to as *frame sizes* or *frame classifications* and apply to circuit breakers that are physically interchangeable with each other. For example, any circuit breaker smaller than, and including, 100A will fit into the same spot. But a 100A or smaller breaker will not fit into a space designed to accommodate the 225A frame size.

3.2.0 Circuit Breaker Interrupting Capacity Ratings

In most large commercial and industrial installations, it is necessary to calculate the available short circuit current at various points in a system to determine whether the equipment meets the requirements of *NEC Sections 110.9 and 110.10*. There are a number of methods used to determine the short circuit requirements in an electrical system. Some give approximate values, while others require extensive computations and are quite exacting.

The breaker interrupting capacity is based on testing. There are two such tests. UL Solutions conducts one type of test while the National Electrical Manufacturers Association (NEMA) conducts the other type. The NEMA tests are self-certifying, while unbiased witnesses certify the UL Solutions tests. UL Solutions tests had been limited to a maximum of 10,000A in the past, so the emphasis was placed on NEMA tests with higher ratings. UL Solutions tests now include the NEMA tests plus other ratings. Consequently, the emphasis is now being placed on UL Solutions test results.

The interrupting capacity of a circuit breaker corresponds to its rated voltage. Where the circuit breaker can be used on more than one voltage, the interrupting capacity will be shown for each voltage level. For example, the LA-type circuit breaker has 42,000A symmetrical interrupting capacity at 240V, 30,000A symmetrical at 480V, and 22,000A symmetrical at 600V.

Case History

Arc Flash

A serious injury resulted when a journey-level electrician on an industrial service call attempted to install a three-phase, 400A, bolt-in circuit breaker into an energized panel. He miscalculated the positioning of the bolt-in lugs of the breaker with the breaker handle in the On position. One phase terminal contacted an energized busbar, while another terminal contacted the grounded cabinet. The circuit breaker exploded in his face, in addition to releasing toxic vapor. He survived but sustained severe injuries to his eyes and lungs.

The Bottom Line: Working with energized equipment requires the use of properly rated PPE. In addition, NFPA 70E®, *Standard for Electrical Safety in the Workplace®*, provides detailed guidance to ensure the safety of electrical workers. *Article 130* in NFPA 70E®, *Work Involving Electrical Hazards*, requires proper arc flash and shock risk assessments to be conducted before work begins, as well as energized work permits that should reflect the results of those assessments. From this information, safe work procedures can be established and followed to help prevent accidents like this from happening.

3.0.0 Section Review

1. Which of the following is a standard circuit breaker frame size?
   a. 100A
   b. 300A
   c. 750A
   d. 1,500A

2. The interrupting capacity of a circuit breaker corresponds to its rated _____.
   a. capacitance
   b. power factor
   c. resistance
   d. voltage

4.0.0 Sizing and Selecting Overcurrent Devices

Performance Tasks

There are no Performance Tasks in this section.

Objective

Explain how to size and select overcurrent protective devices.

a. Select overcurrent protective devices for various applications.

b. Apply short circuit calculations.

All conductors in a circuit are to be protected against overcurrent events in accordance with their ampacities as set forth in *NEC Section 240.4*. They must also be protected against short circuit current damage as required.

Current ratings of overcurrent protective devices must not be greater than the ampacity of the conductor. There is, however, an exception. *NEC Section 240.4(B)* states that if the conductor rating does not correspond to a standard size overcurrent protective device, the next larger size of device may be used, provided its rating does not exceed 800A. Likewise, the conductor cannot be part of a branch circuit supplying more than one receptacle for cord- and plug-connected portable loads. When the ampacity of a busway or cable bus does not correspond to a standard overcurrent protective device, the next larger standard rating may be used only if the rating does not exceed 800A, per *NEC Sections 368.17(A) and 370.23*.

Know the Code

Overcurrent Protection
NEC Section 368.17

Know the Code

Overcurrent Protection
NEC Section 370.23

Arc Fault Circuit Interrupters (AFCIs)

Arc faults can occur due to damaged or deteriorated wiring and are a common cause of house fires. AFCIs are special breakers that can detect arcs and eliminate current flow almost immediately. The requirements for AFCIs can be found in ***NEC Section 210.12***.

4.1.0 Overcurrent Devices for Various Applications

This section describes some of the applications and *NEC®* requirements for branch circuits, feeders, services, and transformer secondaries.

4.1.1 Overcurrent Devices for Branch Circuits

The branch circuit rating must be classified in accordance with the rating of the overcurrent protective device. Classifications for those branch circuits other than individual loads must be 10A, 15A, 20A, 30A, 40A, and 50A, as specified in *NEC Section 210.18*.

Branch circuit conductors and equipment must be protected by an overcurrent protective device with current ratings that conform to *NEC Section 210.20*. Basically, the branch circuit conductor and overcurrent protective device must be sized for the actual noncontinuous load plus 125% of the continuous load. The overcurrent protection size must not be greater than the conductor ampacity. Branch circuits rated 10A through 50A with two or more outlets (other than receptacle circuits) must be protected at their rating and the branch circuit conductor sized according to *NEC Table 210.24(1)* or *NEC Table 210.24(2)*. This protection is normally installed at the point where the conductors receive their supply.

Know the Code

Rating
NEC Section 210.18

Know the Code

Overcurrent Protection
NEC Section 210.20

Surge Protection

In addition to the plug-in surge strips sold to consumers that are typically used to protect electronic equipment, manufacturers are now offering residential surge protective devices that mount directly on the main breaker panel. ***NEC Section 225.42*** now requires surge protection for feeders supplying dwellings and similar residential units. These devices provide comprehensive protection for AC power in use throughout the residence and are used in conjunction with surge strips. Surge strips are not rated to handle the higher surges caused by induced lightning, and the comprehensive device reduces the voltage surge to a level that can be handled by the surge protector. See ***NEC Article 242***.

4.1.2 Overcurrent Devices for Feeders

Per *NEC Section 215.2(A)* and *NEC Section 430.24*, the feeder overcurrent protective device current rating and feeder conductor ampacity must be as follows:

- *Feeder circuit with no motor load* — The overcurrent protective device size must be at least 125% of the continuous load plus 100% of the noncontinuous load.
- *Feeder circuit with all motor loads* — Size the overcurrent protective device at 125% of the FLA of the highest rated motor plus the sum of the FLAs of all other motors.
- *Feeder circuit with mixed loads* — Size the overcurrent protective device at 125% of the FLA of the highest rated motor plus the sum of the FLAs of all other motors, plus 125% of the continuous nonmotor load, plus 100% of the noncontinuous nonmotor load.

Know the Code

Minimum Rating and Size
NEC Section 215.2

Know the Code

Several Motors or a Motor(s) and Other Load(s)
NEC Section 430.24

4.1.3 Overcurrent Devices for Services

Each ungrounded service-entrance conductor must have an overcurrent protective device with a rating not higher than the ampacity of the conductor in series with the conductor. The service overcurrent devices shall be part of the service disconnecting means or be located immediately adjacent to it, per *NEC Section 230.91*. Where fuses are used as the service overcurrent device, the disconnecting means must be placed upstream of the fuse.

Per *NEC Section 230.71*, the service disconnect must consist of a single disconnecting means unless the requirements in *NEC Section 230.71(B)* are met. When more than one switch is used, the switches must be grouped together, per *NEC Section 230.72*.

Know the Code

Location
NEC Section 230.91

Know the Code

Maximum Number of Disconnects
NEC Section 230.71

4.1.4 Overcurrent Devices for Transformer Secondary Circuits

Field installations indicate nearly 50% of transformers installed do not have secondary protection. The *NEC*® generally requires overcurrent protection for the secondary conductors. This overcurrent protection combined with the selected wiring method provides the required protection from physical damage. *NEC Section 240.4(F)* and *NEC Section 240.21(C)* discuss protection of transformer secondary conductors.

Know the Code

Grouping of Disconnects
NEC Section 230.72

Know the Code

Transformer Secondary Conductors
NEC Section 240.4(F)

Know the Code

Transformer Secondary Conductors
NEC Section 240.21(C)

4.1.5 Overcurrent Devices for Motor Circuit Protection

Motors and motor circuits have unique operating characteristics and circuit components. These circuits must be dealt with differently from other types of loads. Generally, two levels of overcurrent protection are required for motor branch circuits:

- *Overload protection* — Motor running overload protection is intended to protect the system components and motor from damaging overload currents.
- *Short circuit protection (including ground fault protection)* — Short circuit protection is intended to protect the conductors, switches, controllers, overload relays, motor, and other motor circuit components against short circuit currents or grounds. This level of protection is commonly referred to as *motor branch circuit protection*. Dual-element fuses are designed to provide this protection when correctly sized.

There are various ways to protect a motor circuit depending on the objective. The current rating of an overcurrent protective device selected for motor protection depends on whether the device is of the dual-element, time-delay type or the nontime-delay type.

NOTE

Design B energy-efficient motors require special circuit protection consideration.

NOTE

In many cases, the overcurrent device size is less than the maximum allowed by the *NEC®*. This is an additional safeguard adopted by many manufacturers and installers.

In general, nontime-delay fuses can be sized at 300% of the motor full-load current for ordinary motors so that the normal motor starting current does not affect the fuse, per *NEC Table 430.52(C)(1)*. Dual-element, time-delay fuses can withstand normal motor starting current and can be sized closer to the actual motor rating than nontime-delay fuses.

A summary of *NEC®* regulations governing overcurrent protection is provided in *Table 2A* and *Table 2B*, while the table in *Figure 15* gives generalized fuse application guidelines for motor branch circuits. *Table 3A*, *Table 3B*, and *Table 3C* may be used to select dual-element fuses for motor protection.

Type of Motor	Dual-Element, Time-Delay Fuses			Nontime-Delay Fuses
	Desired Level of Protection			
	Motor Overload and Short Circuit *NEC Section 430.32(A)(1)*	Backup Overload and Short Circuit *NEC Section 430.32(C)*	Short Circuit Only Based on Current Ratings from *NEC Table 430.52(C)(1)*	Short Circuit Only Based on Current Ratings from *NEC Table 430.52(C)(1)*
Service factor of 1.15 or greater or with temperature rise not over 40°C	125% or less of motor nameplate current	125% or next standard size (not to exceed 140% of motor nameplate current)	150% to 175%	150% to 300%
Service factor of less than 1.15 or with temperature rise over 40°C	115% or less of motor nameplate current	115% or next standard size (not to exceed 130%)	150% to 175%	150% to 300%
	Fuses give overload and short circuit protection.	Overload relay gives overload protection and fuses provide backup overload protection.	Overload relay provides overload protection and fuses provide only short circuit protection.	Overload relay provides overload protection and fuses provide only short circuit protection.

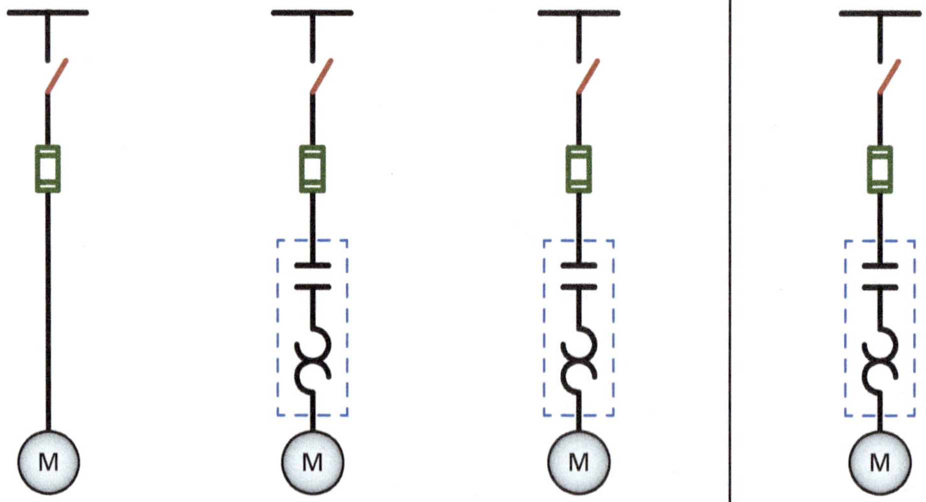

Figure 15 Fuse application guidelines for motor branch circuits.

TABLE 2A *NEC*® Regulations for Overcurrent Protection (1 of 2)

Application	Rule	Code Reference
Scope	Overcurrent protection for conductors and equipment is provided to open the circuit if the current is high enough to cause the temperature of conductors to reach excessive or dangerous levels. See also ***NEC Sections 110.9 and 110.10*** for requirements for interrupting capacity and protection against fault currents.	Informational Notes under ***NEC Section 240.1***
Protection required	Each ungrounded service-entrance conductor must have overcurrent protection in series with it.	***NEC Section 230.90(A)***
Number of devices	Up to six circuit breakers or sets of fuses may be considered as the overcurrent protective device.	Exception No. 3 under ***NEC Section 230.90(A)***
Location	The overcurrent device must be an integral part of the service disconnect or be located adjacent to it. Where fuses are used as service protection, the disconnect must be located upstream of the fuses.	***NEC Section 230.91***
Accessibility	In a multiple-occupancy building, each occupant must have access to their overcurrent protective devices.	***NEC Section 230.72(C)***
Location in circuit	The overcurrent device must protect all circuits and devices, except equipment that may be connected to the supply side, including: (1) Service switch; (2) Special equipment such as surge arresters; (3) Circuits for emergency supply and load management (where separately protected); (4) Circuits for fire alarms or fire pump equipment (where separately protected); (5) Meters with all-metal grounded housing (600V or less); (6) Control circuits for automatic service equipment, if proper overcurrent protection and a disconnecting means are provided.	***NEC Section 230.94*** plus Exceptions No. 1 through 6
Installation and use	Listed or labeled equipment must be applied and per any instructions included in the listing or labeling.	***NEC Section 110.3(B)***
Interrupting rating	Equipment intended to interrupt current at fault levels must have an interrupting rating at least equal to the current available at the line terminals of the equipment within the voltage rating of the protective device.	***NEC Section 110.9***

TABLE 2B *NEC®* **Regulations for Overcurrent Protection (2 of 2)**

Application	Rule	Code Reference
Circuit impedance, short circuit ratings, and other characteristics	The overcurrent protective devices, total impedance, equipment short circuit current ratings, and other characteristics of a circuit must be selected and coordinated to permit the protective devices to clear a fault without extensive damage to the circuit and equipment.	*NEC Section 110.10*
General	Noncurrent-carrying conductive materials that could become energized must be connected to each other and to the electrical source in a way that establishes an effective ground fault current path.	*NEC Section 250.4(A)(4)*
Bonding other enclosures	Regardless of the use of supplemental equipment grounding conductors, bond metal raceways, cable trays, cable armor, cable sheath, enclosures, frames, fittings, and other metal parts not designed to carry current but serve as equipment grounding conductors, to ensure electrical continuity and the ability to safely route fault current. Any nonconductive coating must be removed to improve conductivity or be connected using fittings instead.	*NEC Section 250.96(A)*

TABLE 3A **Selection of Fuses for Motor Protection (1 of 3)**

Dual-Element Fuse Size in Amps	Motor Protection in FLA (Used without properly sized relays)		Backup Motor Protection in FLA (Used with properly sized overload relays)	
	Motor service factor of 1.15 or greater or with temperature rise not over 40°C	Motor service factor of less than 1.15 or with temperature rise not over 40°C	Motor service factor of 1.15 or greater or with temperature rise not over 40°C	Motor service factor of less than 1.15 or with temperature rise not over 40°C
1/10	0.08–0.09	0.09–0.10	0–0.08	0–0.09
1/8	0.10–0.11	0.11–0.15	0.09–0.10	0.10–0.11
5/100	0.12–0.15	0.14–0.15	0.11–0.12	0.12–0.13
2/10	0.16–0.19	0.18–0.20	0.13–0.16	0.14–0.17
1/4	0.20–0.23	0.22–0.25	0.17–0.20	0.18–0.22
3/10	0.24–0.30	0.27–0.30	0.21–0.24	0.23–0.26
4/10	0.32–0.39	0.35–0.40	0.25–0.32	0.27–0.35
1/2	0.40–0.47	0.44–0.50	0.33–0.40	0.36–0.43
6/10	0.48–0.60	0.53–0.60	0.41–0.48	0.44–0.52
8/10	0.64–0.79	0.70–0.80	0.49–0.64	0.53–0.70
1	0.80–0.89	0.87–0.97	0.65–0.80	0.71–0.87
1 1/8	0.90–0.99	0.98–1.08	0.81–0.90	0.88–0.98
1 1/4	1.00–1.11	1.09–1.21	0.91–1.00	0.99–1.09
1 4/10	1.12–1.19	1.22–1.30	1.01–1.12	1.10–1.22
1 1/2	1.20–1.27	1.31–1.39	1.13–1.20	1.23–1.30
1 6/10	1.28–1.43	1.40–1.56	1.21–1.28	1.31–1.39
1 8/10	1.44–1.59	1.57–1.73	1.29–1.44	1.40–1.57
2	1.60–1.79	1.74–1.95	1.45–1.60	1.58–1.74
2 1/4	1.80–1.99	1.96–2.17	1.61–1.80	1.75–1.96
2 1/2	2.00–2.23	2.18–2.43	1.81–2.00	1.97–2.17

TABLE 3B Selection of Fuses for Motor Protection (2 of 3)

Dual-Element Fuse Size in Amps	Motor Protection in FLA (Used without properly sized relays)		Backup Motor Protection in FLA (Used with properly sized overload relays)	
	Motor service factor of 1.15 or greater or with temperature rise not over 40°C	Motor service factor of less than 1.15 or with temperature rise not over 40°C	Motor service factor of 1.15 or greater or with temperature rise not over 40°C	Motor service factor of less than 1.15 or with temperature rise not over 40°C
2 6/10	2.24–2.39	2.44–2.60	2.01–2.24	2.18–2.43
3	2.40–2.55	2.61–2.78	2.25–2.40	2.44–2.60
3 2/10	2.56–2.79	2.79–3.04	2.41–2.56	2.61–2.78
3 1/2	2.80–3.19	3.05–3.47	2.57–2.80	2.79–3.04
4	3.20–3.59	3.48–3.91	2.81–3.20	3.05–3.48
4 1/2	3.60–3.99	3.92–4.34	3.21–3.60	3.49–3.91
5	4.00–4.47	4.35–4.86	3.61–4.00	3.92–4.35
5 6/10	4.48–4.79	4.87–5.21	4.01–4.48	4.36–4.87
6	4.80–4.99	5.22–5.43	4.49–4.80	4.88–5.22
6 1/4	5.00–5.59	5.44–6.08	4.81–5.00	5.23–5.43
7	5.60–5.99	6.09–6.52	5.01–5.60	5.44–6.09
7 1/2	6.00–6.39	6.53–6.95	5.61–6.00	6.10–6.52
8	6.40–7.19	6.96–7.82	6.01–6.40	6.53–6.96
9	7.20–7.99	7.83–8.69	6.41–7.20	6.97–7.83
10	8.00–9.59	8.70–10.00	7.21–8.00	7.84–8.70
12	9.60–11.99	10.44–12.00	8.01–9.60	8.71–10.43
15	12.00–13.99	13.05–15.00	9.61–12.00	10.44–13.04
17 1/2	14.00–15.99	15.22–17.39	12.01–14.00	13.05–15.21
20	16.00–19.99	17.40–20.00	14.01–16.00	15.22–17.39
25	20.00–23.99	21.74–25.00	16.01–20.00	17.40–21.74
30	24.00–27.99	26.09–30.00	20.01–24.00	21.75–26.09
35	28.00–31.99	30.44–34.78	24.01–28.00	26.10–30.43

TABLE 3C Selection of Fuses for Motor Protection (3 of 3)

Dual-Element Fuse Size in Amps	Motor Protection in FLA (Used without properly sized relays)		Backup Motor Protection in FLA (Used with properly sized overload relays)	
	Motor service factor of 1.15 or greater or with temperature rise not over 40°C	Motor service factor of less than 1.15 or with temperature rise not over 40°C	Motor service factor of 1.15 or greater or with temperature rise not over 40°C	Motor service factor of less than 1.15 or with temperature rise not over 40°C
40	32.00–35.99	34.79–39.12	28.01–32.00	30.44–37.78
45	36.00–39.00	39.13–43.47	32.01–36.00	37.79–39.13
50	40.00–47.99	43.48–50.00	36.01–40.00	39.14–43.48
60	48.00–55.99	52.17–60.00	40.01–48.00	43.49–52.17
70	56.00–59.99	60.87–65.21	48.01–56.00	52.18–60.87
75	60.00–63.99	65.22–69.56	56.01–60.00	60.88–65.22
80	64.00–71.99	69.57–78.25	60.01–64.00	65.23–69.57
90	72.00–79.99	78.26–86.95	64.01–72.00	69.58–78.26
100	80.00–87.99	86.96–95.64	72.01–80.00	78.27–86.96
110	88.00–99.00	95.65–108.69	80.01–88.00	86.97–95.65
125	100.00–119.00	108.70–125.00	88.01–100.00	95.66–108.70
150	120.00–139.99	131.30–150.00	100.01–120.00	108.71–130.43
175	140.00–159.99	152.17–173.90	120.01–140.00	130.44–152.17
200	160.00–179.99	173.91–195.64	140.01–160.00	152.18–173.91
225	180.00–199.99	195.65–217.38	160.01–180.00	173.92–195.62
250	200.00–239.99	217.39–250.00	180.01–200.00	195.63–217.39
300	240.00–279.99	260.87–300.00	200.01–240.00	217.40–260.87
350	280.00–319.99	304.35–347.82	240.01–280.00	260.88–304.35
400	320.00–359.99	347.83–391.29	280.01–320.00	304.36–347.83
450	360.00–399.99	391.30–434.77	320.01–360.00	347.84–391.30
500	400.00–479.99	434.78–500.00	360.01–400.00	391.31–434.78
600	480.00–600.00	521.74–600.00	400.01–480.00	434.79–521.74

4.2.0 Applying Short Circuit Calculations

Two essential aspects of overcurrent protection, required by *NEC Sections 110.9 and 110.10*, are adequate interrupting capacity and protection of electrical components. The first step in selecting the proper protective device is to determine the available short circuit current.

Although several methods for determining short circuit currents are in use, the *point-to-point method* is most often recommended. Using this method permits the determination of available short circuit currents with reasonable accuracy for either single-phase or three-phase systems. This method assumes unlimited primary short circuit current (infinite bus) for a facility transformer.

Know the Code

Interrupting Rating
NEC Section 110.9

Know the Code

Circuit Impedance, Short-Circuit Current Ratings, and Other Characteristics
NEC Section 110.10

4.2.1 Basic Point-to-Point Calculation Procedure

The following steps provide a basic calculation procedure for determining short circuit currents:

Step 1 Determine the transformer full-load amperes (I_{FLA}). Transformer full-load kVA ratings are shown in *Table 4*. This can also be determined from the nameplate or by using the following equations.

$$I_{FLA}\text{ (three phase)} = \frac{\text{kVA} \times 1{,}000}{E_{L\text{-}L} \times 1.73}$$

$$I_{FLA}\text{ (single phase)} = \frac{\text{kVA} \times 1{,}000}{E_{L\text{-}L}}$$

Where:

$E_{L\text{-}L}$ = line-to-line voltage

TABLE 4 Three-Phase Transformer Full-Load kVA Ratings

Line-to-Line Voltage	Transformer kVA Rating								
	150 kVA	167 kVA	225 kVA	300 kVA	500 kVA	750 kVA	1,000 kVA	1,500 kVA	2,000 kVA
208	417	464	625	834	1,388	2,080	2,776	4,164	5,552
220	394	439	592	788	1,315	1,970	2,630	3,940	5,260
240	362	402	542	722	1,203	1,804	2,406	3,609	4,812
440	197	219	296	394	657	985	1,315	1,970	2,630
460	189	209	284	378	630	945	1,260	1,890	2,520
480	181	201	271	361	601	902	1,203	1,804	2,406

Step 2 Find the transformer multiplier using the transformer percent impedance (%Z) in the following equation.

$$\text{Transformer multiplier} = \frac{100}{\%Z}$$

Step 3 Determine the transformer let-through short circuit current (I_{SCA}) using *Table 5* or the following equation.

$$I_{SCA} = \text{transformer}_{FLA} \times \text{multiplier}$$

NOTE

The motor short circuit contribution, if significant, may be added to the transformer secondary short circuit current value as determined here. Proceed with the adjusted figure from *Step 4* through *Step 6* to follow. A practical estimate of motor short circuit contribution is to multiply the total load current in amperes by 4.

TABLE 5 Available Short Circuit Currents for Various Transformers

Voltage and Phase	kVA	FLA	% Impedance	Short Circuit Current
120/240V Single Phase	25	104	1.6	10,300
	37½	156	1.6	15,280
	50	209	1.7	19,050
	75	313	1.6	29,540
	100	417	1.6	38,540
	167	695	1.8	54,900
120/208V Three Phase	150	417	2.0	20,850
	225	625	2.0	31,250
	300	834	2.0	41,700
	500	1,388	2.0	69,444
	750	2,080	3.5	59,426
	1,000	2,776	3.5	79,310
	1,500	4,164	3.5	118,965
	2,000	5,552	5.0	111,040
	2,500	6,950	5.0	139,000
277/480V Three Phase	112½	135	1.0	13,500
	150	181	1.2	15,083
	225	271	1.2	22,583
	300	361	1.2	30,083
	500	601	1.3	46,230
	750	902	3.5	25,770
	1,000	1,203	3.5	34,370
	1,500	1,804	3.5	51,540
	2,000	2,406	5.0	48,120
	2,500	3,077	5.0	60,140

Step 4 Calculate the *f* factor using the C values in *Table 6*. For parallel runs, multiply the C value by the number of conductors per phase. Use the following equation.

$$f \text{ (three-phase faults)} = \frac{1.73 \times L \times I}{C \times E_{L\text{-}L}}$$

For single-phase, line-to-line (L-L) faults on single-phase, center-tapped transformers.

$$f = \frac{2 \times L \times I}{C \times E_{L\text{-}L}}$$

For single-phase, line-to-neutral (L-N) faults on single-phase, center-tapped transformers.

$$f = \frac{2 \times L \times I}{C \times E_{L\text{-}N}}$$

Where:

L = length (in feet)
C = constant (from *Table 6*)
I = available short circuit current at the beginning of circuit (in amperes)

NOTE

The C values of other conductors and busways can be found in the manufacturer's literature for the device.

TABLE 6 C Values for Conductors and Busways

AWG or kcmil	Copper, Three Single Conductors				Copper, Three-Conductor Cable			
	Steel Conduit		Nonmagnetic Conduit		Steel Conduit		Nonmagnetic Conduit	
	600V	5 kV	600V	5 kV	600V	5 kV	600V	5 kV
14	389	389	389	389	389	389	389	389
12	617	617	617	617	617	617	617	617
10	981	981	981	981	981	981	981	981
8	1,557	1,551	1,558	1,555	1,559	1,557	1,559	1,558
6	2,425	2,406	2,430	2,417	2,431	2,424	2,433	2,428
4	3,806	3,750	3,825	3,789	3,830	3,811	3,837	3,823
3	4,760	4,760	4,802	4,802	4,760	4,790	4,802	4,802
2	5,906	5,736	6,044	5,926	5,989	5,929	6,087	6,022
1	7,292	7,029	7,493	7,306	7,454	7,364	7,579	7,507
1/0	8,924	8,543	9,317	9,033	9,209	9,086	9,472	9,372
2/0	10,755	10,061	11,423	10,877	11,244	11,045	11,703	11,528
3/0	12,843	11,804	13,923	13,048	13,656	13,333	14,410	14,118
4/0	15,082	13,605	16,673	15,351	16,391	15,890	17,482	17,019
250	16,483	14,924	18,593	17,120	18,310	17,850	19,779	19,352
300	18,176	16,292	20,867	18,975	20,617	20,051	22,524	21,938
350	19,703	17,385	22,736	20,526	22,646	21,914	24,904	24,126
400	20,565	18,235	24,296	21,786	24,253	23,371	26,915	26,044
500	22,185	19,172	26,706	23,277	26,980	25,449	30,028	28,712
600	22,965	20,567	28,033	25,203	28,752	27,974	32,236	31,258
750	24,136	21,386	28,303	25,430	31,050	30,024	32,404	31,338
1,000	25,278	22,539	31,490	28,083	33,864	32,688	37,197	35,748

Step 5 Determine the multiplier (M) using *Table 7.* The multiplier can also be calculated using the f factor with the following equation.

$$M = \frac{1}{1 + f}$$

Step 6 Find the available short circuit current at the fault using the following equation.

$$I_{SCA} \text{ at fault} = I_{FLA} \times M$$

TABLE 7 Multipliers

f	M	*f*	M
0.01	0.99	1.50	0.40
0.02	0.98	1.75	0.36
0.03	0.97	2.00	0.33
0.04	0.96	2.50	0.29
0.06	0.94	3.50	0.22
0.07	0.93	4.00	0.20
0.08	0.93	5.00	0.17
0.09	0.92	6.00	0.14
0.10	0.91	7.00	0.13
0.15	0.87	8.00	0.11
0.20	0.83	9.00	0.10
0.25	0.80	10.00	0.09
0.30	0.77	15.00	0.06
0.35	0.74	20.00	0.05
0.40	0.71	30.00	0.03
0.50	0.67	40.00	0.02
0.60	0.63	50.00	0.02
0.70	0.59	60.00	0.02
0.80	0.55	70.00	0.01
0.90	0.53	80.00	0.01
1.00	0.50	90.00	0.01
1.20	0.45	100.00	0.01

The L-N fault current is lower than the L-L fault current at some distance from the terminals, depending on the wire size. The 1.5 multiplier is an approximation—it will vary from 1.33 to 1.67. These figures are based on the change in turns ratio between the primary and the secondary windings, infinite source available, zero feet from the terminals of the transformer, and 1.2 × %X and 1.5 × %R for L-N versus L-L resistance and reactance values. Begin L-N calculations at the transformer secondary terminals, then proceed the point-to-point calculation.

NOTE

The L-N current is higher than the L-L fault current at the secondary terminals of a single-phase, center-tapped transformer. The short circuit current available (I) for *Step 4* should be adjusted at the transformer terminals as follows: I = 1.5 x L-L short circuit current at transformer terminals.

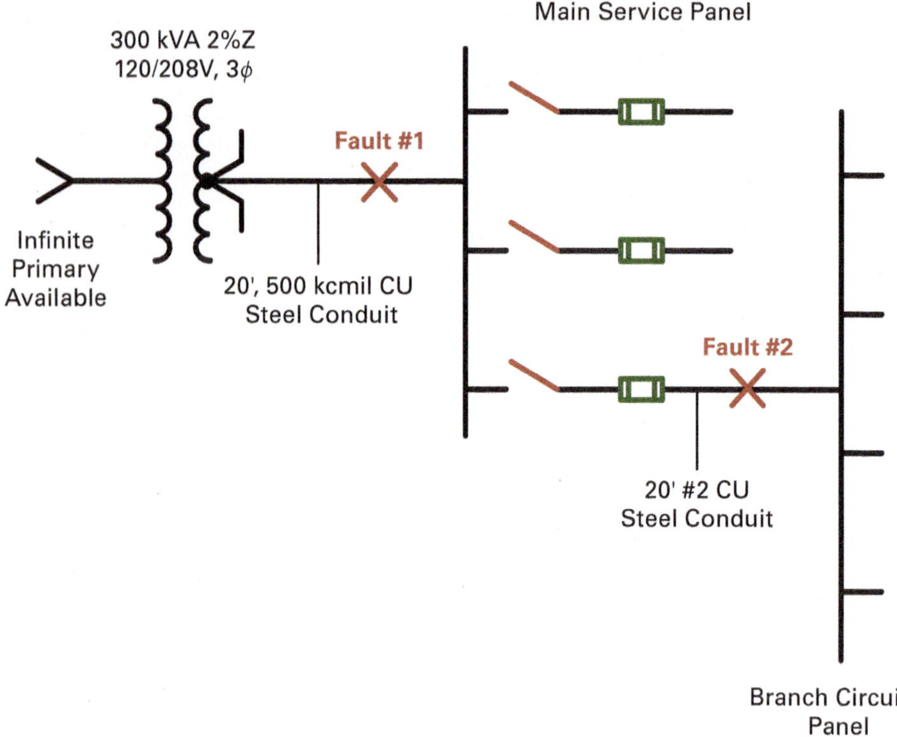

Figure 16 Sample one-line diagram for short circuit calculation.

4.2.2 Using the Point-to-Point Method

This section describes how to use the point-to-point calculation to find the total short circuit current at Fault #1 (I_{SCA1}) and Fault #2 (I_{SCA2}) in the example circuit shown in *Figure 16*. The process outlined here is applied to a practical example using the same formulas shown in *Section 4.2.1*. The process is demonstrated in the following steps:

Step 1 Start by finding the total short circuit current at Fault #1. To do this, first determine the transformer FLA (I_{FLA}) using the following equation.

$$I_{FLA} = \frac{kVA \times 1{,}000}{E_{L\text{-}L} \times 1.73}$$

$$I_{FLA} = \frac{300 \times 1{,}000}{208V \times 1.73}$$

$$I_{FLA} = \frac{300{,}000}{359.84}$$

$$I_{FLA} = 834A$$

Step 2 Find the transformer multiplier using the following equation.

$$\text{Transformer multiplier} = \frac{100}{\%Z}$$

$$\text{Transformer multiplier} = \frac{100}{2} = 50$$

Step 3 Calculate the transformer let-through short circuit current (I_{SCA}). Note that this is the available short circuit current at the beginning of the circuit; you will need this to determine the *f* factor at each fault. Use the following equation.

$$I_{SCA} \text{ (at transformer secondary)} = I_{FLA} \times M$$

$$I_{SCA} = 834A \times 50$$

$$I_{SCA} = 41{,}700A$$

Step 4 Calculate the f factor for Fault #1 using the following equation, along with the C values in *Table 6*.

$$f_1 = \frac{1.73 \times L \times I_{SCA}}{C \times E_{L\text{-}L}}$$

$$f_1 = \frac{1.73 \times 20' \times 41{,}700A}{22{,}185 \times 208V}$$

$$f_1 = \frac{1{,}442{,}820}{4{,}614{,}480} = 0.313$$

NOTE

The C values of other conductors/busways can be found in the manufacturer's literature for the device being sized.

Step 5 Calculate the multiplier (M_1) using the f_1 with the following equation.

$$M_1 = \frac{1}{1 + f_1} = \frac{1}{1 + 0.313}$$

$$M_1 = \frac{1}{1.313} = 0.762$$

Step 6 Find the available short circuit current at Fault #1 (I_{SCA1}) using the following equation.

$$I_{SCA1} = I_{SCA} \times M_1$$

$$I_{SCA1} = 41{,}700A \times 0.762$$

$$I_{SCA1} = 31{,}775A$$

Step 7 Now you have enough information to find the short circuit current at Fault #2 (I_{SCA2}). First, calculate the f factor for Fault #2 (f_2) using the following equation, along with the C values in *Table 6*.

$$f_2 = \frac{1.73 \times L_2 \times I_{SCA1}}{C_2 \times E_{L\text{-}L}}$$

$$f_2 = \frac{1.73 \times 20' \times 31{,}775A}{5{,}906 \times 208V}$$

$$f_2 = \frac{1{,}099{,}415}{1{,}228{,}448} = 0.895$$

Step 8 Determine the multiplier (M_2) using *Table 7*; the multiplier can also be calculated using the f_2 with the following equation.

$$M_2 = \frac{1}{1 + f_2}$$

$$M_2 = \frac{1}{1 + 0.895}$$

$$M_2 = \frac{1}{1.895} = 0.528$$

Step 9 Find I_{SCA2} using the following equation and I_{SCA1} as the available let-through current.

$$I_{SCA2} = I_{SCA1} \times M_2$$

$$I_{SCA2} = 31{,}775A \times 0.528$$

$$I_{SCA2} = 16{,}777A$$

4.2.3 Peak Let-Through Charts

Peak let-through charts (*Figure 17*) allow you to determine both the peak let-through current and the apparent prospective rms symmetrical let-through current. Such charts are commonly referred to as *current limitation curves*. These charts vary with the type of overcurrent protective device and are shown here only to familiarize you with their appearance.

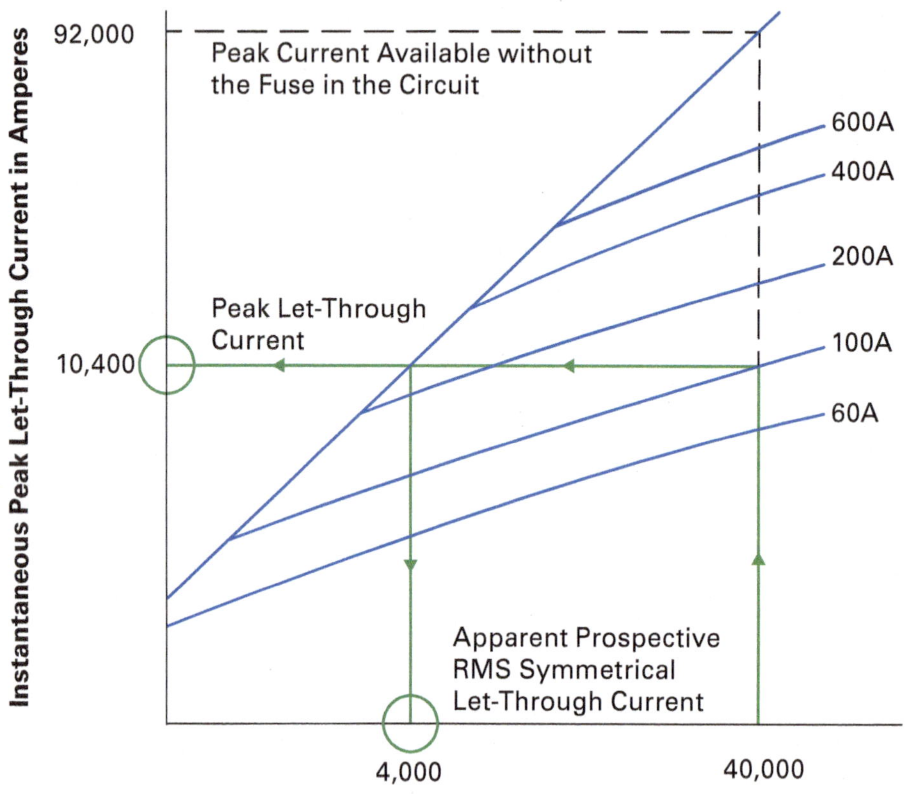

H-CAP® KRP-C Fuses
Instantaneous Peak Let-Through Current in Amperes
400,000 300,000 200,000 100,000 80,000 60,000 40,000 30,000 20,000 10,000 8,000 6,000 4,000 3,000 2,000 1,000
A B
6000 5000 4000 3000 2500 2000 1600 1200 800 601
Ampere Rating
1,000 2,000 3,000 4,000 6,000 8,000 10,000 20,000 30,000 40,000 60,000 80,000 100,000 200,000
Prospective Short Circuit Current in Symmetrical RMS Amperes

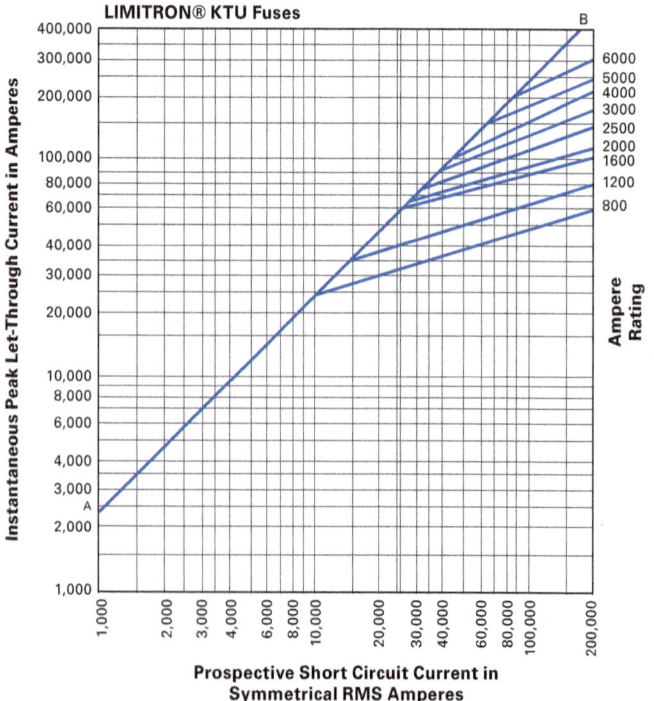

Figure 17 Peak let-through charts.

Source: Schneider Electric

Current-Limiting Circuit Breakers

Most circuit breakers have some current-limiting characteristics. However, they do not limit current at the mid- to low-current levels as required by UL Solutions to qualify as current-limiting breakers. Only those listed as current limiting have published peak let-through charts.

4.0.0 Section Review

1. A feeder circuit with no motor load has a continuous load of 10A and a noncontinuous load of 5A. The feeder overcurrent protective device should be sized at _____.
   a. 10A
   b. 15A
   c. 17.5A
   d. 18.75A

2. The full-load current rating of a 300 kVA, three-phase transformer operating at 480V is _____.
   a. 834A
   b. 722A
   c. 394A
   d. 361A

5.0.0 Testing and Troubleshooting Circuit Breakers and Fuses

Performance Tasks

There are no Performance Tasks in this section.

Objective

Describe how to test and troubleshoot overcurrent protective devices.

a. Describe how to test and troubleshoot circuit breakers.

b. Describe how to test and troubleshoot fuses.

Troubleshooting circuit breakers and fuses is relatively simple. However, tracking down and resolving short circuits and unintentional grounds takes a little more practice to gain proficiency.

5.1.0 Circuit Breakers

The magnetic portion of a circuit breaker provides instantaneous protection against short circuit and other excessive currents. The thermal portion protects the circuit from excessive heat, for instance, the heat developed during a sustained overload.

The first step in troubleshooting is to verify the position of the breaker handle. When tripped, most breaker handles move to a position halfway between On and Off. Some have visual indicators while others do not. To reset a circuit breaker, move the handle to the Off position as far as it will travel, and then move it back to the On position.

WARNING!

Whenever re-energizing a breaker, stand to one side and look away from the breaker. If the breaker trips immediately, stop. Do not re-energize the circuit until the fault is resolved or the circuit is disconnected from the breaker.

5.1.1 Troubleshooting a Short Circuit

An instant trip or a failure to reset is a sign that a short circuit is causing the magnetic portion to trip. This can be verified by turning the breaker Off, checking to ensure there is no voltage on the load side, and then disconnecting the load wires. Then re-energize the breaker and verify the presence of voltage on all load terminals. If the breaker remains On, this likely indicates a short circuit downstream, as opposed to a defective breaker.

5.1.2 Troubleshooting an Overload Condition

If the breaker stays in the On position for a short time before tripping, the problem is likely an overload, a high impedance, or perhaps an arcing fault. A ground fault or true short circuit condition will cause the breaker to trip immediately. Use the following procedure to troubleshoot the problem:

WARNING!

The following procedure must be performed only by qualified individuals operating under the approval of a safe work plan or permit.

Step 1 Test the voltage. First verify the operation of the meter by testing it on a known voltage source. Then check the voltage between the load side of the circuit breaker and the ground terminal. Continue testing by checking the voltage between the phases.

Step 2 If the voltage is as expected, measure the current on each phase present using a clamp-on ammeter. Carefully clamp the meter around each individual conductor. Record the current on each phase in case calculations need to be made. Most importantly, relate the measured current to the breaker rating. Is it low, high, or as you would expect from the load imposed?

Step 3 Check for excess heat. If the breaker remains On for some time (minutes, for example) and the current is within the breaker's range, the problem may be heat related. Outside heat sources can also be a factor in breaker temperature. If the breaker is in a hot area, it may trip at a current below its rating. Loose or corroded connections can also generate additional heat. If an infrared heat detector is available, it can be used to pinpoint and quantify unusual sources of heat. You may also check for heat in the panel by lightly placing the back of your hand on top of the suspected breaker as well as the adjacent breakers. The breaker should feel cool to slightly warm instead of hot.

WARNING!

When testing for this condition, be aware of the voltages present and take the appropriate precautionary measures before proceeding. Touch the breakers very gently at first to avoid burns.

Step 4 Listen for any crackling sounds that may indicate arcing. Electrical arcs generate ozone, which has a distinct sharp odor. You may also detect the odor of burned plastic.

WARNING!

If you suspect or detect arcing, immediately open the breaker.

Step 5 Remove the breaker to visually inspect the connection between the breaker and the busbar. Follow the manufacturer's recommendations for removing the breaker. Inspect the breaker itself at the termination points. If it has been overheated, the terminals will likely be discolored. Check all wiring near the terminations for discoloration or charring of the insulation.

Step 6 Reinstall the breaker.

Breakers are usually stable, reliable, and long-lasting. However, excess current and tripping may damage the internal components and lead to premature wear or unreliable operation. If there is any doubt as to the operation or condition of a breaker, replace it immediately and inspect the connecting bus and nearby components for potential problems.

NOTE

Thermal magnetic breakers are designed not to trip at 100% of the handle rating under normal operating conditions. At slightly over 100% (between 105% and 120%) of the handle rating, the breaker must trip between 2,000 and 10,000 seconds. At 150% of the handle rating, the same molded case breaker must trip between 200 and 900 seconds. This type of breaker will actually sustain a load in excess of its rating for a short time, depending on the size of the load. If the current is less than the handle rating of the breaker but the breaker keeps tripping, it may be due to other factors such as unmeasured harmonic currents, excess heat, or a defective breaker.

5.2.0 Fuses

Remember that a single-element fuse is designed to act immediately on an overcurrent event, which may be either a short circuit or an overload. A dual-element fuse has two fuse elements: a quick-acting element and a thermal time-delay element. The quick-acting element is sized to react to short circuit conditions while allowing smaller overcurrent conditions to continue. The time-delay element will protect against small overcurrent events over time.

WARNING!

Because fuse holders may be energized while testing fuses, you must be especially careful when testing and troubleshooting the circuit.

To troubleshoot a blown fuse using an ohmmeter, use the following steps.

WARNING!

The following procedure must be performed only by qualified individuals operating under the approval of a safe work plan or permit.

CAUTION

Applying voltage to an ohmmeter or a multimeter set to test for continuity will result in a blown meter fuse or significant damage to the meter. Always ensure that a circuit is de-energized before applying an ohmmeter.

Step 1 Disconnect the power to the fuse. This usually involves turning off a local disconnect switch. If the fuse holder is not adjacent to the power switch, follow appropriate lockout/tagout procedures.

Step 2 Verify that the voltage has been removed from all fuses associated with the circuit. After confirming the voltmeter is functional using a reliable voltage source, take readings from the top of the fuse holder to ground and other phases, and then repeat the procedure from the bottom of the fuse holder. Another check between the top and bottom of the fuse holder will verify that no voltage is present.

Step 3 After checking for voltage, remove the fuse with a fuse puller (*Figure 18*). Using an ohmmeter or continuity tester, place one probe at one end of the fuse and the other probe on the opposite end. If the reading shows no resistance, indicating continuity, the fuse is good. If the meter shows no movement or displays "OL," the fuse is open.

Figure 18 Fuse puller.
Source: Photo 19254279 © Dave Willman | Dreamstime.com

To test fuses in an energized circuit, use the following procedure. However, remember that you are working with an energized circuit and must avoid contact with all energized sources.

Step 1 Use a voltmeter to check for voltage between the top of the fuse and ground, and then between the bottom of the fuse and ground. If voltage is present between the top and ground, but not between the bottom and ground, the fuse is open. An intact fuse will show the same voltage between ground (or another phase) and each end of the fuse. If the voltage between phases is not consistent, a fuse other than the one tested is likely open. Test the bottom of each fuse to ground to determine which one is open.

WARNING!

Verify that all voltage has been removed before touching or replacing any fuse.

Step 2 Before removing and replacing a blown fuse, disconnect the power and verify the absence of voltage. Then use a fuse puller to remove and replace the fuse.

Step 3 When a new fuse is installed, verification should follow. After installing the fuse, energize the circuit using the proper safety techniques. Verify the presence of voltage on all load terminals by checking phase-to-phase and phase-to-ground voltages. While the circuit is under load, measure the current on each conductor just below the fuses using a clamp-on ammeter. Ensure that the current is within expectations.For inductive loads like motors, position the ammeter before the circuit is energized and set it to lock in the peak current reached as the load starts. You can then evaluate both the peak current reached as well as the normal operating current.

NOTE

Fuses vary in the amount of time they can sustain an overload. Like a circuit breaker, they are also affected by external heat sources and unmeasured harmonic currents.

Check for excessive heat and other problems using the following procedure:

Step 1 Disconnect all power and verify the absence of voltage with a voltmeter. Visually inspect the fuses and fuse holders for signs of discoloration and burning. If the fuse holder clips are discolored from heat, replacement may be necessary. The metal clips can lose their spring temper with use and overheating. Squeezing them with pliers to improve contact won't likely help; replace questionable fuse holders instead. Also check the wiring for insulation damage or discoloration, which may be accompanied by a distinct odor.

Step 2 Restore power and listen/look for arcing and other abnormal conditions while the circuit is under load.

Fuses are reliable and seldom fail due to internal defects. However, a fuse can be stressed over time and eventually fail. There may be a problem within the circuit that applies a little stress at each event, leading to the slow degradation of the fuse.

Because fuses are good for only one use, it can be difficult to determine the source of the problem. If new fuses continue to blow, do not continue to replace them without conducting thorough troubleshooting downstream. Do not continue to replace fuses, hoping to eventually find one that holds. Do not increase the fuse size or change fuse types without investigating the impact on the protection of the equipment. Be aware of the limitations of single-element fuses when used with inductive loads, since they are rarely the correct choice for motors and similar loads such as HVACR equipment.

5.0.0 Section Review

1. If a circuit breaker has tripped and stays in the On position for a short time before tripping, the problem is *least* likely to be _____.
   a. overheating
   b. an arc fault
   c. a ground fault
   d. an overload

2. Which of the following statements about fuses is *true*?
   a. A fuse that trips repeatedly may be replaced with a larger size as long as it fits in the clip.
   b. Fuse holder clips that have lost their spring temper should be replaced.
   c. Fuses are checked for continuity using a voltmeter.
   d. Fuses are inexpensive devices that often fail due to manufacturing defects.

Module 26305-23 Review Questions

1. Which of the following describes the *maximum* short circuit current that a fuse or circuit breaker will safely interrupt?
   a. CIA
   b. AIC
   c. SOL
   d. ICA

2. A plug fuse is typically used in a _____.
   a. branch circuit
   b. feeder circuit
   c. motor circuit
   d. high-voltage circuit

3. A 600V fuse can be used in _____.
   a. a 240V circuit
   b. a 600V circuit only
   c. any circuit greater than 600V
   d. DC applications only

4. A type of fuse with a grooved ring in one ferrule is a Type _____.
   a. C
   b. H
   c. K
   d. R

5. Each of the following is a basic factor to consider when applying any fuse, *except* _____.
   a. voltage
   b. continuous current-carrying capacity
   c. interrupting rating
   d. manufacturer's brand name

6. For proper coordination, the total clearing time of any downstream protective device must be below a curve representing what percentage of the minimum melting curve of the fuse?
   a. 10%
   b. 20%
   c. 50%
   d. 75%

7. The *minimum* interrupting rating of branch circuit listed fuses is _____.
   a. 5,000A
   b. 10,000A
   c. 15,000A
   d. 20,000A

8. The *minimum* voltage rating of branch circuit fuses is _____.
   a. 24V
   b. 120V
   c. 125V
   d. 240V

9. A current-limiting fuse can typically interrupt high fault currents within _____.
   a. three to four cycles
   b. two cycles
   c. one cycle
   d. the first half cycle

10. An expulsion fuse is *best* described as a _____.
    a. fuse capable of interrupting all currents from the rated interrupting current down to the current that causes melting of the fusible element in an hour
    b. fuse capable of interrupting all currents from the maximum rated interrupting current down to the minimum rated interrupting current
    c. strap-mounted device that vents the gases generated from the arc current from the body of the fuse
    d. vented fuse in which the expulsion effect of gases produced by the arc and lining of the fuse holder extinguishes the arc

11. The *maximum* fuse size for a single-element, nontime-delay fuse serving a 10 hp, three-phase motor is _____.
    a. 40A
    b. 60A
    c. 80A
    d. 100A

12. One characteristic of a dual-element, time-delay fuse is that it _____.
    a. requires the use of larger switches
    b. requires separate overload protection
    c. must be sized at 200% of the motor's full-load current
    d. provides backup protection if other protective devices fail to operate

13. When a common circuit breaker trips, the handle is found in the _____.
    a. Off position
    b. On position
    c. mid position
    d. Reset position

14. A 60A circuit breaker has a frame size of _____.
    a. 100A
    b. 80A
    c. 60A
    d. 40A

15. The amperage range of fuses used to identify groups of circuit breakers that are physically interchangeable with each other is referred to as the _____.
    a. frame size
    b. overload protection class
    c. maximum voltage class
    d. case size

16. A feeder circuit with no motor load has a continuous load of 15A and a noncontinuous load of 10A. The feeder overcurrent device should be sized at _____.
    a. 25A
    b. 27.5A
    c. 28.75A
    d. 31.25A

17. The full-load current rating of a 225 kVA, three-phase transformer operating at 240V is _____.
    a. 362A
    b. 402A
    c. 542A
    d. 722A

18. The full-load current rating of a 150 kVA, three-phase transformer operating at 240V is _____.
    a. 296A
    b. 324A
    c. 362A
    d. 397A

19. The let-through short circuit current of a 500 kVA, 120/208V three-phase transformer is _____.
    a. 10,300A
    b. 20,850A
    c. 69,444A
    d. 139,000A

20. A circuit breaker that trips instantly or fails to reset is *most likely* the result of _____.
    a. a short circuit
    b. corroded connections
    c. a high ambient temperature
    d. loose connections

Answers to odd-numbered Module Review Questions are found in *Appendix A*.

Module 26305-23 Supplemental Exercises

1. A short circuit current is one that flows outside the normal conducting paths and may be either line-to-line or _________.

2. The AIC rating of a circuit breaker or fuse is the _________.

3. If a fault current exceeds the interrupting rating, what can happen?

   ____

   ____

4. Name the factors that must be considered when calculating the required interrupting capacity of an overcurrent protective device.

   ____

   ____

5. The two most common voltage ratings for low-voltage power distribution fuses are _________.

6. True or False? The voltage rating of a fuse can be greater than the circuit voltage.

7. True or False? The rating that defines the capacity of a protective device to maintain its integrity when reacting to fault currents is called its voltage rating.

8. Why is the coordination of protective devices important?

   ____

   ____

9. Class H fuses have an interrupting rating of _________.

10. The basic factors that must be considered when applying any fuse include ________.

11. True or False? Current-limiting fuses produce arc voltages that exceed the system voltage.

12. Circuit breakers are grouped for identification according to given current ranges. Name them.

    ____

    ____

13. If a continuous load on a lighting branch circuit is 16A, the overcurrent device must be sized at an amperage of ________.

14. Service disconnecting means can consist of up to ________ switches or circuit breakers for each service or for each set of service-entrance conductors permitted in Exception No. 3 under *NEC Section 230.90(A)*.

15. In general, nontime-delay fuses can be sized at ________ of the motor full-load current.

Answers to odd-numbered Supplemental Exercises are found in *Appendix B*.

Answers to Section Review Questions

Answer	Section Reference	Objective
Section 1.0.0		
1. a	1.1.0	1a
2. d	1.2.0	1b
3. c	1.3.0	1c
4. a	1.4.0	1d
Section 2.0.0		
1. a	2.1.0	2a
2. c	2.2.1	2b
3. d	2.3.0	2c
Section 3.0.0		
1. a	3.1.0	3a
2. d	3.2.0	3b
Section 4.0.0		
1. c	4.1.2	4a
2. d	4.2.1; *Table 4*	4b
Section 5.0.0		
1. c	5.1.2	5a
2. b	5.2.0	5b

Section Review Calculations

Section 4.0.0

Question 1

For a feeder circuit with no motor load, the protective device size must be at least 125% of the continuous load plus 100% of the noncontinuous load:

Nonmotor feeder load = 125% of continuous load + 100% of noncontinuous load

Nonmotor feeder load = $(1.25 \times 10) + (1 \times 5)$

Nonmotor feeder load = $12.5 + 5 = 17.5$A

The feeder overcurrent protective device should be sized at **17.5A**.

Source: iStock@Croc80

Distribution Equipment

NOTE

NFPA 70®, *National Electrical Code®*, and *NEC®* are registered trademarks of the National Fire Protection Association, Quincy, MA.

Objectives

Successful completion of this module prepares you to do the following:

1. Identify electrical distribution system components.
   a. Identify switchboard components and their installation requirements.
   b. Identify switchgear components and their installation requirements.
   c. Describe the operation and application of medium-voltage limiting switches.
   d. Identify and describe the application of bolted pressure switches.
   e. Identify various transformers.
   f. Identify panelboard arrangements.
2. Describe the installation requirements for distribution equipment.
   a. State the *NEC®* requirements for distribution equipment installation.
   b. Interpret electrical diagrams related to the installation of distribution equipment.
3. Explain how to test and maintain switchgear.
   a. Identify general maintenance guidelines for switchboards and switchgear.
   b. State the guidelines for testing switchboards and switchgear.
   c. Identify and describe devices used to monitor power distribution systems for ground faults.

Performance Tasks

This is a knowledge-based module. There are no Performance Tasks.

Overview

An electrical power system consists of several subsystems on both the utility (supply) side and the customer (user) side. This module discusses switchboards and switchgear, including installation, grounding, and maintenance requirements. It also covers ground fault relay testing.

NCCER Industry-Recognized Credentials

If you are training through an NCCER-accredited sponsor, you may be eligible for credentials from NCCER. The ID number for this module is 26306-23. Note that this module may have been used in other NCCER curricula and may apply to other level completions. Contact NCCER at 1.888.622.3720 or go to **www.nccer.org** for more information.

You can also show off your industry-recognized credentials online with NCCER's digital badges. Transform your knowledge, skills, and achievements into badges that you can share across social media platforms, send to your network, and add to your resume. For more information, visit **www.nccer.org**.

Digital Resources for Electrical

Scan this code using the camera on your phone or mobile device to view the digital resources related to this craft.

1.0.0 Electrical Distribution System Components

Performance Tasks

There are no Performance Tasks in this section.

Objective

Identify electrical distribution system components.

a. Identify switchboard components and their installation requirements.
b. Identify switchgear components and their installation requirements.
c. Describe the operation and application of medium-voltage limiting switches.
d. Identify and describe the application of bolted pressure switches.
e. Identify various transformers.
f. Identify panelboard arrangements.

An electrical power system consists of several subsystems on both the utility side and the consumer side. Electricity generated in power plants is stepped up to transmission voltage and fed into a nationwide grid of transmission lines. The power is then bought, sold, and dispatched as needed.

Local utility companies take power from the grid and reduce the voltage to levels suitable for distribution to the consumer through substations. This may range from the common 200A, 120/240V residential service to voltages as high as 69 kV for an industrial facility that has its own substation.

From the point of service, consumers must control, distribute, and manage the power to supply their electrical needs. This module will discuss how this is done using a typical industrial facility as an example. We will discuss the various components of the distribution system and their relationship. An understanding of single-line diagrams enables the quick analysis of a facility's distribution system.

NOTE

The voltage conventions used in this module are industry standards for distribution systems.

While electrical systems and equipment are often classified by voltage rating, **switchgear** is classified first by the type of construction and then by voltage rating. It is important to note that there is no official industry-wide voltage classification system. The *NEC®* considers *low voltage* as any voltage up to 1,000V, while *medium voltage* refers to systems rated above 1,000V and up to 35,000V. This is the range in which **metal-clad switchgear** and circuit breakers are manufactured in standard configurations. This is also the voltage range in which premolded and shrink-on termination kits are readily available for shielded cable terminations. Above this voltage level, cable is usually run on overhead power lines rather than in raceway or cable tray.

Low-voltage power circuit breaker switchgear, for example, may be rated up to 1,000VAC or 3,200VDC. Metal-clad switchgear or **metal-enclosed switchgear** is applied at voltages over 1,000VAC and up to a maximum of 35,000VAC.

Switchgear: A general term describing a collection of switching and interrupting devices that can also include control, instrumentation, metering, protective, and regulating devices.

Metal-clad switchgear: A subset of metal-enclosed switchgear, with the internal components also enclosed in metal individually. Each component may also be individually insulated and grounded.

Metal-enclosed switchgear: Switchgear that is enclosed on all sides, except for small ventilation and/or viewport openings that are commonly required.

1.1.0 Switchboards

According to the *NEC®*, the term **switchboard** may be defined as a large single panel, frame, or assembly of panels on which switches, protective devices, buses, and instruments can be mounted. Devices can be mounted on the face, on the back, or on both sides. Switchboards are generally accessible from both the rear and front and are not intended to be installed in cabinets.

Switchboard: A large single panel, frame, or assembly of panels on which switches, fuses, buses, and instruments are mounted.

1.1.1 Applications

Switchboards are used in modern distribution systems to subdivide large blocks of electrical power. One common location for switchboards is where the main power enters a building. In this location, the switchboard is referred to as **service-entrance equipment**. Another location common for switchboards is downstream from the service-entrance equipment. In the downstream location, the switchboard is commonly referred to as **distribution system equipment**.

Service-entrance equipment: Equipment located at the service entrance of a structure that provides overcurrent protection to the feeder and service conductors, as well as a means of disconnecting the feeders from the energized service equipment.

Distribution system equipment: Switchboard equipment downstream from the service-entrance equipment.

1.1.2 Switchboard Components

A switchboard is a stationary structure with one or more freestanding units that are mechanically and electrically joined. They take up less space in a plant, are more visually appealing, and eliminate the need for a separate room to protect workers from lethal voltages.

The main portion of the switchboard is formed from heavy-gauge steel with welded members across the top and bottom to provide a rigid enclosure. Most switchboard enclosures are divided into three sections: the front section, the bus section, and the cable section. The three sections are physically separated from one another by metal partitions. This confines any damage that may occur to any one section and keeps it from affecting the other sections.

Typical switchboard components include the following:

- Circuit breakers
- Fuses
- Busbars
- Motor starters
- Ground fault systems
- Instrument transformers
- Control power transformers
- Switchboard metering

Circuit breakers are the only circuit interrupting devices that combine a fault current interruption rating and the ability to be manually or automatically opened or closed. The four general classifications of circuit breakers you will likely encounter are as follows:

- Air circuit breakers
- Oil circuit breakers
- Vacuum circuit breakers
- Gas circuit breakers

The circuit breakers are named based on the medium that surrounds the contacts. Air circuit breakers should now be familiar to you, as they have been presented and discussed at length in previous modules. The other three are found in various utility and industrial applications.

Circuit breakers may conveniently be divided into low-voltage, medium-voltage, and high-voltage classes. Although there is considerable overlap among these classes, each one has certain characteristics.

Circuit breaker ratings are provided on the breaker nameplate. The information from the nameplate must be considered for any replacement breaker selection. The same rating information should be included in the documentation for breakers. The rating information includes the following:

- *Voltage rating* — This is the maximum voltage for which the circuit breaker is designed.
- *Current rating* — This is the continuous current that the circuit breaker can carry without exceeding a standard temperature rise (usually 131°F or 55°C).
- *Interrupting rating* — This is the maximum value of current that a circuit breaker is required to interrupt a limited number of times when operating within its specifications. The term is usually applied to abnormal or emergency conditions.

Electrical ratings include three-phase, three-wire and three-phase, four-wire systems with voltage ratings up to 600V and current ratings up to 6,000A and above.

A switchboard enclosure is described as a dead-front panel, which means that no live parts are exposed on the opening side of the equipment. However, it contains energized breakers. Busbars can be a standard size or customized. Standard sizes are usually made of silver-plated or tin-plated copper or tin-plated aluminum. Conventional bus sizing in the United States is ¼" × 2" through ⅜" × 7". Copper provides an ampacity of 1,000A/in^2 of cross-sectional area. When using aluminum, the ampacity is 750A/in^2.

When two busbars are bolted together using Grade S hardware with the proper torque, the ampacity of the connection is 200A/in^2 of the lapped area for aluminum or copper busing. Busing joints must be bolted together to the specified torque and include Belleville washers or Keps nuts. Aluminum busbars must be tin-plated, and copper busbars over 600A must be plated with tin or silver.

1.1.3 Switchboard Frame Heating

NOTE

Some switchboard frames are engineered differently and will have values other than those shown in *Table 1*.

Table 1 shows guidelines that should be observed to keep the heat losses of iron switchboard frame members to a safe minimum. The dimensions are recommended values and should be adhered to whenever possible.

TABLE 1 Switchboard Frame Heating Guidelines

Amperes	Minimum Distance from Phase Bus to Closest Steel Member	Minimum Distance from Neutral Bus to Closest Steel Member
3,000A	4" (100 mm)	2" (50 mm)
4,000A	6" (150 mm)	3" (75 mm)
5,000A to 6,000A	12" (300 mm)	An aluminum or nonmagnetic material should be used in place of steel frame sections. Where possible, maintain 12" (300 mm) of clearance to steel members and 6" (150 mm) of clearance to aluminum or nonmagnetic members. The neutral spacing can be 6" and 3" (150 mm and 75 mm), respectively. If the main bus is tapered, it is acceptable to use steel frames for the sections containing the tapered bus if working with 4,000A or less.
Over 6,000A	12" (300 mm)	Use an aluminum or nonmagnetic material for frame sections and maintain 12" (300 mm) of clearance to steel members and 6" (150 mm) of clearance to aluminum or nonmagnetic members. Neutral spacing can be 6" and 3" (150 mm and 75 mm), respectively. The use of steel frame members is discouraged. If the main bus is tapered, it is acceptable to use steel frames for the sections containing the tapered bus if working with 4,000A or less.

Note: For currents above 8,000A, the neutral spacing must be 12" (300 mm) wherever possible.

1.1.4 Low-Voltage Spacing Requirements

To minimize tracking or arcing from energized parts to ground, switchboard construction includes spacing requirements. These spacing requirements are measured between live parts of opposite polarity and between live parts and grounded metal parts. *Figure 1* illustrates typical switchboard spacing requirements.

Case History

Oil-Filled Circuit Breaker Testing

An assistant engineer and an electrician had the responsibility of testing the insulation values of the bushings on an oil-filled circuit breaker. The assistant engineer pulled down the handle and opened the breaker, visually determining that the contacts were open and safe. He instructed the electrician to move the test leads from one bushing to the next as the assistant engineer read the values.

On the third and last bushing, the electrician was electrocuted as he placed the test leads on the bushing. The last contact had not opened due to a broken porcelain insulator skirt, which allowed the cap to remain stationary while the rest of the mechanism rotated as the handle was pulled down. This was not noticeable from the assistant engineer's eye level when he determined that all the contacts were open.

The Bottom Line: Never come in contact with switchgear such as this without first testing each and every contact point with the proper meter, and never rely on someone else's word that the power is off. Also remember that, when working with high voltages, contact is not necessary for an injury or electrocution to occur. High voltages can cross a gap, and the higher the voltage, the larger the gap that can be crossed.

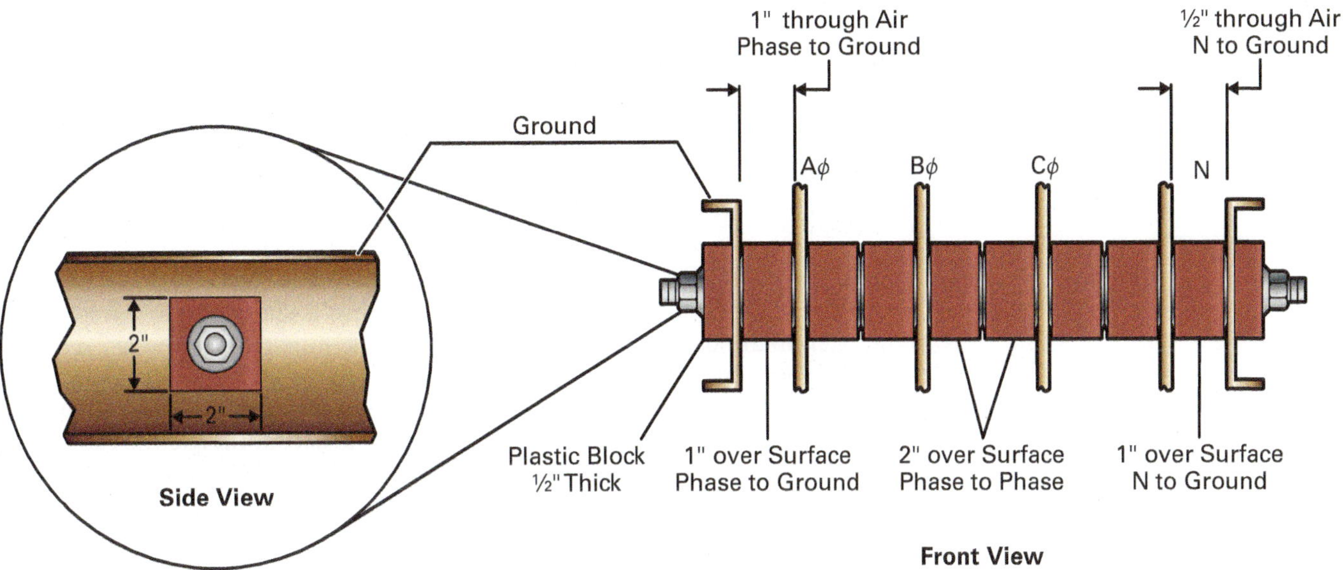

Figure 1 Typical busbar spacing requirements.

Busbars

Busbars have very specific spacing requirements. Note the red spacer blocks between the buses on the switchgear shown in the image.

An isolated metal part, such as a screwhead or washer, situated between uninsulated live parts of opposing polarities or between an uninsulated live part and grounded metal, is considered a reduction in the spacing by the distance equal to its dimension along the measured path.

When measuring over-surface spacing, any slot, groove, or similar characteristic that is 0.013" (0.33 mm) wide or less and in the contour of the insulating material can be ignored.

When measuring spacing, an air space of 0.013" (0.33 mm) or less between a live part and an insulating surface can be disregarded, and the live part must be considered in contact with the insulating material. A wire termination or connector must be prevented from rotating, which might result in less than the minimum acceptable spacings. The means used to ensure terminations do not rotate around their fastener must be reliable, such as a shoulder or boss. A lock washer, for example, cannot physically prevent the connector from rotating.

A means of preventing connector rotation is not needed if air spacings meet the following minimum accepted values:

- When the connector and any connector of opposite polarity have each been turned 30° toward the other.
- When the connector has been turned 30° toward other live parts of opposite polarity and toward grounded dead metal parts.

1.1.5 Cable Bracing

All construction using conductors and having a short circuit current rating greater than 50,000 root-mean-square (rms) symmetrical amperes requires a cable brace positioned as close to the supply lugs as possible. The cable brace is intended to be mounted in the same area allotted for wire bending. It is not necessary to provide additional mounting height to accommodate the cable brace.

The cable brace requirement does not apply to load-side cables, main breakers, or switches. It only applies when cables are connected directly to an unprotected line-side bus. The bus restrictions for a line-side bus are as follows:

- There can be no splice in edgewise bus mounting of 2,100A or less rated at 50,000 rms symmetrical amperes.
- There can be no splice in flatwise bus mounting of 600A or less rated over 50,000 rms symmetrical amperes.

NOTE

This does not apply to connections made from the through bus to a switch or circuit breaker.

The cable restrictions for a line-side bus include the following:

- Busing of 600A or less rated over 50,000 rms symmetrical amperes cannot use cables; it must be bus connected.
- If cabling is required, 800A minimum busing must be used.

Cable bracing requirements may be excluded if the bus is able to fully withstand the total available short circuit current.

1.2.0 Switchgear

Switchgear is a general term used to describe switching and interrupting devices and assemblies of those devices containing control, metering, protective, and regulatory equipment, along with the associated interconnections and supporting structures. Switchgear performs the following two basic functions:

- Provides a means of switching or disconnecting power system apparatus
- Provides power system protection by automatically isolating faulty components

Switchgear can be classified as follows:

- Metal-enclosed switchgear (low voltage)
- Metal-clad switchgear (low and medium voltage)
- Metal-enclosed interrupters
- Unit substations

The low-voltage and medium-voltage switchgear assemblies are completely enclosed on all sides and topped with sheet metal, except for ventilating openings and inspection windows. They contain primary power circuit switching or interrupting devices, buses, connections, and control and auxiliary devices. *Figure 2* shows typical low-voltage, metal-clad switchgear.

The station-type switchgear assemblies, commonly referred to as *cubicles*, consist of indoor and outdoor types with power circuit breakers rated from 14.4 kV to 34.5 kV; 1,200A to 5,000A; and 1,500 kVA with 2,500 kVA interrupting capacity. Equipment can be custom built to accommodate higher kVA ratings.

Figure 2 Typical low-voltage, metal-clad switchgear.

1.2.1 Switchgear Construction

Switchgear consists of a stationary structure that includes one or more freestanding units of uniform height that are mechanically and electrically joined to make a single coordinated installation. Switchgear enclosures are formed from heavy-gauge sheet steel that has been welded or bolted together. Structural members across the top, sides, and bottom provide a rigid structure. Metal-clad switchgear enclosures are divided into three sections: the front section, the bus section, and the cable or termination section.

These three sections are physically separated from one another by metal partitions. This confines any damage that may occur in any one section. It also separates power between the sections for ease of maintenance and safety.

The rigid enclosure provides structural strength and a means to fasten it to a foundation. The strength of the enclosure and its mounting system will vary depending on its intended use. For example, switchgear used in a nuclear application must meet certain seismic qualifications.

The enclosure also provides the required supports and mounts for mounting items and provides for the necessary interconnections between the switchgear and other plant systems. The number of sections and other physical characteristics varies depending on the voltage and current ratings, project specifications, and specific manufacturer.

Figure 3 and *Figure 4* show exterior and interior views, respectively, of medium-voltage, metal-clad switchgear. This equipment is available in voltages from 4.76 kV to 27 kV and current ranges from 1,200A through 3,000A.

1.2.2 Control and Metering Safety Standards

There is a tendency among some people in the industry to use the terms *switchboard* and *switchgear* interchangeably. However, they are not the same. Switchgear is manufactured and tested to more exacting standards and is configured differently than switchboards. For example, in switchgear there are physical barriers between breakers, and between the breakers and the bus. Switchgear is more durable and fault-resistant, and it is commonly selected for larger applications where low-voltage power circuit breakers and selective coordination are applied, such as computer data centers, manufacturing, and process facilities.

Figure 3 Medium-voltage, metal-clad switchgear (exterior view).
Source: Schneider Electric

Figure 4 Medium-voltage, metal-clad switchgear (interior view).
Source: Schneider Electric

1.2.3 Wiring System

The *NEC*® requires wiring to be supported mechanically to keep the wiring in place. Wire harnessing is generally used within the switchboard with the following restrictions:

- Each bundle or cable of wires must be routed vertically or horizontally, securing the harness by means of plastic cable ties or cable clips.
- Plastic wire cable clamps must be located along the harnesses to hold them in place, preventing interference with the control components' required electrical, mechanical, and arcing clearances.
- Wire ties must be used on the harnesses every 3" to 4" (75 mm to 100 mm). Self-adhesive cable ties are spaced at 12" (300 mm).

Some precautions to be observed when wiring the switchboard electrical components are as follows:

- Keep control wires at least ½" (13 mm) from moving parts.
- Avoid running wires across sharp metal edges. To protect the wiring from mechanical damage, use approved cable protectors, such as a nylon clip cable guard, edge-protection wire guards, or edge-protection molding.
- Wires must never touch exposed bare electrical parts of opposite polarity.
- Wires must never interfere with access to the components for adjustment or replacement.
- Keep the wires as straight and short as possible.
- Wires cannot be spliced inside unless they are in a listed wireway.
- To eliminate possible strain on the control wire, a certain amount of slack should be provided in individual or harnessed conductors where they terminate.
- The equipment ground busbar cannot be used as any portion of the control or metering circuits.
- Do not use pliers for bending control wiring. Use your hands or an approved wire bending tool.

No incoming wiring connections may be made directly to door-mounted devices. Wires from door-mounted equipment to the panel terminal block should be a minimum of 19-strand wire.

Wires from the door must be neatly cabled so that the door can be opened easily without straining the wire terminations. In some cases, the cable must be separated into two bundles to accomplish this. Insulated sleeving, tubing, or vinyl tape must be used to bundle and protect the flexible wires.

All control or metering wiring entering or leaving the switchboard should terminate at terminal blocks. One side of the terminal block is left open for the user's connections. No factory connections are allowed on the user's terminals. For factory wiring, allow a maximum of two control wires on the same side of a terminal block. No more than three connections are allowed on the individual terminals of control transformers, meters, meter selector switches, and metering equipment.

Bolted pressure switches or any 100% current-rated, molded-case circuit breaker's line and load power terminals are allowed to have a higher maximum operating temperature than the recommended insulated conductor's temperature rating. Therefore, the control wires cannot be placed directly on the 100% rated disconnect device's line and load connections.

In all cases, control wires cannot touch any exposed part of opposite electrical polarity.

1.2.4 Metering Current and Potential Transformers

Ground connections on a potential transformer (PT) or current transformer (CT) secondary terminal must be connected to the ground bus. CT secondary terminals must be shorted (connected to each other) if no metering equipment is connected to the transformer.

PTs are required to have primary and secondary fusing. If protective circuits, such as ground fault or phase failure protective systems, are placed in the secondary circuit of the PT, no secondary fusing is required.

Metering circuit connections made directly to the incoming bus must be provided with current-limiting fuses that are equal in rating to the available interrupting capacity.

NOTE

CTs and PTs will be discussed in more detail later in this module.

1.2.5 Switchgear Handling, Storage, and Installation

It is important to note that recommendations for the handling of switchgear only supplement the manufacturer's instructions. Manufacturers include instruction books and drawings with their equipment, and it is imperative that you read and understand these documents before handling any equipment. The following are basic guidelines for the handling of switchgear:

- *Switchgear handling* — Immediately upon receipt of switchgear, an inspection for damage during transit should be performed. If any damage is noted, the transportation company should be notified immediately.
- *Switchgear rigging* — Instructions for switchgear should be found in the manufacturer's instruction books and drawings. Verify that the rigging is suitable for the size and weight of the equipment.
- *Switchgear storage* — Indoor switchgear that is not being installed right away should be stored in a clean, dry location. The equipment should be level and protected from the environment if construction is proceeding. The longer equipment is in storage, the more care is required for protection of the equipment. If a temporary cover is used to protect the equipment, this cover should not prevent air circulation. If the building is not heated or temperature controlled, heaters should be used to prevent moisture/condensation buildup. Outdoor switchgear that cannot be installed immediately must be provided with temporary power. This power will allow operation of the space heaters provided with the equipment.
- *Bus connections* — The main bus that is usually removed during shipping should be reconnected. Ensure the contact surfaces are clean and pressure is applied in the correct manner. The conductivity of the joints is dependent on the applied pressure at the contact points. The manufacturer's torque instructions should be referenced.
- *Cable connections* — When making cable connections, verify the phasing of each cable. This procedure is done in accordance with the connection diagrams and the cable tags. When forming and mounting cables, ensure the cables are tightened per the manufacturer's instructions.
- *Grounding* — Any sections of ground bus that were previously disconnected for shipping should be reconnected when the units are installed. In addition, the system must be bonded at this time. The ground bus should be connected to the system ground with as direct a connection as possible. If the system ground is to be run in metal conduit, bonding to the conduit is required. The ground connection is necessary for all switchgear and should be sized per the *NEC®*.

1.3.0 Medium-Voltage Limiting (MVL) and High-Voltage Limiting (HVL) Switches

Figure 5 shows the general appearance of a medium-voltage limiting (MVL) switch. An MVL switch is a switching device for circuits up to the full interrupting current of the switch. The switches are single-throw devices designed for use on 1,001V to 35 kV systems. High-voltage limiting (HVL) switches are for voltages above 35 kV.

HVL and MVL switches can provide both switching and overcurrent protection. They are commonly used as a service disconnect in unit substations and for sectionalizing high-voltage and medium-voltage **feeder** systems. They are designed to conform to ANSI standards for metal-enclosed switchgear.

Feeder: A set of conductors originating at a main distribution center that supply one or more secondary distribution centers, one or more branch circuit distribution centers, or any combination of the two.

1.3.1 Ratings

Switch ratings identify important characteristics, criteria, and limitations for how they can be used. Ratings for switches include the following:

- *Switch (kV)* — The design voltage for the switch. Of course, nominal system voltage is the normal application method. Thus, a 5 kV switch may be used for nominal system voltages of 2.4 kV or 4.16 kV, etc.
- **Basic impulse level (BIL)** — The maximum voltage pulse expressed in kV that the equipment will withstand.
- *Frequency (Hz)* — All HVL and MVL switches may be used in either 50 Hz or 60 Hz power systems.
- *Withstand (kV)* — The maximum 60 Hz voltage that can be applied to the switch for one minute without causing insulation failure.
- *Capacitor switching (kVAR)* — The maximum capacitance expressed in kVAR that can be switched .
- *Fault close current (A)* — The maximum, fully offset fault current that the switch can be closed into without sustaining damage. The term *fully offset* means that the fault current will have a delaying DC component in addition to the AC component.
- *Short time current (A)* — The amount of current that the switch will carry for 10 seconds without sustaining any damage.
- *Continuous current (A)* — The amount of current that the switch will carry continuously.
- *Interrupting current (A)* — The maximum amount of current that the switch will safely interrupt.

Basic impulse level (BIL): A means of expressing the maximum impulse voltage surge that electrical insulation can withstand without damage; also referred to as the *basic insulation level*.

Figure 5 Medium-voltage limiting (MVL) switch.

1.3.2 Variations

There are a variety of switches with different characteristics, making each type suitable for unique applications. These varieties come in the following six main types:

- *Upright* — The upright switch design is the most common type. The upright construction locates the service entry, jaws, and arc chutes near the top of the cubicle. The hinge point is below the jaws and arc chutes.
- *Inverted* — The inverted switch design has the terminals, jaws, and arc chutes located near the bottom of the cubicle. The hinge point is above the jaws and arc chutes. This type of switch is used primarily as a main switch to a lineup of other switches. Its handle operation is identical to that of an upright switch; to close the switch, the handle is moved up, and to open it, the handle is moved down.

- *Fused/unfused* — MVL and HVL switches are available in both fused and unfused models. If equipped with fuses, the entire switch has the fault interrupting capacity of the fuse and therefore provides fault protection. Either current-limiting or boric acid fuses may be used in MVL and HVL switches.
- *Duplex* — A duplex switch is two switches, each in its own bay. The bays are mechanically connected, and the switches are electrically connected on the load side. This switch may be used to supply power to a single load from two different sources.
- *Selector* — A selector allows an HVL or MVL switch to have double-throw characteristics. The selector switch is a single switch with a load connected to the moving or switch mechanism. Throwing the switch to one side connects the load to one source, while throwing it the other way connects it to another source. The selector switch is interlocked with another switch to prevent the selector switch from interrupting current flow. The selector serves a purpose similar to the duplex switch. However, the selector switch is not an interrupter; it is a means of disconnect.
- *Motor operated* — This type of switch is most often used as the major component in an automatic transfer scheme. It can also be used when open and close functions are to be initiated from remote locations.

1.3.3 Opening Operation

In the Closed position, the main switch blade is engaged on the stationary interrupting contacts. The circuit current flows through the main blades.

As the switch operating handle is moved toward the Open position, the stored energy springs are charged. After the springs become fully charged, they toggle over the dead-center position, discharging force to the switch operating mechanism.

The action of the switch operating mechanism forces the movable main blade off the stationary main contacts while the interrupting contacts are held closed, momentarily carrying all the current without arcing. Once the main contacts have separated well beyond the striking distance, the interrupting blade contact that was held captive has charged the interrupter blade hub spring, and the interrupter blade is suddenly forced free and flips open.

The resulting arc drawn between the stationary and movable interrupting contacts is elongated and cooled as the plastic arc chute absorbs heat and generates an arc-extinguishing gas to break up and blow out the arc. The combination of arc stretching, arc cooling, and extinguishing gas causes a quick interruption with only minor erosion of the contacts and arc chutes. The movable main and interrupting contacts continue to the fully Open position and are maintained there by spring pressure.

1.3.4 Closing

When the switch operating handle is moved toward the Closed position, the stored energy springs are being charged and the main blades begin to move. As the main and interrupter blades approach the arc chute, the stored energy springs become fully charged and toggle over the dead-center position.

When the main and movable blades approach the main stationary contacts, a high-voltage arc leaps across the diminishing air gap to complete the circuit before the contacts meet. The arc occurs between the tip of the stationary main contacts and a remote corner of the movable main blades. This arc is short and brief because the fast-closing blades minimize the arcing time.

The spring pressure and momentum of the fast-moving main blades completely close the contacts. The force is great enough to cause the contacts to close even against repelling short circuit magnetic forces if a fault exists. At the same time, the interrupter blade tip is driven through the twin stationary interrupting contacts, definitely latching and preparing them for an interrupting operation when the switch is opened.

WARNING!

Maintenance and testing may only be performed by qualified personnel under the appropriate safe work plan or permit.

No work inside of breakers and switches beyond termination to the conductor should ever be performed by the electrical contractor or electrician without specific oversight and direction from the equipment manufacturer. Ideally, the equipment manufacturer should be hired to perform the work, which may include normal maintenance of the equipment.

1.3.5 Maintenance

Maintenance tasks for an HVL or MVL switch include the following common steps:

Step 1 The switch should be operated several times. Observe the mechanism and check for binding.

Step 2 Inspect the interrupting and main blades every 100 operations for excessive wear or damage; replace as necessary. Also, inspect the arc chutes for damage.

Step 3 Clean the switch and its compartment thoroughly. Use a clean cloth and avoid solvents.

Step 4 Lubricate the switch. The pivot points on the switch should be greased. The switch contacts should also be lubricated with a light film of grease after being cleaned.

Step 5 Final maintenance checks include phase-to-ground and phase-to-phase megger testing. If the results are satisfactory, then a DC high-potential test is performed.

1.3.6 Sluggish Operation

A switch that is operating sluggishly hesitates on the opening cycle. This contrasts with the normal snapping action. Observing the interrupter blade during the opening operation is the proper way to determine sluggish operation. Sluggishness must be resolved to prevent the switch from locking up completely. Perform the following procedure:

Step 1 Close and then open the switch while watching the interrupter blades closely. Sluggishness on close is shown by the main blade's being engaged behind the contacts of the arc chute. On opening, the interrupter blades may hesitate momentarily.

Step 2 Disconnect the links from the operating shaft. Never operate the switch with the links off as this may break the handle crank casting. This is because the main spring energy is absorbed by the handle crank rather than the main blades.

Step 3 Rotate the handle approximately 45° and hold it in this position while trying to operate the switch by hand. Excessive binding will prevent rotation of the shaft.

Step 4 Check the contact adjustment at the jaw and hinge.

Step 5 Check for binding between the interrupter blade and the arc chute.

Step 6 Remove the front panel over the operating mechanism and disconnect the spring yoke from the cam. Check for binding between the spring pivot and the sides of the operator. Ensure the spring is intact and secure.

1.4.0 Bolted Pressure Switches

Bolted pressure switches (*Figure 6*) are used frequently on service-entrance disconnects and feeders in switchgear, such as that shown in *Figure 7*. They are often used instead of circuit breakers because they are inexpensive and can handle a higher available fault current. Bolted pressure switches can be manually operated, shunt-trip operated, or motor operated. However, unlike a circuit breaker, they are tripped by an external signal such as a ground fault relay, phase failure relay, or a blown fuse detector.

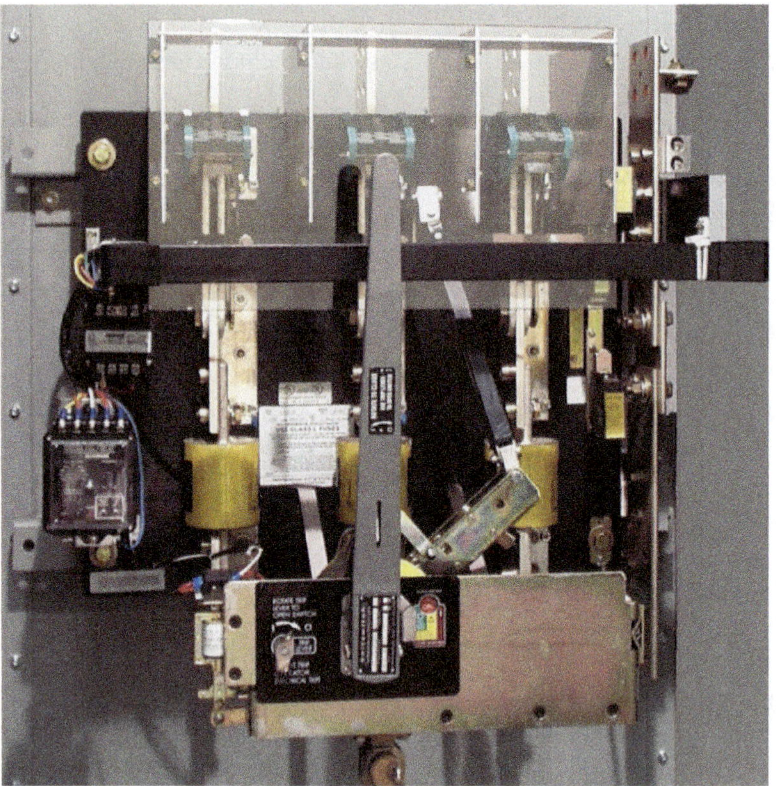

Figure 6 Bolted pressure switch.
Source: Boltswitch

Figure 7 Switchgear.

1.4.1 Ground Fault

Ground faults exist when an unintended current path is established between an ungrounded conductor and ground. These faults occur due to deteriorated insulation, moisture, dirt, rodents, foreign objects such as tools, and careless installation.

Under normal conditions, the currents in all conductors surrounded by the ground fault CT equal zero. When a ground fault occurs, this sensed current increases, eventually reaching the ground fault relay pickup point and causing the bolted pressure switch to trip.

The ground fault system may also be tested. By pressing the Test button, a green test light will illuminate, indicating correct circuit operation. To physically test the switch, press the Test and Reset buttons simultaneously. This sends a trip signal through the current sensor, thus tripping the switch. Whenever a bolted pressure switch trips, a red light or a red flag will appear. The ground fault relay must be reset before the switch can be reclosed.

1.4.2 Phase Failure

If a phase failure relay is installed, it will trip a bolted pressure switch when a phase is lost. This could occur if a falling tree brings a line down, for example. The phase failure relay will sense the lost phase and trip the bolted pressure switch, preventing damage to three-phase loads and overloads on the remaining two phases.

1.4.3 Blown Main Fuse Detector

If one of the in-line main fuses were to blow, the blown main fuse detector would detect it and trip the bolted pressure switch. The trip signal generated comes from a capacitor trip unit. This ensures that power is always available to trip the switch.

1.4.4 Maintenance

These switches have an excessive failure rate due to a lack of maintenance. All manufacturers of bolted pressure switches recommend annual maintenance. A lack of maintenance will eventually result in a switch seizing in the Closed position. Because these switches are often used as service-entrance equipment, a stuck switch poses an immediate safety hazard.

WARNING!

When performing any maintenance, always follow the safety procedures of your company.

Typical annual maintenance includes the following:

Step 1 De-energize the switch, lock and tag it, and perform a preliminary operational check.

Step 2 Record premaintenance digital low-resistance ohmmeter (DLRO) readings. A DLRO is most often referred to as a *megohmmeter.*

Step 3 With the switch open, disassemble the crossbar to free all three phases.

Step 4 Clean off all old grease with denatured alcohol or a similar solvent.

Step 5 Inspect the arc tips and arc chutes for damage.

Step 6 Adjust all pivotal connections on each blade to within the manufacturer's recommended tolerances.

Step 7 Apply a nonconductive grease to the movable blades and the area where the blades initially contact the stationary assembly.

Step 8 Check the pullout torque on each individual blade prior to crossbar reassembly. It should be in accordance with the manufacturer's prescribed limits. Too much torque will result in a switch that will be unable to open under load.

Step 9 Record the megohmmeter readings.

Step 10 Reassemble the crossbar assembly.

Step 11 Close and open the switch manually several times. Ensure that no phases hang up on the arc chute assembly.

Step 12 Meg the switch.

Step 13 Energize and test all accessories, such as the ground fault detector, phase failure detector, and blown main fuse detector.

Remember, if the switch is physically stuck closed, de-energize the switch from the incoming power supply, and take extra precautions when trying to unstick the switch. It may be necessary to pry the blades open but beware of the excessive outward force that will result from a charged opening spring.

Due to the high interrupting capacity of the switch when operated under load, the grease that is used on the movable blades deteriorates over time and eventually turns into an adhesive. Even when the switch is not operated on a recurring basis, the grease still deteriorates due to the high temperatures associated with the current drawn by the phase. The deterioration of this grease has been shown to cause the switch to stick closed. The grease must be cleaned off yearly with denatured alcohol and replaced.

CAUTION

Regular electrical grease cannot be used; use only the grease specifically recommended by the switch manufacturer.

Additionally, infrared scanning of in-service bolted pressure switches has revealed a marked heating concern in switches. A DLRO is used to ensure all three phases carry similar current loads. DLRO readings should never be greater than 75 microohms (μΩ), and there should not be more than a 5% difference between the phases.

Protective Grounding

Remember that even after a circuit has been isolated, de-energized, locked out, and verified without voltage, it still may not be safe to work on. This is because of the possibility that a circuit or conductor may be inadvertently reenergized for any one of the following reasons:

- Induced voltages from other energized conductors
- Static buildup from wind on outdoor conductors
- High voltage from lightning strikes
- Any condition that might bring an energized conductor into contact with the de-energized circuit
- Switching errors causing reenergizing of the circuit
- Capacitive charges in equipment or conductors

When any of these conditions are possible, NFPA 70E®, *Standard for Electrical Safety in the Workplace®*, and NFPA 70B, *Standard for Electrical Equipment Maintenance*, both require that temporary grounds be applied before the circuit or equipment is considered safe. In fact, standard practice in overhead line construction and within open substations is that any conductor without a temporary ground connection is considered energized. While the terms temporary ground, safety ground, and protective ground are often used interchangeably, temporary grounds cover both personal protective grounds and static grounds.

Personal protective grounds consist of cable connected to de-energized lines and equipment by jumpering and bonding with appropriate clamps, to limit the voltage difference between accessible points at a worksite to safe values if the lines or equipment are accidentally reenergized. Protective grounds are sized to carry the maximum available fault current at the worksite for the expected fault duration.

Static grounds include any grounding cable or bonding jumper (including clamps) that has an ampacity less than the maximum available fault current at the worksite or is smaller than No. 2 AWG copper equivalent. Static grounds are used for potential equalizing between conductive parts in grounding configurations that cannot subject them to significant current. Therefore, smaller wire that provides adequate mechanical strength is sufficient (for example, No. 12 AWG).

Low-voltage equipment with only a single source of supply usually does not require temporary grounding for safety. Low-voltage equipment with dual supply and medium-voltage equipment should be grounded at the bus.

High-Resistance Grounding

High-resistance grounding (HRG) is increasingly applied in both medium-voltage and low-voltage distribution systems to limit ground fault energy. Medium-voltage systems have long used low-impedance grounding systems to limit ground fault current. Limiting ground fault current to values of 25A or less increased system reliability by allowing ground faults to be detected and selectively cleared.

Protective Grounding Arrangements

ASTM F855-04, *Standard Specifications for Temporary Protective Grounds to Be Used on De-energized Electric Power Lines and Equipment,* is the national consensus standard covering the equipment making up the temporary grounding system. This standard addresses the parts of a temporary grounding system, which include the clamps, ferrules, cables, or a complete protective ground assembly of clamps, ferrules, and cables. These components work together and must be capable of conducting the maximum available fault current that could occur at a work location if lines or equipment become reenergized from any source, and for the expected duration of the fault. Because the circuit is not safe until grounding is applied, placing and removing temporary grounds is considered work on live parts, and appropriate PPE and safe work practices must be followed.

The image shows a temporary protective ground cluster on incoming medium-voltage feeders at equipment. Not shown is the connection to the permanent system and feeder grounding conductors. This arrangement will provide safety for the connected equipment bus. Notice the phase arrangement is from left to right at the front of the equipment (the back of the equipment is shown in the image).

Source: Jim Mitchem

Bushing: A hollow electrical insulator, usually made from porcelain, that provides safe passage for a conductor through a conductive material, such as the case of a distribution-class transformer or circuit breaker, preventing electrical contact.

These systems used either a large resistor or inductance (transformer primary with a shorted secondary or a relay coil in the secondary) connected between the neutral **bushing** of the service transformer and the system bonding jumper. Detection of a ground fault at certain levels caused a protective relay to open the circuit and clear the fault. The resistors used were mounted in wire cages next to the supply transformers. The resistors were rated at maximum ground fault power for the duration of the fault before it cleared. If the resistors burned up due to failure to clear the fault, the system became an ungrounded wye and extremely dangerous.

In practice, low-resistance grounding for 480V or 600V systems was seldom applied due to the requirements for large resistors and space for enclosures and heat dissipation. The use of high-resistance grounding for low-voltage, high-current systems is becoming common. The high fault currents available on large 480V or 600V systems present a significant arc flash hazard while at the same time the arcing fault may be seen as only an overload by the protective device. High-resistance ground systems allow detection of ground faults and facilitate location of the faulted circuit.

The most common types of faults in power systems are as follows:

- Three phase (balanced)
- Phase to phase
- Phase to phase to ground
- Phase to ground

In industrial facilities it is estimated that 98% or more of all faults begin as a ground fault. If the arcing ground fault current is high enough, the fault develops into a phase-to-phase or three-phase fault. High-resistance grounding in 600V systems limits ground fault current to less than 10A; 5A is a common limit. This level of fault current is too low to present an arc flash hazard or to sustain an arc by itself. This makes development of phase-to-phase faults unlikely. The low fault current also allows continuity of service for a period with little risk of equipment damage. Good practice requires that a fault be located and cleared as quickly as possible. The current allowed in the ground fault must be greater than the capacitive coupled charging current in the system to avoid false alarms. In low-voltage systems, a sensed ground fault is usually indicated and alarmed but does not send a trip signal to controlling equipment. This is done to allow location of the fault using various features available in HRG panels and cabinets.

In the event of a ground fault in an HRG 480V system, the two non-faulted phases will be at 480V (line potential) to ground and to the faulted phase. If a second phase should fault to ground before an existing fault is cleared, the second fault will not be limited and fault currents greater than encountered on a solidly grounded system should be expected. In performing arc hazard analysis, be aware that there is a much lower risk of your actions causing an uncontrolled fault, but you cannot reduce the evaluated incident energy levels or PPE. This is because the risk is not eliminated, and a line-to-line or three-phase fault is still possible. Surveys indicate that human error is responsible for most faults that start as line to line or three phase. It should be noted that resistance or reactance grounded systems may not supply line-to-neutral loads but may supply line-to-line loads.

1.5.0 Transformers

Transformers are used to step voltage up and down in the power transmission and distribution system. The reason for such high transmission voltages is twofold. First, as a transformer increases transmission voltage, the required current decreases in the same proportion; therefore, larger amounts of power can be transmitted and line losses reduced. Second, to send large amounts of power over long distances at a high current and a low voltage requires a very large diameter wire. The reduction in current reduces the conductor size, which results in a cost reduction.

A transformer is an electrical device that uses the process of electromagnetic induction to change the levels of voltage and current in an alternating current (AC) circuit without changing the frequency and with very little loss of power.

1.5.1 Transformer Theory

As current flows through a conductor, a magnetic field is produced around the conductor. This magnetic field begins to form at the instant current begins to flow and expands outward from the conductor as the current increases in magnitude.

When the current reaches its peak value, the magnetic field is also at its peak value. When the current decreases, the magnetic field also decreases.

AC changes direction twice per cycle. These changes in direction or alternation create an expanding and collapsing magnetic field around the conductor.

If the conductor is wound into a coil, the magnetic field expanding from each turn of the coil cuts across other turns of the coil. When the source current starts to reverse direction, the magnetic field collapses, and again the field cuts across the other turns of the coil.

The result in both cases is the same as if a current-carrying conductor is passed through a magnetic field. An electromotive force (EMF) is induced in the conductor. This EMF is called a *self-induced EMF* because it is induced in the conductor carrying the current.

The direction of this induced EMF is always opposite the direction of the EMF that caused the current to flow initially. This principle is known as *Lenz's law* and is explained as follows:

- An induced EMF always has such a direction as to oppose the action that produced it.
- For this reason, the EMF induced is also known as a *counter electromotive force* (CEMF).

The CEMF reaches a value nearly equal to the applied voltage; thus, the primary current is limited when the secondary is open circuited.

The operation of a transformer is based on the principle that electrical energy can be transferred efficiently by mutual induction from one winding to another. When the primary winding is energized from an AC source, an alternating magnetic flux is established in the transformer core. This flux links the turns of the primary with the secondary, thereby inducing a voltage in them. Since the same flux cuts both windings, the same voltage is induced in each turn of both windings. Whenever the secondary of a transformer is left disconnected (or open), there is no current drawn by the secondary winding. The primary winding draws the amount of current required to supply the electromotive force, which produces the transformer core flux. This current is called the *exciting current*, or *magnetizing current*.

The exciting current is limited by the CEMF in the primary and a small amount of resistance, which cannot be avoided in any current-carrying conductor.

When a load is connected to the secondary winding of a transformer, the secondary current flowing through the secondary turns produces a CEMF According to Lenz's law, this electromotive force is in a direction that opposes the flux that produced it. This opposition tends to reduce the transformer flux and is accompanied by a reduction in the CEMF in the primary. Since the primary current is limited by the internal impedance of the primary winding and the CEMF in the winding, whenever the CEMF is reduced, the primary current continues to increase until the original transformer flux reaches a state of equilibrium.

1.5.2 Transformer Nameplate Data

Transformer nameplate data includes both voltage ratings and electrical ratings. The electrical ratings convey information relating to the transformer electrical parameters. The voltage rating identifies the nominal rms voltage value at which the transformer is designed to operate. A transformer can operate within a ±5% range of its rated primary voltage. If the primary voltage is increased to more than +5%, the windings of the transformer can overheat. Operating the transformer at more than −5% decreases its power output proportionally to the percent voltage reduction.

Transformer windings are rated as follows:

- Phase to phase and phase to neutral for wye windings, such as 480Y/277VAC
- Phase to phase for delta windings, such as 480VAC
- Dual-voltage windings, such as 480VAC × 240VAC

When transformers are equipped with a tap changer, the voltage ratings in the nameplate indicate the nominal voltages. The following are some transformer nameplate ratings and classifications:

- *BIL* — This identifies the maximum impulse voltage the winding insulation can withstand without failure.
- *Phase* — The phase information indicates the number of phase windings contained in a transformer tank.
- *Frequency* — The frequency rating of a transformer is the normal operating system frequency. When a transformer is operated at a lower frequency, the reactance of the primary winding decreases. This causes a higher exciting current and an increase in flux density. In addition, there is an increase in core loss, which results in overall heating.
- *Class* — Transformers are classified by the type of cooling they employ.
- *Temperature rise* — The temperature rise rating is the maximum elevation above ambient temperature that can be tolerated without causing insulation damage.
- *Capacity* — The capacity of a transformer to transfer energy is related to its ability to dissipate the heat produced in the windings. The capacity rating is the product of the rated voltage and the current that can be carried at that voltage without exceeding the temperature rise limitation.
- *Impedance* — Impedance identifies the opposition of a transformer to the passage of short circuit current.
- *Phasor diagrams* — Phasor diagrams show phase and polarity relationships of the high and low windings. They can be used with the schematic connection diagram to provide test connection points and to provide proper external system connections.

Transformers can be divided into two main categories: power transformers and **distribution transformers**. Power transformers handle large amounts of power and step down from transmission voltages to distribution voltages. Distribution transformers are designed to handle larger currents at lower voltage levels. Distribution transformers have smaller kVA ratings and are physically much smaller than power transformers.

Distribution transformers: Transformers that are used for reducing the voltage and transferring power from a primary distribution circuit, usually originating at a substation, to the secondary distribution circuit. Distribution transformers are typically rated between 5 kVA and 500 kVA.

Power transformers often have an auxiliary means of cooling, such as fans and radiators. Distribution transformers don't usually have any specific cooling system beyond heat transferring off their surface. Whereas distribution transformers may be pole mounted or mounted on small concrete pads, power transformers are much larger and are mounted on the ground on larger pads.

Although there is some overlap between power and distribution transformers, a transformer that is rated at more than 500 kVA and/or 34.5 kV is generally a power transformer. A transformer rated below these values can be considered a distribution transformer. Remember, there is an overlap in kVA capacity and voltage depending on the system and power requirements.

1.5.3 Dry (Air-Cooled) Transformers

Many transformers do not use an insulating liquid to immerse the core and windings. Dry or air-cooled transformers are used for many jobs where small, low-kVA transformers are required. Large distribution transformers are usually oil filled for better cooling and electrical insulating value. However, for installations in buildings and other locations where the oil in oil-filled transformers would be a serious fire hazard, dry transformers are used. These transformers are generally of the core form. The core and coils are similar to those of other transformers. A three-phase, dry transformer is shown in *Figure 8*.

The case is made of sheet metal and provided with ventilating louvers for the circulation of cooling air. To increase the output, fans can be installed to move air through the coils at a faster rate than is possible with natural circulation.

Figure 8 Dry transformer.
Source: Electro-Mechanical Corp./Line Power/Federal Pacific

Either Class B or Class H insulation is used for the windings. Class B insulation may be operated safely at a hot-spot temperature of 266°F (130°C). Class H insulation may be operated safely at a hot-spot temperature of 356°F (180°C). The use of these materials makes it possible to manufacture smaller transformers. Both Class B and Class H insulation consist of mica, asbestos, fiberglass, and similar inorganic material. Temperature-resistant organic varnishes are used as the binder for Class B insulation. Silicone or fluorine compounds or similar materials are used as the binder for Class H insulation. Such transformers use high-temperature insulation only in locations where the high temperature requires such insulation.

1.5.4 Sealed Dry Transformers

Hermetically sealed dry transformers are constructed in large sizes for voltages above 15 kV. They are used for installations in buildings and other locations where oil-filled transformers would be a serious fire hazard, but they may also be used for lower voltages and kVA ratings and for water-submersible transformers in locations subject to floods. Nitrogen is typically used for the insulation and cooling of sealed dry transformers.

When inspecting the inside of a dry air-cooled transformer case, look for the following:

- Temporary shipping supports or guards
- Bent, broken, or loose parts
- Debris on the floor or in the coils
- Corrosion of any part
- Worn or frayed insulation
- Shifted core members
- Damaged tap changer mounts or mechanisms
- Misaligned core spacers and loose coil elements
- Broken or loose blocking

Upon the completion of the inspection, replace the covers and bolt securely. All information should be recorded on appropriate inspection sheets.

All transformers require regular testing and maintenance. The tests listed below are the recommended minimum tests that should be included as part of a maintenance program. These tests are conducted to determine and evaluate the present condition of the transformer. From the results of these tests, a determination is made as to whether the transformer is suitable for service. The following tests should be performed using the standards and procedures provided by the transformer manufacturer:

- *Continuity and winding resistance test* — There should be a continuity check of all windings. If possible, measure the winding resistance and compare it to the factory test values. An increase of more than 10% could indicate loose internal connections.
- *Insulation resistance test* — A 1,000V insulation resistance test should be conducted to ensure no grounding of the windings exists.
- *Ratio test* — A turns ratio test should be made to ensure proper transformer ratios and to ensure all connections were made. If equipped with a tap changer, all positions should be checked.
- *Core ground* — This test is performed in the same way as the insulation resistance test, except the measurement is made from the core to the frame and ground bus. Remove the core ground strap before the test.
- *Heat scanning* — After the transformer is energized, a heat scan test should be done to detect loose connections. This test is performed using an infrared scanning device that shows or indicates hot spots.

1.5.5 Instrument Transformers

For all practical purposes, the voltages and currents used in the primary circuits of substations are much too large to be used to provide operating quantities to relaying or metering circuits. To reduce voltage and currents to usable levels, instrument transformers are employed. Instrument transformers are used to perform the following tasks:

- Protect personnel and equipment from the high voltages and/or currents used in electric power transmission and distribution.
- Provide reasonable use of insulation levels and current-carrying capacity in relay and metering systems and other control devices.
- Provide a means to combine voltage and/or current phasors to simplify relaying or metering.

Instrument transformers are manufactured with a multitude of different ratios to provide a standard output for the many different system primary voltage levels and load currents. There are two types of instrument transformers: potential transformers (PTs) and current transformers (CTs).

PTs are designed to reduce primary system voltages down to usable levels for metering and are often referred to as *voltage transformers*, or *VTs*. PTs are often used where the system's primary voltage exceeds 600V and sometimes on 240V and 480V systems.

A PT (*Figure 9*) is used to a supply voltage signal to devices such as voltmeters, frequency meters, power factor meters, watt-hour meters, and protective relays. The voltage is proportional to the primary voltage, but it is small enough to be safe for the test instrument. The secondary of a PT may be designed for several different voltages, but most are designed for 120V. The PT is primarily a distribution transformer especially designed for voltage regulation so that the secondary voltage (under all conditions) will be as close as possible to a specified percentage of the primary voltage.

The standard secondary circuit voltage level for a PT circuit is 120V for circuits below 25 kV and 115V for circuits above 25 kV at the PT's rated primary voltage. These voltages correspond to typical transformation ratios of standard transmission voltages. The current flowing in the secondary of the PT circuit is very low under normal operating conditions, typically less than 1A. PTs are constructed to be lightly loaded with the design emphasis on winding ratio accuracy rather than current rating.

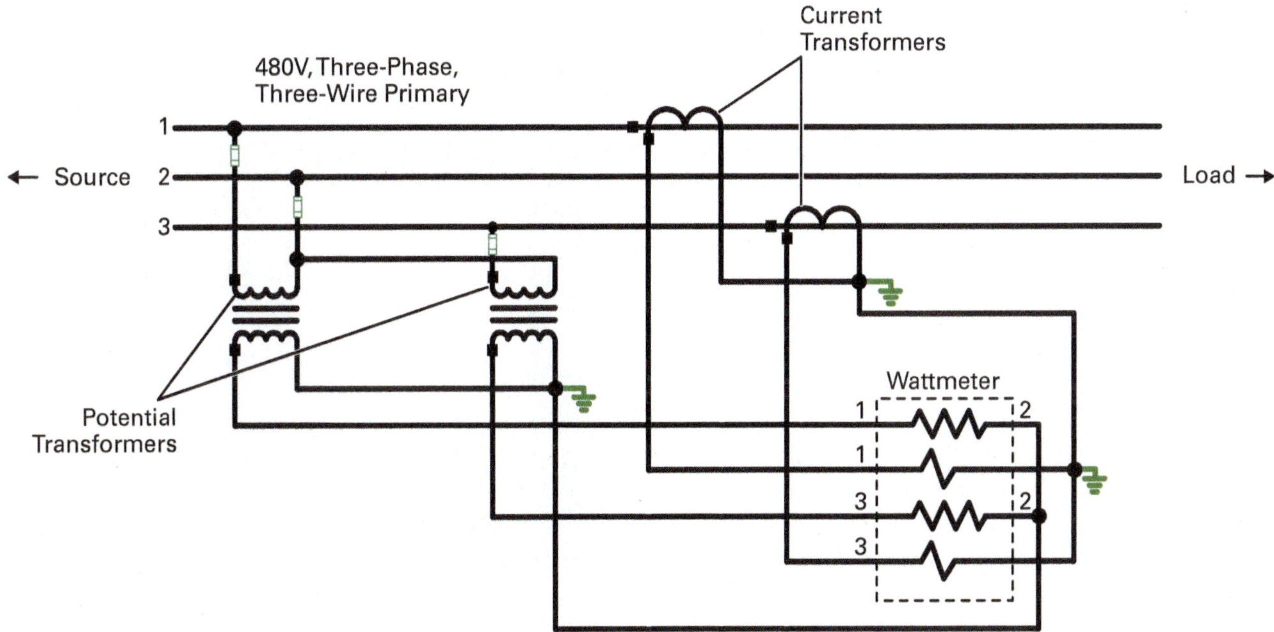

Figure 9 CTs and PTs connected for power metering of a three-phase circuit.

Sulfur hexafluoride (SF_6): A stable, synthetic, odorless gas that is used as an insulator in electrical apparatus.

PTs can be dry, case epoxy insulated, oil filled, or **sulfur hexafluoride (SF_6)** insulated, depending upon the primary circuit voltage level.

The standard output voltage of PTs is either 120V or 69.3V, depending on whether its primary winding uses phase-to-phase or phase-to-neutral connections. Understanding the operation of a potential or voltage transformer is simplified by the inspection of its equivalent circuit.

PTs must have their secondary circuits grounded in case a short circuit develops between the primary and secondary windings and to negate the effects of parasitic capacitance between the primary and the secondary. *Figure 10* shows the connection of an ideal PT circuit.

A CT is used to supply current to an instrument connected to its secondary with the current being proportional to the primary current but small enough to be safe for the instrument. The secondary of a CT is usually designed for a rated current of 5A.

A CT operates in the same way as any other transformer in that the same relationship exists between the primary current and the secondary current. A current transformer uses the circuit conductors as its primary winding. The secondary of the CT is connected to current devices such as ammeters, wattmeters, watt-hour meters, power factor meters, some forms of relays, and the trip coils of some types of circuit breakers.

When no instruments or other devices are connected to the secondary of the CT, a short circuit device or shunt is placed across the secondary to prevent the secondary circuit from being opened while the primary winding is carrying current.

WARNING!

The voltage level across a CT's secondary terminals can rise to a very dangerous level if the secondary circuit opens while the primary circuit is energized. If the secondary circuit is open, there will be no secondary ampere turns to balance the primary ampere turns, so the total primary current becomes exciting current and magnetizes the core to a high flux density. This produces a high voltage across both the primary and secondary windings. This high voltage endangers the life of anyone coming in contact with the meters or leads. This is why current transformers should never be fused. A CT is the only transformer that may be short-circuited on the secondary while energized.

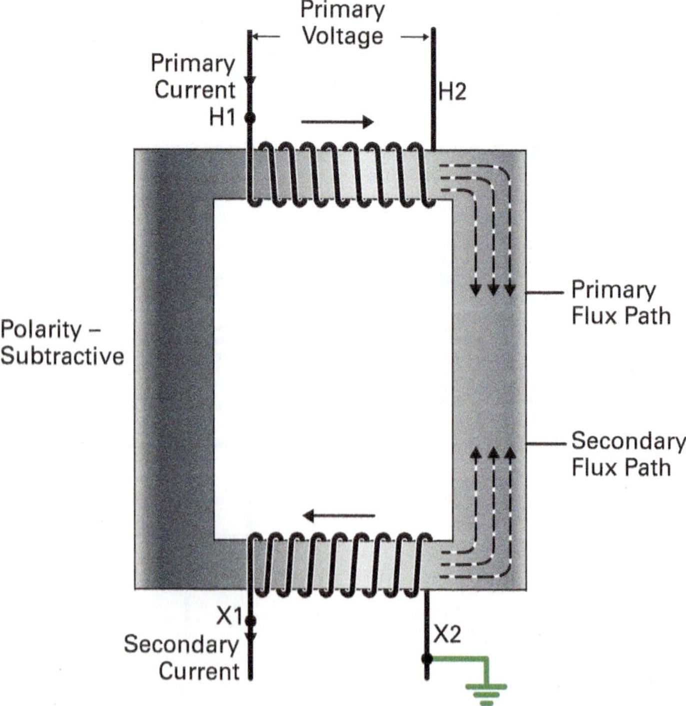

Figure 10 PT construction.

The primary considerations in CT design are the current-carrying capability and saturation characteristics. Insulation systems are of the same generic types as PTs; however, SF_6 insulation is infrequently used in CT construction.

CTs are manufactured in four basic styles: donut, bar, window, and bushing. The bushing-type CT is normally applied on circuit breakers or power transformers. The other types are used for the remaining indoor and outdoor installations. *Figure 11* illustrates some common CT styles.

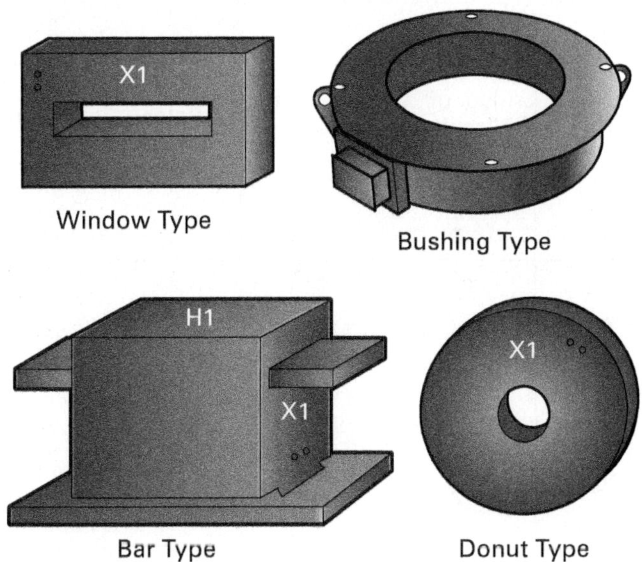

Figure 11 CT construction styles.

The major criteria for the selection of the CT for relaying are its primary current rating, maximum burden, and saturation characteristics. Saturation is particularly important in relaying because many relays are called upon to operate only under fault conditions.

CT circuits operate at a very low voltage. Connected loads, referred to as *burdens*, range from 0.2Ω to 2Ω. These small impedances, together with a maximum continuous current of up to 5A, keep these circuits at low potentials. The voltage can become high momentarily during faults when large secondary currents flow. This voltage is a function of the current, burden, and transformer VA capability.

As with PTs, CTs must also have their secondary windings grounded in the event of an insulation breakdown between the primary and secondary and to negate the effects of parasitic capacitance.

Instrument transformers require regular inspection and maintenance. A typical inspection includes the following steps:

Step 1 Inspect for physical damage and check the nameplate information for compliance with instructions and specification requirements.

Step 2 Verify the proper connection of transformers against the system requirements.

Step 3 Verify the tightness of all bolted connections and ensure adequate clearances exist between the primary circuits and the secondary circuit wiring.

Step 4 Verify that all required grounding and shorting connections provide good contact.

Step 5 Test for proper operation of the PT isolation (PT tip out) compartment and grounding operation when applicable.

Think About It

Transformers

What is the difference between a potential transformer and a control transformer? Why can't one device serve both functions?

1.6.0 Panelboards

Circuit control and overcurrent protection must be provided for all circuits and the power-consuming devices connected to these circuits. Lighting and power panels located throughout large buildings being supplied with electrical energy provide this control and protection. *Figure 12* shows a schedule of 15 panelboards provided in a typical industrial building to feed electrical energy to the various circuits.

Panel No.	Location	Mains	Voltage Rating	No. of Circuits	Breaker Ratings	Poles	Purpose
P-1	Basement N. Corridor	Breaker 100A	208/120V 3ϕ, 4W	19 2 5	20A 20A 20A	1 2 1	Lighting and Receptacles; Spares
P-2	Basement N. Corridor	Breaker 100A	208/120V 3ϕ, 4W	24 2 0	20A 20A	1 2	Lighting and Receptacles; Spares
P-3	2nd Floor N. Corridor	Breaker 100A	208/120V 3ϕ, 4W	24 2 0	20A 20A	1 2	Lighting and Receptacles; Spares
P-4	Basement S. Corridor	Breaker 100A	208/120V 3ϕ, 4W	24 2 0	20A 20A	1 2 1	Lighting and Receptacles; Spares
P-5	1st Floor S. Corridor	Breaker 100A	208/120V 3ϕ, 4W	23 2 1	20A 20A 20A	1 2 1	Lighting and Receptacles; Spares
P-6	2nd Floor S. Corridor	Breaker 100A	208/120V 3ϕ, 4W	22 2 2	20A 20A 20A	1 2 1	Lighting and Receptacles; Spares
P-7	Mfg. Area S. Wall E.	Breaker 100A	208/120V 3ϕ, 4W	5 7 2	20A 20A 20A	1 1 1	Lighting and Receptacles; Spares
P-8	Mfg. Area S. Wall W.	Breaker 100A	208/120V 3ϕ, 4W	5 7 2	20A 20A 20A	1 1 1	Lighting and Receptacles; Spares
P-9	Mfg. Area S. Wall E.	Breaker 100A	208/120V 3ϕ, 4W	5 7 2	50A 20A 20A	1 1 1	Lighting and Receptacles; Spares
P-10	Mfg. Area S. Wall W.	Breaker 100A	208/120V 3ϕ, 4W	5 7 2	50A 20A 20A	1 1 1	Lighting and Receptacles; Spares
P-11	Mfg. Area East Wall	Lugs Only 225A	208/120V 3ϕ, 4W	6	20A	3	Blowers and Ventilators
P-12	Boiler Room	Breaker 100A	208/120V 3ϕ, 4W	10 4	20A 20A	1 1	Lighting and Receptacles; Spares
P-13	Boiler Room	Lugs Only 225A	208/120V 3ϕ, 4W	6	20A	3	Oil Burners and Pumps
P-14	Mfg. Area E. Wall	Lugs Only 400A	208/120V 3ϕ, 4W	3 2 1	175A 70A 40A	3 3 3	Chillers Fan Coil Units Fan Coil Unit
P-15	Mfg. Area W. Wall	Lugs Only 600A	208/120V 3ϕ, 4W	5	100A	3	Trolley Busway and Elevator

Figure 12 Example schedule of electric panelboards for an industrial building.

1.6.1 Panelboard Construction

In general, panelboards are constructed so that the main feed busbars run the height of the panelboard. The buses to the **branch circuit** protective devices are connected to the alternate main buses. In an arrangement of this type, the connections directly across from each other are on the same phase, and the adjacent connections on each side are on different phases. As a result, multiple protective devices can be installed to serve the 208V equipment. An example of a panelboard is shown in *Figure 13*.

Branch circuit: A set of conductors that extends beyond the last overcurrent device in the low-voltage system of a given building.

Figure 13 Typical panelboard.

1.6.2 Identification of Conductors

The ungrounded conductors may be any color except green (or green with a yellow stripe), white, or gray. Green is reserved for grounding purposes only, while white or gray are reserved for the grounded circuit conductor. Refer to *NEC Section 200.6* for the proper identification of grounded conductors.

NEC Section 210.5(C) requires that where different voltages exist in a building, the ungrounded conductors for each system must be identified at each accessible location. Identification may be by color-coding, marking, tape, tagging, or other approved means. The means of identification must be permanently posted at each branch circuit panelboard or must be readily available.

Know the Code

Means of Identifying Grounded Conductors
NEC Section 200.6

Know the Code

Identification of Ungrounded Conductors
NEC Section 210.5(C)

For example, this situation may occur when the building is served with 277/480V, and step-down transformers are used to provide 120/208V for lighting and receptacle outlets. Examples of panelboard wiring connections are shown in the following figures. A single-phase lighting and appliance branch circuit panelboard is shown in *Figure 14*, and a three-phase panelboard is shown in *Figure 15*. A bakery panelboard is shown in *Figure 16*. A four-wire system is shown in *Figure 17*.

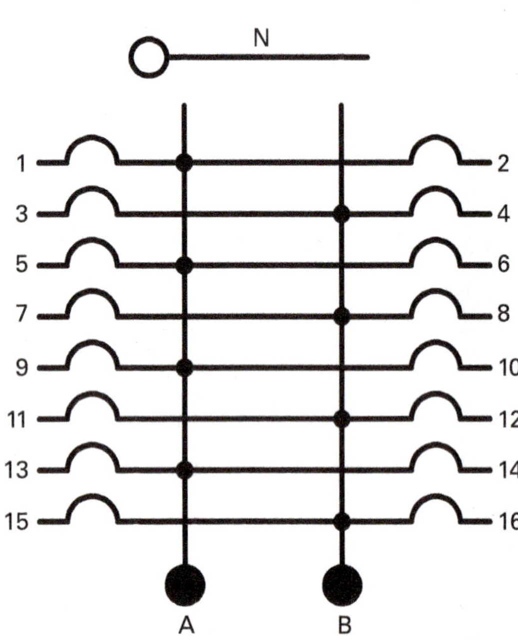

Figure 14 Lighting and appliance branch circuit panelboard with single-phase, three-wire connections.

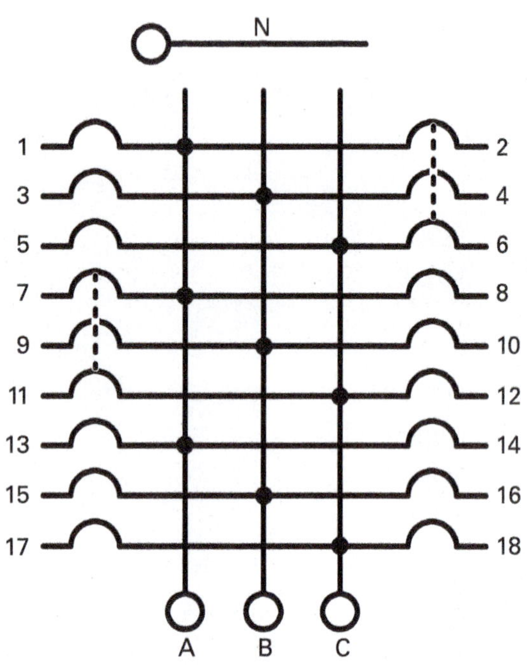

Figure 15 Lighting and appliance branch circuit panelboard with three-phase, four-wire connections.

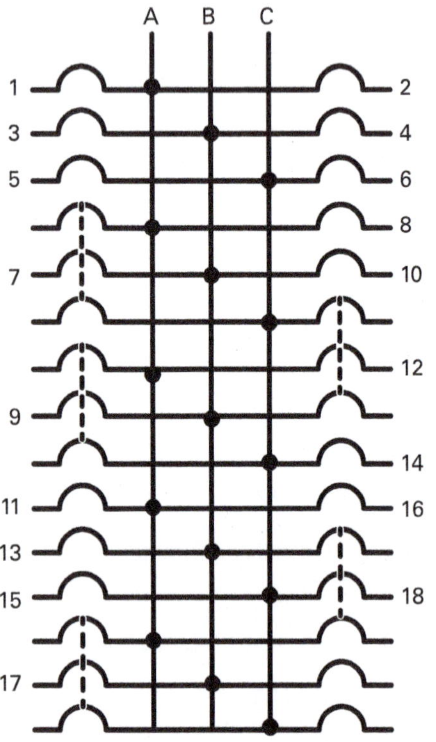

Figure 16 Bakery panelboard circuit showing alternate numbering scheme.

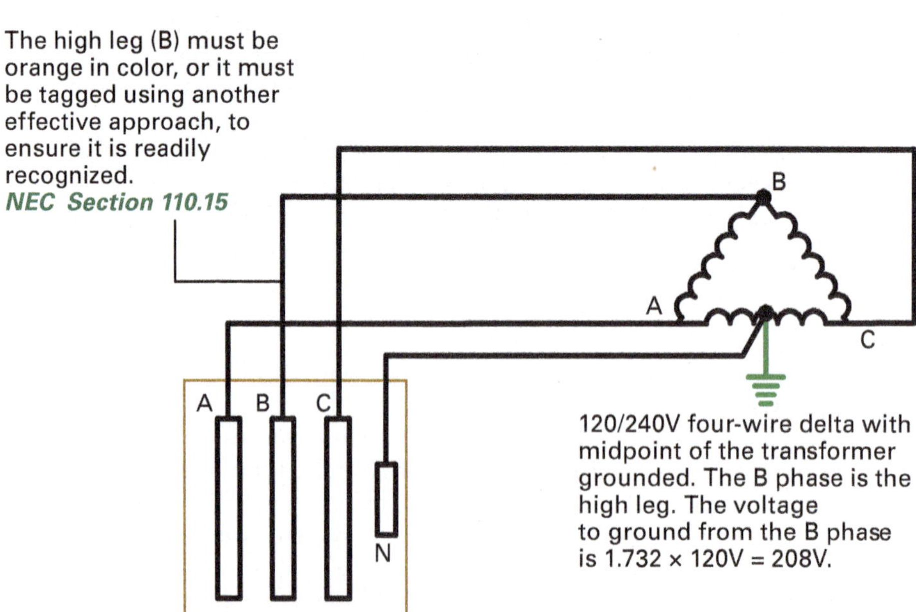

Figure 17 Panelboards and switchboards supplied by four-wire, delta-connected system.

1.6.3 Number of Circuits

The number of overcurrent devices in a panelboard is determined by the needs of the area being served. See *NEC Section 408.54*. Using the bakery panelboard in *Figure 16* as an example, there are 13 single-pole circuits and 5 three-pole circuits. This is a total of 28 poles. When using a three-phase supply, the incremental number is six (a pole for each of the three phases on both sides of the panelboard). This means that poles can only be specified in multiples of six. The minimum number of poles that could be specified for the bakery is 30. This would limit the power available for growth and would not permit the addition of a three-pole lead. The reasonable choice is to go to 36 poles, which provides flexibility for growth. For new construction, it is recommended to provide at least 20% spare poles.

Know the Code

Maximum Number of Overcurrent Devices
NEC Section 408.54

1.6.4 Panelboard Protective Devices

The main protective device for a panelboard may be either a fuse or a circuit breaker. This section describes the use of circuit breakers. The selection of a circuit breaker should be based on the following criteria:

- Proper overload protection
- Suitable voltage rating
- Sufficient interrupting current rating
- Short circuit protection
- Coordination of breaker(s) with other protective devices

The choice of the overload protection is based on the rating of the panelboard. The trip rating of the circuit breaker cannot exceed the amperage capacity of the busbars in the panelboard. The number of branch circuit breakers is generally not a factor in the selection of the main protective device, except in a practical sense. It is a common practice to have the total amperage of the branch breakers greatly exceed the rating of the main breaker; however, it makes little sense for a single branch circuit breaker to be the same size as, or larger than, the main breaker.

The voltage rating of the breaker must be higher than that of the system. Breakers are usually rated at 250V to 600V.

The importance of the proper interrupting rating cannot be overstressed. Remember that, if there is ever any question regarding the exact value of the short circuit current available, the circuit breaker with the higher interrupting rating should be selected.

Many circuit breakers used as the main protective device are provided with an electronic trip unit (*Figure 18*). Adjustments of this trip determine the degree of protection provided by the circuit breaker if a short circuit occurs. The manufacturer of this device provides exact information about the adjustments to be made. In general, a low setting may be 10 or 12 times the overload trip rating.

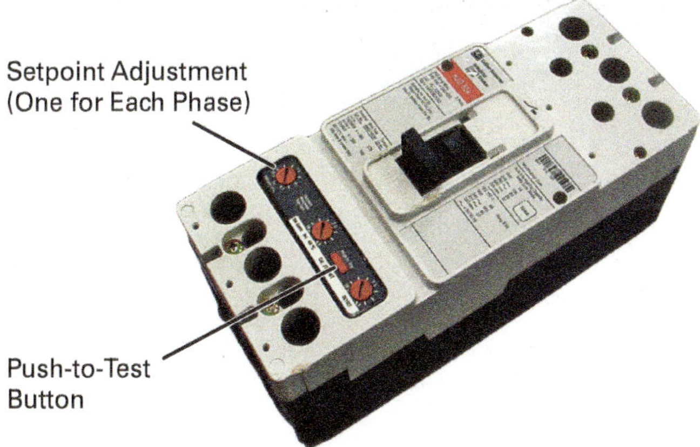

Figure 18 Circuit breaker with electronic trip unit.
Source: Tim Ely

These two rules should be followed whenever the trip is set:

- The trip must be set to the minimum practical setting.
- The setting must be lower than the value of the short circuit current available at that point.

Know the Code

Terminals
NEC Section 110.14(A)

Know the Code

Feeder Taps
NEC Section 240.21(B)

Know the Code

Temperature Limitations
NEC Section 110.14(C)

If lugs are used, ensure the lugs are suitable for making multiple conductor connections, as required by *NEC Section 110.14(A)*. In general, this means that a separate lug is to be provided for each conductor being connected.

If taps are made to the subfeeder, they can be reduced in size according to *NEC Section 240.21(B)*. This specification is very useful in cases such as that of panel P-12 in *Figure 12*. For this panel, a 100A main breaker is fed by a 350 MCM conductor. Within the 10' distance given in *NEC Section 240.21(B)(1)*, a conductor with a 20A rating may be tapped to the subfeeder and connected to the 20A overcurrent device in the panel.

Per *NEC Section 110.14(C)*, the temperature rating of conductors must be selected so as not to exceed the lowest temperature rating of any connected termination, conductor, or device.

1.6.5 Branch Circuit Protective Devices

The schedule of panelboards for the industrial building (*Figure 12*) shows that lighting panels P-1 through P-6 have 20A circuit breakers, including double-pole breakers to supply special receptacle outlets. A double-pole breaker requires the same installation space as two single-pole breakers. Common air breakers are shown in *Figure 19*.

Figure 19 Branch circuit protective devices.

1.0.0 Section Review

1. Which of the following is *true* about switchboards?
   a. Switchboards are always installed in cabinets.
   b. Switchboards are only used at the service entrance.
   c. Switchboards are used to subdivide large blocks of power.
   d. Switchboards are designed to prevent access to the panel.

2. When installing wiring within switchgear, use self-adhesive cable ties every _____.
   a. 12" (300 mm)
   b. 16" (400 mm)
   c. 24" (600 mm)
   d. 36" (900 mm)

3. The *maximum* voltage that a piece of equipment can withstand is known as its _____.
   a. interrupting capacity
   b. basic impulse level (BIL)
   c. current limit
   d. frequency

4. Bolted pressure switches are often used in place of _____.
   a. circuit breakers
   b. push-button switches
   c. main fuses
   d. MVL switches

5. A transformer's temperature rise rating refers to _____.
   a. the normal operating temperature of the transformer
   b. the maximum elevation above ambient temperature that the transformer can withstand without insulation damage
   c. the difference between the temperature of the de-energized transformer and the ambient temperature under normal operating conditions
   d. the difference between the temperature of the transformer when it starts and the temperature of the transformer while running under normal operating conditions

6. A service has a three-phase supply with 14 single-pole circuits and 4 three-pole circuits. The *minimum* number of poles required is _____.
   a. 18
   b. 22
   c. 26
   d. 30

2.0.0 Installation Requirements

Performance Tasks

There are no Performance Tasks in this section.

Objective

Describe the installation requirements for distribution equipment.

a. State the *NEC*® requirements for distribution equipment installation.

b. Interpret electrical diagrams related to the installation of distribution equipment.

This section provides a brief description of the *NEC*® articles that are applicable to switchboard and switchgear construction, installation, and accessories.

2.1.0 *NEC*® Requirements for Distribution Equipment

Know the Code

Switchboards, Switchgear, and Panelboards
NEC Article 408

Distribution equipment requirements can be found in several articles of the *NEC*®. In addition to the specific requirements listed in *NEC Article 408*, these installations must follow the *NEC*® general requirements for electrical installations, as well as those for conductors, tap rules, and grounding.

2.1.1 Electrical Installation Requirements

NEC® has many requirements for electrical installations. These requirements include the following:

- *Interrupting rating* — The interrupting rating refers to the maximum current a device is capable of interrupting without suffering damage, if it is operating within its voltage specifications. *NEC Section 110.9* requires the interrupting rating of an overcurrent protective device to be the same as or higher than the available fault current at the equipment wiring connections.
- *Deteriorating agents* — *NEC Section 110.11* provides for the protection of equipment and conductors from environments that could cause deterioration, such as gases, vapors, liquids, or moisture, unless the equipment is specifically designed for those applications.
- *Mechanical execution of work* — *NEC Section 110.12* states that electrical equipment is to be installed in a skillful and professional manner. Any unused openings provided by the equipment manufacturer or added during installation must be sealed in a way that is as effective as the cabinet wall. This section also forbids the use of damaged electrical parts that could affect the safe operation or integrity of the equipment.
- *Mounting and cooling* — *NEC Section 110.13* states that electrical equipment shall be securely fastened with mechanical fasteners, excluding wooden plugs driven into concrete, masonry, plaster, or similar materials. Equipment must be positioned so cooling airflow is not restricted.
- *Electrical connections* — Due to the resistive oxidation created when dissimilar metals are placed together, splicing devices and pressure connectors must be identified for use with the conductor (*NEC Section 110.14*). Dissimilar metals cannot be used in terminations or splices. Inhibitor compounds to reduce oxidation must be made for the purpose cannot negatively affect conductors, installation, or equipment. Terminals that can be used with more than one conductor metal must be marked to reflect it.
- *Markings* — The manufacturer's trademark or logo and system ratings, including voltage, current, and wattage, must be durable enough to withstand the installed environment. Per *NEC Section 110.24*, service equipment must be marked to indicate the available fault current and the date the relevant calculations were performed. Modifications to equipment that affect the available fault current must be accompanied by new calculations to ensure the ratings are appropriate, and any field markings must be updated.

- *Disconnect identification* — Each disconnecting device, such as circuit breakers, fused switches, feeders, or unfused disconnects must be clearly marked to indicate its purpose unless it is positioned in a manner that makes its purpose clear (*NEC Section 110.22*). In other than one- or two-family residences, the source and location of power disconnects such as circuit breakers must be identified to the occupants.
- *Working space* — Suitable access and working space must be maintained around the electrical equipment to permit safe operation and maintenance (*NEC Section 110.26*). Minimum clearances in front of electrical enclosures must conform to *NEC Section 110.26*. In all cases, doors or hinged parts must be able to open to a 90° angle. In differing conditions, the distances in *NEC Table 110.26(A)(1)* must be adhered to. Storage of any kind is not permitted within the clearance area. Per *NEC Section 110.26(C)(1)*, at least one entrance of sufficient size to enter and exit the work area must exist. In cases of services over 1,200A and over 6' (1.8 m) wide, two entrances are required. The workspace must also have adequate lighting.
- *Flash protection* — *NEC Section 110.16* states that electrical equipment such as switchboards, industrial control panels, and motor control centers (MCCs) in non-dwelling facilities that are likely to need service or maintenance while energized must be marked to warn of the potential arc flash hazards. The markings must be clearly visible to everyone. The markings cannot be handwritten and must be durable enough to stand up to the environment.

2.1.2 Requirements for Conductors

NEC Section 200.6 covers requirements associated with identifying grounded conductors. It includes the following rules:

- *Neutrals* — Grounded conductors (neutrals) must be color-coded with a solid white or gray marking, or with three white or gray stripes along their complete length. Conductors size 4 AWG and larger can be color-coded with a white or gray marking tape where and when terminated. The markings must encircle the entire conductor or its insulation. Where different electrical systems are installed together, each system's grounded conductor must be clearly identified, per *NEC Section 200.6(D)*.
- *Protection* — Branch circuit conductors must be protected by overcurrent protective devices, per *NEC Sections 240.4 and 240.21*.
- *Loading* — *NEC Section 210.19(A)(1)* states that the conductor ampacity must accommodate the noncontinuous load plus 125% of the continuous load. The minimum conductor size must be based on this load after any derating or adjustment factors are applied per *NEC Section 210.19(A)(2)* and *NEC Tables 310.15(B)(1)(1) and 310.15(C)(1)*.
- *Marking* — All conductors and cables must be permanently marked to show their manufacturer, voltage, AWG size, and insulation type (*NEC Section 310.8*). Equipment grounding conductors can be bare wire. Insulated equipment grounding conductors require a continuous green marking along their full length. Of course, fully green insulation is acceptable and common. Larger conductors may be marked at each end and at every point where the conductor can be accessed. Ungrounded conductors, such as phase wires, must be recognizable from grounded or equipment grounding conductors with colors other than white, gray, or green used, since those colors are reserved for grounded conductors and equipment grounds. The typical ungrounded conductor colors are black, red, blue, brown, orange, and yellow. However, the only code-specified colors are for the high leg in a delta system, direct current, and isolated power systems as specified by *NEC Section 110.15*, *NEC Section 210.5(C)(2)*, and *NEC Section 517.160(A)(5)* respectively. In switchboards fed by a four-wire delta system where one transformer phase winding is grounded at its midpoint, the phase having the higher voltage of the group must be marked with an orange color according to *NEC Section 110.15*.

Know the Code

Means of Identifying Grounded Conductors
NEC Section 200.6

Know the Code

Protection of Conductors
NEC Section 240.4

Know the Code

Location in Circuit
NEC Section 240.21

Know the Code

Conductors — Minimum Ampacity and Size
NEC Section 210.19

Know the Code

Ampacity Tables
NEC Section 310.15

Know the Code

Identification for Branch Circuits - Identification of Ungrounded Conductors
NEC Section 210.5(C)

Know the Code

Isolated Power Systems
NEC Section 517.160

Case History

How *Not* to Build a Better Mousetrap

The expensive switchgear shown in the image had an unused opening that was left uncovered after the installation. A rodent entered the compartment, shorting out one of the busbars and causing extensive damage.

The Bottom Line: *NEC Section 110.12(A)* requires that any unused openings must be sealed in way that makes it equivalent to the wall of the structure.

Source: Tri-City Electrical Contractors Inc.

2.1.3 Tap Rules

Tap conductors in switchboards are tapped onto the line-side bus of the switchboard to feed control circuits, control power transformers, and metering devices. Overcurrent protective devices (typically fuses) are usually connected where the protected conductor receives power. However, per *NEC Section 240.21(B)(1)*, tap conductors do not require protection at the tap location if the following conditions are met:

- The length of the conductor is not over 10' (3 m).
- The ampacity of the conductor is greater than or equal to the combined loads it serves. The ampacities of conductors are determined by the tables referenced in *NEC Sections 310.16 through 310.21* or with engineering support per *NEC Section 310.15(A)*.
- The conductors do not extend beyond the switchboard for control devices they serve.
- The conductors are enclosed in a raceway except where they connect to the bus.
- For field installations where the tap conductors leave the enclosure where the tap is made, the rating of the overcurrent protective device on the line side of the tap does not exceed 10 times the tap conductor's ampacity.

Know the Code

Ampacity Tables
NEC Section 310.15

2.1.4 Grounding Requirements

NEC Section 250.20(B) states that AC systems between 50V and 1,000V must be grounded when any of the following conditions are met:

- Where the system can be grounded in such a way that the maximum phase-to-ground voltage does not exceed 150V
- When the system is three-phase, four-wire, wye-connected and the neutral is used as a circuit conductor
- When the system is three-phase, four-wire, delta-connected and the midpoint of a phase is used as a conductor (developed neutral)

Know the Code

Grounding and Bonding
NEC Article 250

NEC® grounding requirements include the following:

- *Grounding electrode conductor* — *NEC Sections 250.24 and 250.66* cover the requirements of grounding electrode conductors, including proper sizing of the equipment grounding conductors to the service equipment enclosures. *NEC Section 250.24(C)* states that for grounded systems (delta or wye), an unspliced main bonding jumper in the service equipment must be used to connect the grounding conductor and the service disconnect enclosure to the grounded conductor of the system within the enclosure.
- *Grounding electrodes* — *NEC Sections 250.52 and 250.53* require that when rod or pipe electrodes are used, they must extend a minimum of 8' (2.44 m) into the soil. The electrode must be no less than ¾" (MD 21) in diameter for pipe and ⅝" (15.87 mm) in diameter for rods. It must be galvanized metal or copper-coated to resist corrosion. Underground structures, such as water piping systems, may also be used as electrodes. Underground gas piping systems and aluminum electrodes cannot be used. If a single rod, pipe, or plate electrode has a resistance to ground less than or equal to 25Ω, then a supplemental electrode isn't necessary.
- *Bonding of grounding electrode* — *NEC Section 250.64(E)* states that a grounding electrode conductor or its enclosure must be securely mounted to the surface along which it runs. In cases where the conductor is enclosed, the enclosure must be electrically continuous and firmly bonded.
- *Ground connection surfaces* — Nonconducting coatings, such as paint, enamel, or insulating materials, must be thoroughly removed at any point where a grounding or bonding connection is made (*NEC Section 250.12*).

NOTE

While most metric conversions in the *NEC®* are rounded to the nearest tenth or whole unit, the *NEC®* conversions for grounding are very specific. Always refer to applicable *NEC®* requirements when making conversions.

2.1.5 Requirements for Switchboards and Panelboards

NEC Article 408 covers switchboards, switchgear, and panelboards. Some of the *NEC®* requirements for switchboards and panelboards include the following:

- *Dedicated space* — *NEC Section 110.26(E)(1)* states that panelboards and switchboards can only be installed in spaces specifically designed for that purpose. No other piping or ducts may be installed in the spaces or pass through them, except equipment that is relevant and needed to support the electrical equipment and system.
- *Inductive heating* — *NEC Section 408.3(B)* indicates that busbars and conductors must be arranged to prevent overheating due to induction.
- *Phasing* — *NEC Section 408.3(E)(1)* states that AC phasing in switchboards must be arranged A, B, C from front to back, top to bottom, or left to right, relative to facing the front of the board or switchgear. In systems containing a high leg, the B phase must be the one with the higher voltage to ground.
- *Wire bending space* — *NEC Section 408.3(G)* requires that the wire bending space allowed for conductors must be as shown in *NEC Tables 312.6(A) and (B)(2)*.
- *Minimum spacing* — *NEC Section 408.56* requires the spacing provided between bare metal parts and conductors to follow *NEC Table 408.56*. Conductors entering switchboards from the bottom must maintain the clearances outlined in *NEC Table 408.5*.
- *Conductor insulation* — Insulated conductors inside switchboards must be listed as flame-retardant and rated for the voltage applied to them or any adjacent conductors with which they could potentially come into contact (*NEC Section 408.19*).

Know the Code

Dedicated Equipment Space
NEC Section 110.26(E)

Know the Code

Support and Arrangement of Busbars and Conductors
NEC Section 408.3

Know the Code

Minimum Spacings
NEC Section 408.56

Know the Code

Conductor Insulation
NEC Section 408.19

2.2.0 Electrical Diagrams of Distribution Equipment

NOTE

Before looking at actual plant diagrams, it is necessary to understand the symbols used to condense electrical drawings. The designer uses symbols and abbreviations as a type of shorthand. This section will present standard symbols, abbreviations, and device numbers that make up the designer's shorthand. For IEEE device numbers, see *Appendix 26306-23A*. Note, however, that symbol sets used by different engineers and engineering firms can vary.

Every line, symbol, and letter in a diagram must have a specific purpose and be presented accurately and concisely. For example, when the rating of a current transformer is given, a transformer symbol is shown, and the device abbreviation (CT in this case) is not needed; the symbol alone is sufficient. Writing the unit of measure (amperes in this case) is also unnecessary since a CT is always rated in amperes. Thus, the number and the transformer symbol alone provide the needed information.

The key to reading and interpreting electrical diagrams is often the electrical legend. The legend reflects the symbols used in the diagram along with general notes and other important information. Most electrical legends are very similar. However, there are some variations between the legends developed by different companies. Only the legend shown on a given set of drawings should be used to interpret those drawings. The legend eliminates the need to memorize all the symbols presented on a diagram. It can be used as a reference for any unfamiliar symbols.

The legend is usually found in the bottom right corner of a drawing or on its own page in a large drawing set. In addition to symbols, abbreviations are an important part of the designer's shorthand. For example, a circle can be used to symbolize a meter, relay, motor, or indicating light. A circle's application can generally be distinguished by its location in the circuit; however, the designer uses a set of standard abbreviations to make the distinction clear. The following abbreviations are used to represent meters:

A — Ammeter
AH — Ampere-hour meter
CRO — Oscilloscope
DM — Demand meter
F — Frequency meter
GD — Ground detector
OHM — Ohmmeter
OSC — Oscillograph
PF — Power factor meter
PH — Phase meter
SYN — Synchroscope
TD — Transducer
V — Voltmeter
VA — Volt-ammeter
VAR — VAR meter
VARH — VAR hour meter
W — Wattmeter
WH — Watt-hour meter

As mentioned earlier, indicating lamps may also be represented by a circle. The following abbreviations are used to represent indicating lamps:

A — Amber
B — Blue
C — Clear
G — Green
R — Red
W — White

Relays are another component commonly represented by a circle. The following abbreviations are used for relays:

CC — Closing coil
CR — Closing/control relay
TC — Trip coil
TR — Trip relay
TD — Time-delay relay
TDE — Time-delay energize
TDD — Time-delay de-energize
X — Auxiliary relay

Still another component that is commonly represented by a circle is the motor. Motors usually have the horsepower rating in or near the circle representing them. Any other piece of equipment represented by a circle will be identified in the legend or notes, or will be spelled out on the diagram itself.

Contacts and switches are also identified using standard abbreviations. Common abbreviations are listed as follows:

A — Breaker A contact
B — Breaker B contact
BAS — Bell alarm switch
BLPB — Backlighted push button
CS — Control switch
FS — Flow switch
LS — Limit switch
PB — Push button
PS — Pressure switch
PSD — Differential pressure switch
TDO — Time-delay open
TDC — Time-delay closed
TS — Temperature switch
XS — Auxiliary switch

The following figures illustrate examples of these abbreviations and symbols. *Figure 20* shows A and B contacts in their de-energized state. This is common among wiring diagrams unless they are specifically modified to display a sequence of operation or for a similar purpose. In *Figure 20,* if relay CR is de-energized, contact A is open and contact B is closed. When relay CR is energized, contact A is closed and contact B is open.

Figure 21 illustrates a control switch and its associated contacts. Contacts 1 through 4 open and close as control switch 1 (CS1) is repositioned.

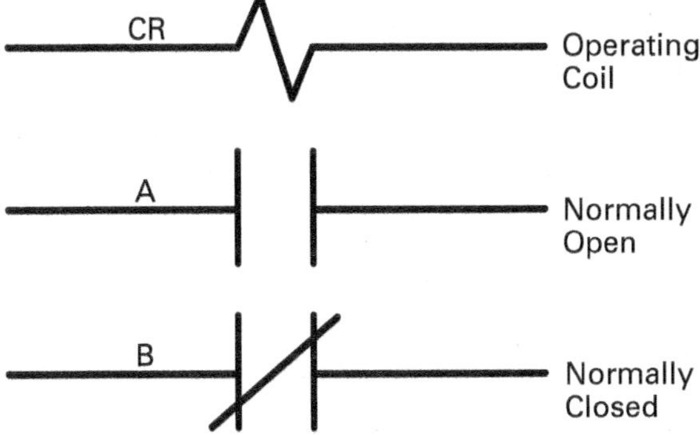

Figure 20 Contact symbols.

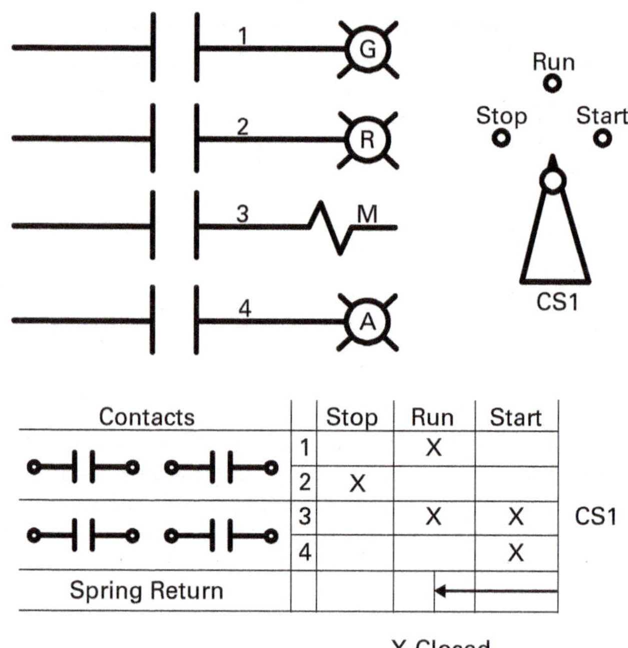

Figure 21 Switch development.

In the Stop position, contact 2 is closed, and the red indicating lamp is lit. In the Start position, contacts 3 and 4 are closed, energizing the M coil and the amber indicating lamp, respectively. When the switch handle is released, the spring returns to the Run position, and contact 4 opens, de-energizing the amber lamp and closing contact 1 to energize the green lamp.

There are many abbreviations used in electrical drawings. Designers do try to use standard abbreviations, but you will encounter nonstandard abbreviations occasionally. Nonstandard abbreviations should be defined in the diagram notes or legend. *Figure 22* defines abbreviations commonly used in wiring prints and specifications. The symbols also help illustrate descriptions of the abbreviations.

2.2.1 Single-Line Diagrams

Analyzing and reading complex electrical circuits can be challenging. Some diagrams can be simplified to single-line diagrams, also known as *one-line diagrams,* to aid in interpreting the prints. A single-line diagram uses single lines and standard symbols to indicate the paths, interconnections, and component parts of an electrical circuit or system. A single line also represents all conductors and phases of a system.

SPST NO		SPST NC		SPDT		Terms
Single Break	Double Break	Single Break	Double Break	Single Break	Double Break	SPST = Single Pole Single Throw
						SPDT = Single Pole Double Throw
DPST NO		DPST NC		DPDT		DPST = Double Pole Single Throw
Single Break	Double Break	Single Break	Double Break	Single Break	Double Break	DPDT = Double Pole Double Throw
						NO = Normally Open NC = Normally Closed

Figure 22 Supplementary contact symbols.

All components of power circuits are represented by symbols and notations. One-line diagrams are valuable tools for system visualization during planning, installation, operation, and maintenance, and they provide a basic understanding of how a portion of the electrical system functions in terms of the physical components of the circuit.

Types of single-line diagrams for industrial facilities include summary (overall facility) diagrams and detailed single-line diagrams. The summary diagrams show each bus and disconnecting device from the point of supply to the line side connection of MCCs. The drawings typically include all voltage levels and power transformers down to the voltage level for three-phase power usage (575V or 480V). The detailed single-line diagrams identify the components and disconnect devices all the way to downstream users. Individual motors or panelboards fed from the MCC will be shown, as well as lighting transformers and lighting panels. The single line of branch circuits and end devices from panelboards is usually called the *panel schedule*. An example of a summary single-line diagram is shown in *Figure 23*.

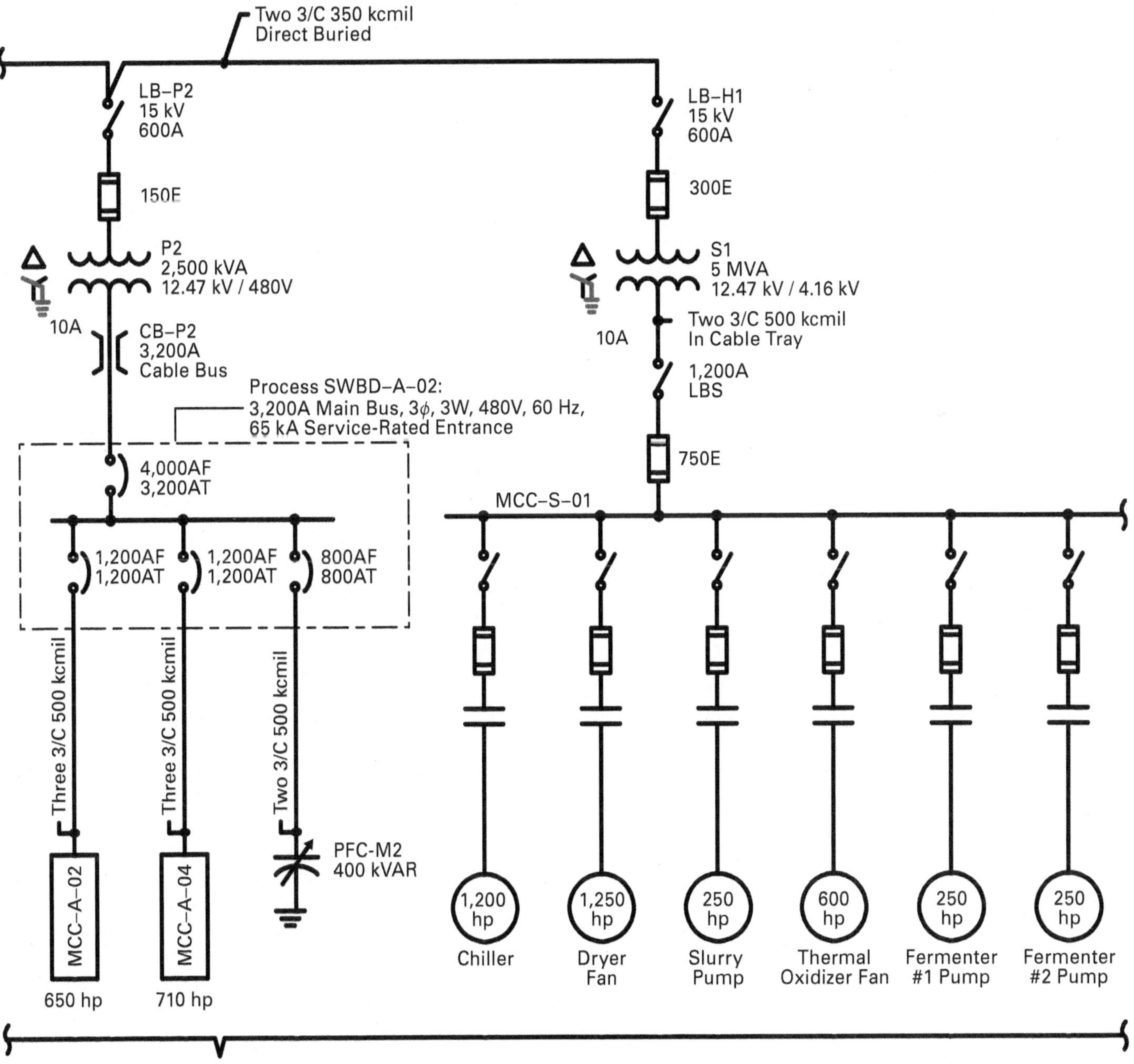

Figure 23 Single-line diagram.

2.2.2 Elementary Diagrams

An elementary diagram is a drawing that falls between single-line diagrams and schematics in terms of complexity. An elementary diagram shows how each individual conductor is connected. *Figure 24* is an example of an elementary schematic diagram with the circuit powered by the phase voltage between L1 and L2.

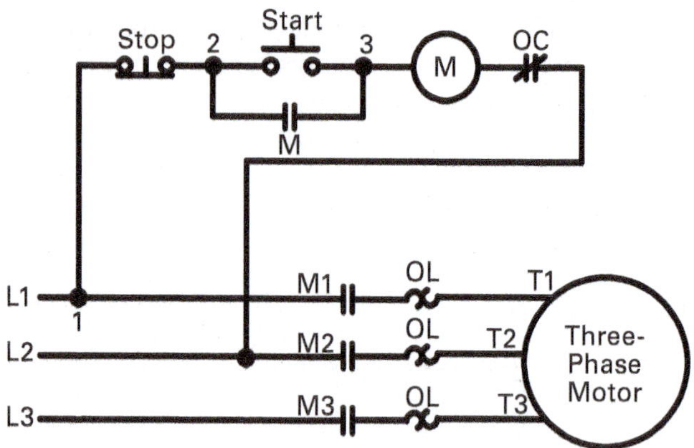

Figure 24 Elementary schematic diagram.

Elementary diagrams, interconnection diagrams, and connection diagrams all illustrate individual conductors. Elementary diagrams are used to show the wiring of instruments and electrical control devices in an elementary ladder or schematic form. The elementary diagram reflects the control wiring required to support the operation and sequence of operations described in the logic diagram. When presented in ladder form, the vertical lines in each ladder diagram represent the power sources in a control circuit. The vertical lines are continuous from the top to the bottom of the ladder, unless there is also a control transformer included in the scheme. The power conductor is always shown on the left side of the ladder and the ground or neutral wire on the right. A ground symbol is not normally shown on the neutral wire in a ladder diagram.

The control power wire numbers are shown above the vertical lines on a ladder diagram. The circuit identification number and source of the control circuit are shown at the top center of each ladder. If two or more control circuits are represented in a single ladder, the vertical lines are broken, and the wire numbers and circuit identification are entered at the top of each ladder segment. Each horizontal line in a ladder diagram represents a circuit path. All devices shown on a single horizontal line represent a series circuit path. Parallel circuit paths are shown on two or more horizontal lines.

2.2.3 Interconnection Diagrams

When troubleshooting electrical circuits, you may use an elementary circuit diagram to determine the cause of a failure. However, since elementary diagrams are drawn without regard to physical locations, connection diagrams should be used to aid in locating faulty components. Interconnection and connection diagrams are structured in such a way that they present all the wires that were shown in the elementary drawing in their actual locations. These drawings show all electrical connections within an enclosure, with each wire labeled to indicate where each end of the wire is terminated.

The interconnection diagram is made to show the actual wiring connections between unit assemblies or equipment. Internal wiring connections within unit assemblies or equipment are usually omitted. The interconnection diagrams will appear adjacent to the schematic diagram or on a separate drawing, depending on the format chosen when making the schematic diagram. The development of the interconnection diagram is integrated with that of the schematic diagram and only the equipment, terminal blocks, and wiring pertinent to the accompanying schematic diagram appear in the interconnection diagram. A typical interconnection diagram will contain the following:

- An outline of the equipment involved in its relative physical location
- Terminal blocks in the equipment that are concerned with the wiring illustrated on the schematic
- Wire numbers, cable sizes, cable numbers, cable routing, and cable tray identification (should not be repeated on the interconnection diagram except where necessary)
- Wiring between equipment (normally shown as individual cables but may be combined on complex drawings)
- Equipment identification information

2.2.4 Connection Diagrams

The connection diagram shows the internal wiring connections between the parts that make up an apparatus. It will contain as much detail as necessary to make or trace any electrical connections involved. A connection diagram generally shows the physical arrangement of component electrical connections. It differs from the interconnection diagram by excluding external connections between two or more assemblies or pieces of equipment.

The schematic diagram shows the arrangement of a circuit with the components represented by conventional symbols. Its intent is to show the function of a circuit. The schematic, like the elementary drawing, is not laid out with respect to physical locations.

A wiring diagram can also show the physical locations of electrical equipment and/or components along with the interconnecting wiring. It shows the actual connection point of every wire and the color of the wires connected to each terminal. It allows the electrician to easily locate terminals and wires. A wiring diagram in conjunction with a schematic greatly aids in troubleshooting a given piece of equipment. Connection diagrams can be shown in various forms.

Two types of connection diagrams are point-to-point connection diagrams and cable connection diagrams.

The point-to-point method is used for the simpler diagrams where sufficient space is available to show each individual wire without sacrificing the clarity of the diagram. Point-to-point diagrams provide accurate information to terminate and troubleshoot the wiring. *Figure 25* is an example of an internal point-to-point connection diagram.

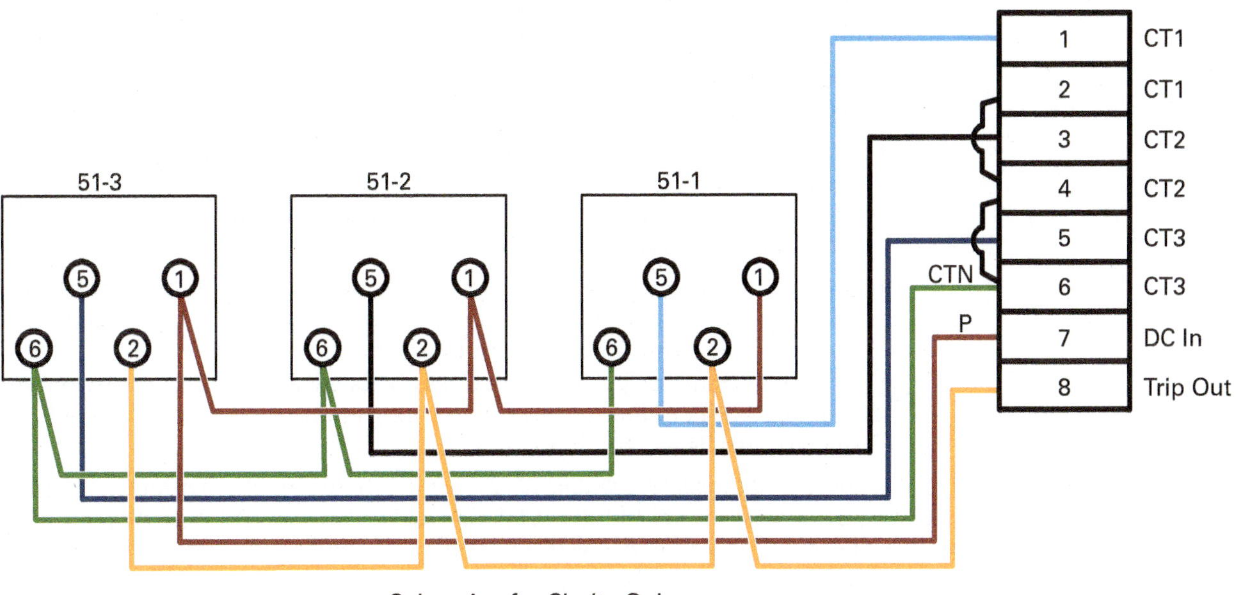

Figure 25 Point-to-point connection diagram.

In complex diagrams, only individual cables are shown between devices or terminal strips. Lines from the cable go to each termination point at the device or strip, but individual connections are not identified. Connection diagrams provide adequate information to route the cable, but not to terminate the conductors. *Figure 26* is an example of an internal cable connection diagram.

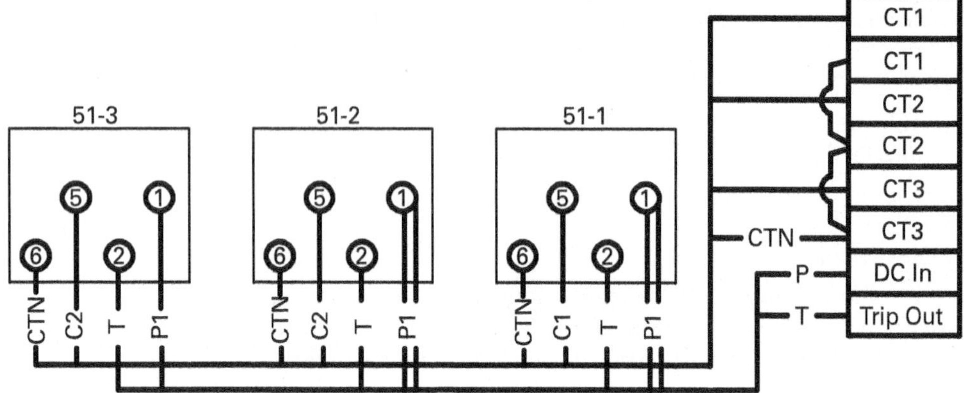

Figure 26 Cable connection diagram.

2.2.5 Manufacturer Drawings

When a project is proposed, several different types of drawings are involved. The engineer and designer work from manufacturer drawings to determine specific details of the equipment and incorporate them into the design. The electrical worker must also be familiar with the manufacturer drawings because the design drawings don't always include all the manufacturer's information.

This section discusses the information found in manufacturer drawings. *Appendix 26306-23B* includes a package of manufacturer's drawings for a new switchboard to distribute power to downstream equipment such as MCCs, panelboards, and major utilization equipment. In this case, the switchboard may be the service equipment, or it may be fed from the facility power transformer.

A brief description of each drawing sheet follows:

- *Title Sheet* — This sheet (not shown) includes the manufacturer and purchaser reference numbers, location, contact persons, dates, and other identifying information. A table of contents is also included so that the job requirements can be easily navigated.
- *Sheet 1, General Notes* — *Figure 27* shows the front view of the switchboard along with the dimensions. Descriptive notes are included to describe the equipment. Electrical ratings are noted, as well as physical data such as the enclosure type and weight. References are also made to relevant literature such as instruction manuals.
- *Sheet 2, Floor Plan* — *Figure 28* provides a top view, side view, and floor plan view. This gives the installer a footprint of the space needed for the installation. The installer must make sure to include the minimum working space around the equipment as required by the *NEC*®.
- *Sheet 3, Single-Line Diagram* — *Figure 29* shows that the project consists of a three-section switchboard. The single-line diagram for each section is shown, along with how they are interconnected with the main bus. The sizes of overcurrent protective devices are listed, along with their trip characteristics. Where interconnecting wiring is to be field-installed, details of that are also listed.
- *Sheet 4, Equipment Schedule* — *Figure 30* shows all the equipment that is installed in the switchboard. All the breakers are listed along with their size, frame designation, trip setting, location, function, and lug size. Additional accessories and information are also listed, including a legend.

NOTE

The figures in *Section 2.2.5* are representative of the drawings and are not large enough for practical use. *Appendix 26306-23B* contains a QR code that directs you to the full-size drawings for review online.

- *Sheet 5, Wiring Diagram* — *Figure 31* shows the wiring diagram for each device installed in the switchboard. Any interconnection between devices is also documented. Any notes for wiring are also included; these may include the wire size and type. This shows both the factory and field wiring.
- *Sheet 6, Catalog Sheet* — *Figure 32* shows catalog data for the equipment and is sometimes referred to as a *bill of materials* (*BOM*). It provides catalog numbers and additional details on the components included in the order.

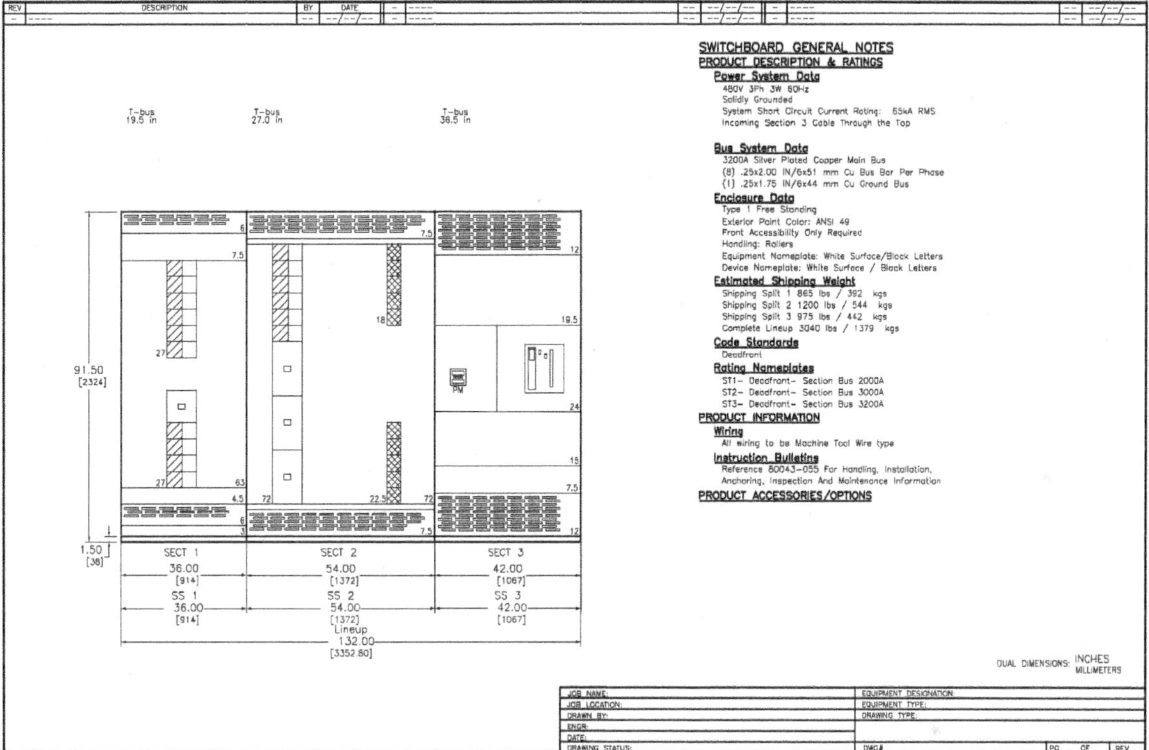

Figure 27 Sheet 1, general notes.

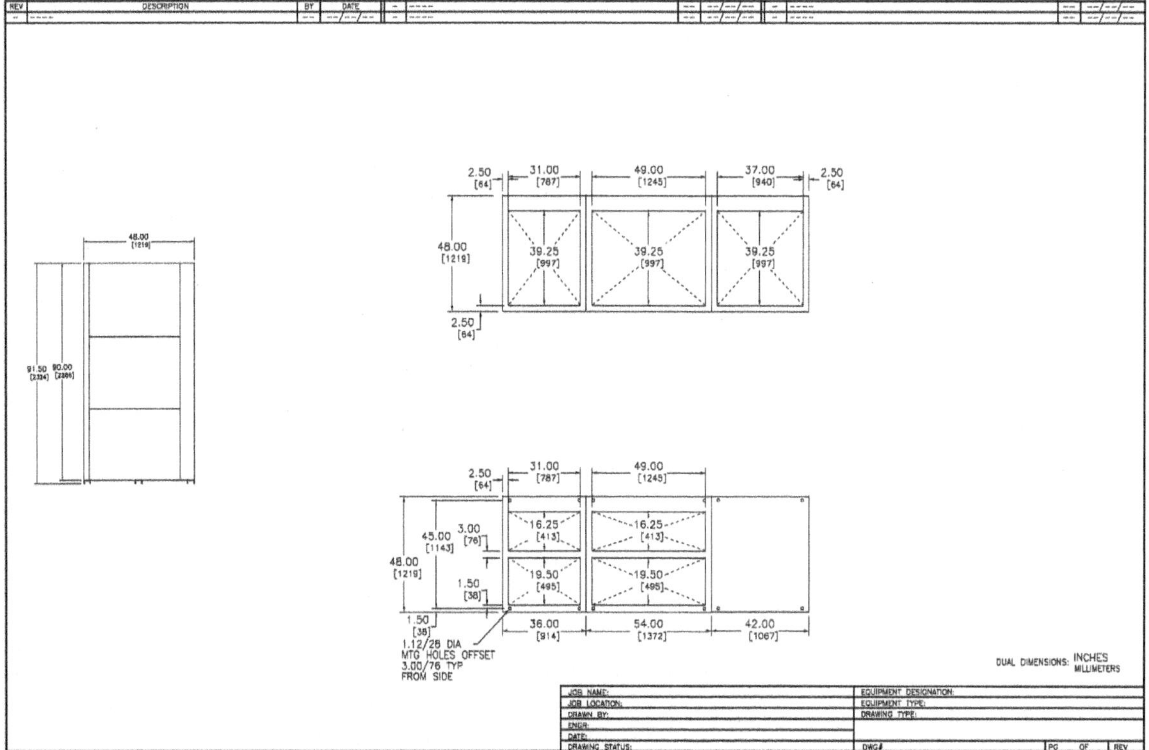

Figure 28 Sheet 2, floor plan.

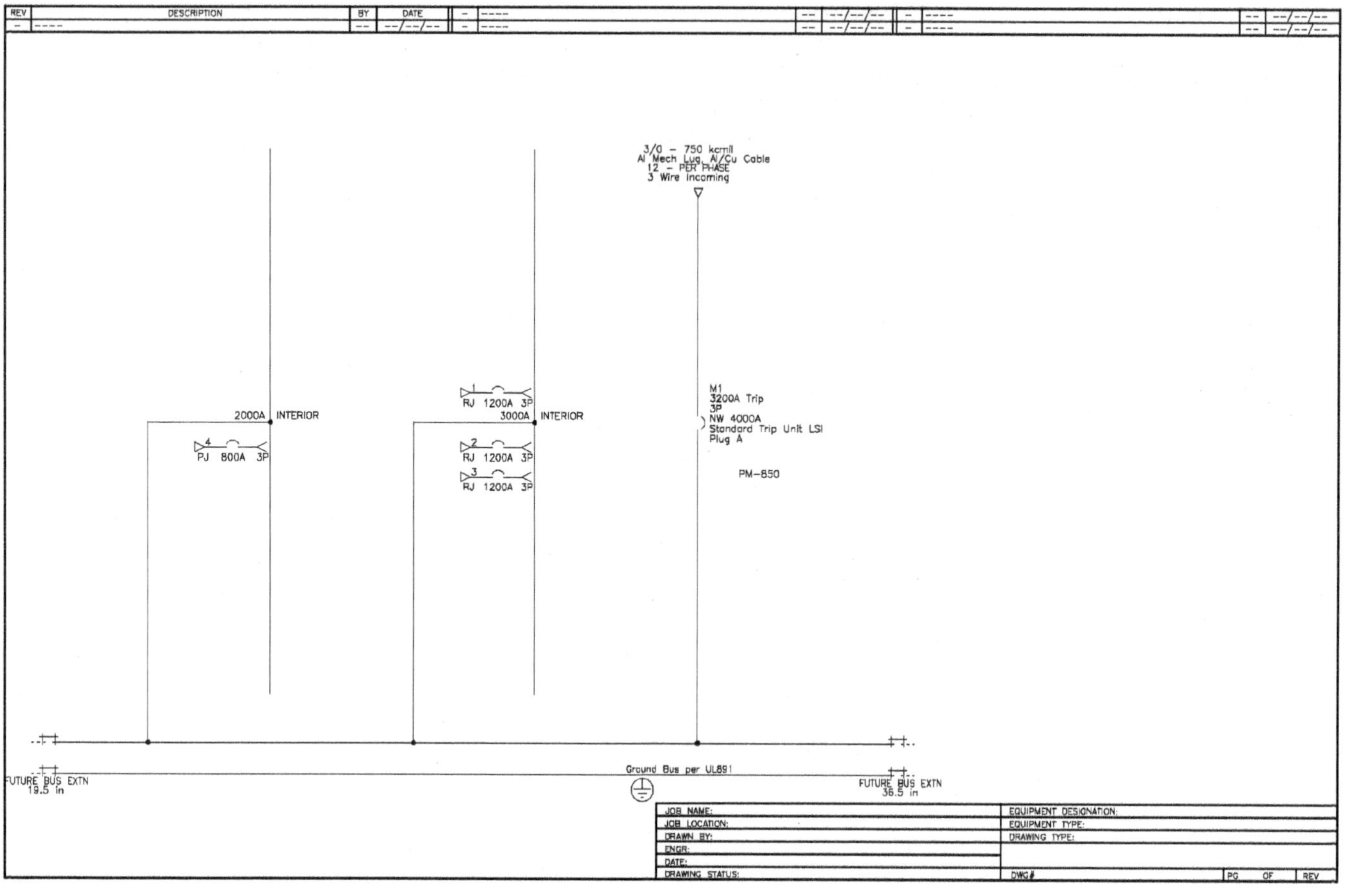

3/0 – 750 kcmil
Al Mech Lug, Al/Cu Cable
12 – PER PHASE
3 Wire Incoming

M1
3200A Trip
3P
NW 4000A
Standard Trip Unit LSI
Plug A

PM-850

2000A INTERIOR
4 PJ 800A 3P

1 RJ 1200A 3P
3000A INTERIOR
2 RJ 1200A 3P
3 RJ 1200A 3P

FUTURE BUS EXTN 19.5 in
Ground Bus per UL891
FUTURE BUS EXTN 36.5 in

JOB NAME:
JOB LOCATION:
DRAWN BY:
ENGR:
DATE:
DRAWING STATUS:
EQUIPMENT DESIGNATION:
EQUIPMENT TYPE:
DRAWING TYPE:
DWG#
PG OF REV

Figure 29 Sheet 3, single-line diagram.

POWER STYLE QED-2 SWITCHBOARD

SECT NO	CKT NO	GMD HEIGHT	DEVICE/FRAME RATING	TRIP AMP	FUSE/TRIP	#P	DESIGNATION	N/P	LUG INFORMATION				ACCESSORIES
									QTY	PHASE WIRE RANGE	QTY	NEUT. WIRE RANGE	
1	4	9 in	PJ 800A	800A	LI	3P	PFC	Yes	3	3/0 – 500 kcmil	–	–	PLA
2	1	15 in	RJ 1200A	1200A	LI	3P	SPARE	Yes	4	3/0 – 600kcmil	–	–	PLA
2	2	15 in	RJ 1200A	1200A	LI	3P	MCC	Yes	4	3/0 – 600kcmil	–	–	PLA
2	3	15 in	RJ 1200A	1200A	LI	3P	MCC	Yes	4	3/0 – 600kcmil	–	–	PLA
3	M1	–	NW 4000A	3200A	LSI	3P	MAIN BREAKER	Yes	12	3/0 – 750 kcmil	–	–	PM/W23008090-008-01,OC,OF4,PLA,

LEGEND	
OC	Over Current Trip Switch
OF4	4 Form C Breaker Aux Contacts
PLA	Padlock Attachment-Fixed
PM	Power Meter (PM-850)

JOB NAME:
JOB LOCATION:
DRAWN BY:
ENGR:
DATE:
DRAWING STATUS:
EQUIPMENT DESIGNATION:
EQUIPMENT TYPE:
DRAWING TYPE:
DWG#
PG OF REV

Figure 30 Sheet 4, equipment schedule.

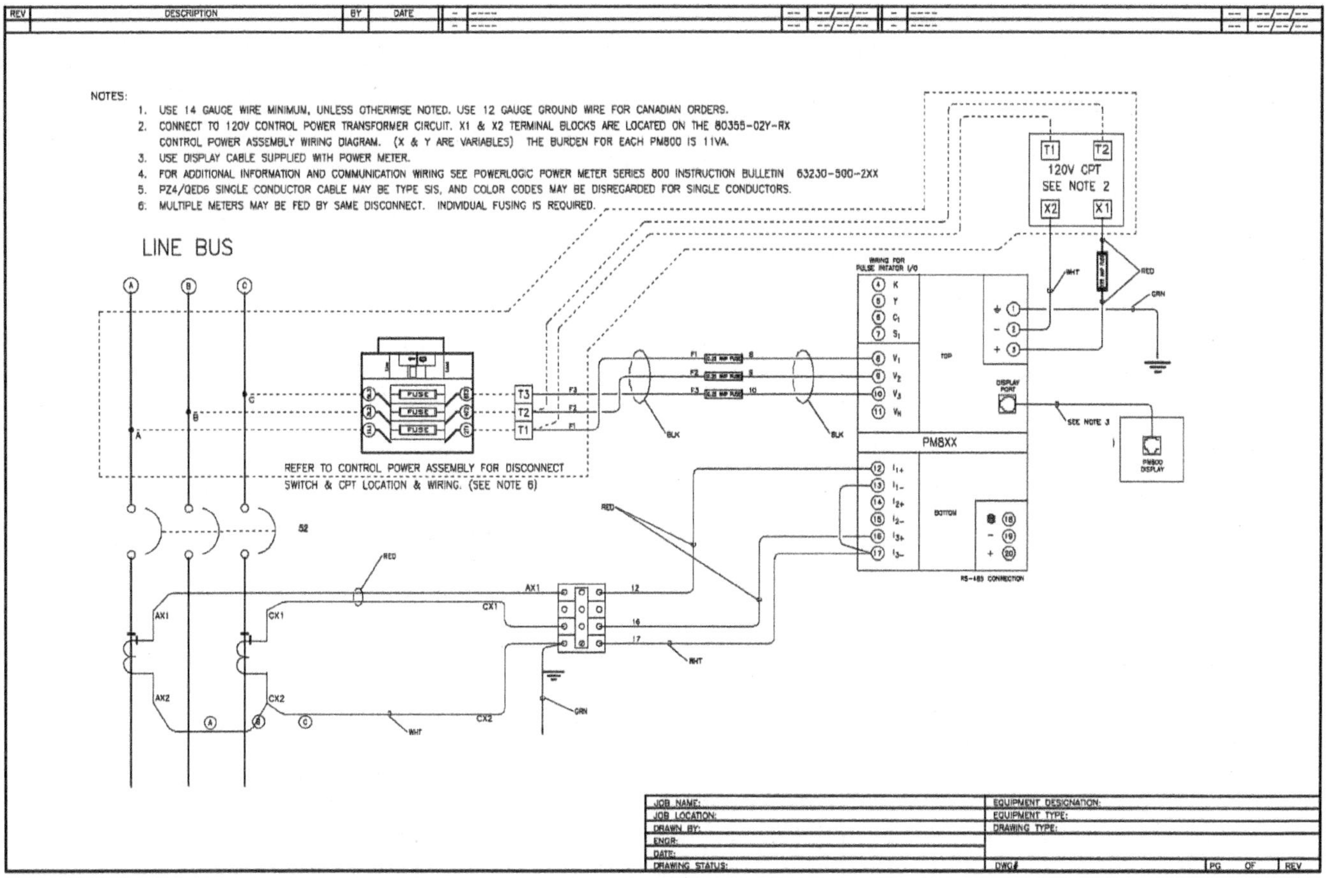

REV
DESCRIPTION
BY
DATE
NOTES:
1. USE 14 GAUGE WIRE MINIMUM, UNLESS OTHERWISE NOTED. USE 12 GAUGE GROUND WIRE FOR CANADIAN ORDERS.
2. CONNECT TO 120V CONTROL POWER TRANSFORMER CIRCUIT. X1 & X2 TERMINAL BLOCKS ARE LOCATED ON THE 80355-02Y-RX CONTROL POWER ASSEMBLY WIRING DIAGRAM. (X & Y ARE VARIABLES) THE BURDEN FOR EACH PM800 IS 11VA.
3. USE DISPLAY CABLE SUPPLIED WITH POWER METER.
4. FOR ADDITIONAL INFORMATION AND COMMUNICATION WIRING SEE POWERLOGIC POWER METER SERIES 800 INSTRUCTION BULLETIN 63230-500-2XX
5. PZ4/QED6 SINGLE CONDUCTOR CABLE MAY BE TYPE SIS, AND COLOR CODES MAY BE DISREGARDED FOR SINGLE CONDUCTORS.
6. MULTIPLE METERS MAY BE FED BY SAME DISCONNECT. INDIVIDUAL FUSING IS REQUIRED.
LINE BUS
T1
T2
120V CPT
SEE NOTE 2
X2
X1
WIRING FOR PULSE INITATOR I/O
FUSE
T3
T2
T1
PM8XX
REFER TO CONTROL POWER ASSEMBLY FOR DISCONNECT SWITCH & CPT LOCATION & WIRING. (SEE NOTE 6)
SEE NOTE 3
AX1
CX1
AX2
CX2
RS-485 CONNECTION
JOB NAME:
JOB LOCATION:
DRAWN BY:
ENGR:
DATE:
DRAWING STATUS:
EQUIPMENT DESIGNATION:
EQUIPMENT TYPE:
DRAWING TYPE:
DWG#
PG
OF
REV

Figure 31 Sheet 5, wiring diagram.

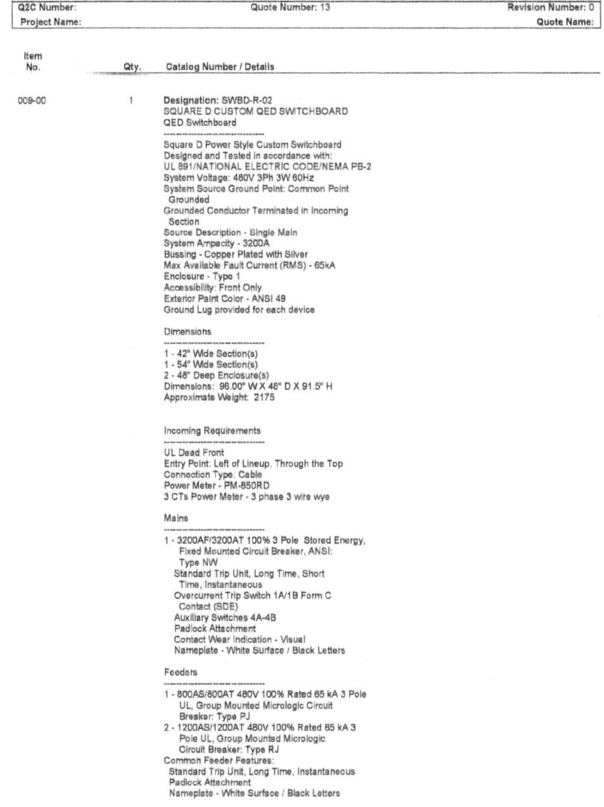

Q2C Number:	Quote Number: 13	Revision Number: 0
Project Name:		Quote Name:

Item No.	Qty.	Catalog Number / Details
009-00	1	**Designation: SWBD-R-02**

SQUARE D CUSTOM QED SWITCHBOARD
QED Switchboard

Square D Power Style Custom Switchboard
Designed and Tested in accordance with:
UL 891/NATIONAL ELECTRIC CODE/NEMA PB-2
System Voltage: 480V 3Ph 3W 60Hz
System Source Ground Point: Common Point Grounded
Grounded Conductor Terminated in Incoming Section
Source Description - Single Main
System Ampacity - 3200A
Bussing - Copper Plated with Silver
Max Available Fault Current (RMS) - 65kA
Enclosure - Type 1
Accessibility: Front Only
Exterior Paint Color - ANSI 49
Ground Lug provided for each device

Dimensions
----
1 - 42" Wide Section(s)
1 - 54" Wide Section(s)
2 - 48" Deep Enclosure(s)
Dimensions: 96.00" W X 48" D X 91.5" H
Approximate Weight: 2175

Incoming Requirements

UL Dead Front
Entry Point: Left of Lineup, Through the Top
Connection Type: Cable
Power Meter - PM-850RD
3 CTs Power Meter - 3 phase 3 wire wye

Mains
----
1 - 3200AF/3200AT 100% 3 Pole Stored Energy, Fixed Mounted Circuit Breaker, ANSI: Type NW
Standard Trip Unit, Long Time, Short Time, Instantaneous
Overcurrent Trip Switch 1A/1B Form C Contact (SDE)
Auxiliary Switches 4A-4B
Padlock Attachment
Contact Wear Indication - Visual
Nameplate - White Surface / Black Letters

Feeders
----
1 - 800AS/800AT 480V 100% Rated 65 kA 3 Pole UL, Group Mounted Micrologic Circuit Breaker: Type PJ
2 - 1200AS/1200AT 480V 100% Rated 65 kA 3 Pole UL, Group Mounted Micrologic Circuit Breaker: Type RJ
Common Feeder Features:
Standard Trip Unit, Long Time, Instantaneous
Padlock Attachment
Nameplate - White Surface / Black Letters

Figure 32 Sheet 6, catalog sheet.

2.2.6 Shop Drawings

When large pieces of electrical equipment are needed, such as high-voltage switchgear and MCCs, most are custom built for each individual project. In doing so, shop drawings are normally furnished by the equipment manufacturer prior to shipment to ensure the equipment will fit the location at the shop site and to instruct the workers about preparing for such equipment as rough-in conduit and cable trays. The drawing in *Figure 33* is one page of a shop drawing showing a pictorial view of the enclosure. Shop drawings will also usually include connection diagrams for all components that must be field wired or connected.

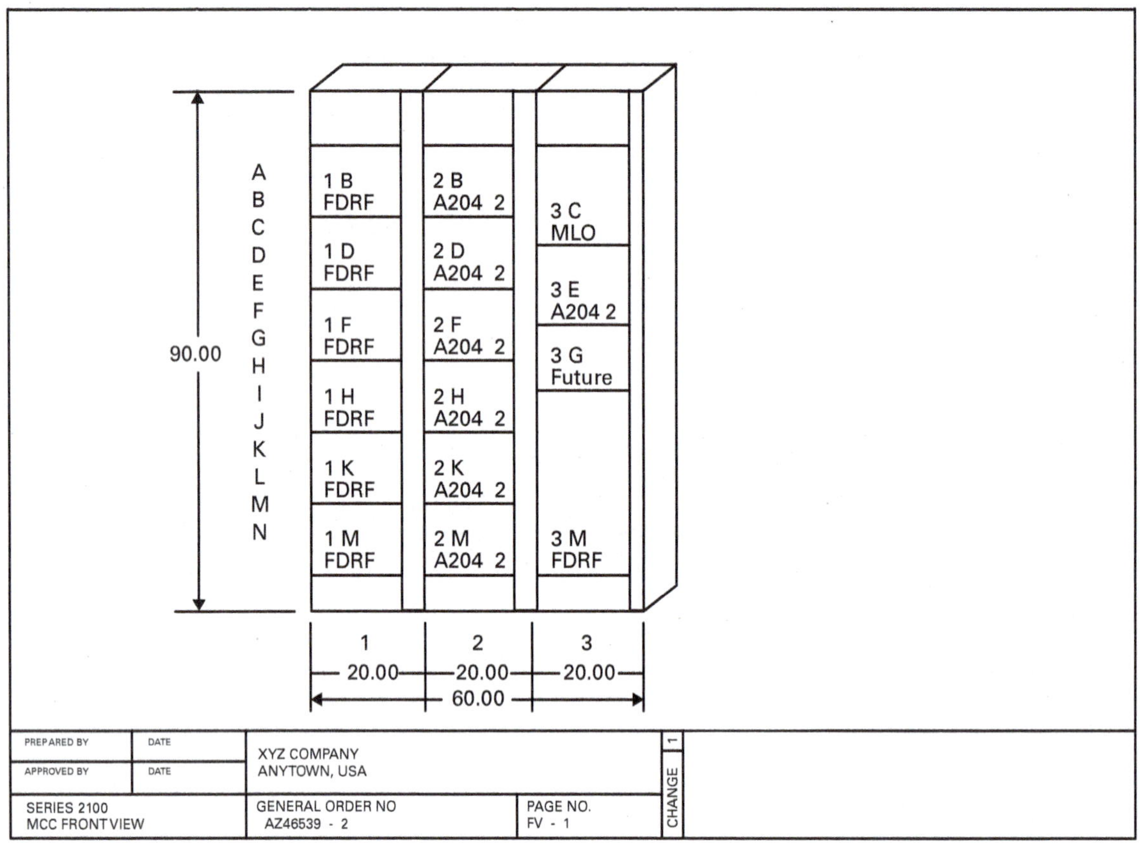

Figure 33 View of an MCC.

2.2.7 As-Built Drawings

As-built drawings, including detailed factory connection diagrams, are also included to assist workers and maintenance personnel in making the final connections, and in troubleshooting problems once the system is in operation.

Typical drawings of this type are shown in *Figure 34* and *Figure 35*.

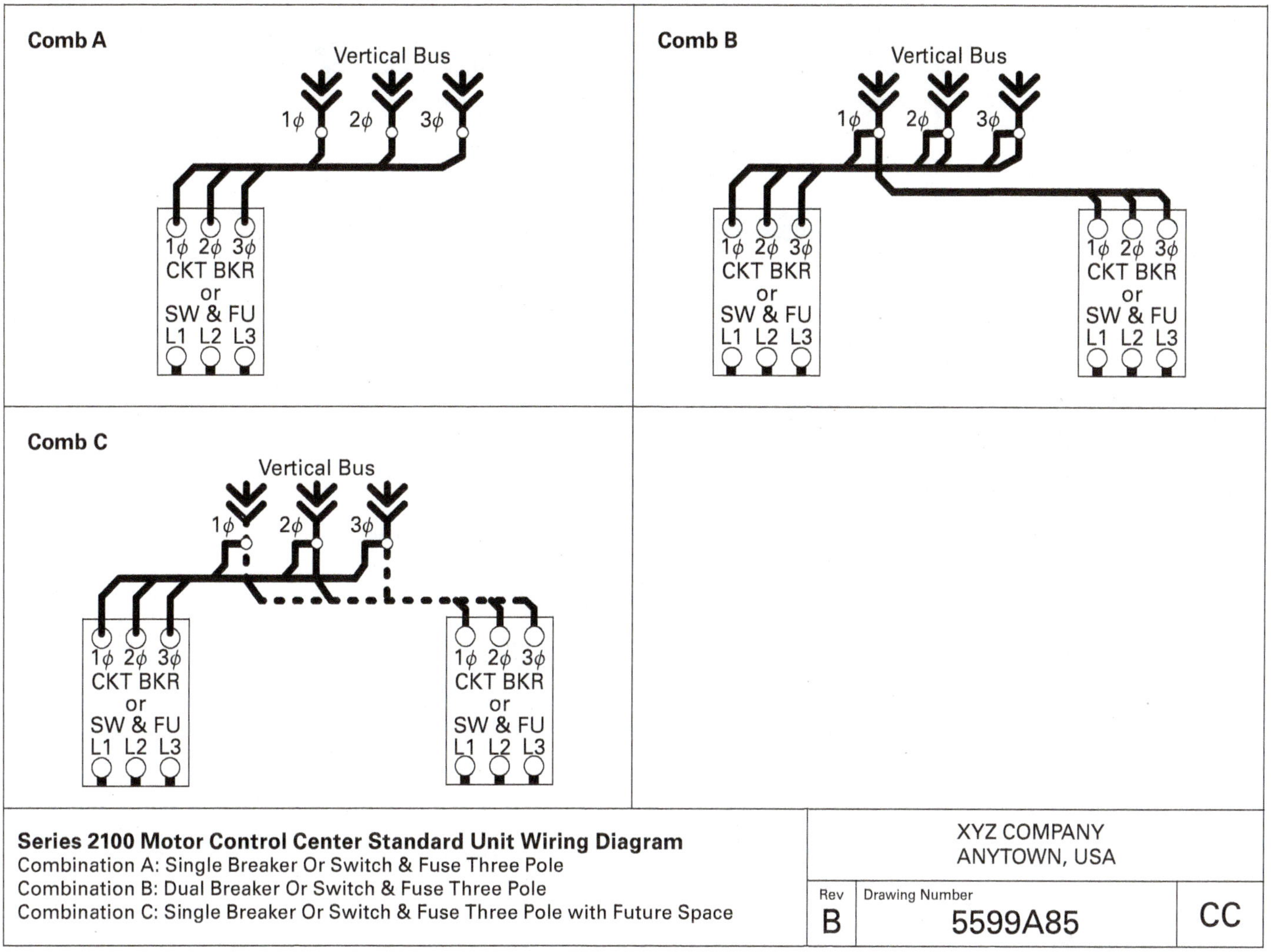

Comb A
Vertical Bus
1ϕ 2ϕ 3ϕ
1ϕ 2ϕ 3ϕ
CKT BKR
or
SW & FU
L1 L2 L3
Comb B
Vertical Bus
1ϕ 2ϕ 3ϕ
1ϕ 2ϕ 3ϕ
CKT BKR
or
SW & FU
L1 L2 L3
1ϕ 2ϕ 3ϕ
CKT BKR
or
SW & FU
L1 L2 L3
Comb C
Vertical Bus
1ϕ 2ϕ 3ϕ
1ϕ 2ϕ 3ϕ
CKT BKR
or
SW & FU
L1 L2 L3
1ϕ 2ϕ 3ϕ
CKT BKR
or
SW & FU
L1 L2 L3
Series 2100 Motor Control Center Standard Unit Wiring Diagram
Combination A: Single Breaker Or Switch & Fuse Three Pole
Combination B: Dual Breaker Or Switch & Fuse Three Pole
Combination C: Single Breaker Or Switch & Fuse Three Pole with Future Space
XYZ COMPANY
ANYTOWN, USA
Rev
B
Drawing Number
5599A85
CC

Figure 34 MCC standard unit wiring diagram.

MCC Labels

NEC Section 430.99 requires the available fault current to be determined, placed on a label, dated, and attached to the MCC. This type of information also cannot be handwritten to ensure readability.

WARNING
Arc Flash and Shock Risk
Appropriate PPE Required
Maximum available fault current:
29,049 Amps
Date: 9/12/16
ACME Electric
MCC #10-5A

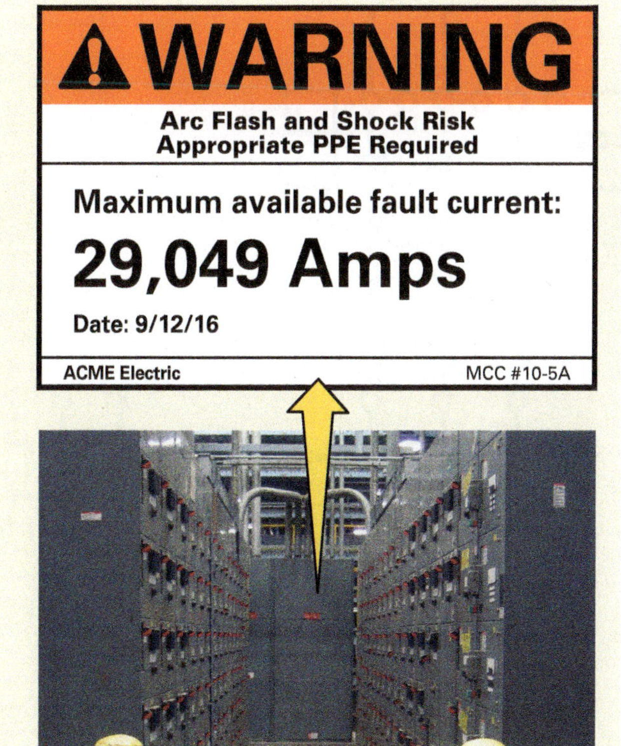

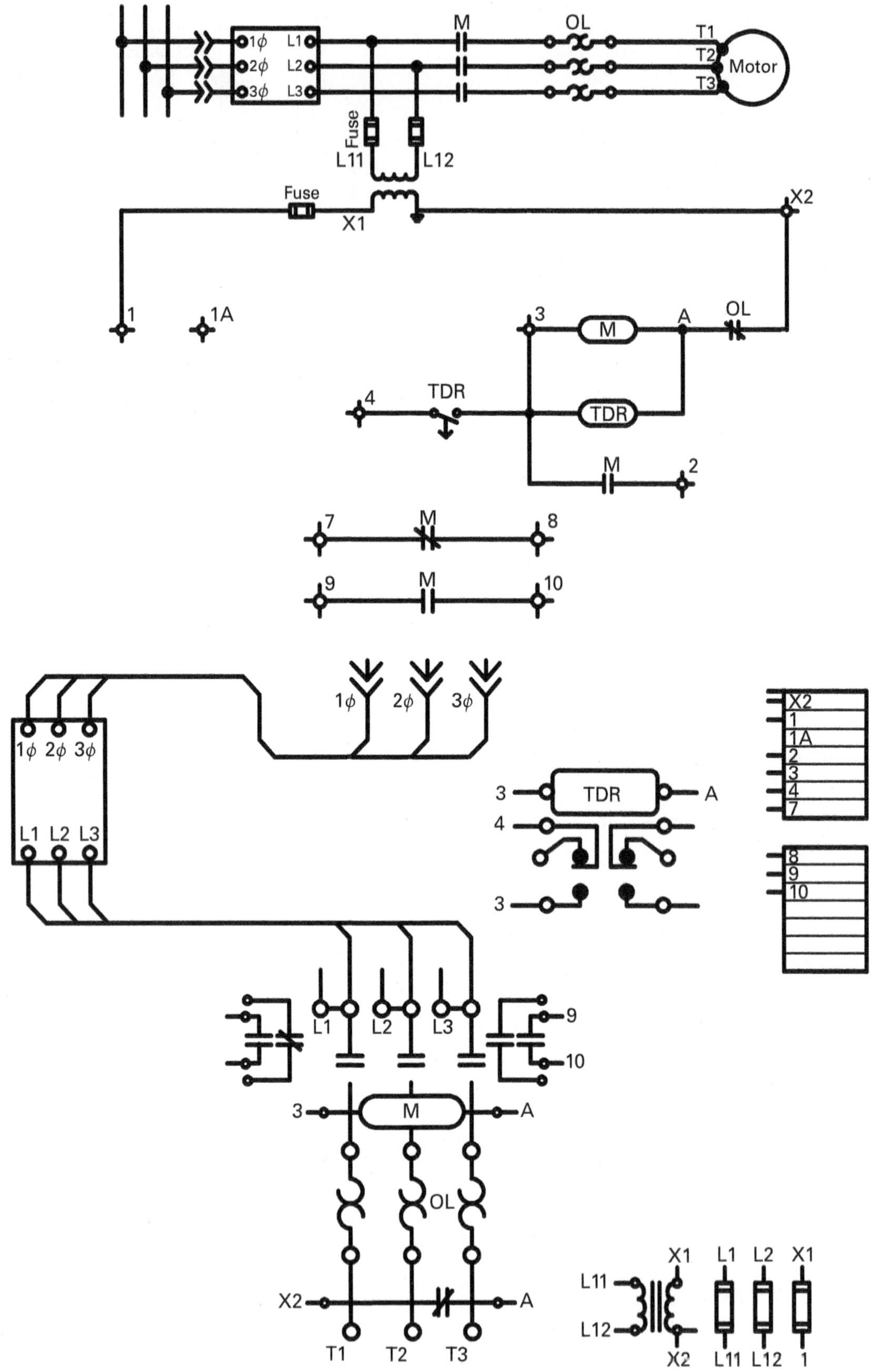

Figure 35 Unit diagrams for MCC.

2.0.0 Section Review

1. Per *NEC Section 408.3(E)(1)*, AC phasing in switchboards *must* be labeled _____.
   a. C, B, A from front to back, top to bottom, or left to right when read facing the front of the switchboard
   b. C, A, B from front to back, top to bottom, or left to right when read facing the front of the switchboard
   c. B, C, A from front to back, top to bottom, or left to right when read facing the front of the switchboard
   d. A, B, C from front to back, top to bottom, or left to right when read facing the front of the switchboard

2. On a typical electrical diagram, the standard abbreviation for a phase meter is _____.
   a. PM
   b. PH
   c. 3M
   d. S/TM

3.0.0 Testing and Maintaining Switchgear

Objective

Explain how to test and maintain switchgear.

a. Identify general maintenance guidelines for switchboards and switchgear.
b. State the guidelines for testing switchboards and switchgear.
c. Identify and describe devices used to monitor power distribution systems for ground faults.

Performance Tasks

There are no Performance Tasks in this section.

This section introduces general testing and maintenance procedures for switchgear. Note that these procedures may only be performed by qualified and authorized personnel operating under a safe work plan or permit.

WARNING!

When working on switchgear or any piece of electrical equipment, you must be aware of and follow all applicable safety procedures. You must also understand the construction and operation of the equipment. You must be specifically trained and qualified to work on or near energized electrical circuits and equipment. National consensus standards such as NFPA 70E® and NFPA 70B provide specific guidance for achieving an electrically safe working condition.

Article 120 in NFPA 70E®, *Standard for Electrical Safety in the Workplace®*, provides a step-by-step procedure for implementing the appropriate safety measures.

Chapter 7 in NFPA 70B, *Standard for Electrical Equipment Maintenance*, provides guidelines for personnel safety for qualified electrical workers, while other chapters provide specific direction for maintenance and troubleshooting of various types of equipment.

3.1.0 General Maintenance Guidelines

To perform a visual inspection of switchboards and switchgear, follow this general procedure. However, it is essential to follow the maintenance procedures provided by the equipment manufacturer for the specific equipment being serviced:

Step 1 Check the exterior for the proper fit of doors and covers, paint, etc.

Step 2 Check the interior, particularly the current-carrying parts, including the following items:

- Inspect the busbars for dirt, corrosion, and/or overheating.
- If necessary, perform an infrared or thermographic test. Note any discoloration that would represent a poor bus joint.
- Check the busbar supports for cracks.
- Check for correct electrical spacing.
- Verify the integrity of all bolted connections.

To clean the switchboard, proceed as follows:

Step 1 Vacuum the interior (do not use compressed air).

Step 2 Wipe down the interior using a clean, lint-free cloth. Use a nonconductive, residue-free solution such as contact cleaner or denatured alcohol.

To check equipment operation, proceed as follows:

Step 1 Manually open and close circuit breakers and switches.

Step 2 Electrically operate all components, such as ground fault detectors, sure trip metering, CTs, test blocks, ground lights, blown main fuse detectors, and phase failure detectors.

To perform a megger test, proceed as follows:

Step 1 Isolate the bus by opening all circuit breakers and switches.

Step 2 Disconnect any devices, such as relays and transformers, that may be connected to the busbars.

Step 3 Make sure all personnel are clear of the switchboard.

Step 4 Use a 1,000V megger to check the phase-to-phase and phase-to-ground resistance. Megger readings should reflect the values listed in the equipment manufacturer's instructions. Typical values are shown in *Table 2*.

TABLE 2 Typical Insulation Resistance Tests on Electrical Apparatus and Systems at 68°F (20°C)

Minimum Voltage Rating of Equipment	Minimum Test Voltage	Recommended Minimum Insulation Resistance
2V to 250V	500VDC	50 MΩ
251V to 600V	1,000VDC	100 MΩ
601V to 5,000V	2,500VDC	1,000 MΩ
5,001V to 15,000V	2,500VDC	5,000 MΩ
15,001V to 39,000V	5,000VDC	20,000 MΩ

3.2.0 Test Guidelines

This section provides typical guidelines for testing distribution equipment.

3.2.1 Thermographic Survey

A thermographic (infrared) survey (*Figure 36*) involves checking items such as switches, busways, open buses, switchgear, cable and bus connections, and circuit breakers. This is only a partial list of the many devices that are inspected through thermography.

WARNING!

The following test is performed while the equipment is energized, and the covers are removed. This test may only be performed by qualified personnel under the appropriate safe work plan or permit.

Figure 36 Infrared camera used in thermographic surveys.
Source: Reproduced with Permission, Fluke Corporation.

Infrared surveys are best performed when maximum loads are present. The highest temperatures are generated under maximum load. At least 40% of the rated load should be active for a meaningful analysis. However, there is no reason that a simple thermographic survey cannot be applied under any operating conditions.

Thermographers must understand that the instrument sees the temperature at the surface. In many cases, the actual source of heat is deeper inside or behind that surface. Thus, hot spots on the surface may only reflect heat transfer from a hotter source. A thermal gradient represents the temperature difference between the two points. Analyzing the gradients observed can help identify places where something hotter is beneath the surface.

Meggers

To test for potential insulation breakdown, phase-to-phase shorts, or phase-to-ground shorts in switchgear, you need to apply a much higher potential than that supplied by the battery of an ohmmeter. A megohmmeter, or *megger*, is commonly used for these tests. The megger is a portable instrument consisting of a hand-driven DC generator (which supplies the level of voltage for making the measurement) and the instrument portion (which indicates the value of the resistance being measured).

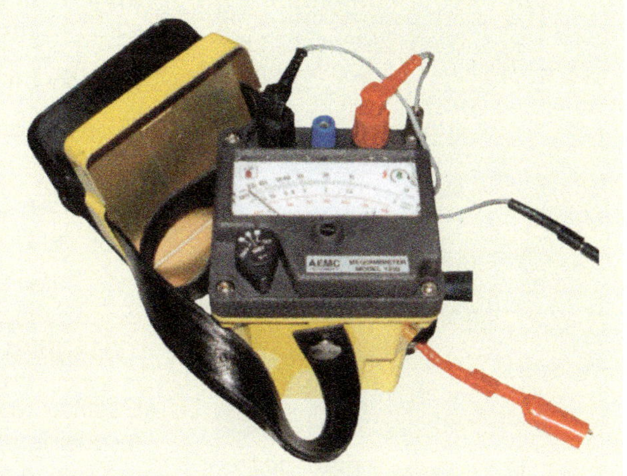

The gradient is often subtle and doesn't have to be large to indicate a problem. Negative test results can include the following:

- Temperature gradients of 2°F to 5°F (1°C to 3°C) may be worthy of further investigation.
- Temperature gradients of 7°F to 27°F (4°C to 15°C) indicate a deficiency. Repair as time permits.
- Temperature gradients of 30°F (16°C) and higher indicate a major deficiency. De-energize and secure the power source and resolve the problem.

3.2.2 Metal-Enclosed Switchgear and Switchboards

To perform a visual and mechanical inspection, proceed as follows:

WARNING!

You must be qualified and authorized to perform these tests. Ensure there is no voltage present prior to testing.

Keyed Interlocks

Keyed interlocks, such as the one shown in the image, ensure qualified personnel perform operations in the required sequence. This is achieved by preventing or allowing the operation of one part only when another part is locked in a predetermined position. These devices can be used for a variety of safety applications, such as preventing personnel from accessing a high-voltage compartment before opening the disconnect switch.

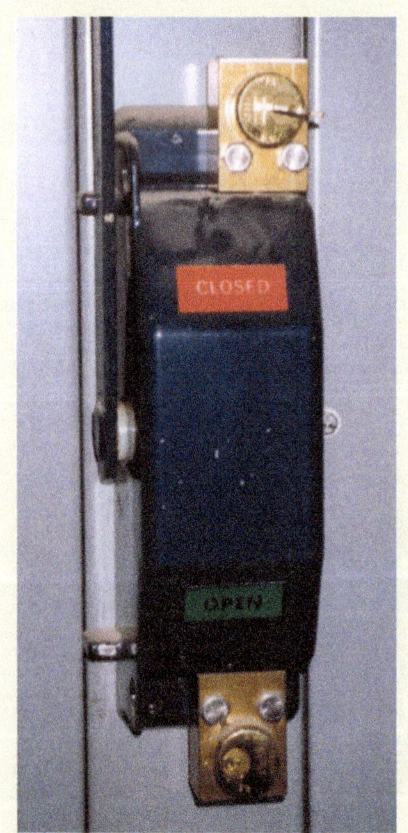

Source: Tri-City Electrical Contractors Inc.

Step 1 Inspect the physical, electrical, and mechanical condition of the equipment.

Step 2 Compare the equipment nameplate information with the latest single-line diagram and report any discrepancies.

Step 3 Check for proper anchorage, required area clearances, physical damage, and proper alignment.

Step 4 Inspect all doors, panels, and sections for missing paint, dents, scratches, fit, and missing hardware.

Step 5 Inspect all bus connections for high resistance. Use a low-resistance ohmmeter or check tightness of bolted bus joints using a calibrated torque wrench.

Step 6 Test all electrical and mechanical interlock systems for the following proper operation and sequencing:

- A closure attempt must be made on all locked-open devices. An opening attempt must be made on all locked-closed devices.
- A key exchange must be made with all devices operated in normally off positions.

Step 7 Clean the entire switchgear using the manufacturer's approved methods and materials.

Step 8 Inspect insulators for evidence of physical damage or contaminated surfaces.

Step 9 Inspect the lubrication as follows:

- Verify manufacturer-approved contact lubricant on moving current-carrying parts.
- Verify appropriate lubrication of moving and sliding surfaces.
- Exercise all active components.
- Inspect all indicating devices for proper operation.

WARNING!

Electrical testing may produce hazardous voltages and may only be performed by qualified personnel under the appropriate safe work plan or permit. Prepare the area to avoid any accidental contact with the system under test and wear appropriate PPE.

To perform electrical testing, proceed as follows:

Step 1 Perform ratio and polarity tests on all current and voltage transformers.

Step 2 Perform ground resistance tests.

Step 3 Perform insulation resistance tests on each bus section (phase to phase and phase to ground) for one minute. Refer to the specific manufacturer's guidelines, an example of which is shown in *Table 2*.

Step 4 Perform an overpotential test on each bus section (phase to ground) for one minute. Refer to specific manufacturer's guidelines, an example of which is shown in *Table 3*.

Step 5 Perform an insulation resistance test on the control wiring. Do not perform this test on wiring connected to solid-state components.

Step 6 Perform a phasing check on double-ended switchgear to ensure proper bus phasing from each source.

NOTE

The values shown in *Table 2* and *Table 3* are typical acceptance values. Maintenance values will vary by manufacturer.

TABLE 3 Overpotential DC Test Voltages for Electrical Apparatus Other than Inductive Equipment

Nominal Voltage Class	DC Test Voltage Maximum New	DC Test Voltage Maximum Used
250V	2,500VDC	1,500VDC
600V	3,500VDC	2,000VDC
5,000V	18,000VDC	11,000VDC
15,000V	50,000VDC	30,000VDC

Any values of insulation resistance less than those listed in the manufacturer's literature should be investigated. Overpotential tests should not proceed until insulation resistance levels are raised above minimum values.

Overpotential test voltages must be applied in accordance with the manufacturer's literature. Test results are evaluated on a go/no-go basis by slowly raising the test voltage to the required value. The final test voltage is applied for one minute.

3.2.3 Low-Voltage Cables (600V Maximum)

To perform a visual and mechanical inspection, proceed as follows:

Step 1 Inspect cables for physical damage and proper connection in accordance with the single-line diagram.

Step 2 Verify the integrity of all bolted connections.

Step 3 Check color coded cable against the applicable engineer's specifications and *NEC*® standards.

To perform electrical testing, proceed as follows:

Step 1 Perform an insulation resistance test on each conductor with respect to ground and adjacent conductors. The applied potential should be 1,000VDC for one minute.

Step 2 Perform a continuity test to ensure proper cable connection. The minimum insulation resistance values must not be less than two MΩ.

3.2.4 Medium-Voltage Cables (15 kV Maximum)

To perform a visual and mechanical inspection, proceed as follows:

Step 1 Inspect exposed sections for physical damage.

Step 2 Inspect for shield grounding, cable support, and termination.

Step 3 Inspect for proper fireproofing in common cable areas.

Step 4 If cables are terminated through window-type CTs, verify that neutrals and grounds are properly terminated for normal operation of the protective devices.

Step 5 Visually inspect the jacket and insulation condition.

Step 6 Inspect for proper phase identification and arrangement.

3.2.5 Metal-Enclosed Busways

To perform a visual and mechanical inspection, proceed as follows:

Step 1 Inspect the bus for physical damage.

Step 2 Inspect for proper bracing, suspension, alignment, and enclosure.

Step 3 Check the tightness of bolted joints using a calibrated torque wrench.

Step 4 Check for proper physical orientation per the manufacturer's labels to ensure proper cooling. Perform continuity tests on each conductor to verify that proper phase relationships exist.

Step 5 Check outdoor busways for removal of weephole plugs if applicable and check for the proper installation of a joint shield.

To perform electrical testing, proceed as follows:

Step 1 Perform an insulation resistance test. Measure the insulation resistance on each bus run (phase to phase and phase to ground) for one minute.

Step 2 Perform AC or DC overpotential tests on each bus run, both phase to phase and phase to ground.

Step 3 Perform a contact resistance test on each connection point of the uninsulated bus. On an insulated bus, measure the resistance of the bus section and compare values with adjacent phases.

Step 4 Insulation resistance test voltages and resistance values must be in accordance with the manufacturer's specifications.

Step 5 Apply overpotential test voltages in accordance with the manufacturer's specifications.

Manufacturer's Data

Never assume anything when it comes to equipment operation, testing, or maintenance. Always refer to the manufacturer's installation, operating, and maintenance instructions for the equipment in use. These materials provide important data that explain the warranty requirements, appropriate test procedures, and specific maintenance and test points.

3.2.6 Metering and Instrumentation

To perform a visual and mechanical inspection, proceed as follows:

Step 1 Examine all devices for broken parts, indication of shipping damage, and wire connection tightness.

Step 2 Verify that meter connections are in accordance with appropriate diagrams.

To perform electrical testing, proceed as follows:

Step 1 Check the calibration of meters at all cardinal points.

Step 2 Calibrate watt-hour meters to one-half of 1% (0.5%).

Step 3 Verify all instrument multipliers.

3.3.0 Ground Faults

Ground faults can create significant arcs while the current remains relatively low, which conventional breakers might not detect. Ground fault protection is used to protect equipment and cables against these low-level faults. *NEC Section 230.95* addresses ground fault protection of equipment.

The *NEC*® requires ground fault protection on solidly grounded wye services of more than 150V to ground but not over a phase-to-phase voltage of 1,000V with each service disconnect rated 1,000A or more.

Know the Code

Ground-Fault Protection of Equipment
NEC Section 230.95

3.3.1 Ground Fault Monitoring Systems

Ground fault relays are used to protect equipment against ground faults. The three basic methods of sensing ground faults include the following:

- Ground-return method
- Residual method
- Zero-sequence method

The *ground-return method* incorporates a sensing coil around the grounding electrode conductor. The *residual method* uses three individual sensing coils to monitor the current on each phase conductor. The *zero-sequence method* requires a single, specially designed sensor to monitor all the phases and the neutral conductor of a system at the same time (*Figure 37*).

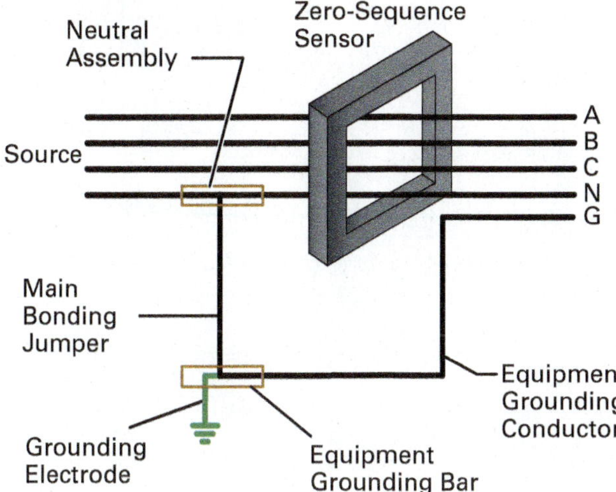

Figure 37 Zero-sequencing detection diagram.

3.3.2 Sensing Operation

When circuit conditions are normal, the currents from all the phase and neutral (if used) conductors add up to zero, and the CT sensor produces no signal. When any ground fault occurs, the currents add up to equal the ground fault current, and the sensor produces a signal proportional to the ground fault. This signal provides power to the ground fault relay, which trips the circuit breaker.

A ground fault lasting for less than the time-delay period will not energize the ground trip coil, thus eliminating nuisance tripping of self-clearing faults.

A ground fault relay is a high-reliability device due to its solid-state construction. The use of redundant, self-protecting, and high-reliability components further improves performance. Self-protection against failure is provided through an internal fuse that will blow and result in a tripping function if the solid-state circuitry fails during a ground fault situation.

3.3.3 Zero-Sequencing Sensor Mounting

The CT-style sensor should be mounted so that all phase and neutral (if present) conductors pass through the core window once. The grounding conductor (if present) must not pass through the core window. The neutral conductors must be free of all grounds after passing through the core window, as shown in *Figure 37*.

When specified by the system design engineer, the sensor may be mounted so that only the conductor connecting the neutral to ground at the service equipment passes through the core window. In such cases, the sensor must provide power to the specific ground fault relay that is associated with the main circuit breaker.

Transformer Grounding

This is the secondary termination compartment of a 2.5 MVA pad-mount transformer that shows the connection and arrangement of parallel feeders serving downstream 480V switchgear. This is a solidly grounded wye transformer.

Source: Jim Mitchem

Maintain at least 2" of clearance from the iron core of the sensor to the nearest busbar or cable to avoid false trips. Cable conductors should be bundled securely and braced to hold them at the center of the core window. The sensor should be mounted within an enclosure and protected from mechanical damage.

3.3.4 Relay Mounting

The ground fault relay should be mounted in a vertical position within an enclosure with the terminal block at the lower end. The location of the relay should be such that the trip setting knob is accessible without exposing the operator to contact with live parts or arcing from disconnect operations.

3.3.5 Connections

Connections for standard applications should be made in accordance with the wiring diagrams in the manufacturer's literature. An example of one circuit is shown in *Figure 38*. Wires from the sensor to the ground fault relay should not be longer than 25' (7.6 m) and no smaller than No. 14 AWG wire. Wires from the ground fault relay to the trip coil should not be longer than 50' (15.2 m) and no smaller than No. 14 AWG wire. All wires should be protected from arcing fault and physical damage by barriers, conduit, armor, or location in an equipment enclosure. Do not disconnect or short circuit wires to the circuit breaker trip coil at any time when the power is turned on.

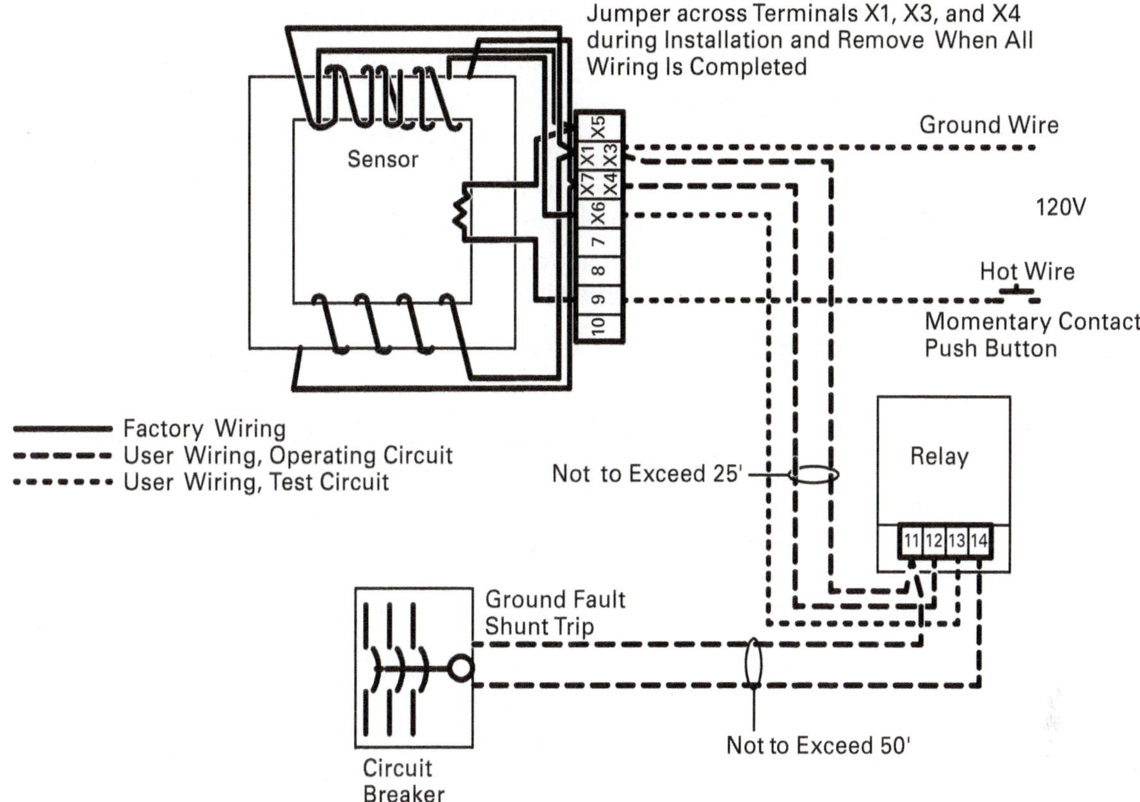

Figure 38 Typical wiring diagram.

3.3.6 Relay Settings

The ground fault relay has an adjustable trip setting. The amount of time delay is factory set and is available in nominal time delays of 0.1, 0.2, 0.3, and 0.5 seconds. When ground fault protection is used in downstream steps, the feeder should have the next lower time-delay curve than the main, the branch the next lower curve than the feeder, and so on. This supports proper sequencing.

High trip settings on main and feeder circuits are desirable to avoid nuisance tripping. High settings usually do not reduce the effectiveness of the protection if the ground path impedance is reasonably low. Ground faults usually reach a value of 40% or more of the available short circuit current very quickly.

It is recommended that the magnetic trip setpoints of any downstream circuit breakers that are not equipped with ground fault protection be set as low as possible. Likewise, the ground fault relay trip settings for main or feeder circuits should be higher than the magnetic trip settings for unprotected downstream breakers where possible. This will minimize nuisance tripping of the main or feeder breaker for ground faults occurring on downstream circuits.

Standard ground-powered ground fault relays have a built-in instantaneous trip feature. This instantaneous trip has a fixed time delay of approximately $1\frac{1}{2}$ cycles, and the fixed trip setting is higher than found on most feeder or branch breakers to avoid nuisance tripping. Its purpose is to interrupt very high-current ground faults on main disconnects as quickly as possible and to protect the ground fault relay components.

3.3.7 Ground Fault System Test

This section provides an overview of a generic visual inspection and electrical test for ground faults. Always follow the procedures specified by the equipment manufacturer for the system being tested.

Ground Fault Trip Settings

Even minor ground faults will usually create an arc and cause immediate damage. Major damage can occur in a matter of a second. Ideally, the ground fault relay responds in less than 30 cycles (0.5 seconds in a 60 Hz system) to clear the fault.

To perform a visual inspection, proceed as follows:

Step 1 Inspect the components for physical damage.

Step 2 Determine if a ground sensor was located properly around the appropriate conductor(s) as follows:

- Zero-sequence sensing requires all phases and the neutral to be encircled by the sensor(s).
- Ground-return sensing requires the sensor to encircle the main bonding jumper.

Step 3 Inspect the main bonding jumper to ensure the following:

- Proper size
- Termination on the line side of the neutral disconnect link
- Termination on the line side of the sensor on zero-sequence systems

Step 4 Inspect the grounding electrode conductor to ensure the following:

- Proper size
- Correct switchboard termination

Step 5 Inspect the ground fault control power transformer for proper installation and size. When the control transformer is supplied from the line side of the ground fault protection circuit interrupting device, overcurrent protection and a circuit disconnecting means must be provided.

Step 6 Visually inspect the switchboard neutral bus downstream of the neutral disconnect line to verify the absence of ground connections.

Know the Code

Performance Testing
NEC Section 230.95(C)

Perform the following electrical tests as required by *NEC Section 230.95(C)*:

Step 1 Check for proper ground fault system performance, including correct response of the circuit interrupting device confirmed by primary/secondary ground sensor current injection, as follows:

- Measure the relay pickup current.
- Ensure the relay time delay is measured at two values above the pickup current.

Step 2 Test system operation at 57% of the rated voltage.

Step 3 Functionally check the operation of the ground fault monitor panel by performing the following:

- Trip test
- No-trip test
- Nonautomatic reset

Step 4 Verify proper sensor polarity on the phase and neutral sensors for residual systems.

Step 5 Measure the system neutral insulation resistance downstream of the neutral disconnect link to verify the absence of grounds.

Step 6 Test systems (zone interlock / time coordinates) by simultaneous ground sensor current injection and monitor for the proper response.

Evaluate the test results for the following:

- The system neutral insulation resistance should be above 100Ω and preferably one megohm or greater.
- The maximum pickup setting of the ground fault protection shall be 1,200A and the maximum time delay shall be one second for ground fault currents equal to or greater than 3,000A, according to *NEC Section 230.95(A)*.
- The relay pickup current should be within 10% of the manufacturer's calibration marks or fixed setting.
- The relay timing should be in accordance with the manufacturer's published time-current characteristics.

Know the Code

Setting
NEC Section 230.95(A)

3.0.0 Section Review

1. The typical insulation resistance of equipment rated for 480V is _____.
   a. 10 MΩ
   b. 50 MΩ
   c. 100 MΩ
   d. 1,000 MΩ

2. A typical overpotential test voltage for a used piece of equipment with a nominal voltage class of 600V is _____.
   a. 600VDC
   b. 1,000VDC
   c. 1,500VDC
   d. 2,000VDC

3. A ground fault relay should be mounted in an enclosure _____.
   a. in a vertical position with the terminal block at the lower end
   b. in a horizontal position with the terminal block at the lower end
   c. in a vertical position with the terminal block at the upper end
   d. in a horizontal position with the terminal block at the upper end

Module 26306-23 Review Questions

1. In industrial applications, medium voltage may refer to systems rated over _____.
   a. 50V up to 480V
   b. 600V up to 120,000V
   c. 1,000V up to 35,000V
   d. 2,000V up to 69,000V

2. The term *interrupting rating* refers to the _____.
   a. trip setting of a circuit breaker
   b. voltage rating of a fuse
   c. highest voltage level a device can withstand
   d. maximum current a device will safely interrupt at its rated voltage

3. Which of the following is *not* a general classification of circuit breakers?
   a. Fused circuit breaker
   b. Air circuit breaker
   c. Oil circuit breaker
   d. Gas circuit breaker

4. A 4,000A switchboard requires a *minimum* distance from the phase bus to the closest steel member of _____.
   a. 2" (50 mm)
   b. 4" (100 mm)
   c. 6" (150 mm)
   d. 12" (300 mm)

5. When an unintended path is established between an ungrounded conductor and ground, it is called a(n) _____.
   a. phase fault
   b. open circuit
   c. ground fault
   d. overload

6. A transformer rated at more than 500 kVA is considered a(n) _____.
   a. power transformer
   b. control transformer
   c. distribution transformer
   d. isolation transformer

7. The term *capacity* on a transformer nameplate refers to _____.
   a. its voltage rating
   b. its ability to transfer energy
   c. the voltage produced by the secondary
   d. the number of secondary windings

8. The term *class* on a transformer nameplate refers to _____.
   a. its use, such as control or power
   b. the type of cooling it uses
   c. whether it is step up or step down
   d. its range of operating frequencies

9. The secondary of a CT is usually designed for a rated current of _____.
   a. 5A
   b. 10A
   c. 15A
   d. 20A

10. Of the following types of CTs, which is normally used with circuit breakers?
    a. Bar
    b. Bushing
    c. Oil filled
    d. Window

11. Which of the following is a color that can be used to designate an ungrounded conductor?
    a. Green
    b. White
    c. Gray
    d. Red

12. The trip rating of a circuit breaker used as the main protective device in a panelboard *cannot* exceed _____.
    a. the total amperage of the branch breakers
    b. the amperage capacity of the busbars in the panelboard
    c. the amperage of the individual branch fuses
    d. 250V

13. What type of diagram shows the actual wiring connections between unit assemblies or equipment with each wire labeled to indicate where to terminate it?
    a. Schematic diagram
    b. Front panel diagram
    c. Interconnection diagram
    d. Block diagram

14. A device that is specifically designed to protect equipment from ground faults using sensors is a _____.
    a. molded-case circuit breaker
    b. dual-element fuse
    c. ground fault relay
    d. ground fault circuit interrupter

Answers to odd-numbered Module Review Questions are found in *Appendix A*.

Module 26306-23 Supplemental Exercises

1. Name the four general classifications of circuit breakers.

____

____

2. What are the two basic functions performed by switchgear?

____

____

3. When wiring switchboard electrical components, keep the control wires at least __________ from moving parts.

4. Specific instructions for handling, storage, installation, testing, and maintenance of distribution equipment are found in the __________.

5. What is the function of HVL and MVL switches?

____

____

6. Common causes of ground faults include __________, __________, __________, __________, __________, and __________.

7. The standard output voltage of a PT is __________ or __________, depending on whether the primary connections are from phase to phase or phase to neutral.

8. The type of conductor that may be any color except green, green with a yellow stripe, white, or gray is a(n) __________.

9. Define the term *interrupting rating*.

____

____

10. Switchboard control and meter wiring standards shall meet the requirements of __________.

11. What are the standard abbreviations for a time-delay open, pressure switch, auxiliary switch, and backlighted push button?

____

____

12. What is the purpose of a single-line (one-line) diagram?

____

____

13. The three methods of sensing ground faults include ________, ________, and ________.

14. Wires from the sensor to the ground fault relay should be no longer than ________ and no smaller than ________ AWG.

15. The instantaneous trip feature of a standard ground fault relay has a fixed time delay of approximately ________ cycles.

Answers to odd-numbered Supplemental Exercises are found in *Appendix B*.

Answers to Section Review Questions

Answer	Section Reference	Objective
Section 1.0.0		
1. c	1.1.1	1a
2. a	1.2.3	1b
3. b	1.3.1	1c
4. a	1.4.0	1d
5. b	1.5.2	1e
6. d	1.6.3	1f
Section 2.0.0		
1. d	2.1.5	2a
2. b	2.2.0	2b
Section 3.0.0		
1. c	3.1.0; *Table 2*	3a
2. d	3.2.2; *Table 3*	3b
3. a	3.3.4	3c

User Update

Did you find an error? Submit a correction by visiting **https://www.nccer.org/olf or** by scanning the QR code using your mobile device.

Source: iStock@Алексей Кравчук

Transformers

Objectives

Successful completion of this module prepares you to do the following:

1. Describe the construction and operation of transformers.
   a. Describe the operation of a typical transformer.
   b. Describe the construction characteristics of a typical transformer.
   c. Explain how to make transformer connections for various applications.
2. Identify *NEC*® requirements related to transformers and capacitors.
   a. Identify the *NEC*® requirements for transformers.
   b. Identify the *NEC*® requirements for capacitors.
   c. Identify the *NEC*® requirements for resistors and reactors.
3. Explain how to troubleshoot and maintain transformers.
   a. Describe common transformer problems.
   b. Explain how to perform transformer testing and maintenance.

Performance Tasks

This is a knowledge-based module. There are no Performance Tasks.

Overview

When the AC voltage needed for an application is lower or higher than the source voltage, a transformer is used to raise or lower the voltage. The main components of a transformer are the primary winding, which is connected to the power source, and the secondary winding, which is connected to the load. This module introduces the various characteristics and applications of transformers as well as relevant *NEC*® requirements.

NOTE

NCCER Industry-Recognized Credentials

If you are training through an NCCER-accredited sponsor, you may be eligible for credentials from NCCER. The ID number for this module is 26307-23. Note that this module may have been used in other NCCER curricula and may apply to other level completions. Contact NCCER at 1.888.622.3720 or go to **www.nccer.org** for more information.

You can also show off your industry-recognized credentials online with NCCER's digital badges. Transform your knowledge, skills, and achievements into badges that you can share across social media platforms, send to your network, and add to your resume. For more information, visit **www.nccer.org**.

Digital Resources for Electrical

Scan this code using the camera on your phone or mobile device to view the digital resources related to this craft.

1.0.0 Transformer Construction and Operation

Performance Tasks

There are no Performance Tasks in this section.

Objective

Describe the construction and operation of transformers.

a. Describe the operation of a typical transformer.
b. Describe the construction characteristics of a typical transformer.
c. Explain how to make transformer connections for various applications.

Power transformers: A term often used to identify larger transformers used at transmission and distribution substations, as opposed to smaller transformers found in electrical subsystems and control systems.

The electric power produced in a generating station is transmitted to locations where it can be distributed to users. Many types of transformers play an important role in the distribution of electricity.

The main purpose of a transformer is to change the output voltage. **Power transformers** are located at generating stations to step up the voltage for more economical transmission. Substations with additional power transformers and distribution equipment are installed along the transmission line. Finally, distribution transformers are used to step down the voltage to a level suitable for consumption. Distribution transformers can be found on power poles and mounted on small slabs on the ground.

Transformers are also used extensively in control work to raise or lower the AC voltage for control circuits. They are also used in 480Y/277V systems to reduce the voltage for operating 208Y/120V lighting and other equipment. *Buck-and-boost transformers* are used for maintaining voltage levels in certain electrical systems.

1.1.0 Transformer Operation

It is important for anyone working with electricity to become familiar with transformers—how they work, how they are connected into circuits, their practical applications, and precautions to take when working with them.

A basic transformer consists of two coils (the primary and secondary windings) formed on a single magnetic core as shown in *Figure 1*. Such an arrangement will allow transforming a large alternating current at a low voltage into a small alternating current at a high voltage, or vice versa.

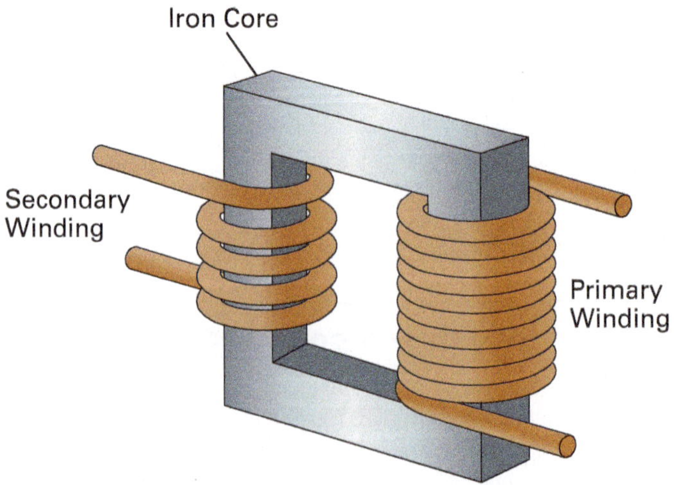

Figure 1 Basic transformer structure.

1.1.1 Mutual Induction

The term **mutual induction** refers to the condition in which two circuits are sharing the energy from one of the circuits. It means that energy is being transferred from one circuit to the other. **Induction** can occur only when there is a change in the magnetic force, which can happen in either a DC or an AC circuit.

Mutual induction: The condition of voltage generation in a second conductor caused by current flow in another conductor.

Induction: The production of an electromotive force (emf) in an electrical conductor when placed in a changing magnetic field.

Magnetic field: The area around a magnet in which the effect of the magnet can be detected.

Consider the diagram in *Figure 2*. Coil A is the primary circuit that obtains energy from the power source. In a DC circuit, when the switch is closed and a voltage is applied to coil A, the current starts to flow, and a **magnetic field** expands from coil A. Coil A then changes the electrical energy of the power source into the energy found in a magnetic field. When the field of coil A is changing, it cuts across coil B, the secondary circuit, inducing a voltage in coil B. The indicator (a *galvanometer*) in the secondary circuit is deflected and shows that a current developed by the induced voltage is flowing in the circuit.

When the voltage applied to coil A reaches its peak value, no changes occur in the field and the induction to coil B ceases, the field collapses, and no voltage is produced in coil B. The induced voltage may be generated again by moving coil B through the **flux** of coil A. As long as motion or change occurs in coil B or coil A and the field changes, voltage can be induced in coil B. Opening and closing the switch will create a change or motion in the field as well.

Flux: A measure of the electric field passing through a surface or area.

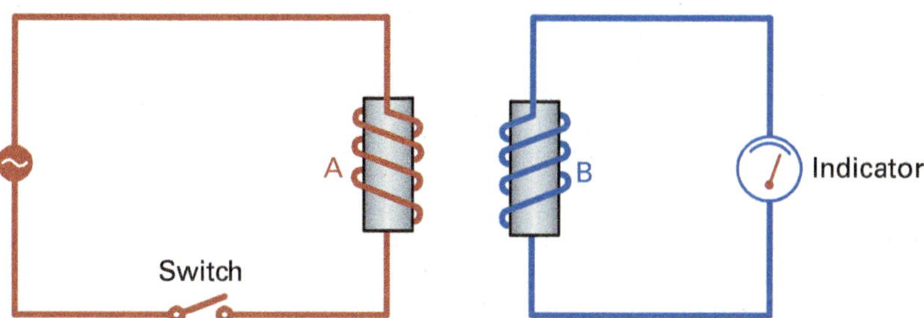

Figure 2 Mutual induction circuits.

In a DC circuit, the induced voltage may be generated by moving coil B through the flux of coil A. However, in an AC transformer, this voltage is induced without moving coil B. When the switch in the primary circuit is open, coil A has no current and no field. As soon as the switch is closed, current passes through the coil and the magnetic field is generated. With each half cycle, this expanding field moves or cuts across the wires of coil B, thus inducing a voltage without the movement of coil B.

The magnetic field expands to its maximum strength and remains constant as long as full current flows. Flux lines stop their cutting action across the **turns** of coil B because the expansion of the field has ceased. At this point, the indicator needle on the meter reads zero because the induced voltage no longer exists. If the switch is opened, the field collapses back to the wires of coil A. As it does so, the changing flux cuts across the wires of coil B, but in the opposite direction. The current present in the coil causes the indicator needle to deflect, showing this new direction.

Turns: The basic coil element in a transformer that forms a single conducting loop comprised of one insulated conductor.

Therefore, the indicator shows current flow only when the field is changing, either building up or collapsing. In effect, the changing field produces an induced voltage in the same way as a magnetic field moving across a conductor. This principle of inducing voltage by holding the coils steady and forcing the field to change is used in innumerable applications. The transformer is designed to take advantage of mutual induction. Transformers are the ideal components for transferring and changing AC voltages as needed.

Transformers are generally composed of two coils placed close to each other but not physically connected. Refer again to *Figure 1*. The coil that receives energy from the line voltage source is called the *primary winding*, and the coil that delivers energy to a load is called the *secondary winding*. Even though the coils are not physically connected, they manage to convert and transfer energy as required by mutual induction.

Transformers, therefore, change or convert power from one voltage to another. For example, generators that produce moderately large alternating currents at moderately high voltages use transformers to convert the power to a very high voltage and proportionately smaller current in transmission lines, permitting the use of smaller cable while experiencing less **power loss**.

Power loss: Power expended without accomplishing useful work.

When AC flows through a coil, an alternating magnetic field is generated around the coil. This alternating magnetic field expands outward from the center of the coil and collapses into the coil as the AC through the coil varies from zero to a maximum and back to zero again. Because the alternating magnetic field must cut through the turns of the coil, a self-inducing voltage occurs in the coil, which opposes the change in current flow.

If the alternating magnetic field generated by one coil cuts through the turns of a second coil, voltage will be generated in this second coil just as voltage is induced in a coil that is cut by its own magnetic field. The induced voltage in the second coil is called the *voltage of mutual induction*, and the action of generating this voltage is called *transformer action*. In transformer action, electrical energy is transferred from one coil (the primary winding) to another (the secondary winding) by means of a varying magnetic field.

1.1.2 Induction in Transformers

NOTE

In the electrical craft, the term *primary winding* is often shortened to simply *primary*, while the term *secondary winding* is shortened to *secondary*.

As stated previously, a simple transformer consists of two coils located very close together and electrically insulated from each other. The primary coil generates a magnetic field that cuts through the turns of the secondary coil and generates a voltage in it. The coils are magnetically coupled to each other, and consequently, a transformer transfers electrical power from one coil to another by means of an alternating magnetic field.

Assuming all the magnetic lines of force from the primary winding cut through all the turns of the secondary winding, the voltage induced in the secondary will depend on the ratio of the number of turns in the primary to the number of turns in the secondary. For example, if there are 100 turns in the primary and only 10 turns in the secondary, the voltage in the primary will be 10 times the voltage in the secondary. Because there are more turns in the primary than in the secondary, the transformer is called a *step-down transformer*, since the voltage is being reduced. Transformers are rated in volt-amperes (VA) or kilovolt-amperes (kVA) because these values are independent of the power factor.

Figure 3 shows a diagram of a step-down transformer with a **turns ratio** of 100:10, or 10:1. The induced voltage on the secondary coil can be determined by multiplying the turns ratio by the voltage on the primary. The ratio is first written as a fraction and then simplified. The voltage available from the secondary is calculated as follows:

Turns ratio: The ratio of the number of turns in the transformer primary winding compared to that in the secondary winding. In current transformers, it is the ratio of the turns in the secondary winding to the primary winding.

$$\frac{\text{10 secondary turns}}{\text{100 primary turns}} = 0.1$$

$$0.1 \times 120\text{V} = 12\text{V}$$

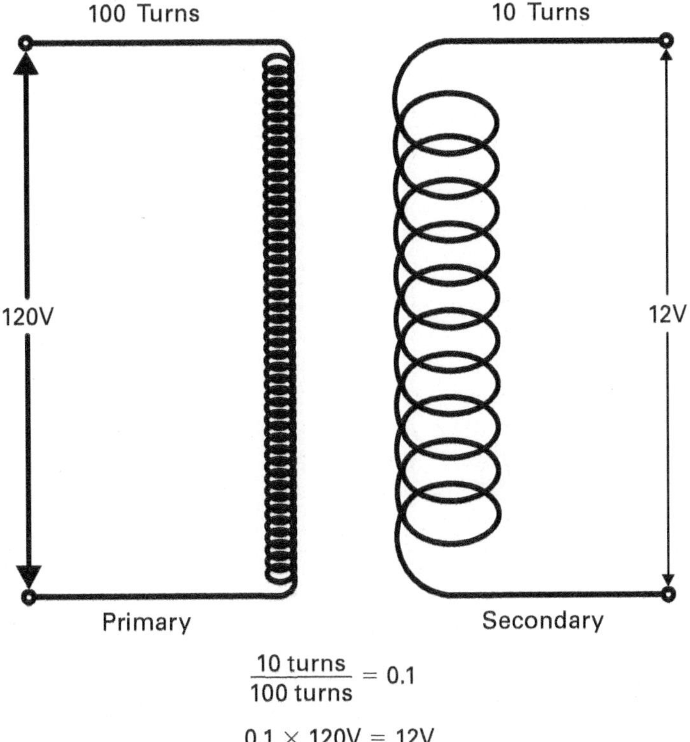

$$\frac{10 \text{ turns}}{100 \text{ turns}} = 0.1$$

$$0.1 \times 120\text{V} = 12\text{V}$$

Figure 3 Step-down transformer with a 10:1 turns ratio.

As long as all the primary magnetic lines of force cut through all the secondary turns, the induced voltage in the secondary will vary with the turns ratio.

If there are more turns in the secondary winding than in the primary winding, the secondary voltage will be higher than that in the primary and by the same proportion. The secondary current, on the other hand, will be proportionately smaller than the primary current.

With fewer turns in the secondary than in the primary, the secondary voltage will be proportionately lower than that in the primary, and the secondary current will be proportionately larger. This is the most common arrangement for simple transformers. Because alternating current continually increases and decreases as a sine wave demonstrates, every change in the primary winding of the transformer produces a similar change of flux in the core. Every change of flux in the core and every corresponding movement of the magnetic field around the core produce a similar changing voltage in the secondary winding, causing alternating current to flow in the circuit that is connected to the secondary.

For example, if there are 100 turns in the secondary and only 10 turns in the primary, the voltage induced in the secondary will be 10 times the voltage applied to the primary (*Figure 4*). Because there are more turns in the secondary than in the primary, the transformer is called a *step-up transformer.* Step-up transformers aren't as common as step-down transformers and have fewer applications. In this example, the voltage on the secondary can be calculated as follows:

$$\frac{100 \text{ secondary turns}}{10 \text{ primary turns}} = 10$$

$$10 \times 120\text{V} = 1{,}200\text{V}$$

Remember, though, that the current produced by the secondary is proportionally smaller in this arrangement. Although the voltage available in the secondary may be much higher than in the primary, its available current would only be 10% of the current in the primary winding.

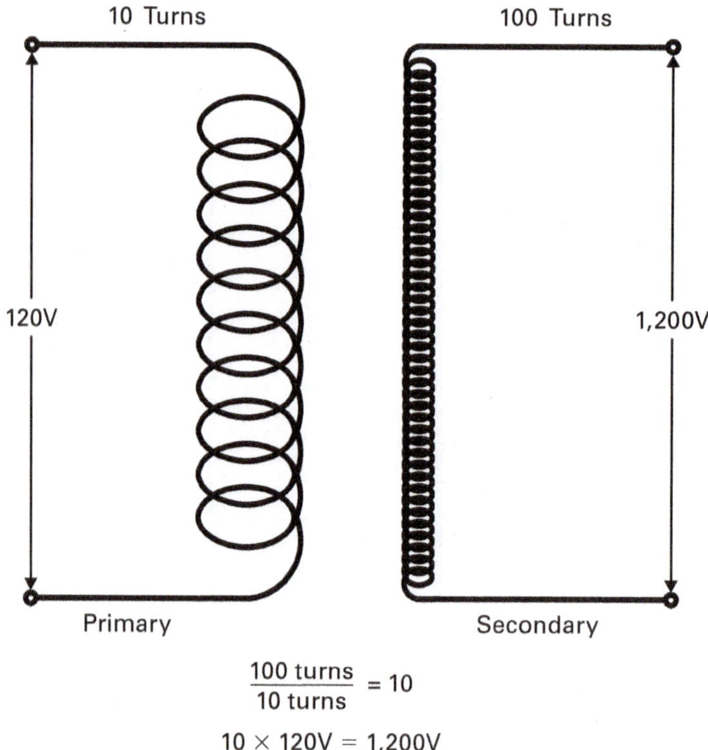

Figure 4 Step-up transformer with a 1:10 turns ratio.

1.1.3 Magnetic Flux in Transformers

Figure 5 shows a cross section of what is known as a *high-leakage flux transformer.* In these transformers, when no load is connected to the secondary or output winding, a voltmeter indicates a specific voltage across the secondary terminals. The voltage drops if a load is applied and would drop to zero if the terminals were shorted.

During these circuit changes, the flux in the core of the transformer would also change. This process, known as *flux leakage,* forces flux out of the transformer core. Flux leakage can be demonstrated with iron filings placed close to the transformer core. As the changes take place, the filings shift their positions, demonstrating the change in the flux pattern.

As the current flows in the secondary, it tries to create its own magnetic field, which is in opposition to the original field. Like a valve in a water system, this action restricts the flux flow that forces the excess flux to find another path, either through the air or adjacent structural steel, such as the transformer housing.

Note that the coils in *Figure 5* are wrapped on the same iron core but are separated from each other, while the transformer in *Figure 6* has its coils wrapped around each other, resulting in a low-leakage transformer design.

1.2.0 Transformer Construction

Transformers that are designed to operate on low frequencies have their coils (windings) wound on iron cores. Because iron offers little resistance to magnetic lines, nearly all the magnetic field of the primary flows through the iron core and cuts the secondary. Iron cores of transformers are constructed in three basic types: *open core, closed core,* and *shell core* (*Figure 7*).

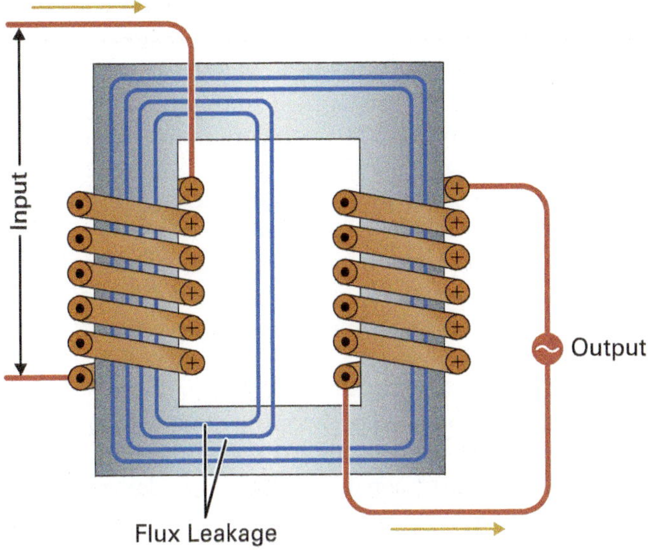

Figure 5 Transformer with high flux leakage.

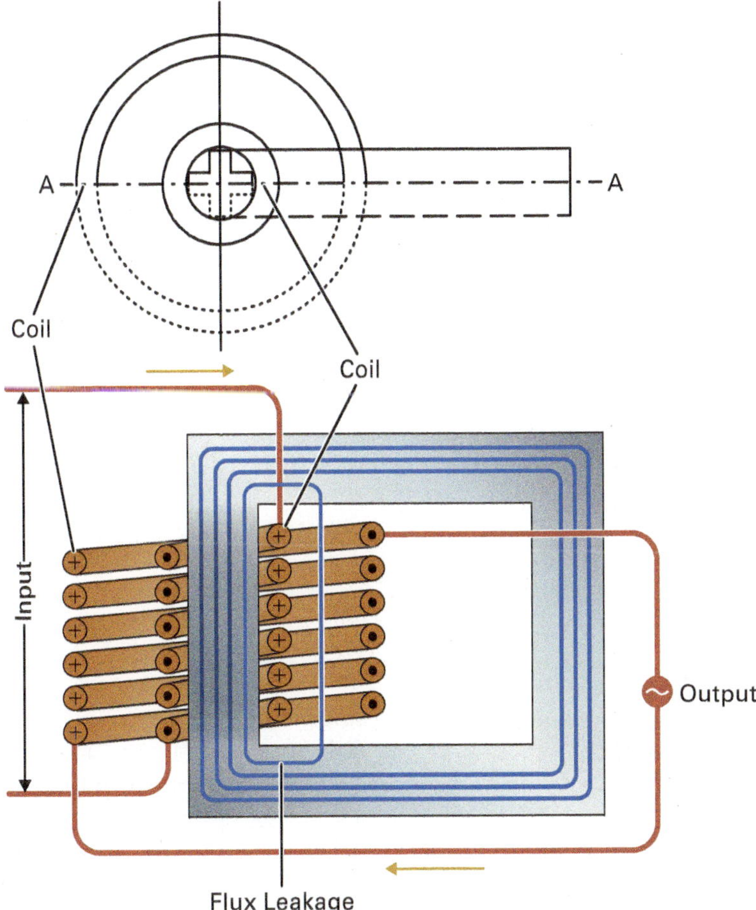

Figure 6 Low-leakage transformer.

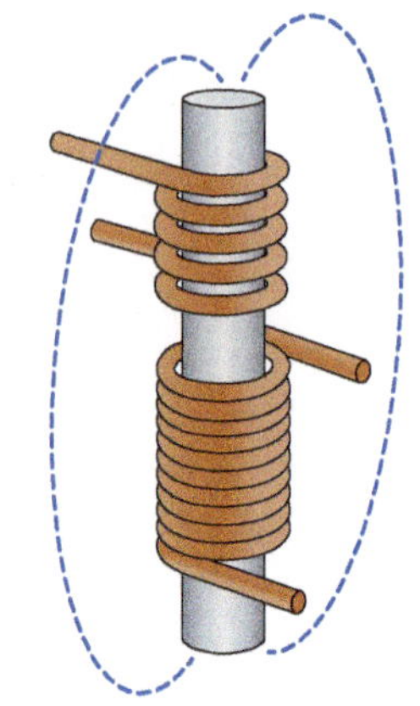

Open Core

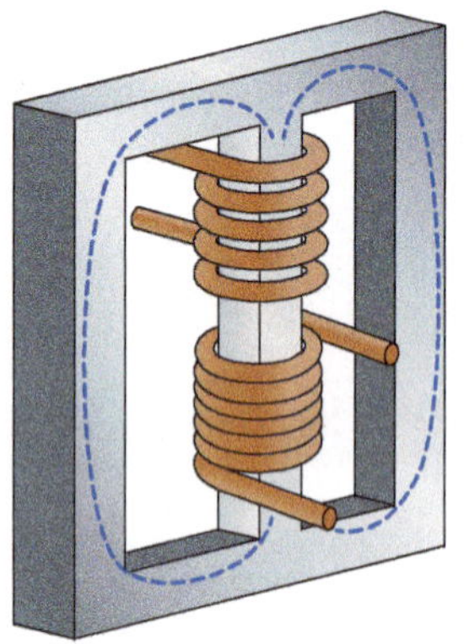

Closed Core

Shell Core

Figure 7 Three types of winding arrangements for iron-core transformers.

The open core is the least expensive to manufacture because the primary and secondary are wound on one cylindrical core. The magnetic path, as shown in *Figure 7*, is partially through the core and partially through the surrounding air. The air path opposes the magnetic field so that the magnetic interaction or linkage is weakened. As a result, the open core is the most inefficient type of transformer.

The closed core improves the transformer efficiency by offering more iron paths and a reduced air path for the magnetic field. The shell core further increases the magnetic coupling and results in greater transformer efficiency. This is achieved by two parallel magnetic paths for the magnetic field which maximum coupling between the primary and the secondary.

1.2.1 Cores

Steel is used to provide a controlled path for the flow of magnetic flux generated in a transformer. In most practical applications, the transformer core is not a solid bar of steel but instead constructed of many layers of thin-sheet steel called *laminations*. Although the specifications of the core steel are of primary interest to a transformer design engineer, the electrical worker should at least have a conversational knowledge of the materials.

The steel used for transformer core laminations will vary with the manufacturer, but a common size is 0.0136" (0.34 mm) thick and referred to as *29-gauge steel*. It is processed from silicon iron alloys containing approximately 3.25% silicon. The addition of silicon to the iron increases its ability to be magnetized and eliminates aging.

The most important characteristic of electrical steel is *core loss*. It is measured in watts per pound at a specified frequency and flux density. The core loss is responsible for the heating in the transformer and contributes to heat buildup in the windings. Much of the core loss is a result of eddy currents that are induced in the laminations when the core is energized. To hold this loss to a minimum, adjacent laminations are coated with an inorganic varnish.

Since the open cores are inefficient, the popular core designs are the closed core (*Figure 8*) and the shell core (*Figure 9*). Of the two, the closed core is favored in dry transformers for the following reasons:

- Only three core legs require stacking, which reduces cost.
- Steel does not encircle the two outer coils, which provides better cooling.
- The required floor space is reduced.

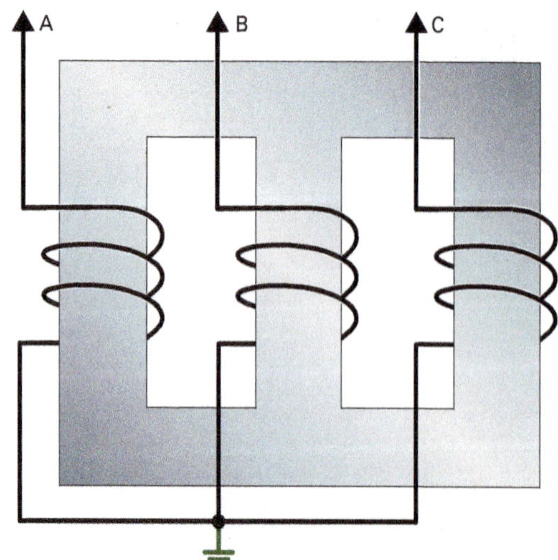

Figure 8 Closed core construction.

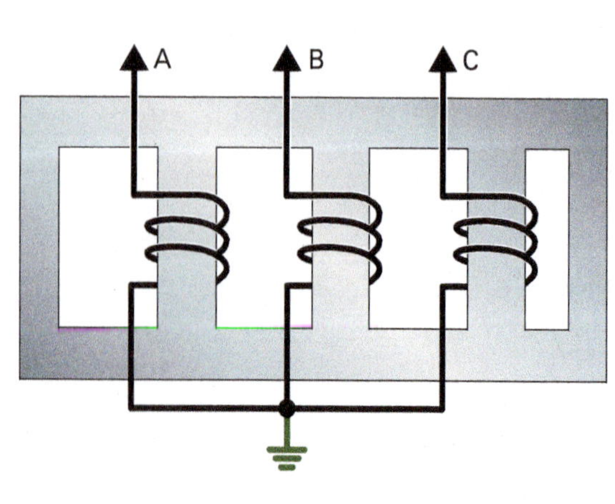

Figure 9 Shell core construction.

1.2.2 Core Assembly

Transformer cores are normally assembled in three different ways, as follows:

- Butt lap
- Wound
- Mitered

The butt-lap core is shown in *Figure 10*. Only two sizes of core steel are needed due to the lap construction shown at the top and right sides. For clarity, the core strips shown are much thicker than the 0.0136" (0.34 mm) thickness mentioned earlier. Each strip is carefully cut so that the air gap indicated in the lower left corner is as small as possible. The permeability of steel to the passage of flux is about 10,000 times as effective as air. Thus, the air gap must be held to the barest minimum to reduce the **ampere turns** necessary to achieve adequate flux density. Also, the amount of sound produced by a transformer is a function of the flux density, making this construction different from other types.

Ampere turns: The product of the current (in amperes) times the number of turns in a transformer coil.

Another phenomenon in core steel is that the flux flows more easily in the direction in which the steel was rolled. This also varies between hot-rolled steel and cold-rolled steel. For example, when flux passes at right angles to the rolling direction, the core loss is almost $1\frac{1}{2}$ times as great in hot-rolled steel and $2\frac{1}{2}$ times as great in cold-rolled steel, compared with that in the direction of rolling. The difference in exciting current is more dramatic, with ratios of 2:1 in hot-rolled steel and almost 40:1 in cold-rolled steel. These are primarily the designer's concern, but you should be aware of this difference.

Eddy currents are restricted from passage from one lamination to another due to the inorganic insulating coating. The magnetic lines of flux easily transfer at adjacent laminations in the lap area. However, in doing so, they are forced to cross at an angle to the preferred direction.

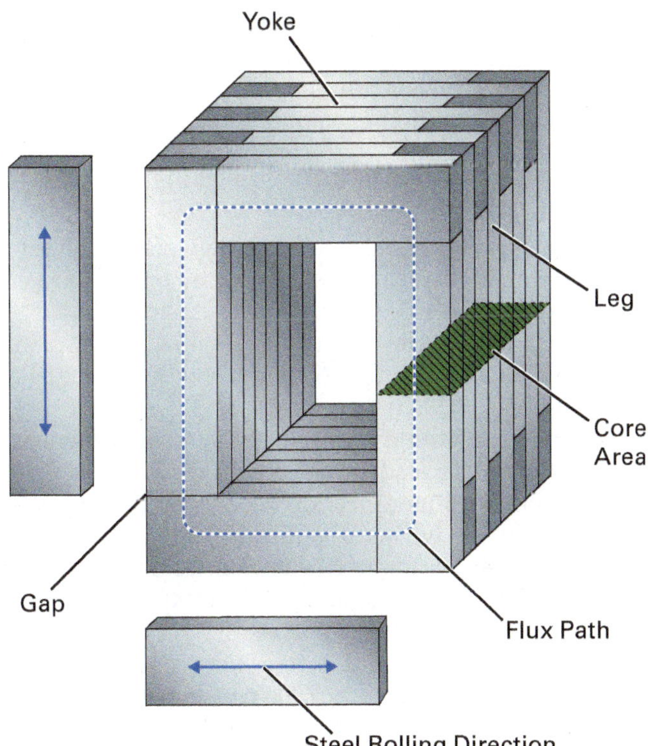

Figure 10 Butt-lap core assembly.

1.2.3 Wound Cores

Some cores are designed to take advantage of the unique characteristics of core steel. One such type is shown in *Figure 11*. The core loops are cut to predetermined lengths so that the gap locations do not coincide. These cuts allow assembly of the core around a prewound coil that passes through both openings.

1.2.4 Mitered Cores

Figure 12 shows a mitered core design. It is basically a butt-lap core with the joints made at 45° angles. Compare the corner construction shown in *Figure 12* with the butt-lap design shown in *Figure 10*.

There are two benefits derived from the mitered joints:

- It eliminates all cross-grain flux, thereby reducing the core loss and exciting current values.
- It reduces the flux density in the air gap, resulting in less noise.

This type of core is normally used only with cold-rolled, grain-oriented steel and permits this steel to be used to its fullest capability.

1.2.5 Transformer Characteristics

In a well-designed transformer, there is very little flux leakage. Leakage causes a decrease in secondary voltage when the transformer is loaded. When a current flows through the secondary in phase with the secondary voltage, a corresponding current flows through the primary in addition to the magnetizing current. The magnetizing effects of the two currents are equal and opposite.

In a perfect transformer—one having no eddy current losses, no resistance in its windings, and no flux leakage—the magnetizing effects of the primary load current and the secondary current would neutralize each other, leaving only the constant primary magnetizing current effective in setting up the constant flux. Obviously, the perfect transformer has yet to be built. The best transformers available today still have very small eddy current losses where the drop in current in the secondary windings is not more than 1% to 3%, depending on the size of the transformer.

1.2.6 Transformer Taps

It is not possible to have the same voltage delivered at every transformer location. Therefore, transformer taps are necessarily provided to either increase or decrease the secondary voltage.

Generally, if a load is very close to a substation or power plant, the voltage will consistently be above the average. Near the end of the line, the voltage may be below the average.

In large transformers, it would naturally be very inconvenient to move the thick, well-insulated primary leads to different tap positions when changes in source voltage levels make a modification necessary. Therefore, taps are used, such as those shown in the wiring diagram in *Figure 13*. In this transformer, the permanent high-voltage leads would be connected to H_1 and H_2; and the secondary leads, in their normal fashion, to X_1 and X_2, and X_3 and X_4.

Note, however, the taps are still available at positions 2 through 7. Until a pair of these taps are interconnected with a jumper wire, the primary circuit is not completed. If this were a typical 7,200V primary, the transformer would normally have 1,620 turns. Assume 810 of these turns are between taps 1 and 6 and another 810 between taps 3 and 2. Then, if taps 6 and 3 are connected using a flexible jumper on which the lugs have already been installed, the primary circuit is completed, and we have a normal ratio transformer that can deliver 120V and 240V from the secondary.

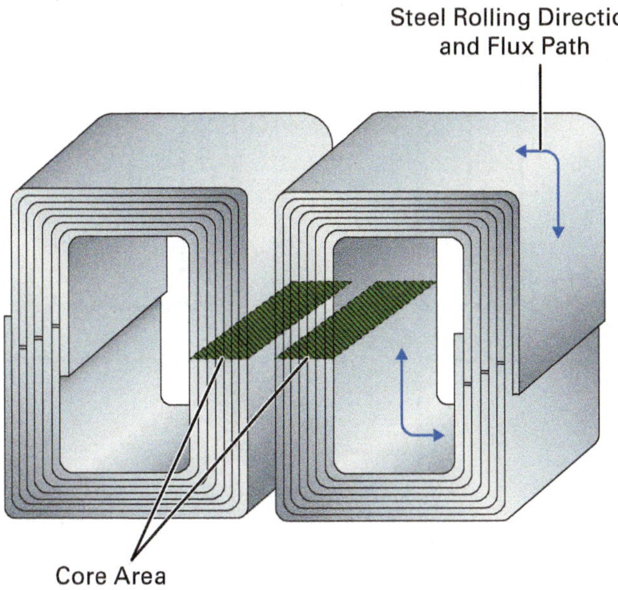

Figure 11 Wound transformer coil.

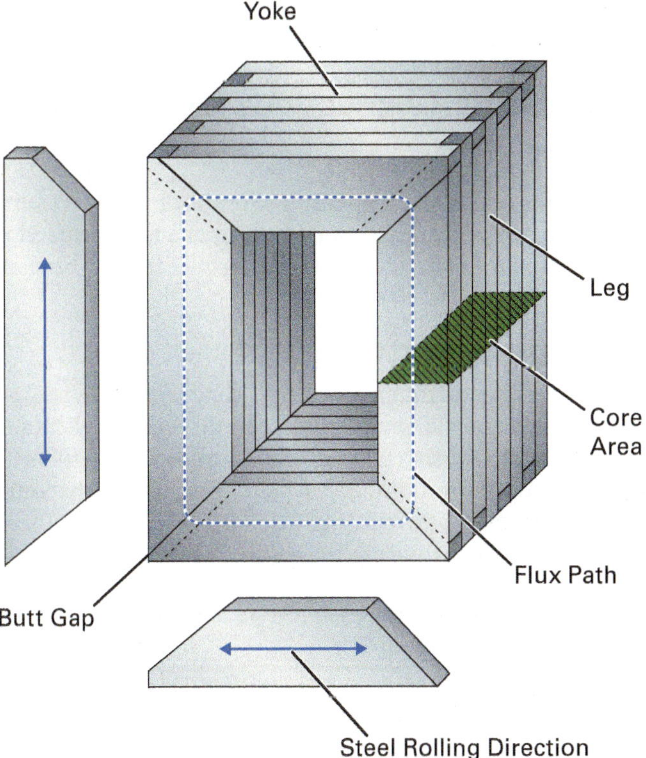

Figure 12 Mitered transformer core.

Between taps 6 and either 5 or 7 exist 40 turns of wire. Similarly, 40 turns are present between taps 3 and either 2 or 4. Changing the jumper from taps 3 and 6 to taps 3 and 7 removes 40 turns from the left half of the primary. The same condition would apply on the right half of the winding if the jumper were between taps 6 and 2. Either connection would boost the secondary voltage by 2.5%. Had taps 2 and 7 been connected, 80 turns would have been omitted, and a 5% boost would result. Placing the jumper between taps 6 and 4 or taps 3 and 5 would reduce the output voltage by 2.5%.

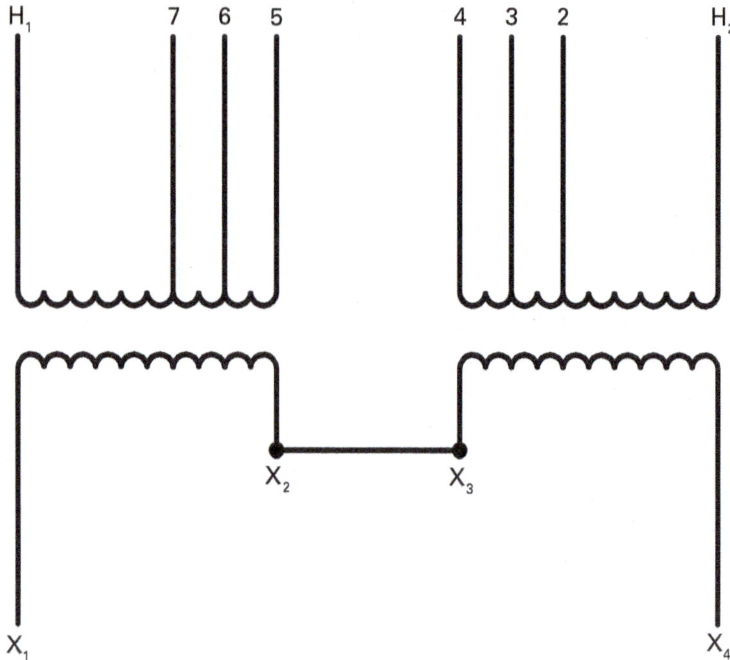

Figure 13 Transformer primary taps to adjust secondary voltage.

1.3.0 Transformer Connections for Various Applications

There are many types of transformer connections, but we won't introduce all of them in this module. However, an understanding of a few connection types will provide the basic requirements and make it possible to use the manufacturer's data.

Think About It

Secondary Neutrals

Why do we ground and bond secondary neutrals?

1.3.1 Single-Phase Light and Power Systems

Figure 14 is a single-phase transformer line diagram showing a connection used primarily for residential and small commercial applications. It is the most common single-phase distribution transformer in use today. It is known as a three-wire, 120/240V, single-phase system. Because of its configuration, it is easy to balance the load between the two coils. The kVA of a single-phase transformer is calculated by dividing the total VA by 1,000.

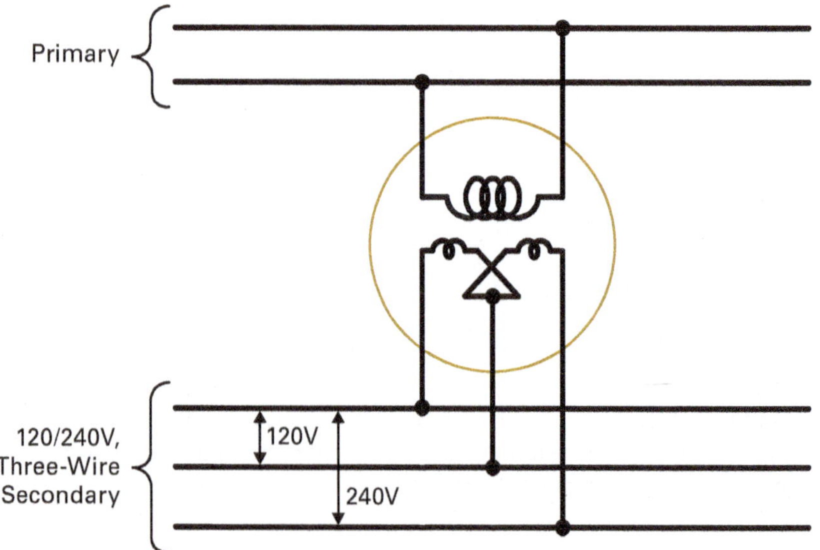

Figure 14 Single-phase transformer connection.

1.3.2 Three-Phase Power Systems

The following factors need to be considered when choosing a three-phase power system:

- The load(s) to be supplied
- The voltage(s) needed
- Future expansion needs

There are advantages and disadvantages associated with the different types of three-phase systems. The wye system can be used for both single-phase and three-phase loads. Its configuration makes it easy to balance the single-phase loads while still allowing it to be used for three-phase loads.

Transformers installed in this type of system are sized by first dividing the total single-phase load in kVA by three and adding this result to the total three-phase load in kVA divided by three. The sum of these two loads identifies the required kVA rating of the transformer.

The nominal voltage levels provided by this type of system are 120/208V or 277/480V. Note that 240V is not an option, which may prove to be a disadvantage in many applications because 240V is a common operating voltage.

One of the most common transformer systems found in today's commercial and industrial settings is the delta primary, wye secondary transformer system (*Figure 15*). This is a three-phase, four-wire system that has the advantage of both providing three-phase power and allowing lighting to be connected between any of the secondary phases and the neutral.

In *Figure 15*, the system is basically made up of three typical single-phase, step-down transformers with the interconnections between each individual transformer's primary and secondary coils determining the output voltage on the secondary side. However, this does not mean that any three single-phase transformers can be developed into a functional three-phase, delta-wye system for power and lighting. The coils must be rated for the loads to be served, as well as the primary voltage level to be connected. Power and light delta-wye transformers are universally found in both commercial and industrial installations as floor-mounted, dry-type transformers.

This transformer provides the convenience of three-phase and single-phase power, as well as the benefit that each phase of the 208V system shares the neutral to supply three legs of 120V lighting circuits. However, remember that the three-phase and single-phase power availability is at the 208V level instead of 240V. If a power requirement specifically calls for 240V and does not permit a 208V supply, this type of transformer is not the best option.

The delta-delta system in *Figure 16* operates a little differently from the delta-wye system. Whereas the wye-connected system is formed as a common terminal by connecting one terminal from each of three equal voltage transformer windings together, the delta-connected system has its windings connected in series, forming a triangle that relates it to the Greek symbol *delta* (Δ).

In *Figure 17*, a center-tapped terminal is used on one winding to ground the system. A 120/240V system has 120V between the center-tapped terminal and each ungrounded terminal on either side, such as phases A and C, and 240V across the full winding of each phase.

Refer to *Figure 17* and note the high leg. This is also known as the *wild leg*. This high leg has a higher voltage to ground than the other two phases. The voltage of the high leg can be determined by multiplying the voltage to ground of either of the other two legs by the square root of 3, which we round to a value of 1.732. Therefore, if the voltage between phase A to ground is 120V, the voltage between phase B to ground may be determined as follows:

$$120V \times 1.732 = 208V$$

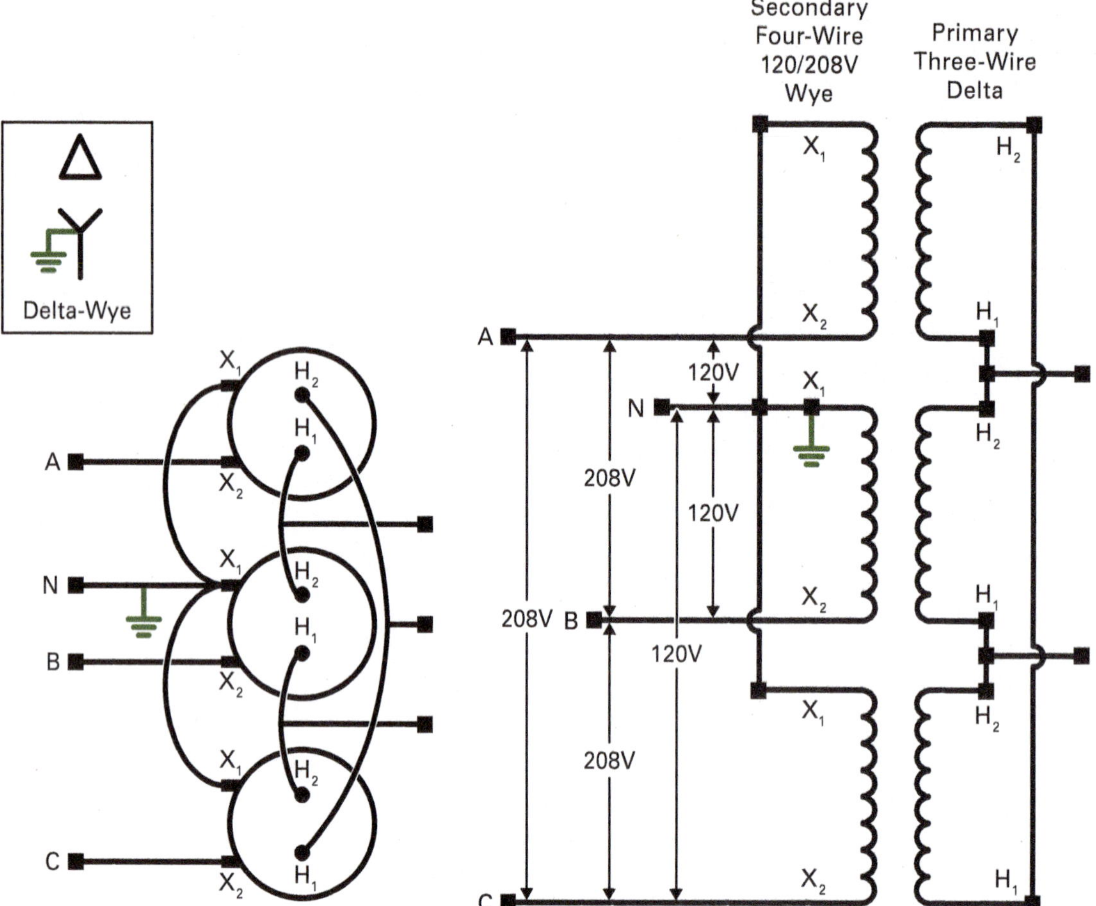

Figure 15 Delta-wye transformer system.

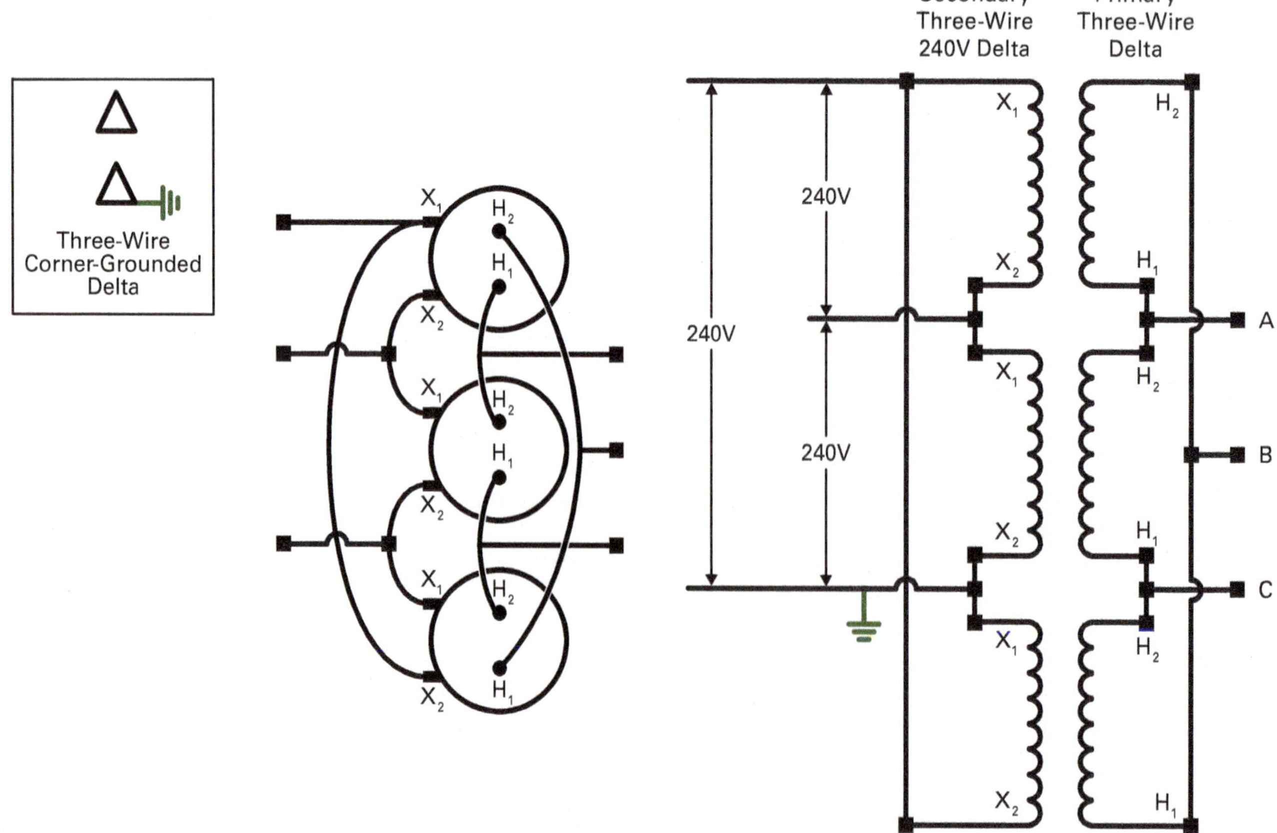

Figure 16 Delta-connected secondary.

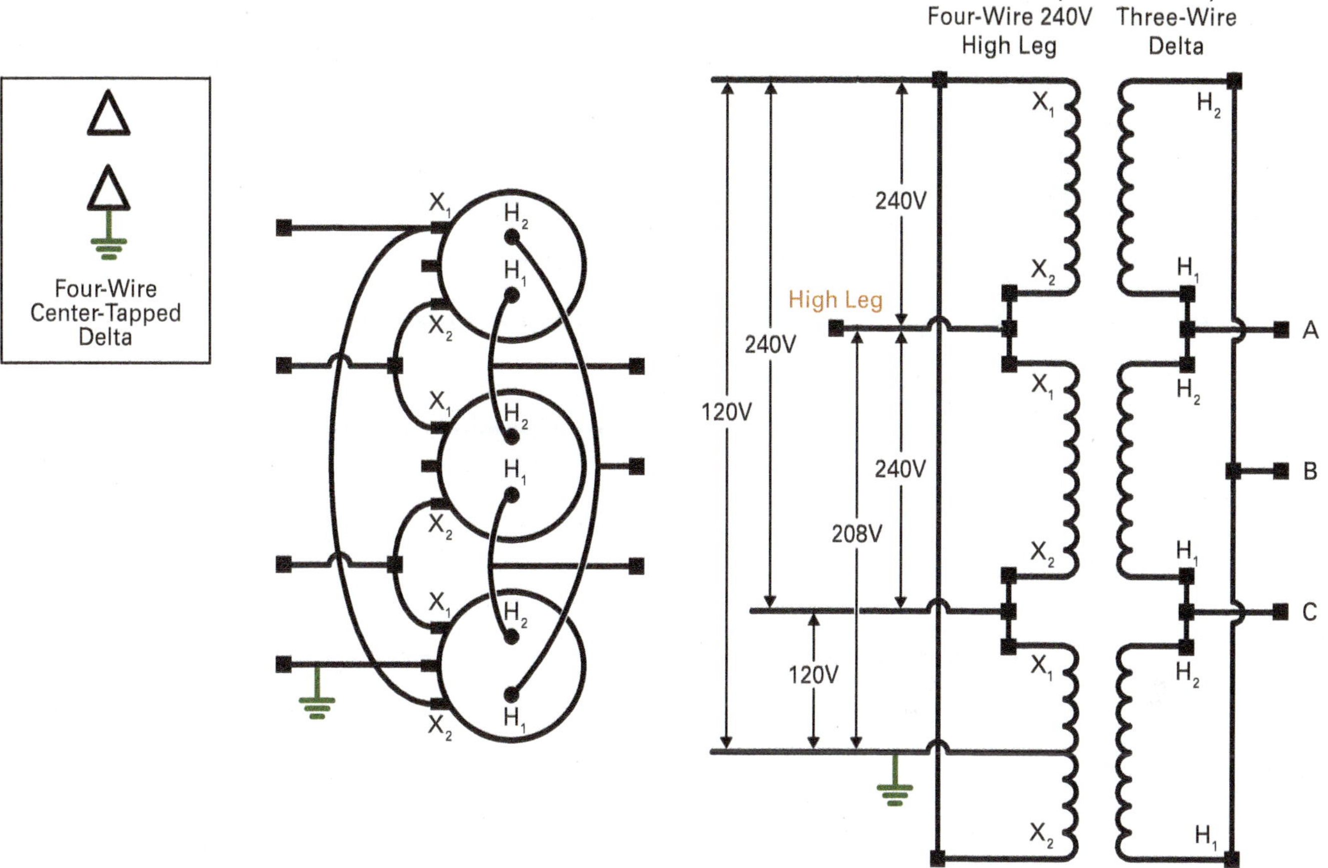

Figure 17 Characteristics of a center-tapped, delta-connected system.

Obviously, no single-pole breakers should be connected to the high leg of a center-tapped, four-wire, delta-connected system. In fact, *NEC Section 110.15* requires that the phase busbar or conductor with the higher voltage to ground be permanently marked by an outer finish that is orange in color, or by an equally effective means. This prevents future workers from connecting 120V single-phase loads to this high leg and possibly damaging any equipment connected to the circuit. Remember, the orange color means no line-to-neutral loads should be connected to this phase.

Know the Code

High-Leg Marking

NEC Section 110.15

WARNING!

Always use caution when working on a center-tapped, four-wire, delta-connected system. Phase B has a higher voltage to ground than phases A and C. Never connect 120V circuits to the high leg. Doing so will result in damage to the connected equipment.

Three-phase, delta-connected systems may be connected so that only two transformers are used. This arrangement is known as an *open delta system*, as shown in *Figure 18*. It is frequently used on a delta system when one of the three transformers becomes damaged. The damaged transformer is disconnected from the circuit, and the remaining two transformers carry the load. In doing so, the three-phase load carried by the open delta bank is only 86.6% of the combined rating of the remaining two equally sized units. This is only 57.7% of the normal full-load capability of a full bank of transformers. In an emergency, this provides single-phase and three-phase power to a location where one transformer has failed without a replacement readily available. The load, however, must be reduced and carefully controlled to avoid a second failure before repairs are completed.

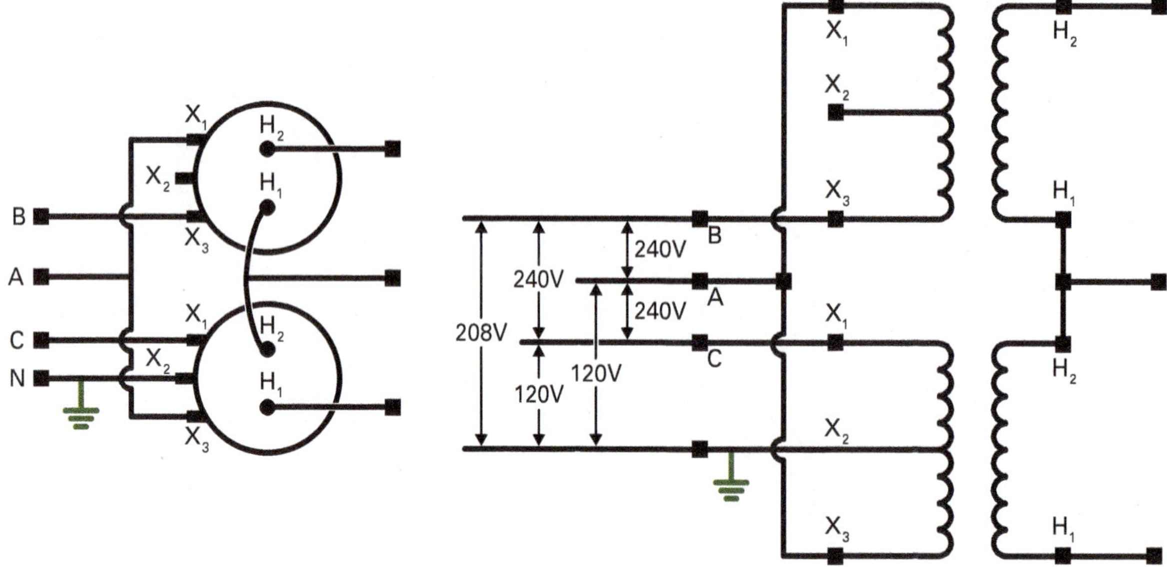

Figure 18 Open delta system.

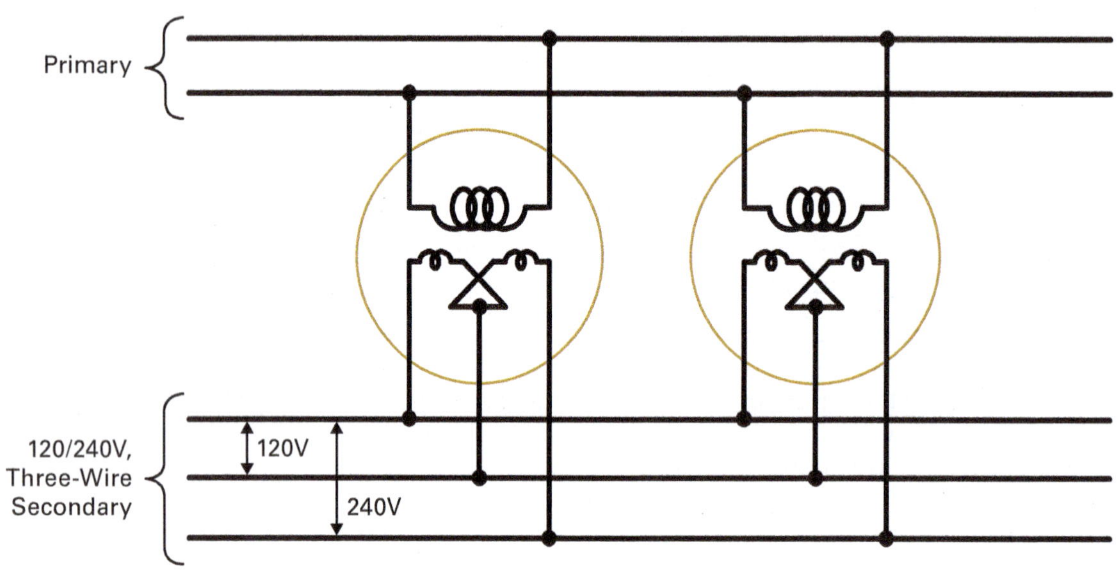

Figure 19 Parallel operation of single-phase transformers.

1.3.3 Parallel Operation

Transformers may operate in parallel under certain conditions (*Figure 19*). Generally, satisfactory operation occurs only when applying transformers with equivalent winding configurations, voltages, and impedances. Small differences between transformers can cause circulating currents to flow, resulting in excess heat and reduced efficiency.

Reactance: The opposition to AC due to capacitance and/or inductance.

If transformers of unequal kVA ratings must be connected in parallel, they must have equal X/R ratios (impedance to resistance or **reactance**) to divide the load proportionately between the transformers.

Impedance plays an important role in the successful operation of transformers connected in parallel. The impedance of the transformers must make sure the voltage drop from no load to full load be the same in all transformer units in both magnitude and phase. In most transformers, reactance causes a larger voltage drop than resistance does. For our purposes, load sharing of similar transformers can be calculated using total impedance instead of reactance. If the full-load impedances of two similar transformers are the same, the load will be divided equally between them.

The following equation may be used to obtain the division of loads between two transformer banks operating in parallel on single-phase systems:

$$P = \frac{kVA_1 + Z_1}{(kVA_1 + Z_1) + (kVA_2 + Z_2)} \times kVA$$

Where:

P = power

kVA_1 = kVA rating of transformer 1

kVA_2 = kVA rating of transformer 2

Z_1 = percent impedance of transformer 1

Z_2 = percent impedance of transformer 2

kVA = total kVA load

In this equation, you can assume that the ratio of resistance to reactance is the same in all units because the error introduced by differences in this ratio is usually small enough to be negligible.

The preceding equation can also be applied to more than two transformers operated in parallel by adding the denominator of the fraction with the kVA of each additional transformer divided by its percent impedance.

Three-phase transformers, or banks of single-phase transformers, can be connected in parallel, provided each of the three primary leads in one three-phase transformer is connected in parallel with a corresponding primary lead of the other transformer. The secondaries are then connected in the same way. The corresponding leads are those that consistently have the same potential and polarity. Furthermore, the transformers must have the same voltage ratio and the same impedance voltage drop.

When three-phase transformer banks operate in parallel and the three units in each bank are similar, the division of the load can be determined by the same method previously described for single-phase transformers connected in parallel on a single-phase system.

In addition to the requirements of polarity, ratio, and impedance, paralleling of three-phase transformers also requires that the angular displacement between the voltages in the windings be taken into consideration when they are connected.

Single-Phase and Three-Phase Transformers

These photos show single-phase and three-phase transformers. Note the orange marking on the high leg of the three-phase transformer in accordance with ***NEC Section 110.15***.

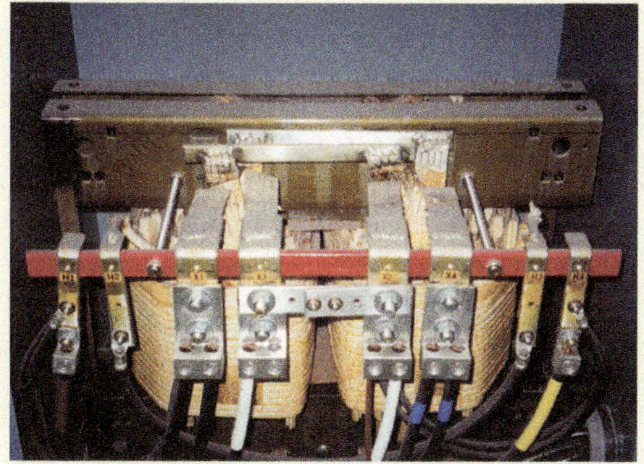

100 kVA Single-Phase Transformer
480V Primary, 120V/240V Secondary

300 kVA Three-Phase Center-Tapped Delta Transformer
480V Primary, 120V/240V Secondary

Source: Dan Lamphear

Phasor diagrams of three-phase transformers that are to be paralleled greatly simplify matters. With these, all that is required is to compare the two diagrams to make sure they consist of phasors that can be made to coincide, and then to connect the terminals corresponding to coinciding voltage phasors. If the phasor diagrams can be made to coincide, leads that are connected will always have the same potential. This is one of the fundamental requirements for paralleling.

1.3.4 Autotransformers

Autotransformer: A transformer in which primary and secondary connections are made to a single winding.

An **autotransformer** is a transformer in which primary and secondary circuits share part of a winding. Thus, the two circuits are not isolated from each other, as shown in *Figure 20*. An autotransformer is a good choice where a 480Y/277V or 208Y/120V, three-phase, four-wire distribution system is used. The advantages include the following:

- Lower purchase price
- Lower operating cost due to lower losses
- Smaller size, making it easier to install
- Better voltage regulation
- Lower sound levels

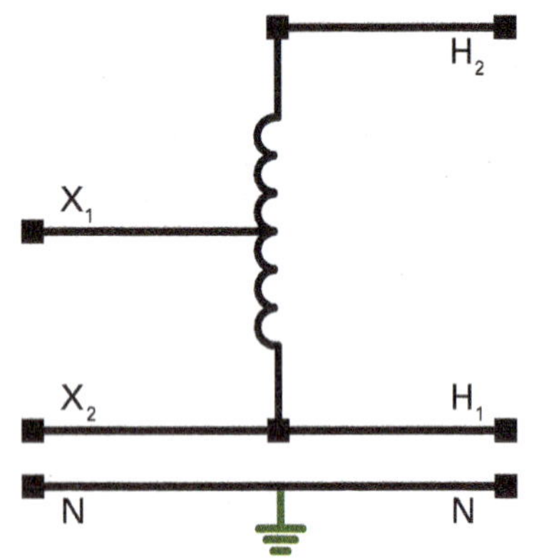

Figure 20 Step-down autotransformer.

For example, when the ratio of transformation from the primary to the secondary voltage is small, the most economical way of stepping down the voltage is using autotransformers, as shown in *Figure 21*. For this application, it is necessary that the neutral of the autotransformer bank be connected to the system neutral, similar to a wye connection.

> **Autotransformers**
>
> Autotransformers are not separately derived systems and must not be bonded on the secondary side.

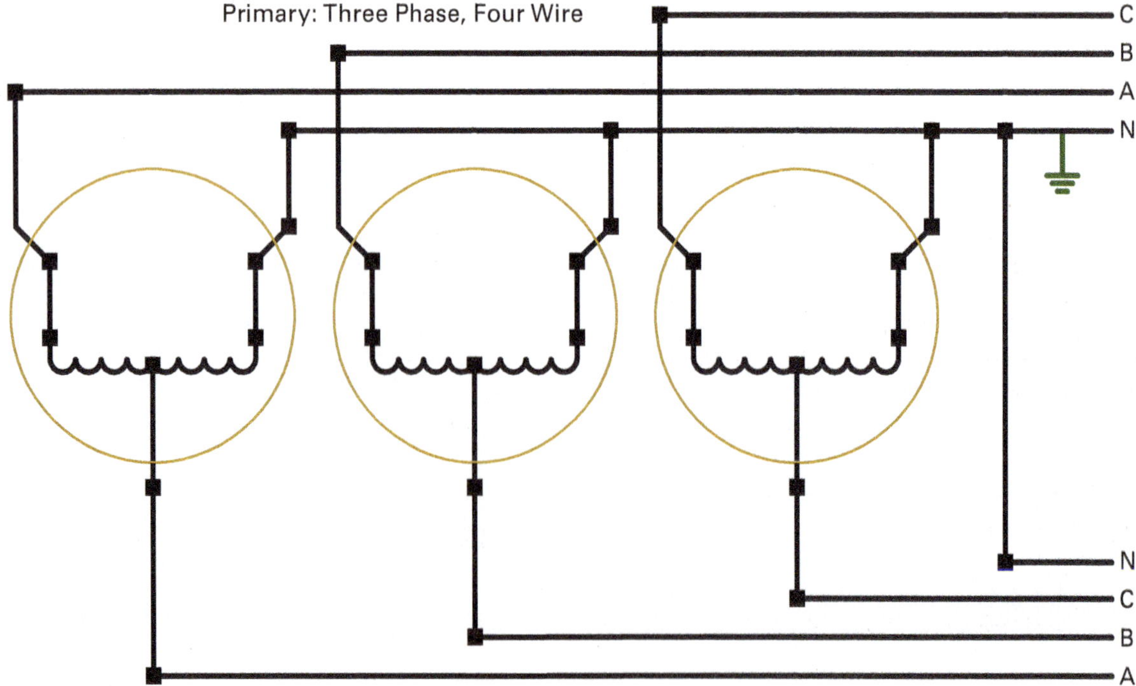

Figure 21 Autotransformers supplying power from a three-phase, four-wire system.

Liquid-Filled Transformers

This 2,500 kVA liquid-filled transformer has cable bus secondary conductors and a single protection and control interface cabinet. Larger transformers may have two or more interface cabinets, depending on the number of heaters, fans, CTs, and other devices.

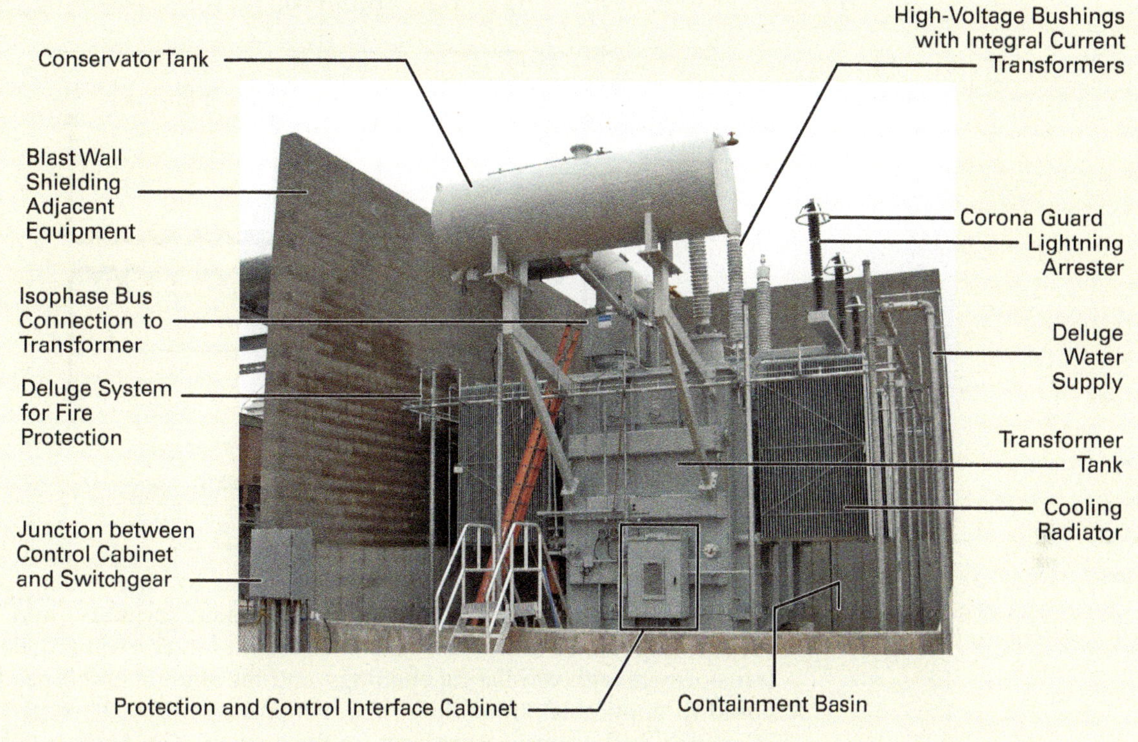

Source: Jim Mitchem

1.3.5 Control Transformers

Most control transformers are dry-type, step-down units with the secondary control circuit isolated from the primary line circuit to maximize safety (*Figure 22*). Industrial control transformers are designed to accommodate the momentary current inrush caused when electromagnetic components, such as relays and contactors, are energized without sacrificing secondary voltage stability beyond practical limits.

Other types of control transformers, sometimes referred to as *control and signal transformers*, are constant-potential, air-cooled units. Their purpose is to supply the proper reduced voltage for control circuits of electrically operated switches, signal circuits, or other equipment. These transformers do not normally require the industrial regulation characteristics found in other transformers. Some are of the open type with no protective casing over the windings; others are enclosed within a metal casing.

When choosing control transformers for any application, the loads must be calculated and completely analyzed before the proper transformer selection can be made. This analysis must consider every electrically energized component in the control circuit.

To select an appropriate control transformer, first determine the voltage and frequency of the supply circuit. Next, determine the total inrush VA of the control circuit. When doing so, do not neglect the current requirements of indicating lights and timing devices, which do not experience inrush current but are energized at the same time as the other components in the circuit. Their total VA should be added to the total inrush VA.

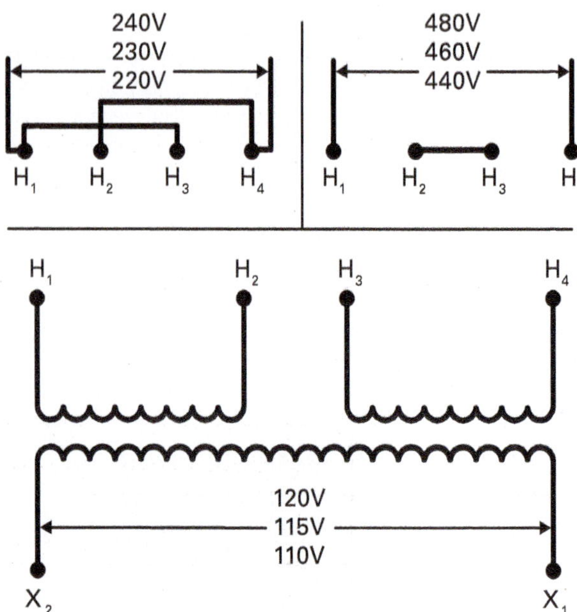

Figure 22 Typical control transformer wiring diagram.

1.3.6 Transformer Nameplate and Wiring Data

NEC Section 450.11 requires each transformer to have a nameplate listing the manufacturer, rated kVA, rated frequency, primary and secondary voltage, impedance of transformers 25 kVA and larger, required clearances for transformers with ventilating openings, and the amount and type of insulating liquid (if applicable). In addition, the nameplate of each dry-type transformer must include the temperature class for the insulation system.

In addition, most manufacturers include a wiring diagram and a connection chart, as shown in *Figure 23*, for a 480V delta primary to 208Y/120V wye secondary. It is recommended that all transformers be connected as shown on the manufacturer's nameplate.

In general, this wiring diagram and accompanying table indicate that the 480V, three-phase, three-wire primary conductors are connected to terminals H_1, H_2, and H_3 respectively, regardless of the desired voltage on the primary. A neutral conductor, if required, is carried from the primary through the transformer to the secondary.

Two variations are possible on the secondary side of this transformer. One is a 208V, three-phase, three-wire or four-wire arrangement. The other is a 120V, single-phase, two-wire arrangement. To connect the secondary side of the transformer as a 208V, three-phase, three-wire system, the secondary conductors are connected to terminals X_1, X_2, and X_3. The neutral is carried through with conductors usually terminating at a solid neutral bus in the transformer.

Another common transformer arrangement is the 480V primary to 240V delta/120V secondary. This configuration is shown in *Figure 24*. Again, the primary conductors are connected to transformer terminals H_1, H_2, and H_3. The secondary connections for the desired voltages are made as indicated in the table.

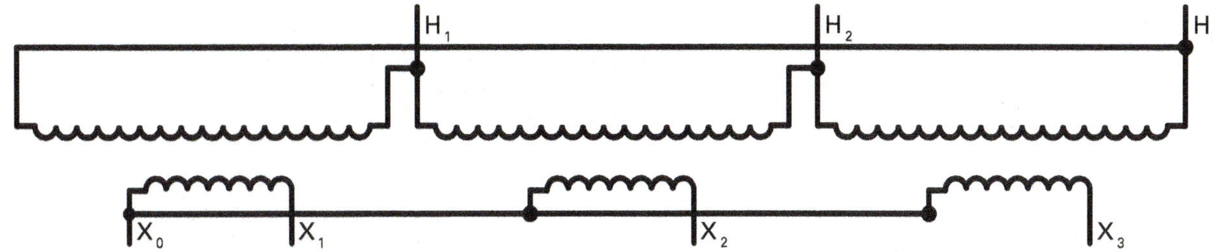

Primary Volts	Connect Primary Lines to	Connect Secondary Lines to
480V	H_1, H_2, H_3	———
Secondary Volts	**Connect Primary Lines to**	**Connect Secondary Lines to**
208V	———	X_1, X_2, X_3
120V Single-Phase	———	X_1 to X_0 X_2 to X_0 X_3 to X_0

Figure 23 Typical manufacturer's wiring diagram for a delta-wye transformer.

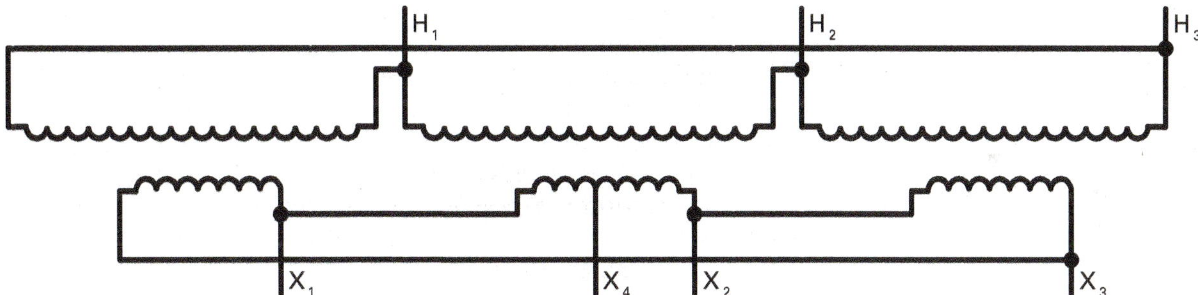

Primary Volts	Connect Primary Lines to	Connect Secondary Lines to
480V	H_1, H_2, H_3	———
Secondary Volts	**Connect Primary Lines to**	**Connect Secondary Lines to**
240V	———	X_1, X_2, X_3
120V	———	X_1, X_4 or X_2, X_4

Figure 24 480V delta to 240V delta transformer connections.

1.3.7 Calculating a System's Power Factor

The concept of power factor was introduced in NCCER Module 26201-23, *Alternating Current*. A brief review of the subject will be provided here. The equation for power factor is:

$$\text{Power factor } (pf) = \text{kW} \div \text{kVA}$$

Where:

kW = kilowatts
kVA = kilovolt-amperes

Calculating the power factor of an electrical system requires determining true power, inductive reactance, and capacitive reactance of the system. An analogy should enhance your understanding of these terms.

Imagine a farm wagon to which three horses are hitched, as shown in *Figure 25*. Horse #1 is pulling straight ahead. We name this horse *true power* because all the effort is in the direction that the work should be done. Horse #2 wants to nibble at the grass growing along the side of the road. This horse does not contribute an ounce of pull in the desired direction and actually causes a problem by pulling the wagon toward the ditch. We will call horse #2 *inductance*. Horse #3 has about the same strength as horse #2 and enjoys the grass on the opposite side of the road, which causes this horse to pull in the opposite direction of horse #2. Horse #3 also contributes nothing to the forward motion. This horse is called **capacitance**.

Capacitance: The storage of electricity in a capacitor that produces an opposition to voltage change. The unit of measurement is the *farad* (*F*) or *microfarad* (*μF*).

Going Green

Increasing Power Factors

Industrial facilities with large inductive loads that must rely on commercial power to operate are required to pay a rate based on power factor and consumption. The rate increases when the power factor is low because the utility must supply additional kVARs not metered by the watt-hour meter. To decrease the cost of electricity caused by these inductive loads, capacitors are often installed to increase the power factor.

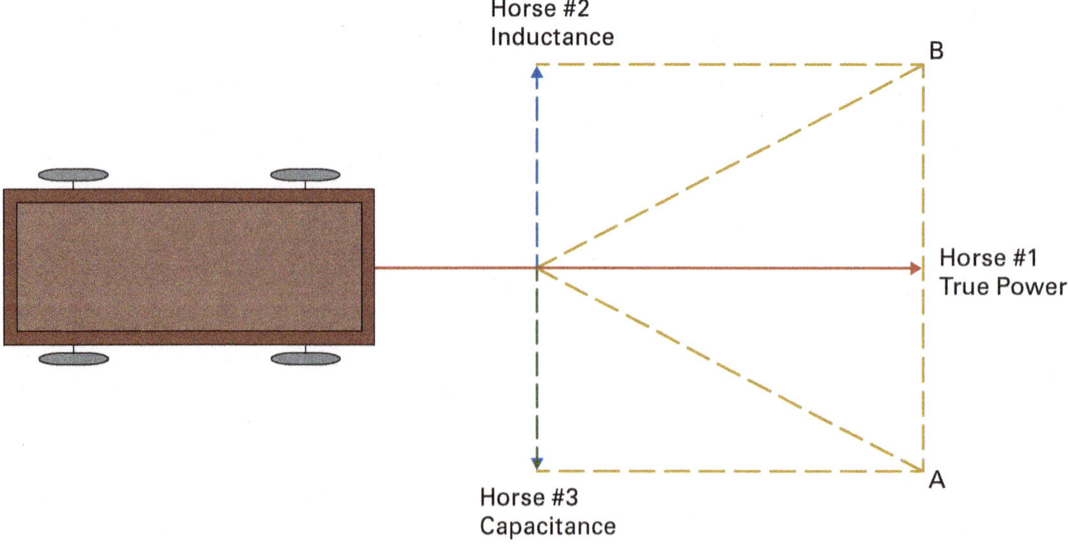

Figure 25 Depiction of power factor.

Now forget about horse #3 for the moment. If only horse #1 and horse #2 were pulling, the wagon would go in the direction of dotted line B. Notice that the length of that line is greater than the line to #1, so horse #2 (inductance) influences the final result. The direction and length of line B might well be called *apparent power* and happens to be the hypotenuse of a right triangle.

If horse #2 were unhitched, then horses #1 and #3 would cause the wagon to move in the direction of line A, whose length would also be apparent power. If all three horses were pulling, horses #2 and #3 would cancel one another out, and the only useful animal would be horse #1, true power.

In an AC circuit, there are always three forces working in varying lengths. Inductance is present in every magnetic circuit and always works at a 90° angle with true power. There is a power factor of less than 100% because the wagon does not move straight ahead but instead travels toward the right due to the pull of horse #2. However, we can improve the power factor by adding horse #3, capacitance, which tends to cancel out the pull of inductance, enabling the wagon to travel straight ahead.

The primary function of capacitors is to improve the power factor of an electrical installation or an individual piece of electrically operated equipment. In general, this efficiency lowers the cost of power.

Because capacitors may store an electrical charge and hold a voltage that is present even when disconnected from a circuit, capacitors must be enclosed, guarded, or located so that persons cannot accidentally contact the terminals. In most installations, capacitors are installed out of reach or placed in an enclosure accessible only to qualified persons. The stored charge of a capacitor must be drained by a discharge circuit—either permanently connected to the capacitor or automatically connected when the line voltage of the capacitor circuit is removed. The windings of a motor or a circuit consisting of resistors and reactors will serve to drain the capacitor charge.

Capacitor circuit conductors must have an ampacity of not less than 135% of the rated current of the capacitor. This current is determined from the VA rating of the capacitor as for any other load. For example, a 100 kVA (100,000VA), three-phase capacitor operating at 480V has a rated current determined as follows:

$$I = \frac{100{,}000\text{VA}}{(1.732 \times 480\text{V})} = 120.3\text{A}$$

The minimum conductor ampacity is then calculated as follows:

$$I = 1.35 \times 120.3\text{A} = 162.4\text{A}$$

When a capacitor is switched into a circuit, a large inrush current results as you start to charge it to the circuit voltage. Therefore, an overcurrent protective device for the capacitor must be rated or set high enough to allow the capacitor to charge. Although the exact setting is not specified in the *NEC*®, typical settings vary between 150% and 250% of the rated capacitor current.

In addition to overcurrent protection, a capacitor must have a disconnecting means rated at not less than 135% of the rated current of the capacitor, unless the capacitor is connected to the load side of the motor running the overcurrent device. In this case, the motor disconnecting means would serve to disconnect the capacitor and the motor.

A capacitor connected to a motor circuit serves to increase the power factor and reduce the total kVA required by the motor capacitor circuit. As stated earlier, the power factor (*pf*) is defined as the true power kW in kilowatts divided by the total kVA, or:

$$\text{Power factor } (pf) = \text{kW} \div \text{kVA}$$

A power factor of less than one represents a lagging current for motors and inductive devices. The capacitor introduces a leading current that reduces the total kVA and raises the power factor to a value closer to unity (one). If the inductive load of the motor is completely balanced by the capacitor, a maximum power factor of unity results, and all the input energy serves to perform useful work.

Know the Code

Conductors
NEC Section 460.8

According to *NEC Section 460.8*, the capacitor circuit conductors for a power-factor correction capacitor must have an ampacity of not less than 135% of the capacitor's rated current. In addition, the ampacity must not be less than one-third of the ampacity of the motor circuit conductors.

The connection of a capacitor reduces the current in the feeder up to the point of connection. If the capacitor is connected on the load side of the motor running the overcurrent device, the current through this device is reduced and its rating must be based on the actual current, not on the full-load current of the motor.

1.3.8 Diodes and Rectifiers

Rectifiers: Devices used to change alternating current to direct current.

Diodes and **rectifiers** are the simplest types of electronic components and are used in control circuits. The major difference between the two is their current rating. A component rated less than 1A is called a *diode*; a similar component rated above 1A is called a *rectifier*. Both devices serve to convert or rectify alternating current to direct current.

Diodes and rectifiers are composed of two types of semiconductor materials and classified as either P or N type. One has extra electrons; the other has a shortage of electrons. When the two types of material are bonded together, a solid-state component is produced, allowing electrons to flow in one direction and act as an insulator when the voltage is reversed.

Figure 26 shows an AC voltage supplying a control transformer that, in turn, must supply a DC electronic controller. Consequently, two rectifiers are installed on the secondary side of the transformer to change AC to DC. The rectifiers allow current to flow in one direction but will not allow the reversal of normal alternating current, simulating direct current.

Types N and P Diodes or Rectifiers

Type N semiconductor diodes or rectifiers are called *donor diodes* because they have an abundance of free electrons to give away. Type P semiconductor diodes or rectifiers are referred to as *acceptor diodes* because they have a shortage of free electrons and will receive electrons. To help you remember, the word donor has the letter *n* in it, while the word acceptor has the letter *p* in it.

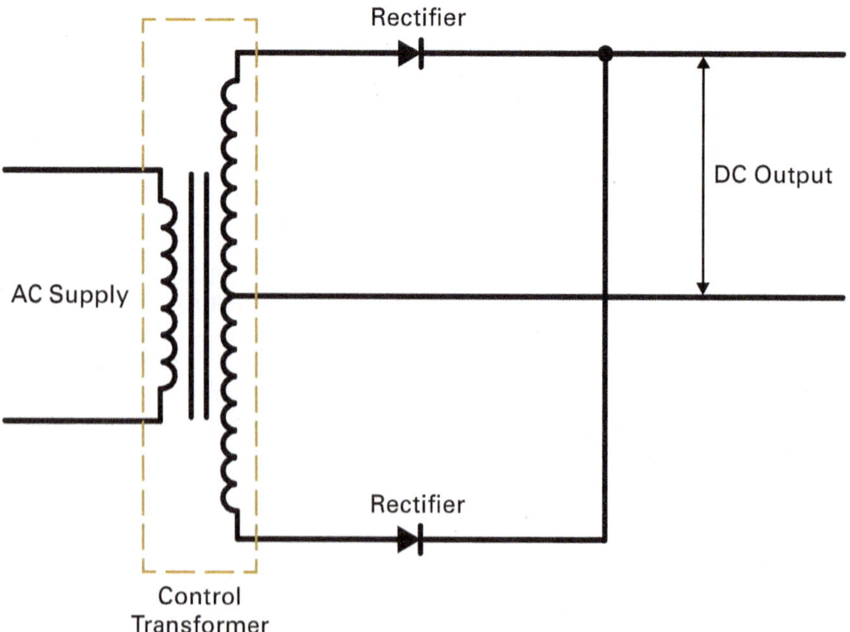

Figure 26 Rectifiers used in a control circuit to change AC to DC.

1.3.9 Using Vectors to Identify the Operating Characteristics of Transformers

The theoretical study of transformers includes the use of vectors (phasor diagrams) that graphically represent voltages and currents in transformer windings.

A vector or phasor diagram is a line with direction and length. Reading a vector diagram is like reading a road map. It is not much more difficult, just a bit more refined. For example, it's simple to understand road directions that instruct you to go east 40 miles and then south 30 miles to get to your destination. In other words, you must drive 70 miles to get there. However, had you been able to drive directly from point to point, the distance would have been shorter—only 50 miles, as shown in *Figure 27* (A). If the miles were converted into electrical terms, this same triangle would be the classic 3-4-5 right triangle or the 80% power factor relationship, as shown in *Figure 27* (B).

Referring to *Figure 27* (B), if the working current (line A-B) is 4A and the inductive current (line B-C) is 3A, then the diagonal line A-C, or hypotenuse, will be 5A. This may be obtained mathematically using the Pythagorean theorem, which states that the hypotenuse is equal to the square root of the sum of the other two sides' squares:

$$AC = \sqrt{AB^2 + BC^2}$$

$$AC = \sqrt{4^2 + 3^2} = \sqrt{16 + 9} = \sqrt{25}$$

$$AC = 5$$

In this example, the working current is related to kW, and the total current is related to kVA. The ratio between the two is 4 ÷ 5 = 0.80. Because power factor = kW ÷ kVA, the power factor of this circuit is 80%.

When three single-phase transformers are used as a three-phase bank, the direction of the voltage in each of the six phase windings may be represented by a voltage phasor. A voltage phasor diagram of the six voltages involved provides a convenient way to study the relative directions and amounts of the primary and secondary voltages.

The same is true for one three-phase transformer, which also has six voltages to be considered. Its phase windings on the high-voltage and low-voltage sides are connected in the same way as the phase windings of three single-phase transformers.

To show how a phasor diagram is drawn, consider the three-phase transformer shown in *Figure 28*. It is a wye-connected/delta-connected, three-phase transformer with three legs, with each leg carrying a high-voltage and a low-voltage winding. The high-voltage windings are connected in wye (with a common neutral point at N) with the leads to the high-voltage terminals designated H_1, H_2, and H_3. The three low-voltage windings are connected in delta. The junction points of the three windings serve as low-voltage terminals X_1, X_2, and X_3.

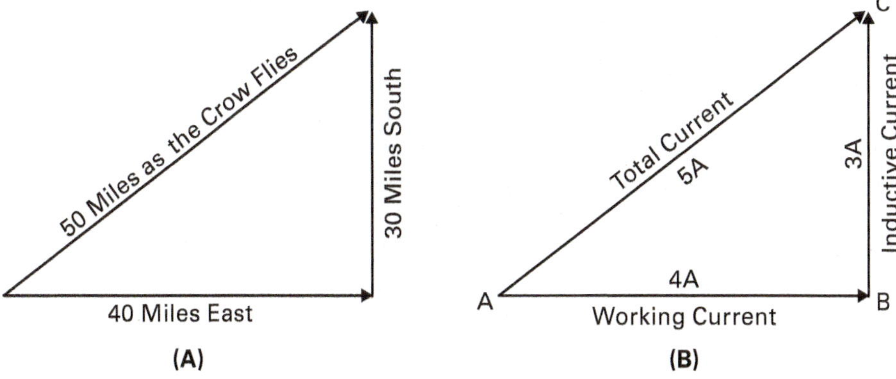

Figure 27 Typical vectors.

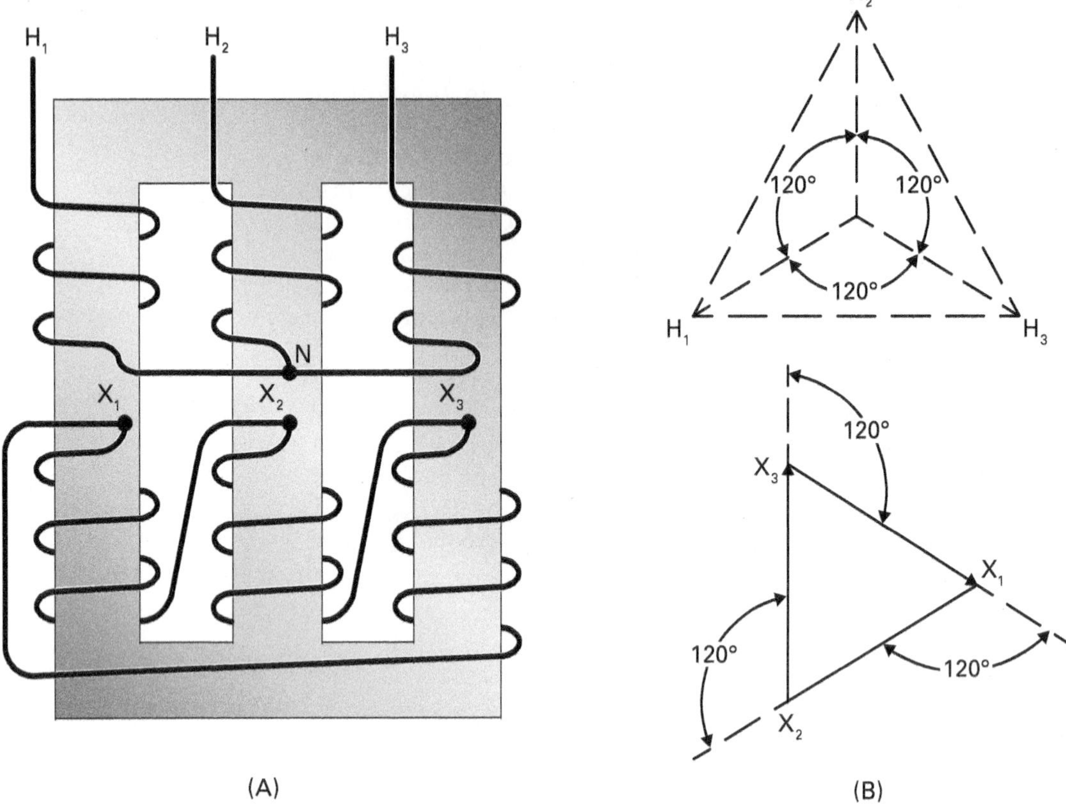

Figure 28 Windings and phasor diagrams of a three-phase transformer.

The voltages in the low-voltage windings are assumed to be equal but displaced from each other by 120°.

When drawing the phasor diagram for the low-voltage windings, phasor X_1X_2 is drawn first in any selected direction and to any convenient scale. The arrowhead indicates the instantaneous direction of the alternating voltage in winding X_1X_2, and the length of the phasor represents the amount of voltage in the winding. The broken lines that extend past the arrowheads represent reference lines for phase angles.

Because winding X_2X_3 is physically connected to the end of X_2 in the X_1X_2 winding, phasor X_2X_3 will be started at point X_2, 120° out of phase with phasor X_1X_2 in the clockwise direction. The length of phasor X_2X_3 is equal to the length of phasor X_1X_2. Winding X_3X_1 is drawn in a similar manner.

The high-voltage phasor comes next. Because the low-voltage winding X_1X_2 is wound on the same leg as the high-voltage winding NH_1, the voltages in these two windings are in phase and represented by parallel voltage phasors. Therefore, the phasor NH_1 is drawn from a selected point N parallel to X_1X_2. Note that all three high-voltage windings are physically connected to a common point N. Therefore, the phasors representing the voltages, NH_1, NH_2, and NH_3, all start at the common point (point N in the phasor diagram). Again, the high-voltage phasors are all the same length but displaced from each other by 120° as shown.

The high-voltage phasors are the same length as the low-voltage phasors. Each phase of the high voltage is, however, proportional to the low voltage in the same phase according to the turns ratio. This results in the line voltage being 1.732 times higher than the phase voltage in any single winding. It can be geometrically proven that, for example, $H_1H_2 \times NH_1 = 1.732 \times NH_2$, and so forth for each line voltage.

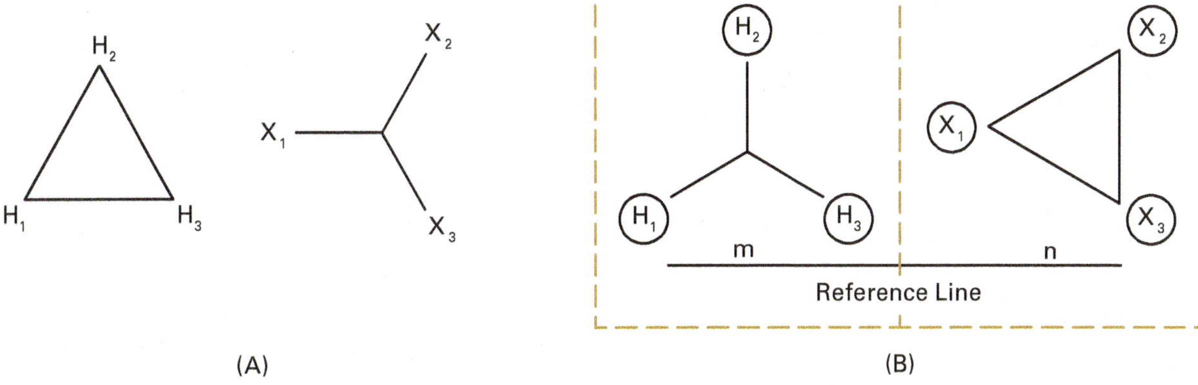

Figure 29 Phasor diagrams for three-phase transformer banks in parallel.

When comparing three-phase transformers for possible operation in parallel, draw the voltage phasor diagram for transformer bank A and mark the terminals as shown in *Figure 29* (A). Next, draw the phasor diagram for bank B on drafting or other transparent paper. Then draw a reference line (m-n) as shown in *Figure 29* (B). Cut the transparent diagram into two parts as indicated by the dashed lines in the drawing. The two parts of the reference line are now marked m and n. Place diagram m on the high-voltage diagram in *Figure 29* (A) so that the terminals that need to be connected coincide. Place diagram n on the low-voltage diagram in *Figure 29* (A) so that the reference line of n is parallel to the reference line of m. If the terminals of n can be made to coincide with the low-voltage terminals, the terminals that coincide can be connected for parallel operation. If the low-voltage terminals cannot be made to coincide, parallel operation is not possible with the assumed high-voltage connection.

Vector diagrams can also be used to illustrate voltage drop. *Figure 30* shows a simple 100% power factor circuit in which a 1Ω resistance appears between the source of power and the load. The current in this series circuit is 10A, and the voltage at the source is 120V. Because Ohm's law states that $E = IR$, there must be a voltage drop equal to 10 times 1, or 10V, across the resistance. With the voltage and current in phase, only 110V will be available at the load because you must subtract the resistance drop from the source voltage.

Voltage drop, however, becomes a little more complicated when inductance is introduced into a circuit, as shown in *Figure 31*. In this case, the load voltage will be equal to the source voltage minus the voltage drops through R and X, but the drops cannot be summed. The vector diagram in *Figure 32* shows that the dotted line A-C is the combination of the two voltages across R and X and represents the voltage drop in the line only.

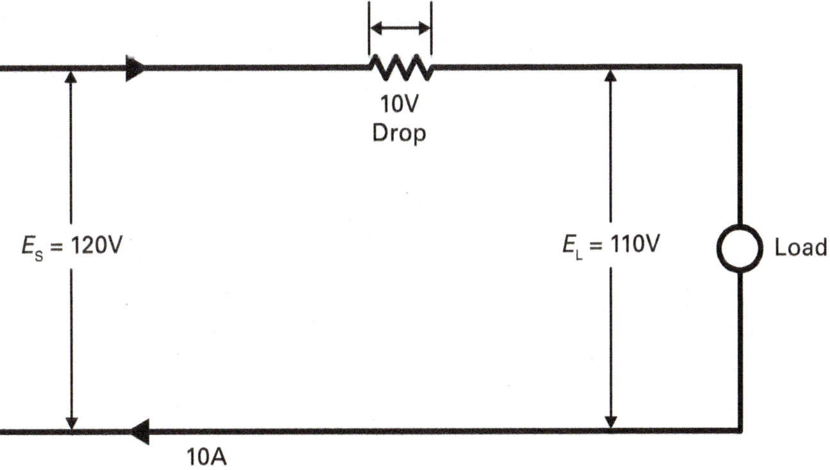

Figure 30 Circuit containing resistance only.

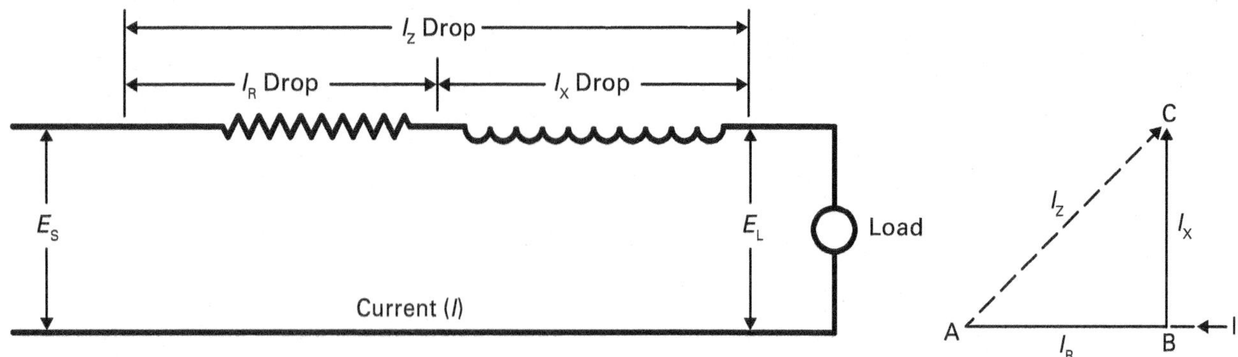

Figure 31 Electrical circuit with both resistance and inductance.

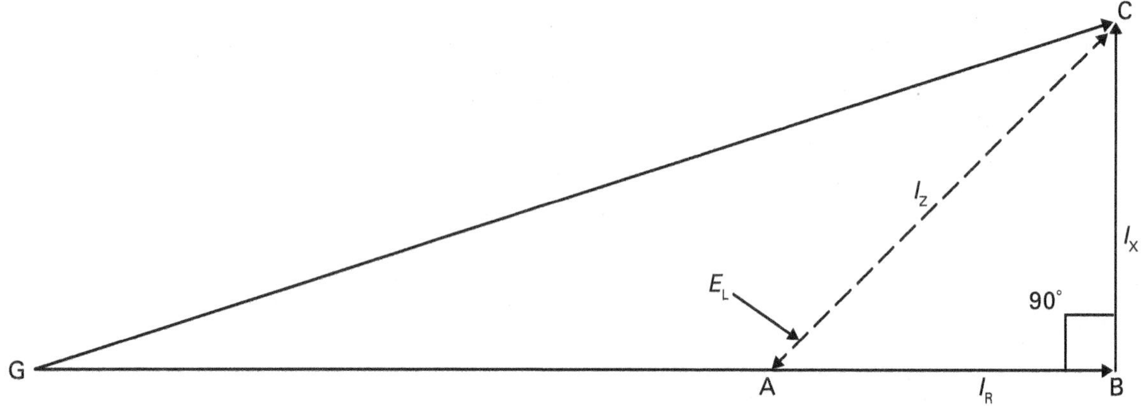

Figure 32 Effect of voltage drop in an AC circuit.

Figure 32 shows that line G-A, which is the load voltage E_L, is less than the source voltage G-C due to the voltage drop in line A-C. Because of the effect of the reactance in the line, this voltage drop cannot be subtracted arithmetically from the source voltage to obtain the load voltage. Instead, vectors must be used.

Calculations of impedance can be simplified by using equivalent circuits. The reactance voltage drop is governed by the leakage flux, and the voltage regulation depends on the power factor of the load. To determine transformer efficiency at various loads, it is necessary to first calculate the core loss, hysteresis loss, eddy current loss, and load loss.

However, for all practical purposes, a transformer's efficiency can be considered 100%. Therefore, a transformer may be defined as a device that transfers power from its primary circuit to the secondary circuit without any significant loss.

Apparent power (*VA*) equals voltage (*E*) times current (*I*). Therefore, if $E_P I_P$ represents the primary apparent power and $E_S I_S$ represents the secondary apparent power, then $E_P I_P = E_S I_S$, where P stands for primary and S stands for secondary (*Figure 33*). If the primary and secondary voltages are equal, the primary and secondary currents must also be equal. Assume that E_P is twice as large as E_S. For $E_P I_P$ to equal $E_S I_S$, I_P must be $\frac{1}{2}$ of I_S. Therefore, a transformer that steps voltage down always steps current up. Conversely, a transformer that steps voltage up always steps current down. However, transformers are classified as step-up or step-down only in relation to their effect on voltage.

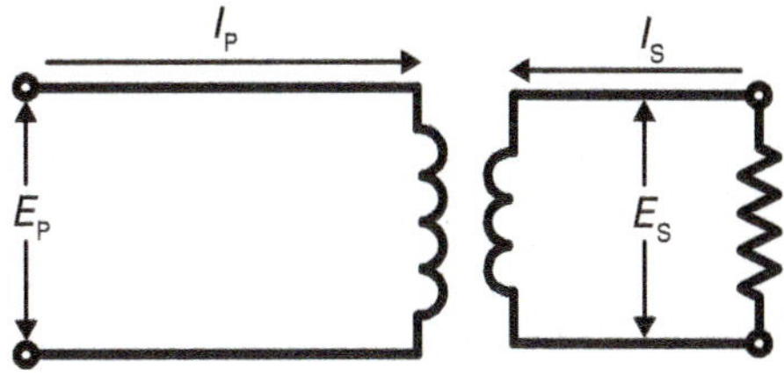

(A) Low Voltage, High Current

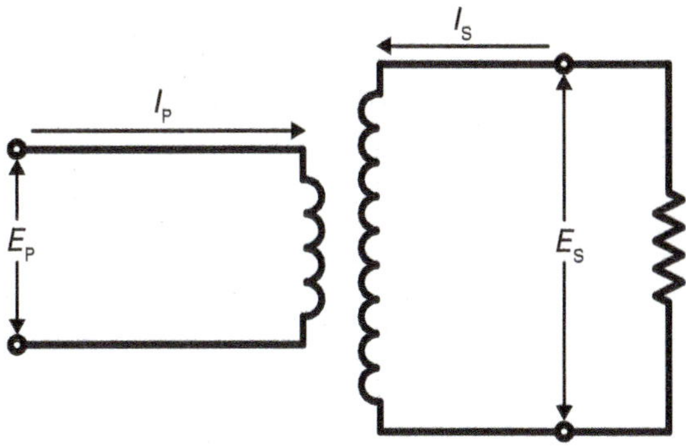

(B) High Voltage, Low Current

Figure 33 Voltage-current relationship in a transformer.

1.0.0 Section Review

1. If there are 20 turns in the primary and only 10 turns in the secondary and the transformer operates at 120V, the voltage in the secondary is _____.
   a. 480V
   b. 240V
   c. 120V
   d. 60V

2. The *least* efficient transformer core design is the _____.
   a. open core
   b. closed core
   c. shell core
   d. mitered core

3. Capacitor circuit conductors must have an ampacity of *not* less than _____.
   a. 135% of the rated current of the capacitor
   b. 150% of the rated current of the capacitor
   c. 175% of the rated current of the capacitor
   d. 200% of the rated current of the capacitor

2.0.0 *NEC*® Requirements for Transformers and Capacitors

Performance Tasks

There are no Performance Tasks in this section.

Objective

Identify *NEC*® requirements related to transformers and capacitors.

a. Identify the *NEC*® requirements for transformers.
b. Identify the *NEC*® requirements for capacitors.
c. Identify the *NEC*® requirements for resistors and reactors.

Know the Code

Disconnecting Means
NEC Section 450.14

Per *NEC Section 450.14*, transformers other than low-voltage Class 2 or Class 3 models must have a disconnecting means. The disconnect location must be either in sight of the transformer or in a remote location. If it is mounted remotely, you must be able to lock the disconnect in the open position. Its location is then marked on the transformer.

2.1.0 *NEC*® Requirements for Transformers

Know the Code

Dry-Type Transformers Installed Indoors
NEC Section 450.21

Know the Code

Dry-Type Transformers Installed Outdoors
NEC Section 450.22

Know the Code

Less-Flammable Liquid-Insulated Transformers
NEC Section 450.23

Know the Code

Nonflammable Fluid-Insulated Transformers
NEC Section 450.24

Know the Code

Askarel-Insulated Transformers Installed Indoors
NEC Section 450.25

Know the Code

Oil-Insulated Transformers Installed Indoors
NEC Section 450.26

Know the Code

Oil-Insulated Transformers Installed Outdoors
NEC Section 450.27

The *NEC*® requirements for transformer installations are based on the type of insulating fluid (if any), the transformer rating, and whether they are installed indoors or outdoors. There are five basic classifications:

- *Dry-type transformers* — These transformers can be ventilated, air-cooled, or sealed. See *NEC Sections 450.21 and 450.22*. They are not required to be installed in a vault unless their rating is greater than 35 kV.
- *Less-flammable liquid-insulated transformers* — These transformers contain a liquid with a fire point not less than 572°F (300°C) . Refer to *NEC Section 450.23*. They do not need to be installed in a vault unless their rating is greater than 35 kV and additional conditions are met. Otherwise, the indoor and outdoor installation requirements are the same as for oil-insulated transformers.
- *Nonflammable fluid-insulated transformers* — These transformers contain a nonflammable dielectric fluid that does not have a flash or fire point as a liquid or in any concentration with air. See *NEC Section 450.24*. They are not required to be installed in a vault unless their rating is greater than 35 kV.
- *Askarel-insulated transformers* — These transformers contain Askarel fluid, which is a generic name for a synthetic, nonflammable liquid that often contains polychlorinated biphenyls (PCBs). Because of the possible presence of PCBs, these transformers have special safety requirements. See *NEC Section 450.25*. They are not required to be installed in a vault unless their rating is greater than 35 kV.
- *Oil-insulated transformers* — These transformers do contain flammable oil and must be installed in vaults. Exceptions to this requirement can be found in *NEC Section 450.26*. The requirements for oil-insulated transformers installed outdoors can be found in *NEC Section 450.27.*

Transformers must normally be accessible for inspection, except for dry-type transformers under certain specified conditions. Certain types of transformers with a high voltage or kVA rating are required to be enclosed in transformer rooms or vaults when installed indoors. The construction of these vaults is covered in *NEC Article 450, Part III* and described in *Figure 34* and *Table 1*.

TABLE 1 Summary of *NEC*® Transformer Installation and Overcurrent Protection Requirements

Application	Code Requirement	Code Reference
Location	Transformers must be easily accessed by qualified personnel to provide service.	*NEC Section 450.13*
	Transformers other than low-voltage Class 2 or Class 3 must have a disconnecting means. The disconnect location must be either within sight of the transformer or in a remote location. When mounted remotely, the disconnect must be lockable in the open position, with its location marked on the transformer.	*NEC Section 450.14*
	Dry-type transformers rated at 112½ kVA or less can be located out in the open if they are separated from combustible material by at least 12" (300 mm) or an appropriate fire/heat barrier.	*NEC Section 450.21(A)*
	Dry-type transformers rated at more than 112½ kVA must be installed in a fire-resistant transformer room.	*NEC Section 450.21(B)*
	Dry-type transformers not exceeding 1,000V and 50 kVA do not need to be readily accessible and can be placed in fire-resistant hollow spaces in a building under conditions specified by the *NEC*®.	*NEC Sections 450.13(A) and (B)*
	Dry-type transformers installed outdoors must have a weatherproof enclosure.	*NEC Section 450.22*
	Liquid-filled transformers must be installed as specified in the *NEC*® and usually in vaults when installed indoors.	*NEC Section 450.23(A)(3)*
Overcurrent protection	The primary protection for a transformer rated at 1,000V or less must be as follows: If the primary current (line side) is 9A or more, an overcurrent protective device of the next higher standard size than 125% of the primary current is used. For example, if the primary current is 15A, 125% of 15A = 18.75A. The next standard circuit breaker available is 20A. Conductors on the secondary side of a single-phase transformer with a two-wire secondary may be protected by the primary overcurrent protective device under certain *NEC*® conditions.	*NEC Table 450.3(B)*
Transformers used in motor control circuits	Special rules apply to transformers in motor control circuits.	*NEC Section 430.72(C)*
Over 1,000V	Special *NEC*® rules apply to transformers operating at over 1,000V.	*NEC Table 450.3(A)*

In general, the *NEC*® specifies that the walls and roofs of vaults must be constructed of materials that have adequate structural strength for the environment and a minimum fire resistance of three hours. However, where transformers are protected with an automatic sprinkler system or other fire suppression system, the fire resistance of the structure may be lowered to only one hour.

Vault floors that are in contact with the earth must be concrete and at least 4" (100 mm) thick. If the vault is built with a vacant space or other floors below it, its floor must have sufficient structural strength for the total load and a minimum fire resistance of three hours. Again, if fire suppression is provided as outlined above, the fire resistance of the structure can be reduced to one hour. Note that the *NEC*® does not permit the use of wooden studs and wallboard construction in transformer vaults.

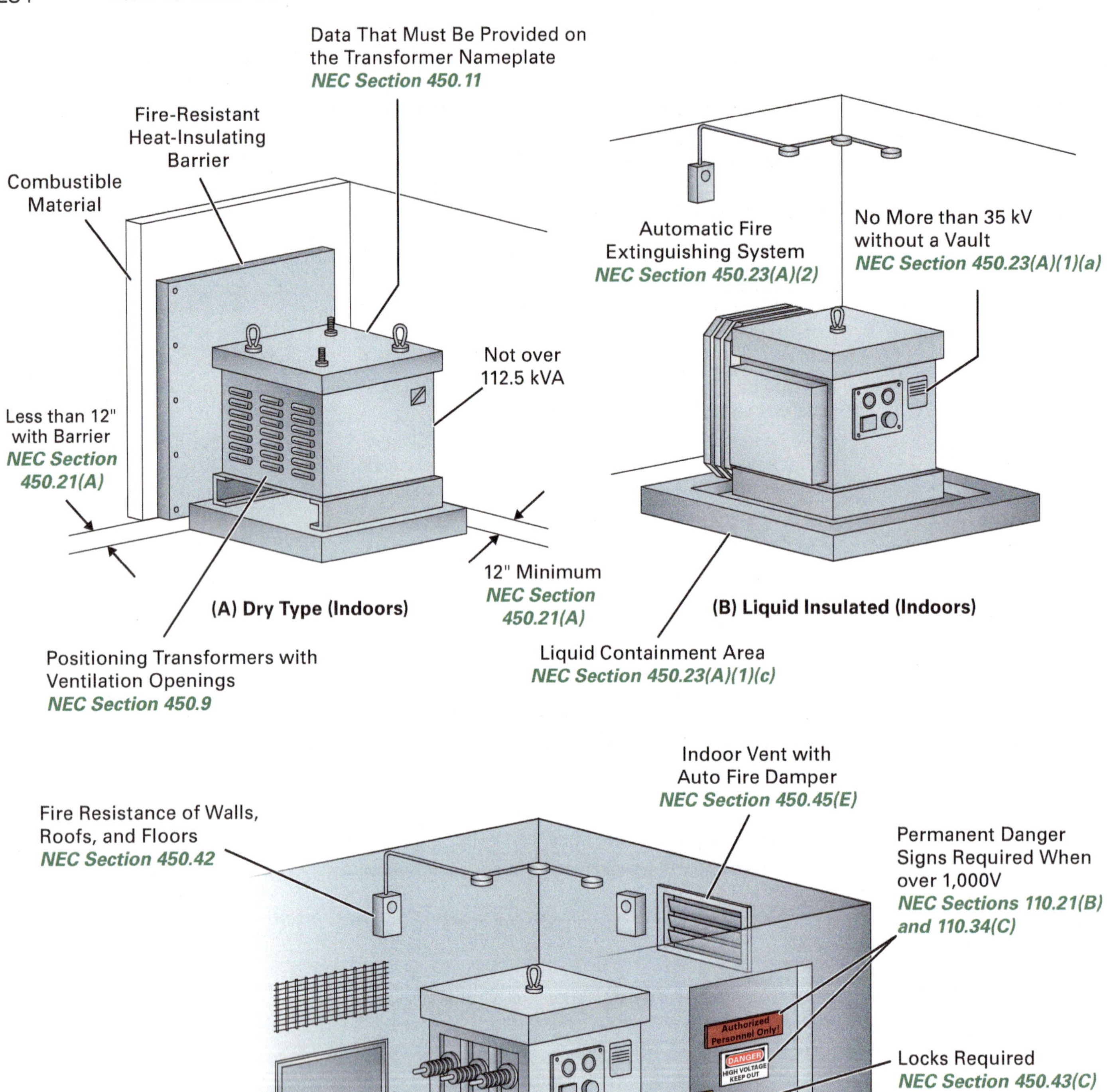

Figure 34 Transformer installation requirements.

Transformer Classifications

Transformers are classified by four-letter cooling designations. These designations allow you to identify the cooling approach employed and how cooling fluids are circulated.

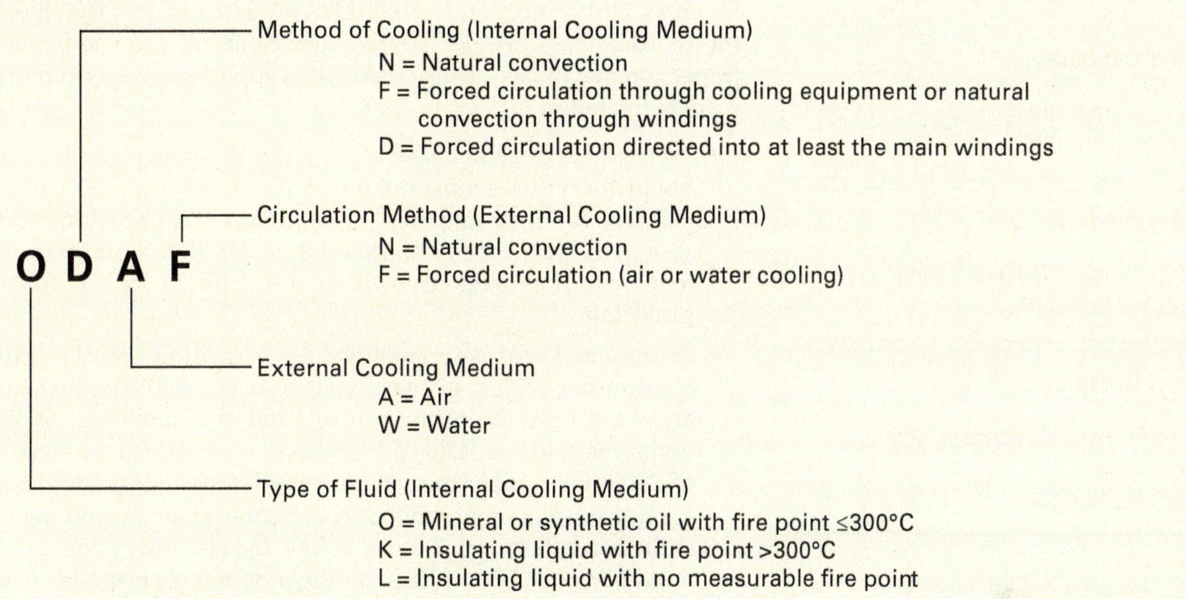

2.1.1 Overcurrent Protection for Transformers (1,000V or Less)

The overcurrent protection for transformers is based on their rated current, not on the load to be served. The primary circuit may be protected by a device rated or set at not more than 125% of the rated transformer primary current for transformers rated at 9A and higher.

Instead of individual protection on the primary side, the transformer may be protected only on the secondary side if all the following conditions are met:

- The overcurrent protective device on the secondary side is rated or set at not more than 125% of the rated secondary current.
- The primary feeder overcurrent protective device is rated or set at not more than 250% of the rated primary current.

For example, if a 12 kVA transformer has a primary voltage rating of 480V, calculate the amperage as follows:

$$12 \text{ kVA} = 12{,}000\text{VA}$$

$$12{,}000\text{VA} \div 480\text{V} = 25\text{A}$$

If the secondary voltage is 120V, the amperage will be higher:

$$12{,}000\text{VA} \div 120\text{V} = 100\text{A}$$

The individual primary protection is then determined as follows:

$$1.25 \times 25\text{A} = 31.25\text{A}$$

In this case, a standard 30A cartridge fuse rated at 600V could be used, as could a circuit breaker approved for use on 480V. However, when certain conditions are met, individual primary protection for the transformer is not necessary if the feeder overcurrent protective device is rated at not more than the result of the following calculation:

$$2.5 \times 25\text{A} = 62.5\text{A}$$

In addition, the protection of the secondary side is determined as follows:

$$1.25 \times 100\text{A} = 125\text{A}$$

In this case, a standard 125A circuit breaker could be used.

NOTE

The example is for the transformer only, not the secondary conductors. The secondary conductors must be provided with overcurrent protection as outlined in ***NEC Section 240.4***.

Know the Code

Overcurrent Protection
NEC Section 450.3

Know the Code

Overcurrent Protection
NEC Article 240

Know the Code

Conductors for General Wiring
NEC Article 310

Know the Code

Switchboards, Switchgear, and Panelboards
NEC Article 408

NOTE

The use of primary protection only does not waive the requirements for overcurrent protection compliance found in *NEC Articles 240 and 408*. See Informational Note No. 1 in *NEC Section 450.3*, which references *NEC Sections 240.4, 240.21, 245.26, and 245.27*, for proper protection of secondary conductors.

2.1.2 Overcurrent Protection for Other Components

The requirements of *NEC Section 450.3* cover only transformer protection. In practice, other components must be considered when applying circuit overcurrent protection. Circuits with transformers must meet the requirements for conductor protection in *NEC Articles 240 and 310*. Panelboards must meet the requirements of *NEC Article 408*. Some of the components or types of protection that must be considered when applying circuit overcurrent protection include the following:

- *Primary fuse protection only* — If secondary fuse protection is not provided, the primary fuses must not be rated at more than 125% of the transformer primary full-load amperes (FLA), unless the transformer primary FLA is shown in *NEC Table 450.3(B)* (*Figure 35*). Individual transformer primary fuses are not necessary where the primary circuit fuse provides this protection.
- *Primary and secondary protection* — According to *NEC Table 450.3(A)*, a transformer with a primary voltage over 1,000V located in unsupervised areas can have the primary fuse sized at a maximum of 300%. If the secondary is also over 1,000V, the secondary fuses can be sized at a maximum of 250% for transformers with impedances not greater than 6%, or 225% for transformers with impedances greater than 6% and not more than 10%. If the secondary is 1,000V or below, the secondary fuses can be sized at a maximum of 125%. If these settings do not correspond to a standard fuse size, the next higher standard size is allowed. The maximum settings for supervised locations are as shown in *Figure 36*, except for secondary voltages of 1,000V or below where the secondary fuses can be sized at a maximum of 250%.
- *Primary protection only* — In supervised locations, the primary fuses can be sized at a maximum of 250% or the next larger standard size if 250% does not correspond to a standard fuse size.

2.1.3 Overcurrent Protection for Small Power Transformers

Low-amperage, E-rated, medium-voltage fuses are general-purpose, current-limiting fuses. The E rating defines the melting time-current characteristic of the fuse and permits electrical interchangeability of fuses that have the same E rating.

For a general-purpose fuse to have an E rating, the current-responsive element must melt in 300 seconds at a rms current within 200% to 240% of the continuous current rating of the fuse, according to *ANSI C37.46*.

Low-amperage, E-rated fuses are designed to provide primary protection for potential, small service, and control transformers. These fuses offer a high level of fault current interruption in a self-contained, non-venting package that can be mounted indoors or in an enclosure.

As for all current-limiting fuses, the basic application rules found in the *NEC*® and the manufacturer's literature must be obeyed. In addition, potential transformer fuses must have sufficient inrush capacity to successfully pass through the magnetizing inrush current of the transformer. If the fuse is not sized properly, it will open before the load is energized. The maximum inrush currents to the transformer at system voltage, as well as the duration of the inrush current, varies with the transformer design. Magnetizing inrush currents are usually denoted as a percentage of the transformer full-load current, such as 10X, 12X, 15X, and so on.

The inrush current duration is usually measured in seconds. Where this information is available, an easy check can be made on the appropriate minimum melting curve to verify proper fuse selection. In lieu of transformer inrush data, the rule of thumb is to select a fuse size rated at 300% of the primary FLA or the next larger standard size.

Fuse must not be rated at more than 125% of the transformer primary FLA (note exceptions in the table).

No Secondary Protection

Primary 1,000V or Less

Secondary 1,000V or Less

Transformer

Primary Current	Primary Fuse Rating
9A or more	125% or next higher standard rating if 125% does not correspond to a standard fuse size
2A to 9A	167% maximum
Less than 2A	300% maximum

Figure 35 Transformer circuit with primary fuse only.

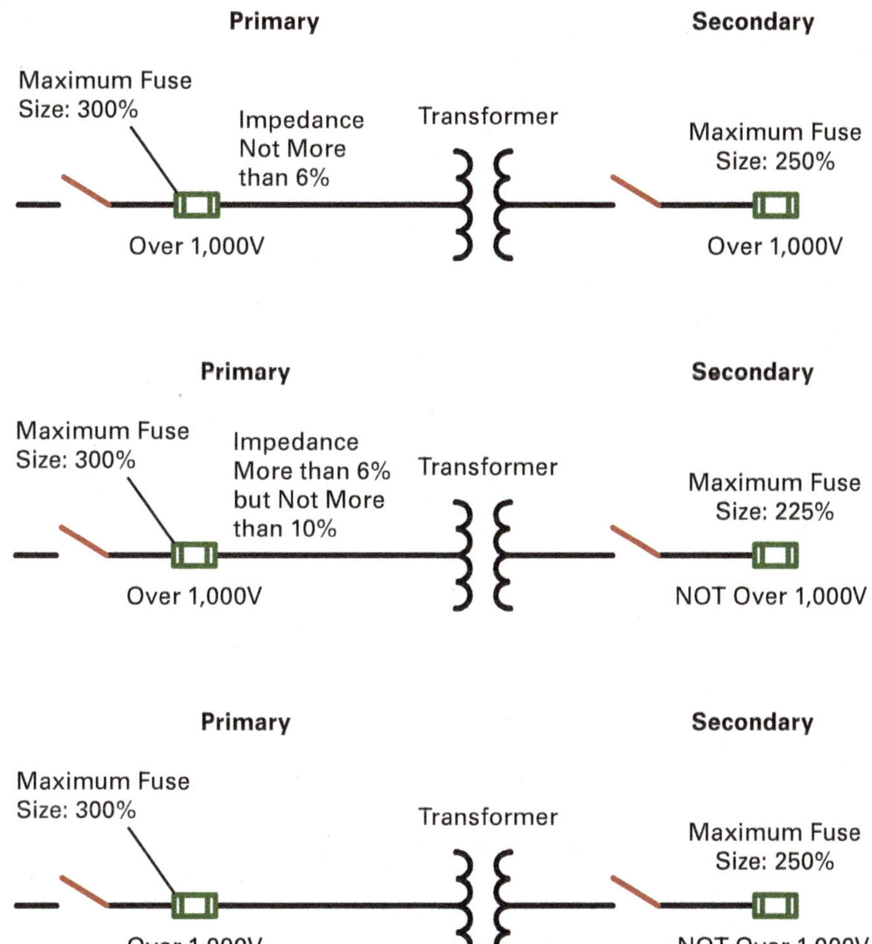

Figure 36 Overcurrent protection for transformers in supervised locations.

For example, a transformer manufacturer states that an 800VA, 2,400V, single-phase potential transformer has an inrush current of 12X lasting for 0.1 seconds. The inrush current can be calculated as follows:

$$I = 800\text{VA} \div 2{,}400\text{V} = 0.333\text{A}$$

$$\text{Inrush Current} = 12 \times 0.333\text{A} = 4\text{A}$$

Because the voltage is 2,400V, you can use either a JCW or a JCD fuse. Using the 300% rule of thumb to determine the fuse rating results in the following equation:

$$300\% \times 0.333\text{A} = 0.999\text{A}$$

Therefore, choose a JCW-1E or JCD-1E fuse.

Typical potential transformer connections can be grouped into two categories:

- Connections that require the fuse to pass the inrush of only one potential transformer (*Figure 37*)
- Connections that require the fuse to pass the inrush of more than one potential transformer (*Figure 38*)

Fuses for medium-voltage transformers and feeders are E-rated, medium-voltage fuses. They are also general-purpose, current-limiting fuses. The fuses carry either an E or an X rating, which defines the melting time-current characteristic of the fuse. The ratings are used to allow electrical interchangeability among different manufacturers' fuses.

For a general-purpose fuse to have an E rating, the following conditions must be met:

- Current-responsive elements with ratings 100A or below must melt in 300 seconds at a rms current within 200% to 240% of the continuous current rating of the fuse (*ANSI C37.46*).
- Current-responsive elements with ratings above 100A must melt in 600 seconds at a rms current within 220% to 264% of the continuous current rating of the fuse (*ANSI C37.46*).

A fuse with an X rating does not qualify the electrical interchangeability for an E-rated fuse, but it offers the user other ratings that may provide better protection for a given application.

Transformer protection is the most popular application of E-rated fuses. The fuse is applied to the primary of the transformer and is solely used to prevent destruction of the transformer due to short circuits. It is essential that the fuse be sized so that it does not open during the inrush current or permissible overload currents. Inrush must also be considered when sizing a fuse. Power transformers usually have a magnetizing inrush current of 12X, which is the full-load rating for a duration of 0.1 seconds.

Transformers

Transformers are not considered separately derived systems when they have a direct electrical connection, including a solidly grounded neutral, from the primary through the secondary.

2.1.4 Transformer Grounding

Grounding is necessary to remove static electricity and act as a precautionary measure in case the transformer windings accidentally contact the core or the enclosure. All transformers should be grounded and bonded to meet *NEC®* requirements as well as local codes where applicable.

The enclosure of every power transformer should be grounded to eliminate the possibility of obtaining static shocks from it or being injured by accidental grounding of the winding to the case. A grounding lug is provided on the base of most transformers.

The *NEC®* specifically states grounding requirements that should be followed in every respect. Furthermore, certain advisory rules recommended by manufacturers provide additional protection beyond that of the *NEC®*. In general, the code requires that separately derived AC systems be grounded as stated in *NEC Section 250.30*.

Know the Code

Grounding Separately Derived Alternating-Current Systems
NEC Section 250.30

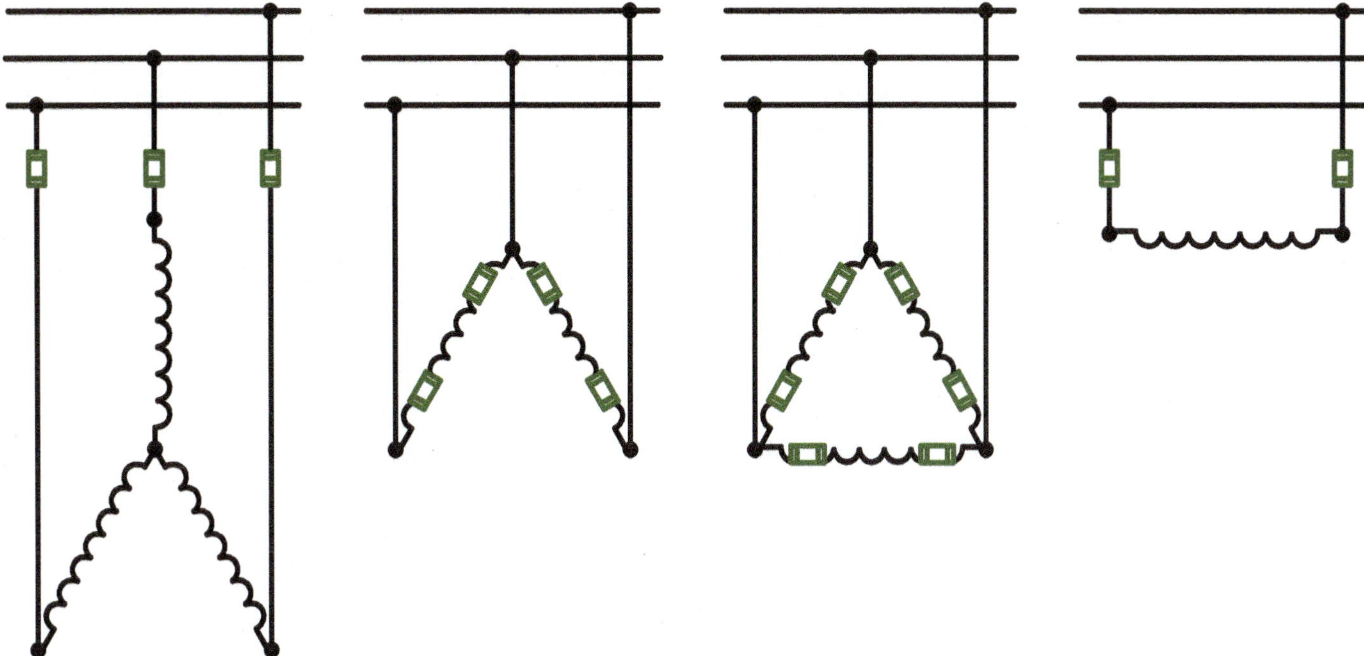

Figure 37 Connections requiring fuses to pass the inrush of only one transformer.

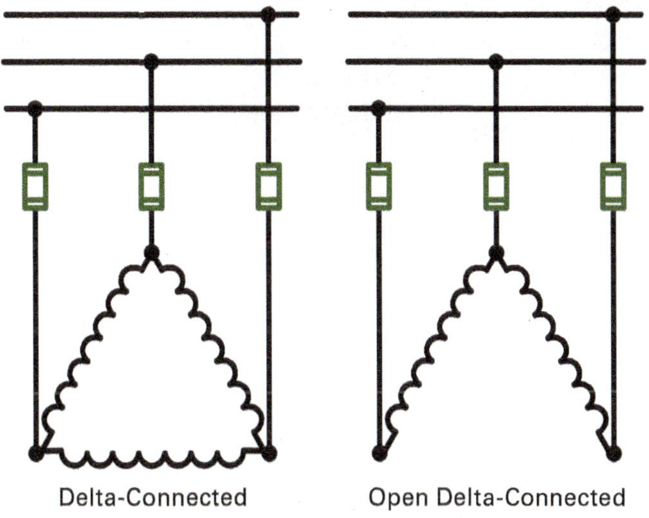

Figure 38 Connections requiring fuses to pass the inrush of more than one transformer.

Generator Step-Up (GSU) Transformers

This GSU steps up the voltage from a power generation facility to a transmission voltage of 230 kV. It has a rated load capacity of 185 MVA, or 185,000,000 watts. This installation includes a deluge system for fire protection, a huge containment basin to protect the surrounding area from oil leaks, and a reinforced blast wall to shield adjacent equipment in the event of an explosion.

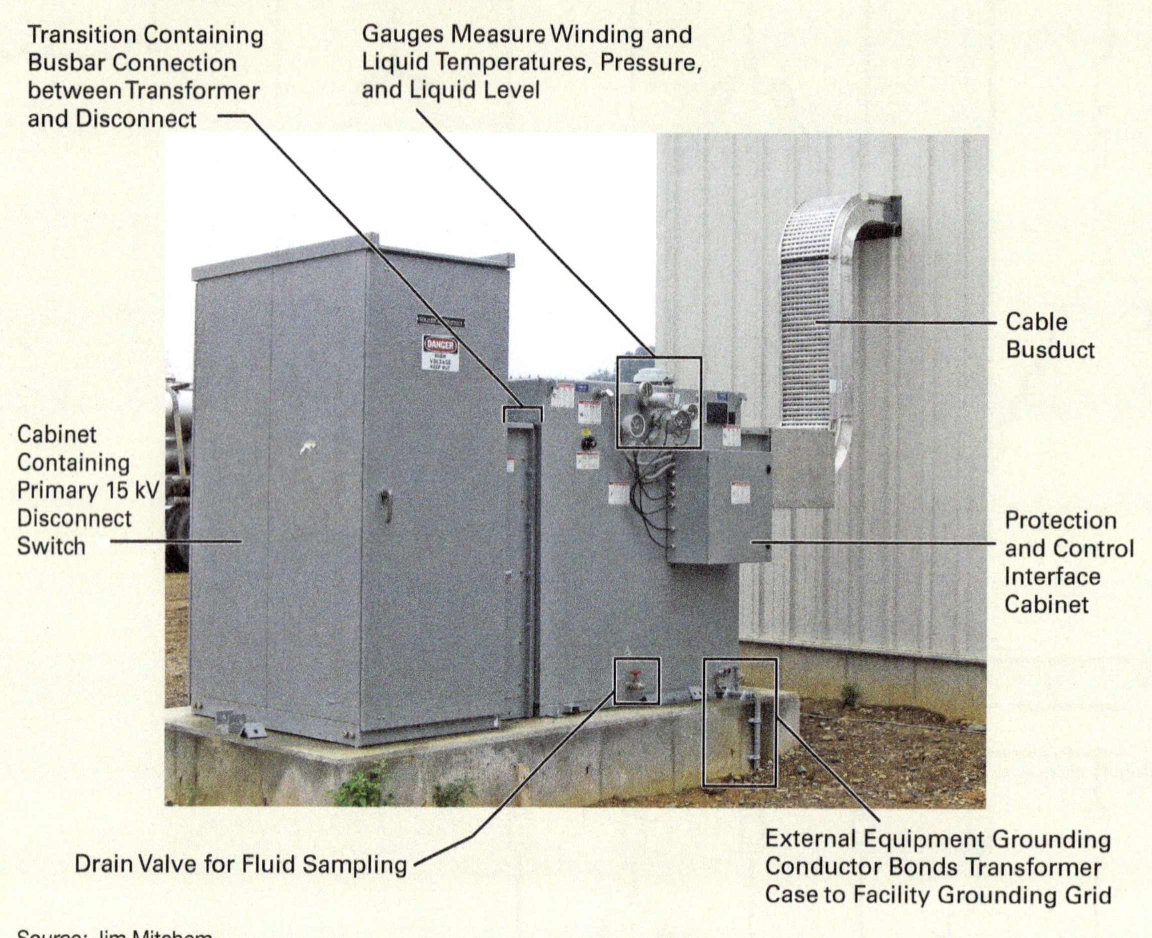

Source: Jim Mitchem

2.2.0 *NEC®* Requirements for Capacitors

NEC Article 460 states specific rules for the installation and protection of capacitors other than surge capacitors or capacitors that are part of other equipment. The *NEC®* requirements for capacitors operating under 1,000V are summarized in *Table* 2.

Know the Code

Capacitors
NEC Article 460

TABLE 2 *NEC®* **Capacitor Installation Requirements**

Application	Code Regulation	Code Reference
Enclosing and guarding	Capacitors must be enclosed, located, or guarded so that accidental contact with exposed energized parts, terminals, and buses by workers or conductive materials is prevented. However, no additional guards are required for enclosures that can only be accessed by qualified persons.	*NEC Section 460.3*
Stored charge	Capacitors must be provided with a way to discharge the stored charge. Once the discharge circuit is de-energized, it must either be permanently connected to the terminals of the capacitor or capacitor bank or be provided with an automatic means of connecting it to those terminals. Connecting the discharge circuit to the capacitors manually is not permitted.	*NEC Section 460.6*
Capacitors on circuits over 1,000V	Special *NEC®* regulations apply to capacitors operating at over 1,000V.	*NEC Articles 460, Part II and 495*
Conductor ampacity	The ampacity of capacitor circuit conductors must not be less than 135% of the rated capacitor current.	*NEC Section 460.8(A)*
Capacitors on motor circuits	The ampacity of conductors that connect a capacitor to motor terminals or motor circuit conductors must not be less than one-third the ampacity of the motor conductors and in no case less than 135% of the rated capacitor current.	*NEC Section 460.8(A)*
Overcurrent protection	Overcurrent protection is required in each ungrounded conductor unless the capacitor is connected on the load side of a motor overcurrent device. The setting must be as low as practicable.	*NEC Section 460.8(B)*
Disconnecting means	A disconnecting means is required for a capacitor unless it is connected to the load side of a motor controller. The rating must be at least 135% of the rated capacitor current.	*NEC Section 460.8(C)*
Overcurrent protection for improved power factor	If the power factor is improved, the motor overcurrent device must be selected based on the reduced current draw, not the FLA of the motor.	*NEC Section 460.9*
Grounding	Capacitor cases must be grounded except when the system is designed to operate at other than ground potential.	*NEC Sections 460.10 and 460.27*

2.3.0 *NEC®* Requirements for Resistors and Reactors

NEC Article 470 covers the installation of separate resistors and reactors on electric circuits. However, this article does not cover such devices that are component parts of other machines and equipment.

In general, *NEC Section 470.2* requires resistors and reactors be installed where they are not exposed to physical damage. Therefore, these devices are normally installed in a protective enclosure, such as a controller housing or other type of cabinet. When these enclosures are constructed of metal, they must be grounded as specified in *NEC Article 250*. Furthermore, a thermal barrier must be provided between resistors and/or reactors and any combustible material that is less than 12" (305 mm) away. A space of 12" (305 mm) or more between the devices and combustible material is considered a safe distance.

Insulated conductors used for connections between resistors and motor controllers must be rated at not less than 194°F (90°C) except for motor starting service. In this case, other conductor insulation is permitted, provided other sections of the *NEC®* are not violated.

Know the Code

Resistors and Reactors
NEC Article 470

Know the Code

Grounding and Bonding
NEC Article 250

2.0.0 Section Review

1. Dry-type transformers rated at $112\frac{1}{2}$ kVA or less may be located out in the open if they are separated from combustible material by a *minimum* distance of _____.
   a. 6" (150 mm)
   b. 8" (200 mm)
   c. 10" (250 mm)
   d. 12" (300 mm)

2. If the power factor is improved by the addition of a capacitor, the motor overcurrent device *must* be selected based on the _____.
   a. reduced current
   b. full-load current
   c. 125% of the locked-rotor current
   d. 200% of the full-load current

3. Except for motor starting service, insulated conductors used for connections between resistors and motor controllers must be rated at *not* less than _____.
   a. 86°F (30°C)
   b. 140°F (60°C)
   c. 167°F (75°C)
   d. 194°F (90°C)

3.0.0 Troubleshooting and Maintaining Transformers

Performance Tasks

There are no Performance Tasks in this section.

Objective

Explain how to troubleshoot and maintain transformers.

a. Describe common transformer problems.
b. Explain how to perform transformer testing and maintenance.

Because transformers are an essential part of every electrical installation, electricians must know how to properly maintain transformers. In addition, they must understand how to test and locate problems that develop in transformers.

3.1.0 Common Transformer Problems

Common transformer problems include open circuits, ground faults or short circuits, or a change in winding resistance. When troubleshooting, think before acting, study the problem thoroughly, and then ask yourself these questions:

- What were the warning signs preceding the trouble?
- What previous repair and maintenance work has been done?
- Has a similar problem occurred before?
- If the circuit, component, or piece of equipment still operates, is it safe to continue operation before further testing is conducted?

The answers to these questions can usually be obtained by:

- Questioning the owner of the equipment
- Taking time to think through the problem
- Looking for additional symptoms
- Consulting troubleshooting charts
- Checking the simplest possibilities first
- Referring to repair and maintenance records
- Testing with calibrated instruments
- Double-checking all conclusions before beginning any repair

NOTE

Always check the easiest and most obvious things first. Following this simple rule will save time and trouble.

The source of many problems can be traced to the relationship of one part with another. For instance, a tripped circuit breaker may be reset to restart a piece of equipment, but what caused the breaker to trip in the first place? It could have been caused by an energized conductor vibrating and momentarily contacting a ground source or simply a loose connection. There are many possibilities.

Too often, electrical equipment is completely disassembled in search of the cause of a complaint and all the evidence is destroyed during disassembly. Avoid overlooking an easy solution to a problem by moving toward invasive troubleshooting too quickly.

After an electrical failure has been corrected in any type of electrical circuit, be sure to locate and correct the cause so the same failure will not repeat. Further investigation may reveal other faulty components beyond the first one encountered.

Also be aware that, although troubleshooting charts and procedures greatly help in diagnosing malfunctions, they can never be complete. There are too many variations and possible solutions for any given problem.

To solve electrical problems consistently, you must first understand the basics of electrical circuits, how they function, and what purpose they serve. If you know that a particular part is not performing its job, then the cause of the malfunction must be within this part or series of parts.

Common transformer problems include open circuits, short circuits between winding turns, complete shorts, and grounded windings. These problems are discussed in the following sections.

3.1.1 Open Circuit

Should one of the windings in a transformer develop an open condition, no current can flow and the transformer will not operate. The main symptom of an open circuit in a transformer is that the secondary circuit(s) does not produce a voltage. Use an AC voltmeter or a multimeter to check the voltage at the transformer output terminals, as shown in *Figure 39*. A reading of zero volts on the secondary indicates an open circuit.

Next, take a voltage reading across the terminals where the primary winding receives its power. This is also shown in *Figure 39*. If the voltage reading is as expected, then the conclusion is that one of the windings in the transformer is open. However, if there is no voltage reading on the input terminals either, it means the transformer isn't energized and the problem is on the line side of the circuit. For example, a disconnect switch could be open, or a circuit breaker has tripped.

WARNING!

Be certain that your test instruments are rated for the voltage category of the system and calibrated for the correct voltage. Never test the primary side of any transformer over 600V unless you are qualified and have the correct high-voltage testing instruments and the test is made under proper supervision. Always set the range selector switch on the test equipment to the highest setting and decrease it in increments until the most accurate reading is obtained.

If voltage is present on the primary side and not present on the secondary side, open the switch to de-energize the circuit, and place a warning tag and lock on the disconnect switch so that it is not inadvertently closed again while someone is working on the circuit. Disconnect all the transformer primary and secondary leads, then check each winding in the transformer with an ohmmeter for continuity (*Figure 40*).

Continuity is indicated by a relatively low resistance reading on control transformers. An open winding will be indicated by an infinite resistance reading on the ohmmeter. On many multimeters, the display shows "OL" when there is no continuity, or it is beyond the range of the meter because the resistance is too high. In most cases, small transformers are replaced unless the damage is accessible and can be repaired.

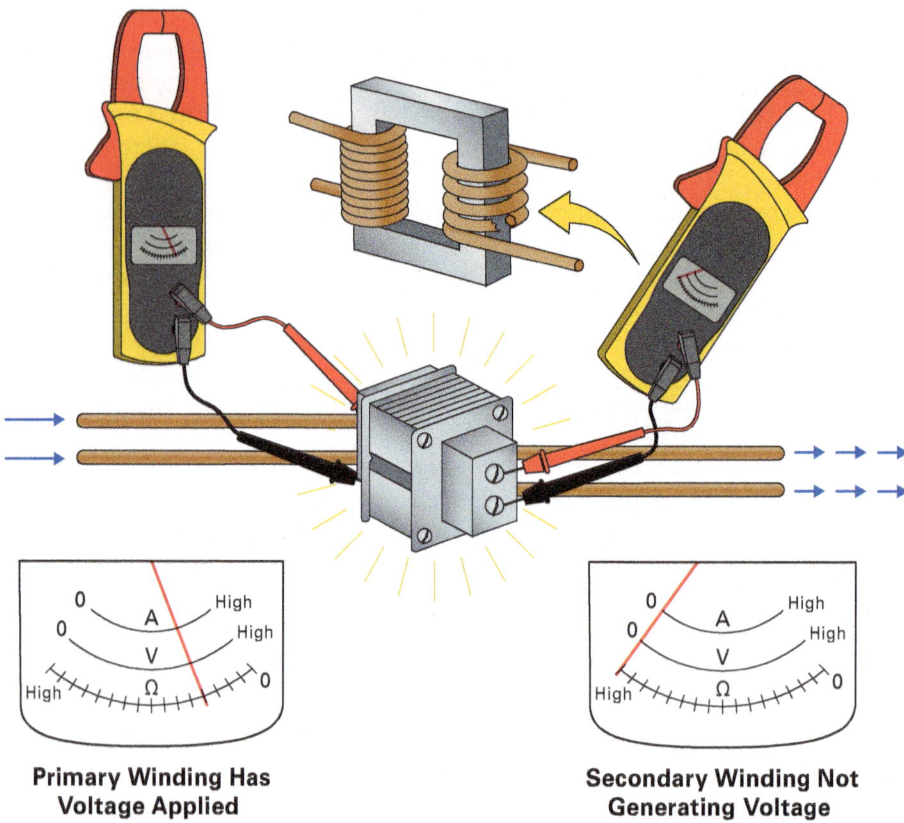

Figure 39 Checking transformer primary and secondary voltages with a multimeter.

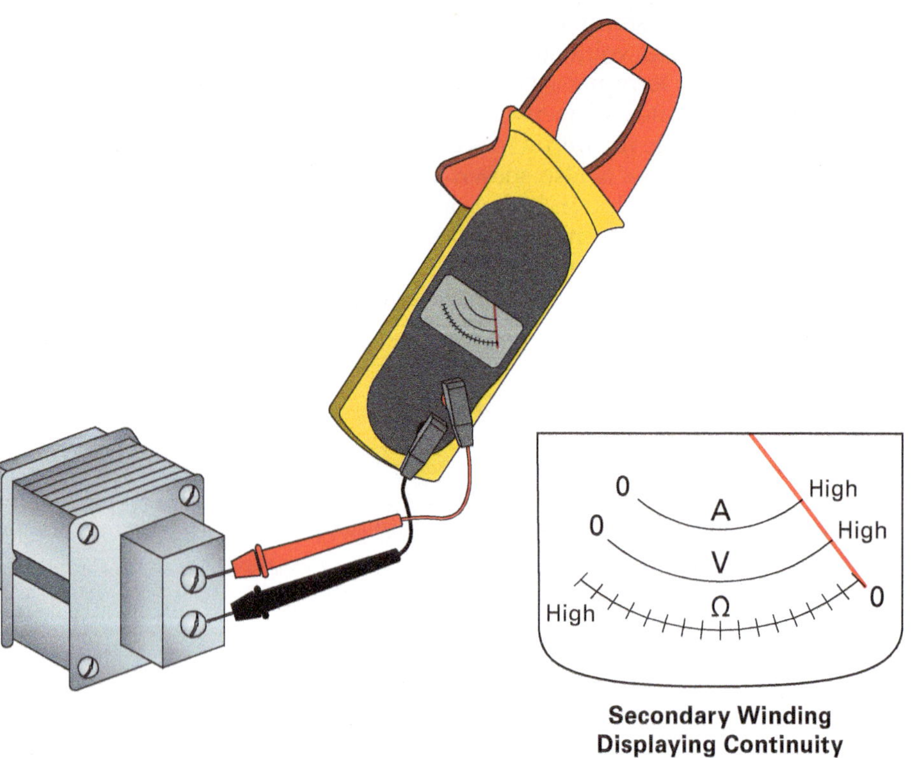

Figure 40 Checking for winding continuity with an ohmmeter.

3.1.2 Shorted Turns

Sometimes a few turns in the secondary winding of a transformer will develop a short circuit between them, which causes a voltage drop across the secondary. The symptom of this condition is usually transformer overheating caused by large circulating currents flowing in the shorted windings.

The most accurate way to check for this condition is with a transformer turns ratio (TTR) tester. Another way to check for this condition is with a multimeter set at the proper voltage range (*Figure 41*). Take a reading on the primary side of the transformer first to make certain the expected voltage is present. Then take a reading on the secondary side. If the transformer has a partial short or ground fault, the voltage reading should be lower than normal.

Replace the faulty transformer with a new one, and again take a reading on the secondary. If the voltage reading is now normal and the circuit operates properly, leave the replacement transformer in the circuit. Discard or repair the original transformer.

A highly sensitive ohmmeter may also be used to test for this condition when the system is de-energized and the leads are disconnected. A resistance reading lower than normal usually indicates a short circuit between the turns. However, the difference is usually too slight for the average ohmmeter to detect. Therefore, the recommended method is to use the voltmeter test.

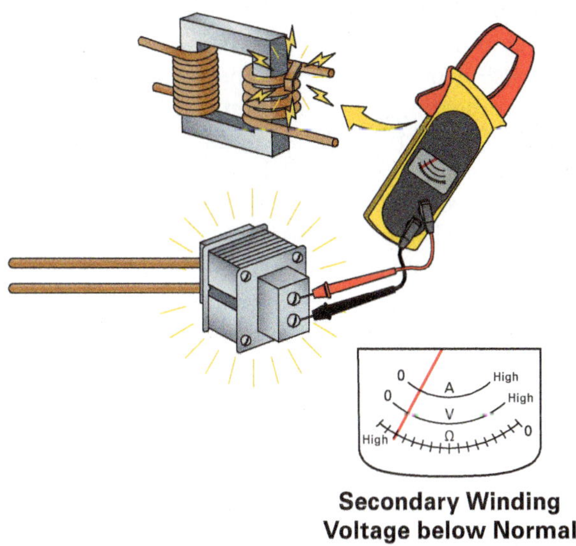

Figure 41 Transformers that overheat usually have a turn-to-turn short in the windings.

Think About It

Open and Unbonded Neutrals

What is the difference between an open neutral and an unbonded neutral?

3.1.3 Complete Short

Occasionally, a transformer winding will become completely shorted. In most cases, this will activate the overload protective device and de-energize the circuit. In a few cases, the transformer may continue trying to operate. Overheating usually accompanies this condition due to the very large circulating current. The added heat will often melt the wax or insulation inside the transformer, which is easily detected by the foul odor. Also, there will be no voltage output across the shorted winding, and the circuit across the winding will be dead.

The short may be in the external load circuit of the secondary winding or in the winding itself. To determine its location, disconnect the load from the secondary winding terminals and take a reading with a voltmeter across the secondary terminals. If the voltage is normal with the external circuit disconnected, then the problem lies within the external circuit. However, if the voltage reading is still zero across the secondary leads, the transformer is the source of the problem.

3.1.4 Grounded Windings

Insulation breakdown is quite common in older transformers, especially those that have been overloaded. At some point, the insulation breaks or deteriorates, exposing the windings beneath. The exposed winding may contact or arc to the housing, causing a short to ground.

If a winding develops a ground, and a point in the external circuit connected to this winding is also grounded, part of the winding will be bypassed. The symptoms include overheating, which is often accompanied by a burning smell, and a low resistance reading from the secondary winding to the enclosure, as indicated on the multimeter in *Figure 42*. In most cases, transformers with this condition must be replaced.

A *megohmmeter* (*megger*) is the best test instrument to check for this condition. Disconnect the leads from both the primary and secondary windings. Tests can then be performed on either winding by connecting the megger negative test lead to a dependable ground point and connecting the positive test lead to the winding to be measured.

WARNING!

Do not use a megger unless you are qualified and properly supervised.

The insulation resistance should then be measured on the windings. This is accomplished by connecting one test lead to the primary winding terminal and the second test lead to the secondary winding and ground. After that, connect the first lead to the secondary winding and the second test lead to the primary winding and ground. Record the results of these tests carefully.

Testing External Circuits

Always retest the external circuits supplied by the transformer before energizing a replacement transformer. This is done to avoid damaging or destroying the replacement due to a shorted external circuit, which may have caused the original transformer to burn out.

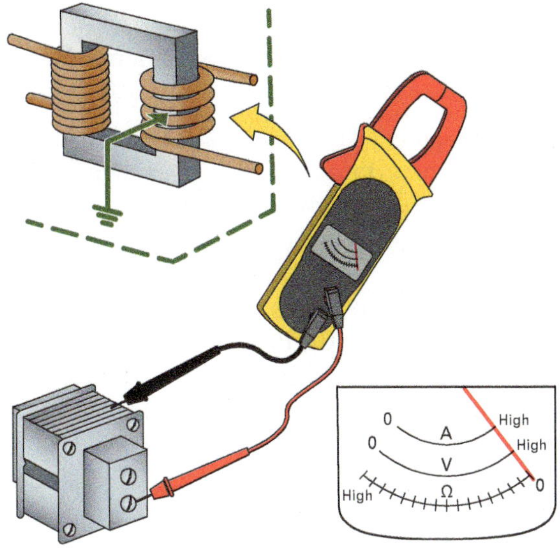

Figure 42 Testing a transformer for a grounded condition by measuring resistance.

3.2.0 Transformer Testing and Maintenance

Regular transformer maintenance can help to reduce energy use and equipment failure. The following sections identify general transformer maintenance tasks. Always refer to the manufacturer's instructions and policies in your facility for specific service and maintenance procedures.

WARNING!

Transformer maintenance may be performed only by qualified individuals operating under an appropriate safe work plan or permit. De-energize, lock out, and tag the transformer circuits and conductors before performing any cleaning or maintenance procedures. Never use solvents or detergents as they may damage components or leave a coating that attracts dust or other contaminants.

3.2.1 Dry Transformers

Consider the following when inspecting and maintaining dry transformers:

- A buildup of dust and grime on cooling surfaces interferes with heat transfer and slows the cooling process, which leads to overheating. Regularly remove dust using a vacuum cleaner or other manufacturer-approved method.
- When access panels are removed for cleaning, check all the internal components for signs of discoloration or heat damage.
- Check for loose or damaged connections and damaged bus bracing. Inspect for physical damage, broken insulation, tightness of connections, defective wiring, and general mechanical and electrical conditions.
- For dry-type transformers above 1,000V, inspect the insulators for cracks. A cracked insulator can allow a high voltage to arc across the gap to a nearby grounded surface.
- Check for the proper clearances to ensure adequate ventilation. If the transformer is in a vault, ensure the ventilation system is working and not blocked.
- Sealed dry transformers use an inert gas blanket under positive pressure to prevent condensation inside. They are equipped with pressure gauges that should be monitored and the results plotted over time. Pressure decreases not associated with outdoor ambient temperature changes indicate leakage.

CAUTION

If a transformer is to remain de-energized for an extended period in a humid atmosphere, it may be necessary to use space heaters to prevent condensation and moisture buildup inside.

3.2.2 Liquid-Filled Transformers

Consider the following when inspecting and maintaining liquid-filled transformers:

- Inspect for physical damage, cracked bushings, leaks, and loose connections. Dust and pollen may adhere to exposed high-voltage bushings, causing **surface tracking** and reduced insulation values.
- Verify the proper operation of auxiliary devices such as fans and indicators. Liquid-filled transformers have indicating gauges to monitor temperature, pressure, and liquid level. These values must be checked consistently and plotted over time for negative trends. Pressures and liquid levels will vary according to ambient conditions and transformer loading. Thus, not every change indicates a problem. Manufacturer guidelines and user experience provide acceptable ranges for the operating conditions.
- Samples of insulating oil or liquid are drawn to test the dielectric strength of the fluid. Most facilities document the test results over time on behalf of the client and provide reports as needed. Gases are generated from electrical corona, the thermal decomposition of insulating oil, and electrical arcing. Trend analysis of dissolved gases in oil and in the gas blanket above the oil reflects the condition of the transformer and its reliability.

Surface tracking: The formation of conductive, carbonized paths along the surface of insulators and similar components that provide a path for current from high-voltage sources.

CAUTION

To prevent contamination and get an acceptable sample of liquid for testing, the sample must be taken using strict procedures and an approved collection vessel. Testing laboratories can provide specific instructions.

Maintenance Testing of Transformers

In commercial and industrial facilities with many significant transformers that are not automatically monitored, periodic maintenance testing and performance trending is normally applied in a manner similar to that for essential motors. These tests include liquid dielectric tests to measure the breakdown voltage of the insulating fluids and DC hi-potential, or *hi-pot*, tests of transformer winding insulation. Some DC hi-pot testers also have a test pot for high-voltage components. The test sets shown here can perform tests with voltages up to 160 kV.

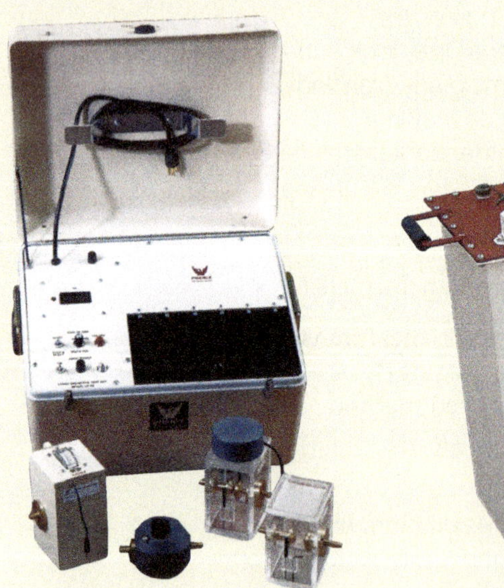

Liquid Dielectric Test Set

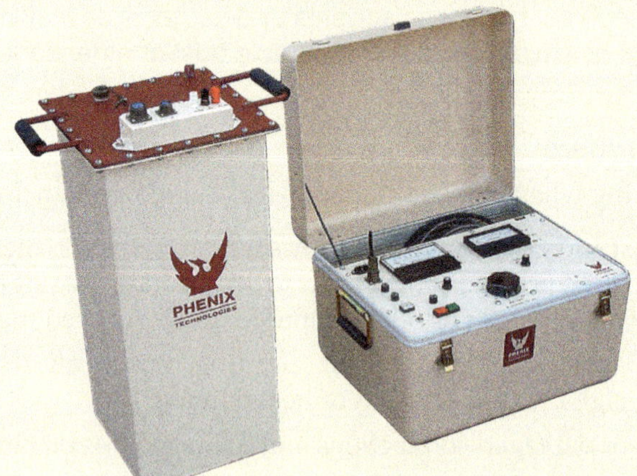

Component Test Pot

DC Hi-Pot Test Set

Source: Phenix Technologies, Inc.

3.0.0 Section Review

1. A transformer reads zero volts across the output terminals while the primary winding is energized. The cause is *most* likely _____.
   a. an open circuit in the transformer
   b. a partially shorted winding
   c. a ground fault
   d. insulation breakdown

2. When performing maintenance on dry transformers, check for cracked insulators on transformers *over* _____.
   a. 240V
   b. 480V
   c. 600V
   d. 1,000V

Module 26307-23 Review Questions

1. The *main* purpose of a transformer is to _____.
   a. modulate the current
   b. improve the power factor
   c. change the output voltage
   d. add impedance to the circuit

2. The three parts of a basic transformer are the _____.
   a. housing, lifting hooks, and base
   b. dry, oil-filled, and gas-filled chambers
   c. shell, open winding, and closed winding
   d. primary winding, secondary winding, and core

3. When AC flows through a transformer coil, the field generated around the coil is a _____.
   a. magnetic field
   b. capacitive field
   c. high-impedance field
   d. rotating field

4. When the field from one coil cuts through the turns of a second coil, _____.
   a. the second coil is not affected
   b. the coils rotate
   c. voltage is generated
   d. no current flows in the circuit

5. What causes voltage to be induced in a transformer?
   a. Transformer taps
   b. Mutual induction
   c. Reluctance
   d. Capacitance

6. In a transformer with a turns ratio of 5:1 (the primary has five times the number of turns as the secondary), what will be the voltage on the secondary if the primary voltage is 120V?
   a. 12V
   b. 24V
   c. 48V
   d. 60V

7. The three basic types of iron-core transformers are _____.
   a. dry, oil-filled, and auto
   b. closed core, open core, and shell core
   c. control, power, and lighting
   d. metal, nonmetallic, and high temperature

8. One effect caused by magnetic leakage in transformers is a _____.
   a. reactance voltage drop
   b. low impedance
   c. higher secondary voltage
   d. lower secondary voltage

9. If one of the transformers is damaged in a delta system, the configuration that may be used to temporarily carry the load is called a(n) _____.
   a. open delta configuration
   b. center-tapped delta configuration
   c. delta-wye configuration
   d. delta-delta configuration

10. The windings in an autotransformer _____.
    a. share part of a winding
    b. are isolated from one another
    c. constitute two separately derived systems
    d. are rarely used in three-phase systems

11. Most control transformers are _____.
    a. dry-type, step-up units
    b. dry-type, step-down units
    c. oil-filled, step-up units
    d. oil-filled, step-down units

12. The power factor of a circuit can be improved by increasing the _____.
    a. voltage
    b. current
    c. wire size
    d. capacitance

13. A dry transformer must be housed in a vault if rated at *more than* _____.
    a. 20 kV
    b. 25 kV
    c. 30 kV
    d. 35 kV

14. A symptom of a transformer with an open circuit is _____.
    a. high voltage on the secondary
    b. excessive overheating
    c. no voltage output on the secondary
    d. voltage drop across the secondary

15. The meter used to measure insulation resistance is called a(n) _____.
    a. ammeter
    b. megger
    c. voltmeter
    d. TTR

Answers to odd-numbered Module Review Questions are found in *Appendix A*.

Module 26307-23 Supplemental Exercises

1. The main purpose of a transformer is to _________.
2. Distribution transformers are used to _________.
3. The basic components of a transformer are _________.
4. Explain the term *mutual induction*.

   ____

   ____

5. If you have a transformer with a turns ratio of 100:10 and a primary voltage of 240V, what will the secondary voltage be?

   ____

   ____

6. Transformers are rated in _________.
7. The three types of transformer core construction are _________.
8. The *NEC*® requires that conductors connected to the high leg of a four-wire delta system be color-coded with _________ insulation or tape.
9. Name some of the advantages of an autotransformer.

   ____

   ____

10. Which conditions must be met in order to protect a transformer only on the secondary?

    ____

    ____

11. E-rated fuses are designed to provide primary protection for _________.
12. Why are transformers grounded?

    ____

    ____

13. What is the main symptom of an open circuit in a transformer?

    ____

    ____

14. Periodic maintenance of dry transformers includes _________.
15. The buildup of dust and pollen on exposed bushings can cause _________.

Answers to odd-numbered Supplemental Exercises are found in *Appendix B*.

Answers to Section Review Questions

Answer	Section Reference	Objective
Section 1.0.0		
1. d	1.1.2	1a
2. a	1.2.0	1b
3. a	1.3.7	1c
Section 2.0.0		
1. d	2.1.0; *Table 1*	2a
2. a	2.2.0; *Table 2*	2b
3. d	2.3.0	2c
Section 3.0.0		
1. a	3.1.1	3a
2. d	3.2.1	3b

Section Review Calculations

Section 1.0.0

Question 1

Write the ratio of turns in the primary to secondary coils as a fraction, simplify, and then multiply the result by the voltage on the primary:

$$\frac{10 \text{ turns}}{20 \text{ turns}} = 0.5$$

$$0.5 \times 120V = 60V$$

The induced voltage on the secondary is **60V**.

User Update

Did you find an error? Submit a correction by visiting **https://www.nccer.org/olf** or by scanning the QR code using your mobile device.

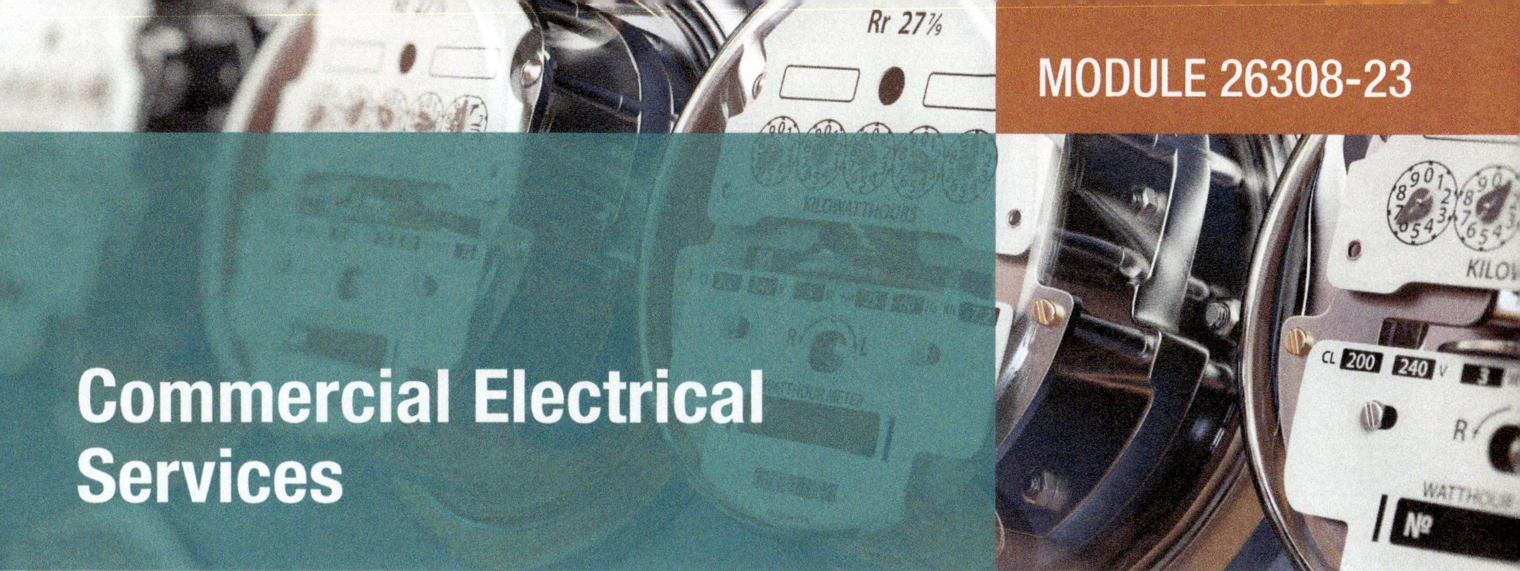

Source: iStock@Bet_Noire

Commercial Electrical Services

Objectives

Successful completion of this module prepares you to do the following:

1. Identify installation considerations for commercial electrical services.
   a. Identify the primary components required for a commercial electrical service.
   b. Identify specific *NEC*® requirements related to commercial electrical services.
2. Explain how to install a commercial electrical service.
   a. Explain how overhead electrical service components are installed.
   b. Explain how underground electrical service components are installed.
   c. Explain how switchgear components are installed.
   d. Explain how an electrical service for multifamily housing is installed.

Performance Tasks

This is a knowledge-based module. There are no Performance Tasks.

Overview

No matter the size or complexity, all electric services have the same purpose: to deliver electrical energy safely. This module covers the components, installation considerations, and *NEC*® requirements for commercial services.

NOTE

NFPA 70®, *National Electrical Code*®, and *NEC*® are registered trademarks of the National Fire Protection Association, Quincy, MA.

NCCER Industry-Recognized Credentials

If you are training through an NCCER-accredited sponsor, you may be eligible for credentials from NCCER. The ID number for this module is 26308-23. Note that this module may have been used in other NCCER curricula and may apply to other level completions. Contact NCCER at 1.888.622.3720 or go to **www.nccer.org** for more information.

You can also show off your industry-recognized credentials online with NCCER's digital badges. Transform your knowledge, skills, and achievements into badges that you can share across social media platforms, send to your network, and add to your resume. For more information, visit **www.nccer.org**.

Digital Resources for Electrical

Scan this code using the camera on your phone or mobile device to view the digital resources related to this craft.

1.0.0 Installation Considerations for Commercial Services

Performance Tasks

There are no Performance Tasks in this section.

Objective

Identify installation considerations for commercial electrical services.

a. Identify the primary components required for a commercial electrical service.

b. Identify specific *NEC*® requirements related to commercial electrical services.

Service: Conductors and equipment for delivering electric energy from the serving utility to the wiring system of the premises served.

NOTE

Electricians install the electric service, but connections to the utility transformer are typically made by the utility company.

Service equipment: The necessary equipment, usually consisting of a circuit breaker or switch along with fuses and their accessories, located near the entrance point of supply conductors to a building and intended to constitute the main control and cutoff means for the electric supply to the building.

Service conductors: The conductors between the point of termination of the overhead service drop or underground service lateral and the main disconnecting device in the building or on the premises.

Service drop: The overhead conductors between the utility electric supply system and the service point.

Service lateral: The underground conductors between the utility main electric supply system and the service point.

Service entrance: All components between the point of termination of the overhead service drop or underground service lateral and the building's main disconnecting device, except for metering equipment.

The electrical contractor is typically responsible for the installation of all electrical systems including the electric **service**. Electric **service equipment** consists of conduit, **service conductors**, panelboard, and overcurrent protective devices. *NEC Article 230* lists the requirements for service conductors and raceways, along with numerous additional requirements for service installations.

Electric services are usually designed by qualified electrical engineers who produce drawings and specifications showing the design and installation requirements for the service. The drawings include the floor plan, one-line diagram, site plan, panelboard schedules, and specifications. A service one-line, or single-line, drawing shows service characteristics including the size and voltage rating, fault current rating, service-entrance conduit and conductor sizes, grounding requirements, transformers, and panelboards. The one-line drawing also shows whether the service is overhead or underground. If the power connects to the electric service overhead, the incoming service supply is referred to as a **service drop**. If the service is supplied underground, the supply is defined as a **service lateral**. The one-line for a small installation typically shows the conduit and conductor sizes for the service. Larger service one-line diagrams often include a schedule or table indicating the conduit and conductor sizes for **service entrance**, distribution, and panelboard feeders.

A building layout shows the location of the electric service and the location of distribution and branch circuit panelboards within the structure. It is important to review all the project drawings and specifications to obtain all the information needed to install the electric service. For example, the building elevation drawing will show the amount of exterior wall space available at the location where the service equipment might be mounted. It is also important to know the type of construction so that the proper anchoring systems can be ordered for the installation.

A qualified electrician should have the ability to work with the supervisor, the project manager, the utility company, and the electrical engineer to install the service components in a professional and efficient manner. Safety considerations come first in every installation. This means that the required personal protective equipment (PPE) must be used to allow the installation to proceed in a safe manner. All equipment must be installed level and plumb and in a manner that meets all *NEC*® requirements and all the requirements presented in the project electrical drawings and specifications. All equipment and materials must be new and installed per the manufacturer's guidelines. The electrician may be required to install the conduit, service-entrance conductors, wireways, bussed gutter, service disconnects, current transformer (CT) can, meter can, and other fittings as needed to comply with *NEC*® requirements, local building department requirements and codes, project specifications, and even the local fire marshal's requirements.

The location of the bus and ground bars and the service-side entrance compartment are important considerations when installing the service-entrance conductors. The service grounding conductors are usually routed to and bonded in the first service disconnecting means. When service-entrance conductors terminate in a wireway or bussed gutter to serve multiple service disconnecting means, the service grounding conductors may be terminated to and bonded to the grounded service conductor in the wireway or bussed gutter enclosure. In many instances, the local Authority Having Jurisdiction (AHJ) will make the determination regarding specific grounding decisions to ensure these connections meet *NEC®* requirements. Another important consideration is whether housekeeping pads are required to mount or raise the switchgear or transformers above the finished floor.

In most cases, electricians are not required to design electrical systems except for design-build projects that require a stamped set of drawings and the oversight of an electrical engineer. Usually, the electrician is only required to interpret the engineer's drawings and apply them to the installation of the project. Other important factors relating to the installation of service equipment are the inspection and acceptance of service equipment from the delivery service and the megging of the conductors and equipment before energizing the equipment.

The following considerations apply to the selection and installation of electric services:

- *Codes and standards* — All services must be designed and installed in accordance with the provisions of the *NEC®* and all local codes. Suitability of service equipment for the intended use is evidenced by listing or labeling.
- *Mechanical protection* — Mechanical strength and durability, including the adequacy of the protection provided, must be considered.
- *Wiring space* — Wire bending and connection space in distribution equipment is provided according to UL Standards. When unusual conditions are encountered requiring additional wiring space, larger wireways and/or termination cabinets must be considered.
- *Classification* — Classification according to type, size, voltage rating, current capacity, interrupting rating, and specific use must be considered when selecting service equipment. Loads may be continuous or noncontinuous, and load calculations per *NEC®* requirements must be completed in order to determine the proper sizing of service equipment. All distribution equipment carries labels showing the maximum voltage and current ratings for the equipment. A typical example is shown in *Figure 1*. The supply voltage and current ratings must not be exceeded. In addition, always observe the minimum and maximum values listed for conductor terminations.
- *Heating effects* — All installations must consider heating effects both under normal operating conditions and under any extreme conditions of use that might reasonably be expected. Ambient heat conditions must be considered when specifying conductor insulation for the specific application.
- *Arcing effects* — The normal arcing effects of overcurrent protective devices must be considered when the application is in or near combustible materials or vapors (classified/hazardous locations). Enclosures and wiring methods must be selected based on *NEC®* requirements.
- *Personal protection* — Every service installation should begin with a review of safety policies and a site visit to determine any special safety requirements relating to the project. Always wear all PPE required for the installation. Safe work practices are extremely important when installing service equipment that may be heavy and awkward to lift and place. Be aware of potential pinch points as well as all other jobsite conditions that may present specific safety hazards. Safety should always be a top priority when installing electric services.

Receiving Equipment

Freight claims for damage caused during shipping must be made by written notation on the freight bill when the equipment is received at the jobsite. If freight damage is not noted on the freight bill, a claim for damages may not be possible later. All boxes should be examined for hidden damage prior to uncrating.

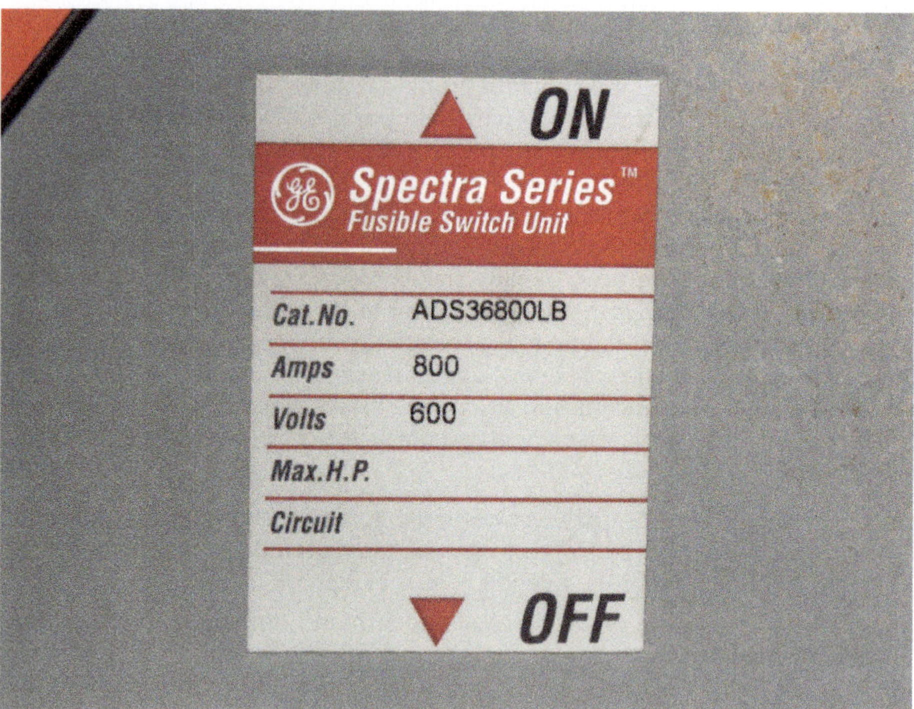

Figure 1 Typical maximum voltage and current rating label.
Source: Al Hamilton

1.1.0 Service Components

All service components must be carefully selected and coordinated to ensure proper system operation. Common service components include the service disconnect, meter, transformers, wireways, busways, weatherhead and service mast, and panelboards.

Know the Code

General
NEC Section 230.70

Know the Code

Marking
NEC Section 230.66

Cold sequence metering: A service installation where the meter disconnect is installed before the metering for power to be disconnected to allow for safe maintenance of the metering.

1.1.1 Service Disconnecting Means

Whether a service is single phase or three phase, overhead or underground, *NEC Section 230.70* requires that a means be provided to disconnect all undergrounded conductors in a building from the service-entrance conductors. The service disconnecting means must be in a readily accessible location, marked as a service disconnect, and listed as suitable for use as a service disconnecting means per *NEC Section 230.66*. The service disconnect can be a fusible switch or a circuit breaker and is typically located at the point where the service-entrance conductors come into the building or structure. In some cases, the utility company requires **cold sequence metering**, in which a disconnect is installed before the electric meter. Cold sequence metering provides the utility company the opportunity to work on the metering equipment with the power disconnected.

1.1.2 Metering

The metering assembly, single meter enclosure, multiple meter enclosures, or CT metering is installed at the point of service entrance. A self-contained meter installation refers to a single meter enclosure with a single watt-hour or demand meter. Typically, the electrician is responsible for the meter enclosure installation and the utility company will provide and install the electric meter. A CT meter installation refers to the installation of a CT enclosure or cabinet. The CT cabinet is a listed piece of manufactured equipment designed to accept the installation of CTs used to measure electric usage. A CT meter enclosure is installed alongside or within a short distance of the CT cabinet. The electrician is usually responsible for the installation of the CT cabinet and the CT meter base. The utility company will then install the CTs, the CT meter, and the wiring between the CT meter enclosure and the CT cabinet.

Most analog watt-hour meters use five dials (*Figure 2*). The dial farthest to the right on the meter counts the kilowatt-hours singly. The second dial from the right counts by tens, the third dial by hundreds, the fourth dial by thousands, and the left-hand dial by ten thousands. The dials may seem a little strange at first, but they are actually very simple to read. The number that the dial has passed is the reading. For example, look at the dial on the far left. Note that the pointer is about halfway between the number 2 and the number 3. Since it has passed the number 2 but has not yet reached 3, the dial reading is 2. The same is true of the second dial from the left—the pointer is between 2 and 3. Consequently, the reading of this dial is also 2. Following this procedure, the complete reading for the meter in *Figure 2* is 22,179 kilowatt-hours.

Although knowing how to read an electric meter is useful, most electricians will be involved only with installing the meter base and making the connections within. Once these connections are made and inspected, the local power company will install and seal the meter.

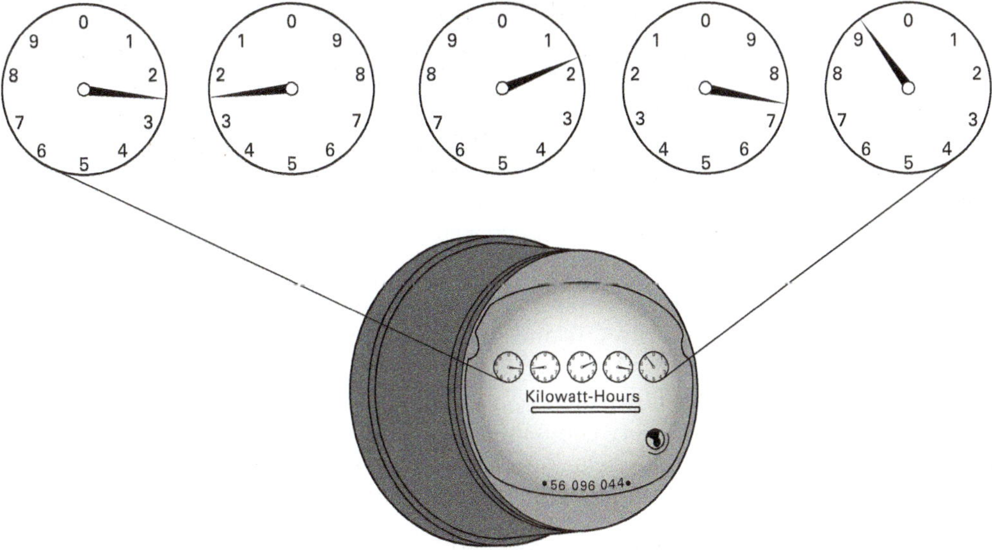

Figure 2 Typical watt-hour meter.

1.1.3 Current Transformers (CTs)

Meters used to record electric usage normally respond to a current level that varies from 0A to 5A. To respond to the actual current, each meter is provided with CTs. If the peak demand is 100A, a 100:5 CT is used. If the peak demand is expected to be 200A, a 200:5 CT is used.

Services rated 400A and above usually have a group of CTs, one for each ungrounded conductor or set of conductors. There are two basic types of CTs: the busbar type and the doughnut type. The doughnut-type CT encircles the ungrounded conductors in the system to measure the current flow, much the same as a clamp-on ammeter (*Figure 3*). The busbar-type CT has each transformer connected in series with a busbar and does not encircle the conductor.

Think About It

Reading a Watt-Hour Meter

What is the reading on this meter?

Dual Services

Some large installations have two or more services. Dual services are provided with separate disconnects, as shown in the image.

Figure 3 Clamp-on ammeters operate on the same principle as doughnut-type CTs.
Source: Tim Dean

CTs are normally enclosed in a CT cabinet. *Figure 4* shows a typical CT cabinet with CTs and their related wiring. In some cases, the CTs may be mounted exposed on overhead conductors, but this arrangement is more of an exception than the rule.

Utility companies have specifications for the location and wiring of CTs and CT cabinets, depending upon the location. Always check requirements for the local utility. The following are requirements for one utility company:

- The meter base and meter may be located on either the side or top of the CT cabinet, or they may be located at a distance if approved by the utility company when the conduit containing the instrument wiring is exposed.
- In no case shall more than one set of conductors terminate in the instrument transformer cabinet. Sub-feeders and branch circuits are to terminate at the customer's distribution panel. The instrument transformer cabinet shall not be used as a junction box.
- When service-entrance conductors enter or leave the back of the cabinet, the size of the CT cabinet must be increased to provide additional working space.
- For services at higher voltages, additional space must be provided in the transformer cabinet for mounting potential transformers. Consult the local utility company for dimensions.
- If recording demand instruments are required, increase the height of the meter mounting by 12" (300 mm) to accommodate the recording instrument.

1.1.4 Metal Wireways

Metal wireways are defined in *NEC Article 100* as sheet metal troughs with hinged or removable covers for housing and protecting electrical wires. As part of a service installation, wireways are used to provide space for splicing service-entrance conductors to conductors supplying individual metering equipment or service disconnects. Follow *NEC Section 376.22* to determine the appropriate size wireway to use in a particular service installation. Wireways installed outdoors must be listed for outdoor use.

Know the Code

Definitions
NEC Article 100

Know the Code

Number of Conductors and Ampacity
NEC Section 376.22

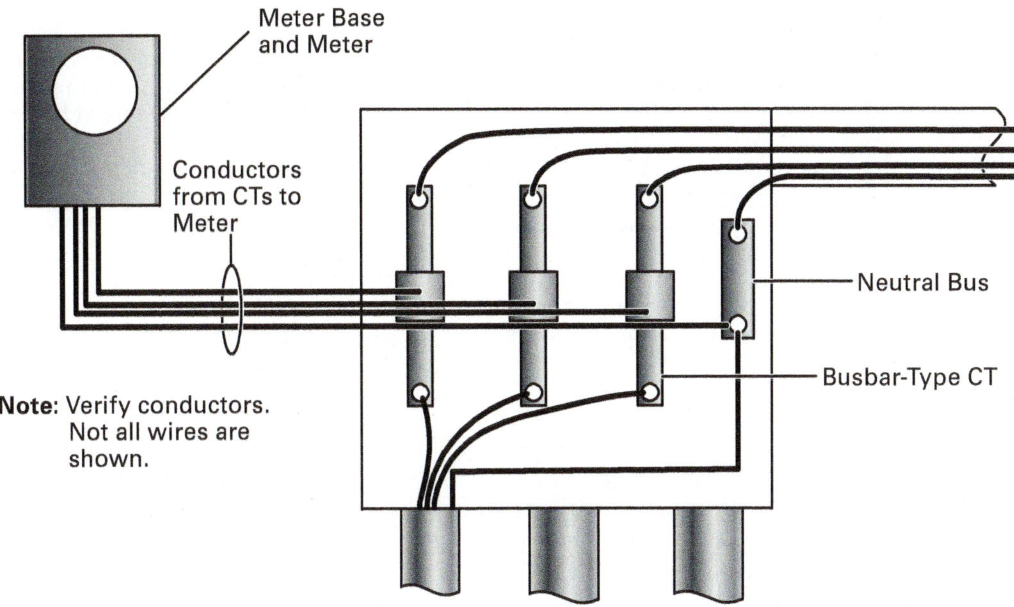

Figure 4 Typical CT cabinet arrangement.

Meter Types

Watt-hour meters are available in both analog and digital forms.

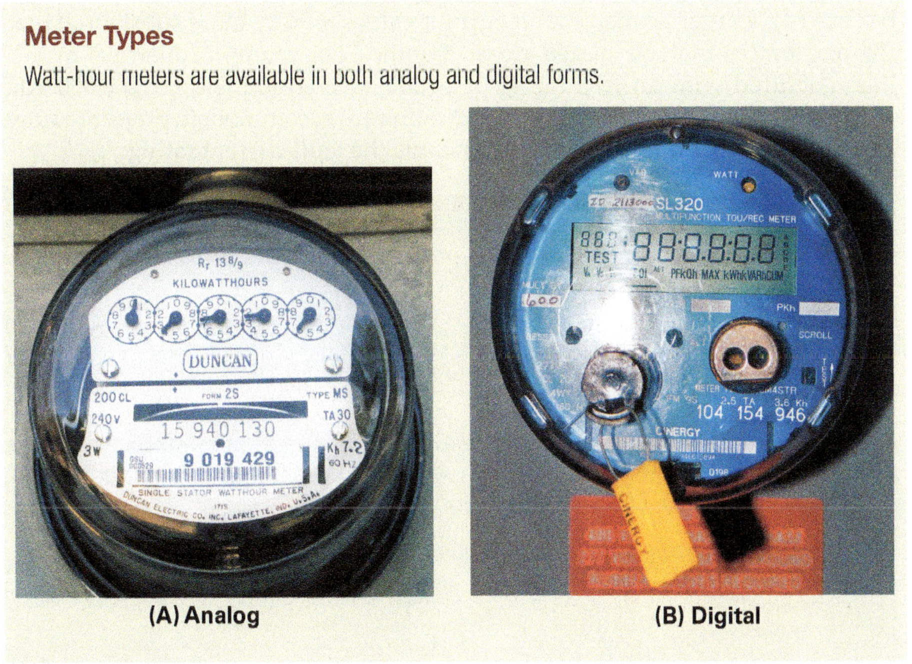

(A) Analog **(B) Digital**

1.1.5 Busways

NEC Article 100 defines a *busway* as a metal enclosure containing factory-mounted bare or insulated conductors, which are usually copper or aluminum bars, rods, or tubes (*Figure 5*). Busways are used to provide a means of tapping the service-entrance conductors or distribution feeders to supply the service disconnecting means, self-contained metering equipment, CT metering equipment, or for other purposes such as disconnects for equipment feeders in an industrial plant or manufacturing facility. Busways are manufactured to meet the project specifications for their intended use. These specifications include the voltage, ampacity, and fault current ratings, along with the NEMA enclosure classification (for example, NEMA 3R for outdoor use) and the gutter length. Busways can be manufactured with copper bus or plated aluminum bus. It may include factory-installed mechanical lugs of specific sizes to accommodate line and load feeders. Busways may be bottom fed or top fed.

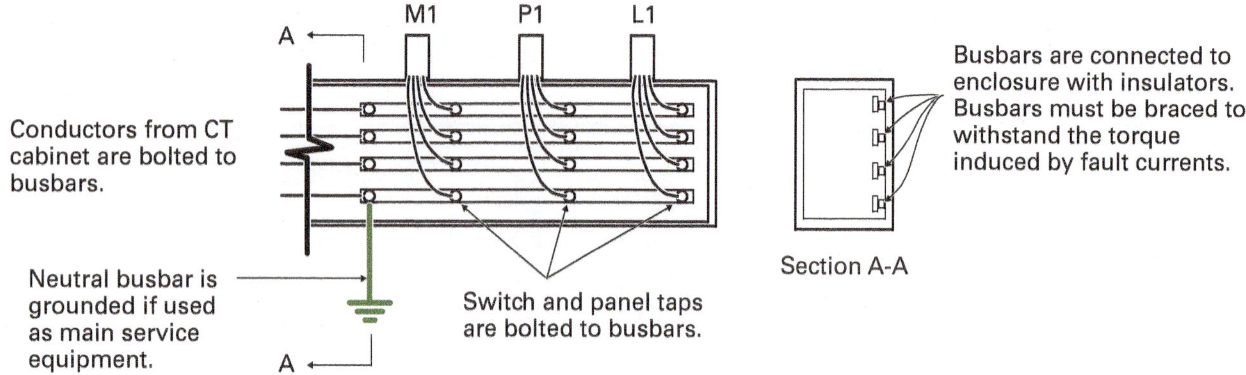

Figure 5 Three-phase busway.

Busways are one of the most common types of wiring methods for use with multiswitch service installations. One advantage of a busway is the ease of installation and modification. Adding disconnect switches or changing switches is relatively easy.

One characteristic of fault currents is an induced torque in conductors carrying the fault. Because of this torque, the busbars must be mounted in the enclosure in a manner that allows them to withstand the fault current that may be imposed. This is accomplished using bracing. *Figure 6* shows typical busbar bracing. Larger available fault current values require more substantial bracing and braces that are placed closer together. For example, busways may be manufactured with an AIC rating of 20,000 AIC; 30,000 AIC; or 200,000 AIC. Manufacturers typically have standard ratings to accommodate project requirements. The listing for busways must include the fault current rating.

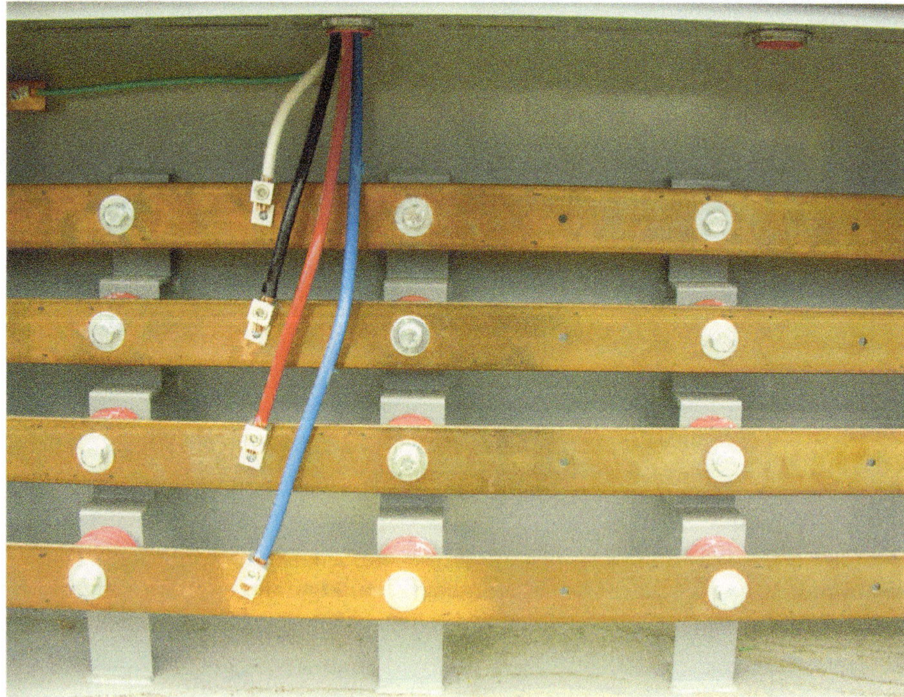

Figure 6 Typical busbar bracing.
Source: Al Hamilton

1.1.6 Weatherhead and Service Mast

Overhead services include a weatherhead and a service mast. Weatherheads may be cast aluminum, galvanized steel, or PVC. Service masts are typically constructed of galvanized steel and supported as required by the *NEC*® and the utility company. The service mast terminates in the hub of the meter housing, CT cabinet, or other service equipment and must use watertight fittings when the components are installed outdoors. The service-entrance riser conductors terminate in the meter housing or the service disconnecting means if the service disconnect is required before the meter (*cold sequence metering*) or after the meter (*hot sequence metering*).

1.1.7 Panelboards

The final component in a commercial electric service is the panelboard. Panelboards are designed to distribute electric energy through branch circuit breakers. In a small commercial service, a panelboard may include the service disconnecting means together with all the branch circuit breakers needed for the building. Panelboards may also be designated as distribution panelboards. Distribution panels are used to provide overcurrent protection for individual panel feeders when the service is large enough to require numerous branch circuit panelboards. The selection of the components used in a panelboard is based on the following criteria:

- Feeder voltage and amperage
- Phase (single or three phase) and number of conductors
- Type of busbar material (copper or aluminum)
- Number of required spaces or blanks
- Whether the panelboard is configured as a main breaker type of panel or main lugs only (MLO) type of panel

NOTE

MLO panels do not have a main circuit breaker.

CAUTION

Always install breakers according to the manufacturer's installation instructions. When installing bolt-in breakers, make sure to tighten them to the correct torque. Failure to do so could result in improper operation of the service equipment.

Panelboards

After the panelboard is assembled, the appropriate number and type of circuit breakers must be ordered separately and installed in the panel.

1.2.0 *NEC®* Requirements

Know the Code

Services
NEC Article 230

In most cases, a building may be supplied by only one service (*NEC Section 230.2*). Special conditions outlined in *NEC Sections 230.2(A) through (D)* permit the use of additional services. Per *NEC Section 230.2(A)*, an additional service is permitted to supply emergency systems and fire pumps, standby systems, interconnected electric power production sources, and systems designed to connect to multiple sources for enhanced reliability. Additional services are permitted for sufficiently large buildings or certain multiple occupancy buildings per *NEC Section 230.2(B)*, capacity requirements as outlined in *NEC Section 230.2(C)*, and for different service characteristics as identified in *NEC Section 230.2(D)*. *NEC Section 230.66* requires service equipment to be marked as suitable for its intended use and listed or field labeled. Distribution panelboards and switchgear assemblies are generally intended to carry and control electric current but are not intended to dissipate or use energy.

In accordance with *NEC Section 230.82(3)*, a disconnect switch connected to the supply side of the meter (cold sequence) must be rated at no more than 1,000V. In addition, it must have a short circuit rating equal to or greater than the available fault current, provided the housings and service enclosures are grounded and bonded in accordance with *NEC Article 250, Parts V and VII*. The switch must be capable of interrupting the load served and be field labeled.

Know the Code

Grounding and Bonding
NEC Article 250

1.2.1 Services Passing through Buildings

Per *NEC Section 230.3*, service conductors supplying one building cannot pass through the interior of another building or structure. *NEC Section 230.6* states that conductors are considered outside of a building if any of the following conditions are met:

- Where installed beneath the building under at least 2" (50 mm) of concrete
- Where buried under at least 18" (450 mm) of earth in conduit
- Located inside the building if encased in at least 2" (50 mm) of concrete or brick
- Installed in a vault that meets the requirements of *NEC Article 450, Part III*
- Where instruments within metal conduit meet clearance requirements of *NEC Section 230.24* and are routed directly through an eave but not a wall of a building

Know the Code

Transformer Vaults
NEC Article 450, Part III

1.2.2 Grounding and Bonding

NEC Article 250 covers the requirements for grounding and bonding electric services. In general, the *NEC®* requires that a premises' AC service be grounded with a grounding electrode conductor connected to a grounding electrode. All the requirements for grounding and bonding identified in *NEC Figure 250.1* must be considered and applied to every electric service.

The grounding electrode conductor must be grounded to the grounded service conductor (neutral) at any accessible point from the load end of the service drop or service lateral up to and including the terminal bus to which the grounded service conductor is connected at the service disconnecting means. A grounding connection must not be made to any grounded circuit conductor on the load side of the service disconnecting means.

NEC Section 250.50 requires all grounding electrodes to be bonded together to create the grounding electrode system. *NEC Section 250.52* lists the grounding electrodes as metal underground water pipe in direct contact with the earth for 10' (3 m) or more, the metal in-ground support structure in direct contact with the earth vertically for 10' (3 m) or more, concrete-encased electrode, ground ring, rod and pipe electrodes, other listed electrodes, plate electrodes, and other metal grounding systems or structures. These requirements apply to all electric services, regardless of the voltage rating of the service or the size (ampacity rating) of the service. *NEC Section 250.53* guides the installation of the grounding electrode system. *NEC Table 250.66* gives the required sizes of the grounding electrode conductor for electric services.

Know the Code

Grounding Electrode Conductor for Alternating-Current Systems
NEC Table 250.66

1.2.3 High-Leg Marking

The conductor or busbar with a higher voltage to ground in a **delta-connected**, four-wire system, also known as the *high leg*, is required to be durably and permanently marked with an outer finish that is orange in color or by other effective means per *NEC Section 110.15*. If the grounded conductor is also present, this marking must appear at each point where a connection is made. In panelboards, the phase having the higher voltage to ground is the B phase per *NEC Section 408.3(E)(1)*.

Delta-connected: A three-phase transformer connection in which the terminals are connected in a triangular shape like the Greek letter delta (Δ).

Know the Code

Support and Arrangement of Busbars and Conductors
NEC Section 408.3

1.2.4 Clearances

NEC Section 230.24 lists the minimum service drop clearances for various applications. Service drop conductors require a vertical clearance of no less than 8'-6" (2.6 m) above roof surfaces per *NEC Section 230.24(A)* and no less than 10' (3.0 m) at the electrical service entrance to buildings, at the lowest point of the drip loop, and over sidewalks accessible to pedestrians only where it does not exceed 150V to ground per *NEC Section 230.24(B)(1)*. Note that this is the minimum for the lowest point of the drip loop, and the weatherhead or tie point for the service drop should be a minimum of 13'-6" (4.1 m) above final grade, based on elevations. This accounts for the height of the weatherhead itself, the insulator that attaches to the mast below the weatherhead, the drip loop, and the conductor sag (lowest point) in a service drop, which varies by conductor size and length of drop. The insulator, whether attached to the mast or to the building, is the point of attachment for the service drop. If the service mast is to support the service drop, it must be well secured per *NEC Section 230.28*. Meter bases should be installed with anchors to hold the weight of the meter as well as the raceway system resting on the meter base.

Know the Code

Clearances
NEC Section 230.24

Service conductor clearances on buildings, porches, and platforms are covered in *NEC Section 230.9*. They may be installed as open conductor or MC cable without an overall outer jacket and must be located at least 3' (900 mm) from locations such as windows, doors, porches, balconies, ladders, stairs, and fire escapes. These clearances do not apply to windows that are not designed to be opened or to locations where the conductor is run above the top level of the window, per the exception in *NEC Section 230.9(A)*.

Know the Code

Clearances on Buildings
NEC Section 230.9

1.0.0 Section Review

1. The meter itself is typically installed and sealed by the _____.
   a. electrical contractor
   b. utility company
   c. general contractor
   d. inspector

2. For a conductor to be considered outside a building, which of the following conditions *must* be met?
   a. Buried under 4" of earth in conduit
   b. Buried under 8" of earth in conduit
   c. Buried under 12" of earth in conduit
   d. Buried under 18" of earth in conduit

2.0.0 Installing Commercial Services

Performance Tasks

There are no Performance Tasks in this section.

Objective

Explain how to install a commercial electrical service.

a. Explain how overhead electrical service components are installed.
b. Explain how underground electrical service components are installed.
c. Explain how switchgear components are installed.
d. Explain how an electrical service for multifamily housing is installed.

The most common service voltage for small commercial and other small loads is a 120/240V, single-phase, three-wire service. Common services for larger commercial installations are 120/240V, three-phase, four-wire; 120/208V, three-phase, four-wire; and 277/480V, three-phase, four-wire installations. To understand the function of each component of an electrical service, we will examine five specific installations of commercial electric services. These include a small overhead service, an underground service utilizing a wireway to connect multiple meters and service disconnects, a bussed gutter installation, a service entrance using switchgear, and a multifamily service.

2.1.0 Overhead Services

Know the Code

Grounding Electrode Conductor for Alternating-Current Systems
NEC Table 250.66

Know the Code

Drip Loops
NEC Section 230.54(F)

Figure 7 shows an overhead 208Y/120V service. The service is rated at 400A, and the grounding requirements are established based on a 400A service. The two service-entrance conductors consist of No. 3/0 THWN copper in 2" (MD 53) conduit. Because these are parallel service-entrance conductors, the GEC is sized at No. 2 copper (or 1/0 aluminum) using *NEC Table 250.66*. Note that there are drip loops formed at the weatherheads as required by *NEC Section 230.54(F)*. The drip loops are used to direct moisture or rainwater away from the meter and disconnect. The parallel service-entrance conductors terminate in a wireway. The service disconnects are connected to the wireway with rigid galvanized steel nipples and the conductors supplying each meter are terminated in the wireway. The terminations in the wireway may be mechanical (split-bolt) type connections or mechanical connections made using a multitap

Figure 7 Overhead service.
Source: Al Hamilton

insulated connector. In this example, the GEC is bonded to the grounded service conductor (neutral) in the wireway. There are two 200A, 240V three-phase meters and service disconnects, and one 100A, 240V single-phase meter and disconnect. Because the building has a 208Y/120V three-phase service, the voltage to the single-phase meters is also 208V. Note the use of weatherheads, drip loops, and conduit supports, and the arrangement of the service equipment. This service illustrates hot sequence metering.

Because this equipment is installed on the exterior of the building, it is rated for outdoor use. In this installation, the point of attachment is bolted through the masonry wall to make it secure enough to support the service drop (also accounting for the wind load and ice weight) and includes a porcelain insulator. Porcelain insulator attachments are preferable and may be required by the local utility. Always check with the local utility before proceeding with any installation. Porcelain insulator attachments provide a separation between the service drop conductors (which may become worn and short over time) and other metal components.

Figure 8 is the one-line diagram for the service just described. Note that the one-line diagram includes sizes for service-entrance conduit and conductors, sizes and fuse requirements for the service disconnects, the size and length of the wireway, and sizes for the grounding electrode conductors.

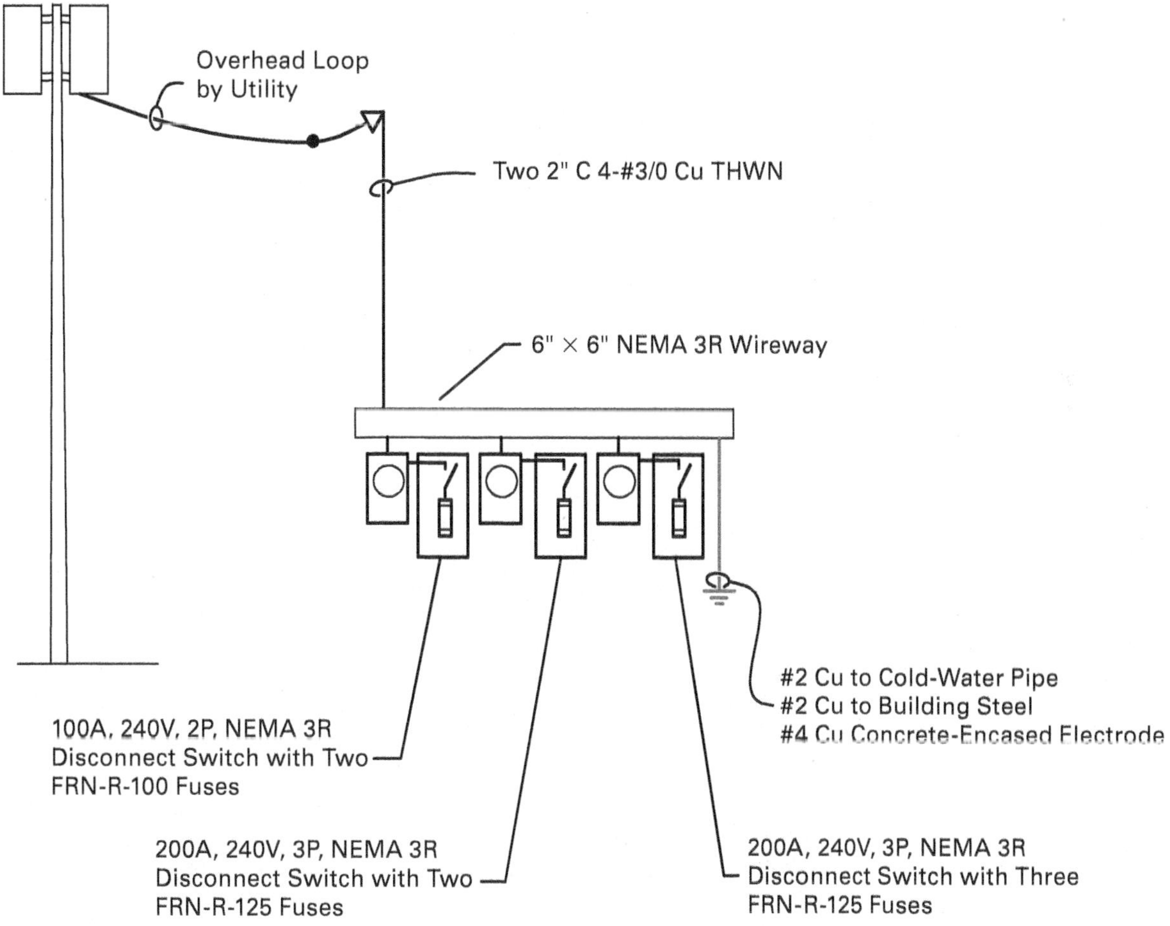

Figure 8 One-line diagram for overhead service.
Source: Al Hamilton

2.2.0 Underground Services

Underground services have special installation requirements. These services can be run in a wireway or a bussed gutter.

2.2.1 Three-Phase Underground Service in a Wireway

Figure 9 shows an underground 208Y/120V three-phase service. The service is rated at 800A, and the grounding requirements are established based on an 800A service. The service-entrance conductors consist of three cables, with each cable being four-conductor, 400 kcmil THWN aluminum. Each cable is in an underground 3" (MD 78) conduit (the fourth conduit is a spare). Because these are parallel service-entrance conductors, the GEC is sized at 2/0 copper (or 4/0 aluminum) using *NEC Table 250.66*. The parallel service-entrance conductors terminate in a main service disconnect. The tenant service disconnects are connected to the wireway with EMT conduit with Myers™ hubs (or similar) used where the conduit terminates into the wireway, and the conductors supplying each tenant disconnect are terminated in the wireway. The terminations in the wireway are mechanical connections made using a multitap insulated connector. Per *NEC Section 310.10(G)(2)(5)*, parallel conductors must be terminated in the same manner. In this example, the GEC is bonded to the grounded service conductor (neutral) in the 800A fused main disconnect switch. In this installation, there is one 100A fusible disconnect ahead of the meter (cold sequence) serving panel HP (house panel), which in turn supplies branch circuit power to the lighting controller for exterior building lighting and parking lot lighting. There is one 200A fusible tenant disconnect ahead of the meter serving a tenant space and one 400A CT cabinet ahead of a 400A fusible disconnect serving a larger tenant space. The vertical line originating at the top of the 200A tenant meter indicates the feeder to the tenant panel located in the tenant space. Note that the feeder to the 400A tenant panel is not shown in the image because it is routed inside the building and terminates into the back of the 400A tenant disconnect.

Know the Code

Uses Permitted
NEC Section 310.10

Figure 9 Underground wireway service.
Source: Al Hamilton

Note that the *NEC*® would not require a main service disconnect for this installation because there are fewer than six switches to disconnect power to the building, and the ampacity of the feeders serving panel HP and the tenant services does not exceed the ampacity of the service-entrance conductors. However, the service for this retail building was designed before any tenants were identified to occupy the building, which provided flexibility to allow a tenant mix that could have required more than six disconnects for tenant services. This design also provides flexibility for future changes in the tenant mix, making it unnecessary to rework the entire service to accommodate changes. Because this equipment is installed on the exterior of the building, all the equipment is rated for outdoor use.

Figure 10 shows the one-line diagram for the service previously described. Note that the one-line diagram includes sizes for service-entrance conduit and conductors, sizes and fuse requirements for the service disconnects, the size and length of the wireway, the sizes of panel feeder conduit and conductors, and sizes for the grounding electrode conductors. The branch circuit panelboard for the lighting contactors is also shown as part of the one-line diagram.

2.2.2 Three-Phase Underground Service in a Bussed Gutter with Multiple Service Disconnects

Figure 11 shows an underground 208Y/120V three-phase service. The service is rated at 800A, and the grounding requirements are established based on an 800A service. The service-entrance conductors consist of two cables, with each cable being a four-conductor, 500 kcmil THWN copper in 3" (MD 78) conduit. Because these are parallel service-entrance conductors, the GEC is sized at 2/0 copper (or 4/0 aluminum) using *NEC Table 250.66*. The parallel service-entrance conductors terminate in a bussed gutter. The service disconnects are connected to the bussed gutter with RGS nipples and the conductors supplying each meter are terminated in the bussed gutter using mechanical lugs. It is important to follow the manufacturer's requirements for torque when terminating conductors using mechanical lugs. Because this equipment is installed on the exterior of the building, all the equipment is rated for outdoor use.

Know the Code

Grounding Electrode Conductor for Alternating-Current Systems
NEC Table 250.66

Figure 12 is the one-line diagram for the service previously described. Note that the one-line diagram includes sizes for service-entrance conduit and conductors, sizes and fuse requirements for the service disconnects, the size and length of the wireway, the sizes of panel feeder conduit and conductors, and sizes for the grounding electrode conductors.

2.3.0 Switchgear

Figure 13 shows a switchgear 480Y/277V three-phase service which is a typical industrial plant installation. The service is rated at 2,000A, and the grounding requirements are established based upon a 2,000A service. The fault current rating for this equipment is 50,000 AIC. The service-entrance conductors consist of seven cables, with each cable being a four-conductor, 500 kcmil THWN copper in 3½" (MD 91) conduit. Because these are parallel service-entrance conductors, the GEC is sized at 3/0 copper (or 250 kcmil aluminum) using *NEC Table 250.66*. The switchgear consists of a network of busbars installed horizontally. The busbars extend through each metal switchgear cabinet to supply power to individual disconnects. If a service is of sufficient size to require more than six disconnects, a main disconnecting means must be included. The service shown in this image has four service disconnects and spaces for two additional future service disconnects. A main disconnecting means is not required in this configuration.

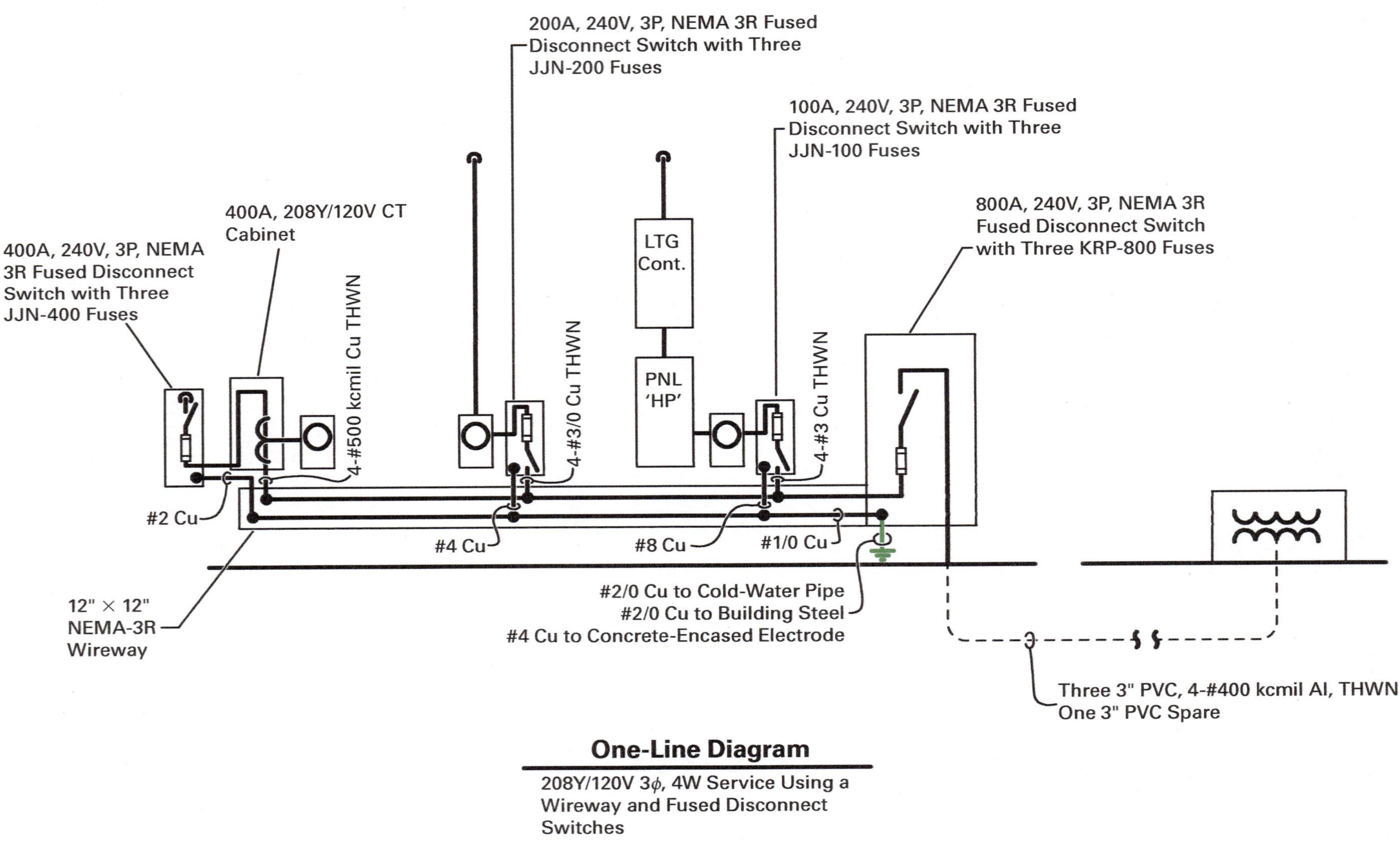

Figure 10 One-line diagram for an underground wireway service.
Source: Al Hamilton

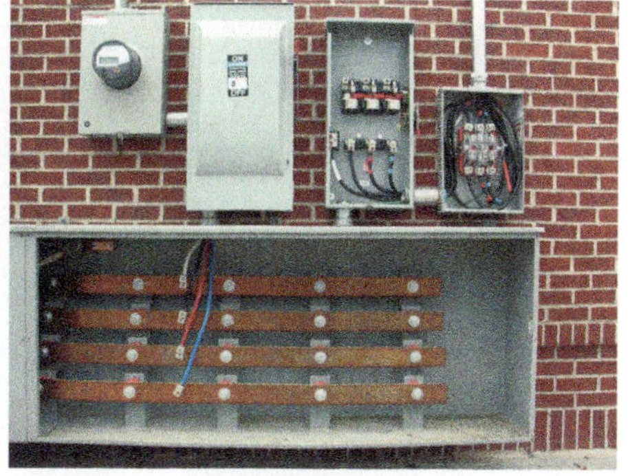

(A) Closed (B) Open

Figure 11 Underground bussed gutter service.
Source: Al Hamilton

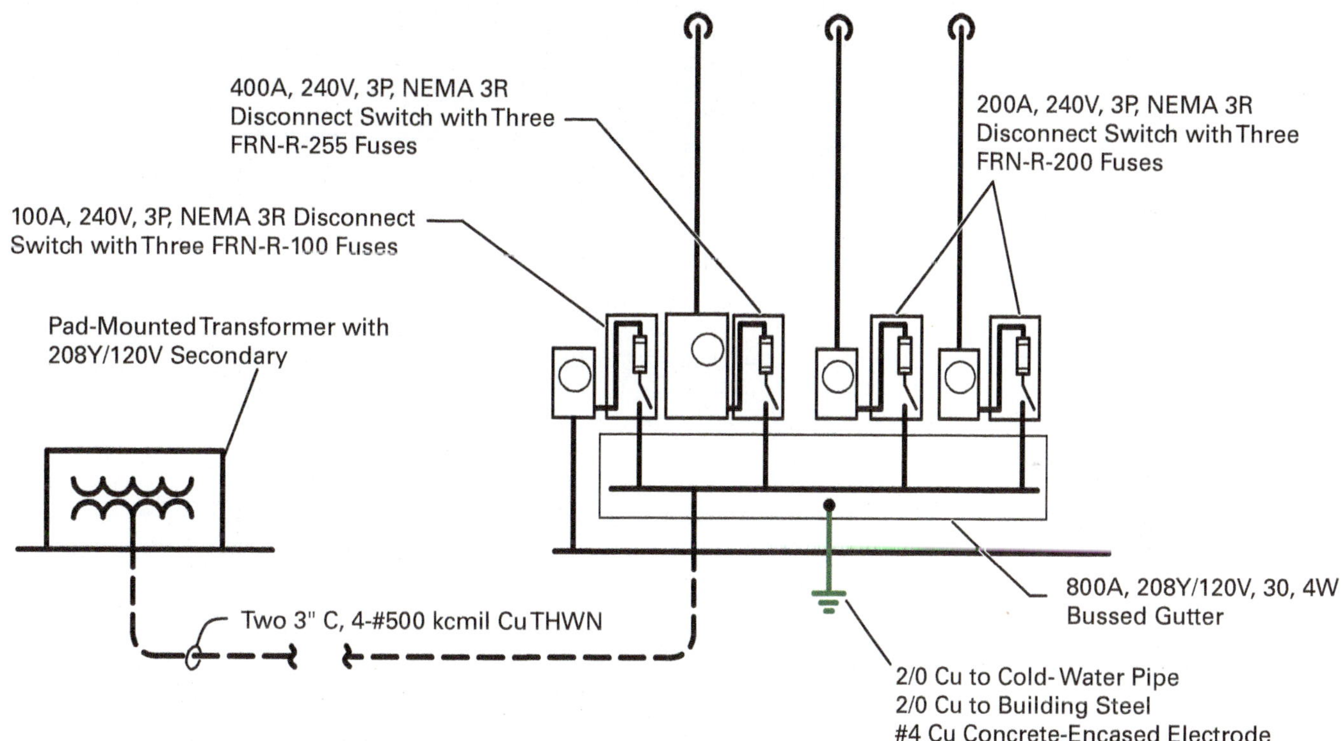

Figure 12 One-line diagram for an underground bussed gutter service.
Source: Al Hamilton

The service-entrance conductors enter underground and are terminated in a CT metering section that is part of the panel main switchboard (MSB). The metering section includes provisions for terminating the service-entrance conductors to the switchgear bus. The bus then carries current through the switchgear to fusible disconnects supplying power to panelboards (*Figure 14*) serving mechanical equipment loads, a lighting panelboard, a transformer, and a shunt-trip circuit breaker supplying an elevator. Because this equipment is installed on the interior of the building, all the equipment is rated for indoor use.

Figure 13 Main switchboard (MSB) and sub switchboard (SSB) switchgear service.
Source: Al Hamilton

Figure 14 Panelboards.
Source: Al Hamilton

Figure 15A and *Figure 15B* are the one-line diagrams for the service previously described. Note that the 600A switch with 500A fuses feeds a 300 kVA transformer to the panel sub switchboard (SSB)—a 1,200A switchgear supplying power for all 208V and 120V loads. This one-line diagram includes sizes for service-entrance conduit and conductors, the size and ratings for the switchgear, transformers, and panelboard feeders. The branch circuit panelboards are also shown as part of the one-line diagram. *Figure 16A* and *Figure 16B* represent panel schedules for two of the panelboards in this installation.

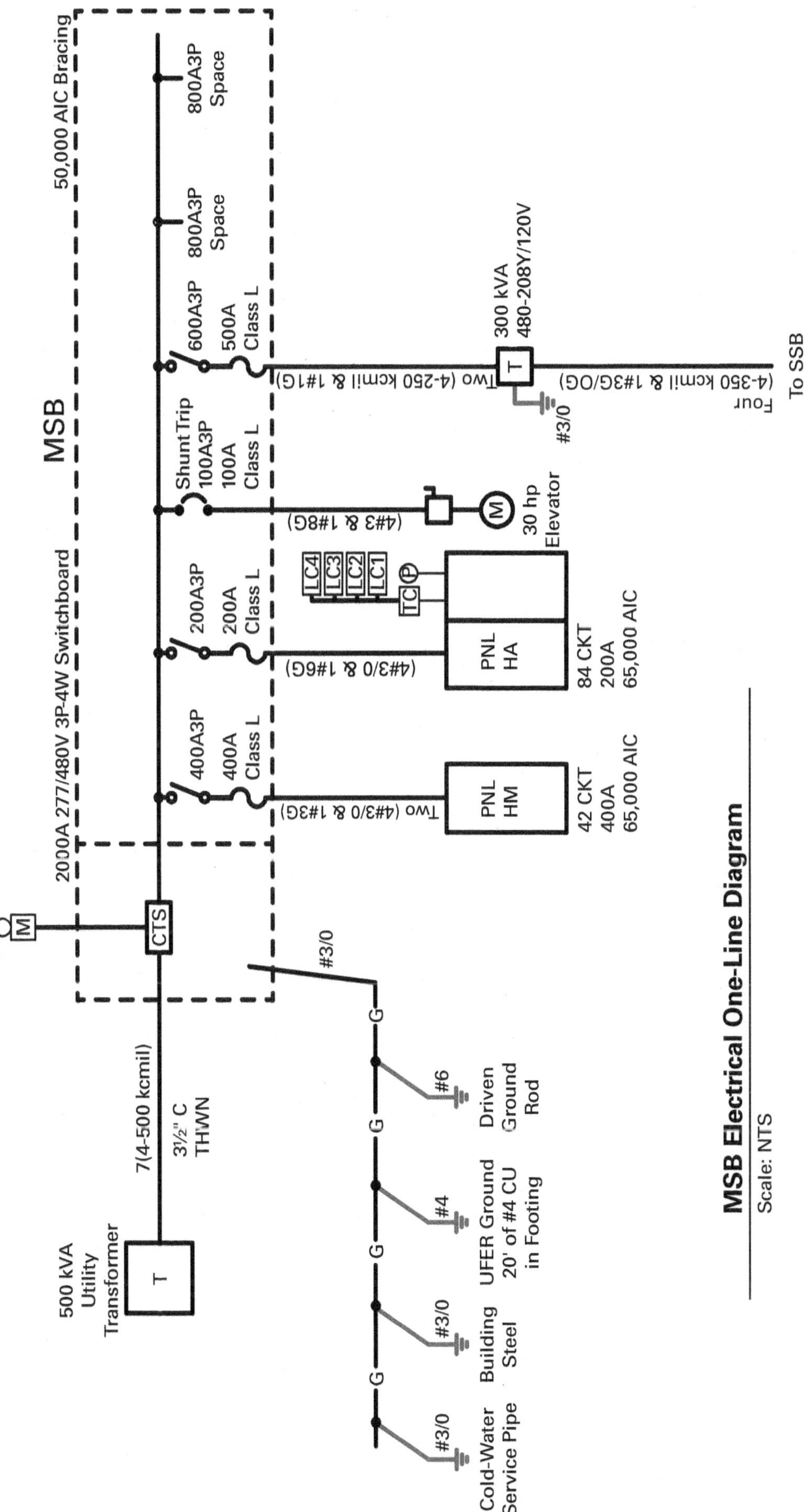

Figure 15A One-line diagram for MSB switchgear service (1 of 2).
Source: Al Hamilton

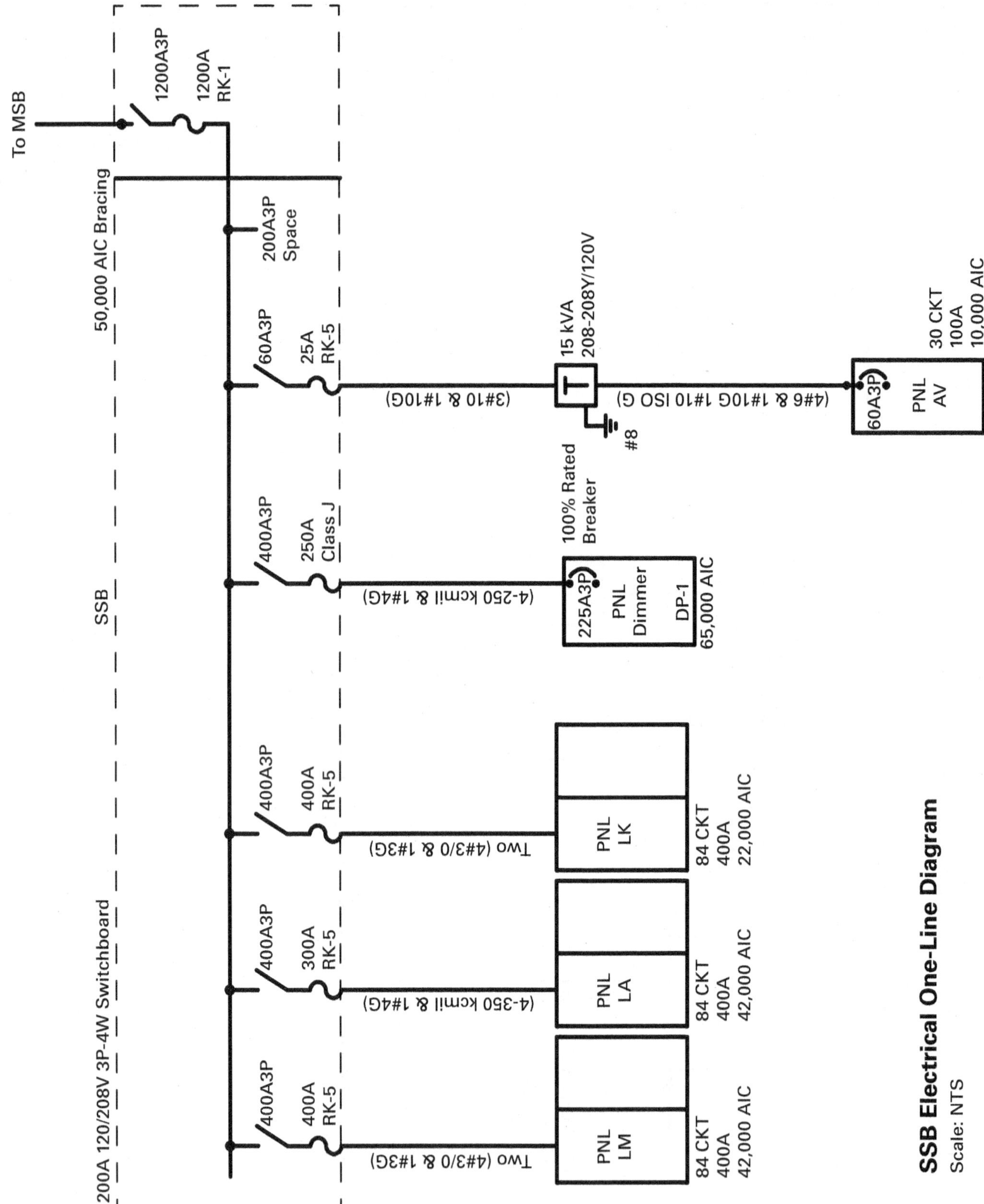

Figure 15B One-line diagram for SSB switchgear service (2 of 2).
Source: Al Hamilton

PANEL: 1H1					480/277V 3 PH 4W
FED:					FEED:
LOAD SERVED	BRK	CIR	CIR	BRK	LOAD SERVED
		1	2		
MS1	20	3	4	20	MS2
		5	6		
		7	8		
VFD2	40	9	10	40	VFD1
		11	12		
		13	14		
VFD3	70	15	16	70	VFD4
		17	18		
		19	20		
BOILER #1	15	21	22	15	BOILER #2
		23	24		
		25	26		
		27	28	20	SPARE
		29	30		
		31	32		
		33	34		
		35	36		
		37	38		
		39	40		
		41	42		

Phase A – Brown
Phase B – Orange
Phase C – Yellow
Neutral – Grey
Ground – Green

Figure 16A Panel schedule for a 480Y/277V panelboard (1 of 2).
Source: Al Hamilton

<table>
<tr><th colspan="3">PANEL:LPG
FEED: BOTTOM</th><th colspan="3">120/208V 3 PH 4W
FED FROM:</th></tr>
<tr><th>LOAD SERVED</th><th>BRK</th><th>CIR</th><th>CIR</th><th>BRK</th><th>LOAD SERVED</th></tr>
<tr><td>RCPT BOYS COACHES OFFICE</td><td>20</td><td>1</td><td>2</td><td></td><td>LTG WEST GYM FLOOR W</td></tr>
<tr><td>RCPT BOYS LOCKER, STORAGE S. ENTRY</td><td>20</td><td>3</td><td>4</td><td></td><td>LTG WEST GYM FLOOR E</td></tr>
<tr><td>RCPT GIRLS COACHES OFFICE</td><td>20</td><td>5</td><td>6</td><td></td><td>LTG WEST GYM E BLEACHERS</td></tr>
<tr><td>RCPT GIRLS LOCKERS, STORAGE, MECH</td><td>20</td><td>7</td><td>8</td><td></td><td>BACKBOARDS & BATTING CAGE N, W GYM</td></tr>
<tr><td>CONCESSIONS POP COOLER</td><td>20</td><td>9</td><td>10</td><td></td><td>BACKBOARDS & BATTING CAGE S, W GYM</td></tr>
<tr><td>CONCESSIONS COUNTERTOP GENERAL</td><td>20</td><td>11</td><td>12</td><td></td><td>BACKBOARD E, W GYM</td></tr>
<tr><td>CONCESSIONS MICROWAVE</td><td>20</td><td>13</td><td>14</td><td></td><td>LTG GIRLS LOCKERS, STORAGE, MECH</td></tr>
<tr><td>CONCESSIONS HOT DOG ROLLER</td><td>20</td><td>15</td><td>16</td><td></td><td>LTG BOYS LOCKERS, STOAGE</td></tr>
<tr><td>CONCESSIONS NACHO WARMER</td><td>20</td><td>17</td><td>18</td><td></td><td>LTG CONCESSIONS, STORAGE</td></tr>
<tr><td>CONCESSIONS POPCORN</td><td>20</td><td>19</td><td>20</td><td></td><td>FREEZER PROTECTION</td></tr>
<tr><td>CONCESSIONS FREEZER</td><td>20</td><td>21</td><td>22</td><td></td><td>RCPT N PITCHING MACHINE</td></tr>
<tr><td>RCPT STORAGE RMS E</td><td>20</td><td>23</td><td>24</td><td></td><td>RCPT S PITCHING MACHINE</td></tr>
<tr><td rowspan="2">HAND DRYER BOYS</td><td rowspan="2">20</td><td>25</td><td>26</td><td></td><td>RCPT A/V RACK</td></tr>
<tr><td>27</td><td>28</td><td rowspan="2">30</td><td rowspan="2">BATTERY INVERTER, W GYM</td></tr>
<tr><td rowspan="2">HAND DRYER GIRLS</td><td rowspan="2">20</td><td>29</td><td>30</td></tr>
<tr><td>31</td><td>32</td><td rowspan="2">20</td><td rowspan="2">SPARE</td></tr>
<tr><td rowspan="2">FCU-4</td><td rowspan="2">15</td><td>33</td><td>34</td></tr>
<tr><td>35</td><td>36</td><td></td><td>SPARE</td></tr>
<tr><td rowspan="2">FCU-5</td><td rowspan="2">15</td><td>37</td><td>38</td><td></td><td>SPARE</td></tr>
<tr><td>39</td><td>40</td><td></td><td>RCPT CKT-EX PANEL G</td></tr>
<tr><td>EF-4</td><td>20</td><td>41</td><td>42</td><td></td><td>RCPT CKT-EX PANEL G</td></tr>
<tr><td colspan="6"></td></tr>
<tr><td colspan="3" rowspan="5"></td><td colspan="3">Phase A – Black</td></tr>
<tr><td colspan="3">Phase B – Red</td></tr>
<tr><td colspan="3">Phase C – Blue</td></tr>
<tr><td colspan="3">Neutral – White</td></tr>
<tr><td colspan="3">Ground – Green</td></tr>
</table>

Figure 16B Panel schedule for a 208Y/120V panelboard (2 of 2).
Source: Al Hamilton

2.4.0 Multifamily Services

Figure 17 shows a 240/120V single-phase service for a multifamily apartment building. This is a 24-unit apartment building with individual tenant meters located on each end of the building. Grounding for this service is sized for three parallel sets of 350 kcmil aluminum service-entrance conductors. The service equipment is designed for a fault current rating of 100,000 AIC. The service-entrance conductors consist of three cables, with each cable being a three-conductor, 350 kcmil THWN aluminum in 3" (MD 78) PVC conduit. Because these are parallel service-entrance conductors, the GEC is sized at 2/0 copper (or 4/0 aluminum) using *NEC Table 250.66* (3 × 350 kcmil = 1,050 kcmil). The service consists of a main fused disconnect switch and meter stacks connected with horizontal busbars to the main disconnect. The busbars extend horizontally from the main disconnect through each meter stack and then the vertical bus is connected to supply power to each meter. Individual meters in the meter stack consist of the meter base and a two-pole, 100A circuit breaker that serves as the overcurrent and short circuit protection for the feeder supplying the apartment. This circuit breaker is sized for the load to be served. The meter breakers may not be equal size—they are sized for the load to be served, and the load for each apartment must be calculated to determine the feeder size. The red sign on the main service disconnect reads, "A DISCONNECT ON OTHER END OF BUILDING." This plaque informs the Fire Department and other authorities that more than one disconnect controls power to the building.

This service complies with *NEC Section 230.72(C)*, which requires that occupants have access to their own service disconnecting means. This exception allows the tenant meters and service disconnecting means to be placed inside the building under the conditions listed. *Figure 18* is a one-line diagram for the multifamily service previously described.

Think About It

Putting It All Together

Examine the service at your school or workplace. Is the service entrance properly installed and protected? Is the service-entrance cable cracked or frayed? What about the electrical panel? Are the breakers properly labeled? Is there a main service disconnect switch?

Know the Code

Access to Occupants
NEC Section 230.72(C)

Arc Flash

The service point typically will deliver the greatest level of arc flash potential to the electrician in the customer's electrical system. However, on-site backup generation may deliver higher energies than the utility point of service.

Figure 17 Multifamily service.
Source: Al Hamilton

Call First before Digging

When installing any underground electrical, a call to the local utility should be the first order of business. This allows the utility company's third-party locating service to come out and mark where the existing underground services are present. These may include existing gas, telephone, or fiber optic lines, or other electrical services. You can call your local utility or 811, which is a national call-before-you-dig number.

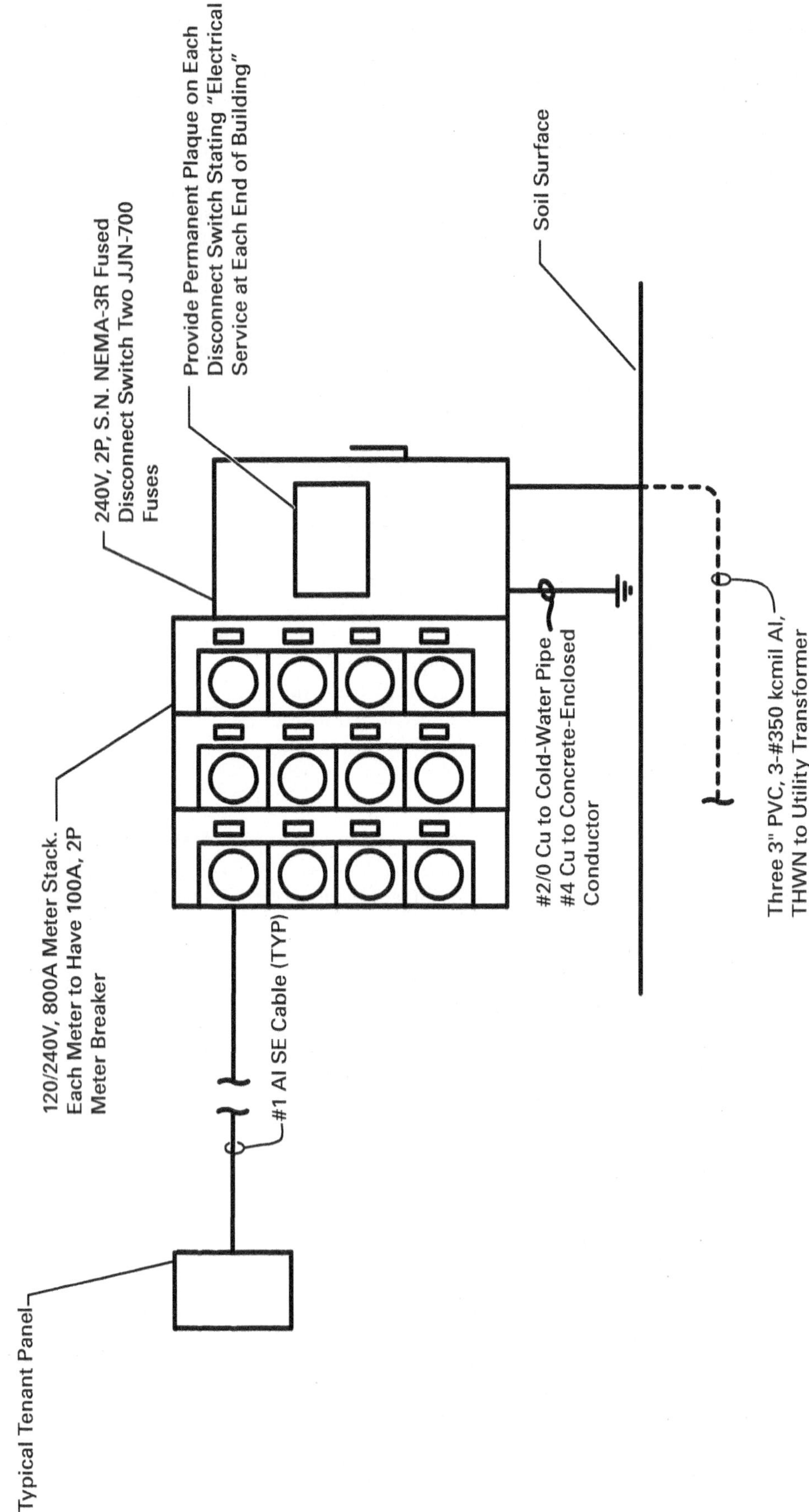

Figure 18 One-line diagram for a multifamily service.
Source: Al Hamilton

2.0.0 Section Review

1. In the overhead service shown in *Figure 7*, the parallel service-entrance conductors terminate in a _____.
   a. bussed gutter
   b. wireway
   c. meter stack
   d. switchgear bus

2. In the underground service shown in *Figure 11*, the parallel service-entrance conductors terminate in a _____.
   a. bussed gutter
   b. wireway
   c. meter stack
   d. switchgear bus

3. In the underground service shown in *Figure 15A*, the parallel service-entrance conductors terminate in a _____.
   a. bussed gutter
   b. wireway
   c. meter stack
   d. CT metering section

4. In the underground service shown in *Figure 18*, the parallel service-entrance conductors terminate in a _____.
   a. bussed gutter
   b. wireway
   c. main disconnect
   d. switchgear cabinet

Module 26308-23 Review Questions

1. The *NEC*® requirements for services can be found in _____.
   a. *NEC Article 225*
   b. *NEC Article 230*
   c. *NEC Article 240*
   d. *NEC Article 250*

2. All electrical service distribution equipment is labeled with _____.
   a. maximum voltage and current ratings
   b. equipment heating effect
   c. mechanical strength data
   d. installation requirements

3. Cold sequence metering is accomplished by _____.
   a. unplugging the meter from the meter base
   b. installing a disconnect after the meter
   c. installing a disconnect before the meter
   d. placing a switch in CT leads to a meter

4. Most analog watt-hour meters have _____.
   a. four dials
   b. five dials
   c. six dials
   d. seven dials

5. A group of CTs, one for each ungrounded conductor or set of conductors, is typically used in services rated _____.
   a. 200A and above
   b. 300A and above
   c. 400A and above
   d. 500A and above

6. The term *busway* is defined in _____.
   a. *NEC Article 100*
   b. *NEC Article 376*
   c. *NEC Article 378*
   d. *NEC Article 392*

7. Bracing of busbars is required primarily to _____.
   a. support the weight of the busbars
   b. allow easier connection of taps
   c. withstand the torque of a fault current
   d. remove heat from the busbar

8. Service conductors passing through a building are considered outside if the conductors are _____.
   a. insulated and fireproofed
   b. in PVC conduit
   c. in conduit encased by 2" (50 mm) of concrete
   d. out of reach

9. The weatherhead for a service drop should be a *minimum* of _____.
   a. 10' (3.0 m) above grade
   b. 13'-6" (4.1 m) above grade
   c. 18' (5.5 m) above grade
   d. 21'-6" (6.6 m) above grade

10. Grounding electrode conductors are sized using _____.
   a. *NEC Table 250.66*
   b. *NEC Table 250.122*
   c. *NEC Table 310.16*
   d. *NEC Table 310.19*

Answers to odd-numbered Module Review Questions are found in *Appendix A*.

Module 26308-23 Supplemental Exercises

1. Service equipment consists of __________, __________, __________, and __________.

2. *NEC Article 230* covers service conductors and __________.

3. The incoming overhead service supply is referred to as the __________.

4. A service __________ drawing indicates service characteristics such as the voltage rating and grounding requirements.

5. Every service installation begins with a review of __________ policies and requirements.

6. The location of the bus and ground bars and the __________ are important in service-entrance conductor installation.

7. True or False? Cold sequence metering allows the utility company to work on metering equipment with the power disconnected.

8. For a typical meter installation, the electrician is responsible for installing the _________.

9. The two basic types of CTs are the busbar type and the _________ type.

10. A characteristic of fault currents is a(n) _________ torque in conductors carrying the fault.

11. True or False? An MLO panelboard has a main circuit breaker.

12. *NEC Section 230.82(3)* requires that a meter disconnect switch connected to the supply side of the service disconnect be rated at no more than _________.

13. According to *NEC Section 230.3*, _________ supplying one building cannot pass through the interior of another building or structure.

14. The most common service voltage for small commercial loads is _________, _________, _________ service.

15. Equipment installed on the exterior of a building must be rated for _________ use.

Answers to odd-numbered Supplemental Exercises are found in *Appendix B*.

Answers to Section Review Questions

Answer	Section Reference	Objective
Section 1.0.0		
1. b	1.1.2	1a
2. d	1.2.1	1b
Section 2.0.0		
1. b	2.1.0	2a
2. a	2.2.2	2b
3. d	2.3.0	2c
4. c	2.4.0	2d

User Update

Did you find an error? Submit a correction by visiting **https://www.nccer.org/olf** or by scanning the QR code using your mobile device.

Motor Calculations

Source: iStock@Supersmario

NOTE

NFPA 70®, *National Electrical Code®*, and *NEC®* are registered trademarks of the National Fire Protection Association, Quincy, MA.

Objectives

Successful completion of this module prepares you to do the following:

1. Identify motor connections and operating characteristics.
   a. Calculate synchronous speed.
   b. Identify stator windings.
2. Size motor circuit conductors.
   a. Calculate conductor ampacities for a typical motor control center.
   b. Calculate conductor ampacities for other motor applications.
3. Size motor protective devices.
   a. Size fuses for short circuit protection.
   b. Size overcurrent protection for oversized motors.
   c. Size motor overload protection.
   d. Size short circuit protection for use with motor starters.
   e. Size protective devices for multimotor branch circuits.
   f. Size equipment grounding conductors.
   g. Explain how to apply capacitors for power factor correction.

Performance Tasks

This is a knowledge-based module. There are no Performance Tasks.

Overview

A basic understanding of motor operation, components, and circuitry is essential when sizing conductors and protective devices. This module presents the calculations required to size conductors and overcurrent protective devices required for motor applications.

Digital Resources for Electrical

Scan this code using the camera on your phone or mobile device to view the digital resources related to this craft.

1.0.0 Motor Connections and Operating Characteristics

Performance Tasks

There are no Performance Tasks in this section.

Objective

Identify motor connections and operating characteristics.

a. Calculate synchronous speed.

b. Identify stator windings.

Electric motors are used everywhere, from residential appliances to heavy industrial machines. Many types of motors are needed, from small shaded-pole motors used for small household fans to huge synchronous motors for use with pumps and other machinery.

However, no motor design has more application potential than the three-phase motor. This is the type of motor electricians encounter most frequently in commercial/industrial settings. Therefore, most of this module is dedicated to three-phase motors. Three-phase motors are also the type most likely to be repaired or refurbished rather than disposed of when they fail or are damaged in some way.

There are three basic types of three-phase motors:

- Squirrel-cage induction
- Wound-rotor induction
- Synchronous

These three motor types are identified by their rotating component (*Figure 1*). The stator winding is basically the same for all three motor types.

The principle of operation for all three-phase motors is the rotating magnetic field. Three factors cause the magnetic field to rotate:

- The voltages of a three-phase electrical system are 120° out of phase with each other.
- The three voltages change polarity at regular intervals.
- The stator windings around the inside of the motor are arranged in a specific manner to induce rotation.

The *NEC*® plays an important role in the installation of electric motors. *NEC Article 430* covers the application and installation of motor circuits and motor control connections, including conductors, short circuit and ground fault protection, controllers, disconnects, and overload protection.

NEC Article 440 contains provisions for motor-driven air conditioning and refrigerating equipment, including the branch circuits and controllers for the equipment. It also accounts for the special considerations of **hermetic** compressors, within which the motor is sealed and cooled by the refrigerant vapor flowing through. When referring to *NEC Article 440*, be aware that the rules in this article are in addition to, or are amendments to, the rules outlined in *NEC Article 430*. Motors are also covered to a lesser degree in *NEC Articles 422 and 424*.

Know the Code

Motors, Motor Circuits, and Controllers
NEC Article 430

Know the Code

Air-Conditioning and Refrigerating Equipment
NEC Article 440

Hermetic: Isolated and sealed airtight.

The rotor of an AC squirrel-cage induction motor (*Figure 2*) consists of a structure of steel laminations mounted on a shaft. Embedded in the rotor is the rotor winding, which is a series of copper or aluminum bars short circuited at each end by a metallic end ring. The stator consists of steel laminations mounted in a frame. Slots in the stator hold stator windings that can be made from either copper or aluminum coils or bars. These are connected to form a circuit.

Energizing the stator coils with an AC supply voltage causes current to flow in the coils. The current produces an electromagnetic field that causes magnetic poles to be created in the stator iron. The strength and polarity of these poles vary as the AC current flows in each direction. This change causes the poles around the stator to alternate between being south and north poles, thus producing a rotating magnetic field.

The rotating magnetic field cuts through the rotor, inducing a current in the rotor bars. This induced current only circulates in the rotor, which in turn creates a magnetic field in the rotor. As with two conventional bar magnets, the north pole of the rotor field attempts to line up with the south pole of the stator magnetic field, and the south pole attempts to line up with the north pole. However, because the stator magnetic field is rotating, the rotor chases the stator field. The rotor field never quite catches up due to the need to furnish **torque** to the mechanical load.

Torque: A twisting force that produces rotary motion, typically measured in foot-pounds or inch-pounds of force.

Figure 1 Basic parts of a three-phase motor.

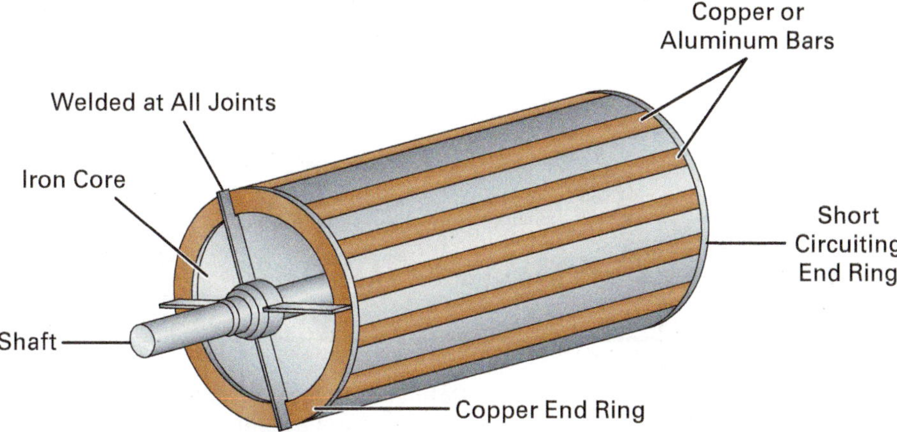

Figure 2 Squirrel-cage rotor.

1.1.0 Synchronous Speed

The speed at which the magnetic field rotates is known as the *synchronous speed*. The synchronous speed of a three-phase motor is determined by two factors:

- Number of stator poles
- Frequency of the AC line in Hz

The synchronous speeds for various 60 Hz motors are as follows:

- Two poles: 3,600 rpm
- Four poles: 1,800 rpm
- Six poles: 1,200 rpm
- Eight poles: 900 rpm

These speeds illustrate that the rpm of a three-phase motor decreases as the number of poles increases. The formula for determining synchronous speed is as follows:

$$\text{Synchronous speed} = 120f \div P$$

Where:

f = frequency

P = number of poles

1.2.0 Stator Windings

The stator windings of three-phase motors are connected in either a wye or delta configuration (*Figure 3*). Some stators are designed to operate both ways, meaning they start as a wye-connected motor to help reduce starting current and are changed to a delta configuration for normal operation.

Many three-phase motors have dual-voltage stators. These stators are designed to be connected to either 240V or 480V. The leads of a dual-voltage stator use a standard numbering system. *Figure 4* shows a dual-voltage, wye-connected stator.

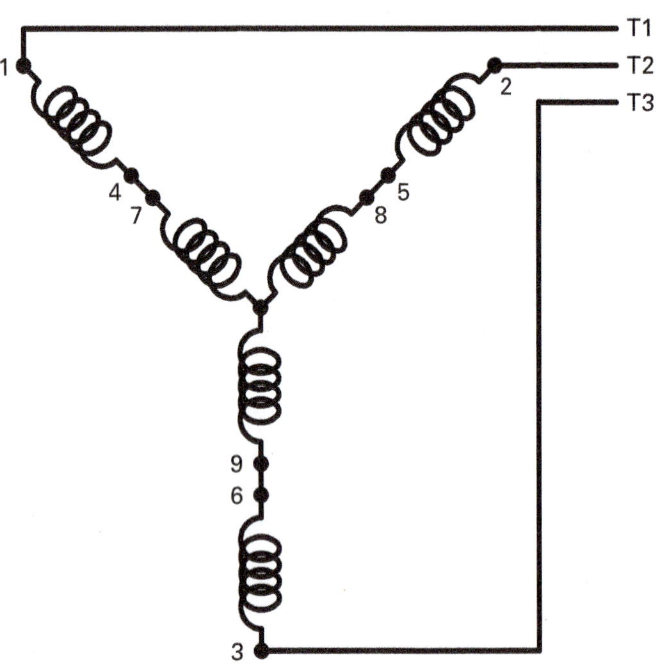

Wye-Connected Motor Windings (Series Connected)

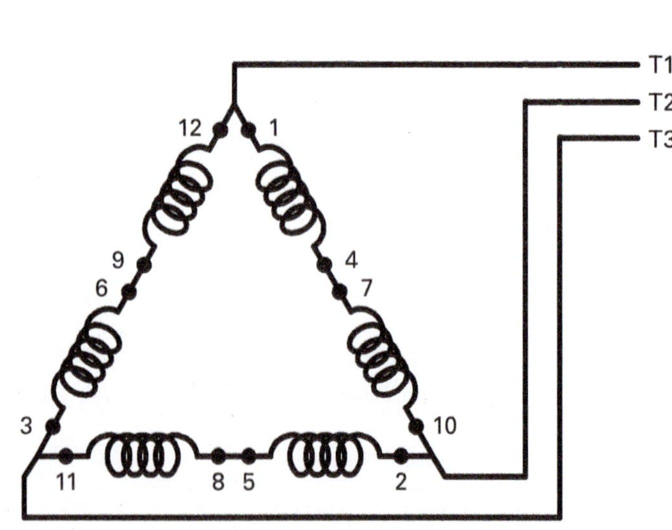

Delta-Connected Motor Windings (Series Connected)

Figure 3 Types of windings found in three-phase motors.

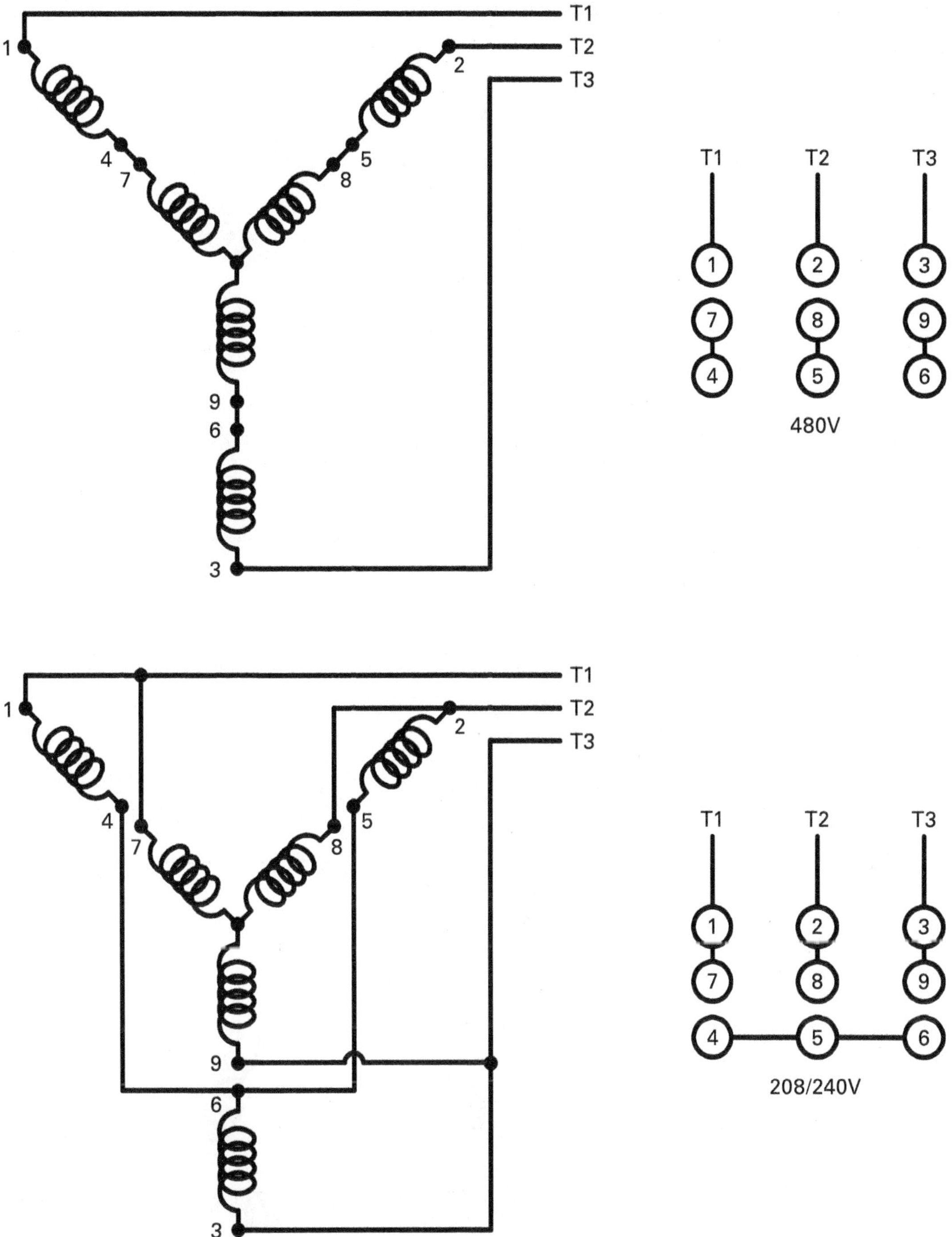

Figure 4 Dual-voltage, wye-connected, three-phase motors.

Note that the nine motor leads are numbered in a spiral. The leads are connected in series for use on the higher voltage and in parallel for use on the lower voltage. Therefore, for the higher voltage, leads 4 and 7, 5 and 8, and 6 and 9 are connected. For the lower voltage, leads 4, 5, and 6 are connected. Finally, further connections 1 and 7, 2 and 8, and 3 and 9 are then connected to the three-phase power source. *Figure 5* shows the equivalent parallel circuit when the motor is connected for use on the lower voltage.

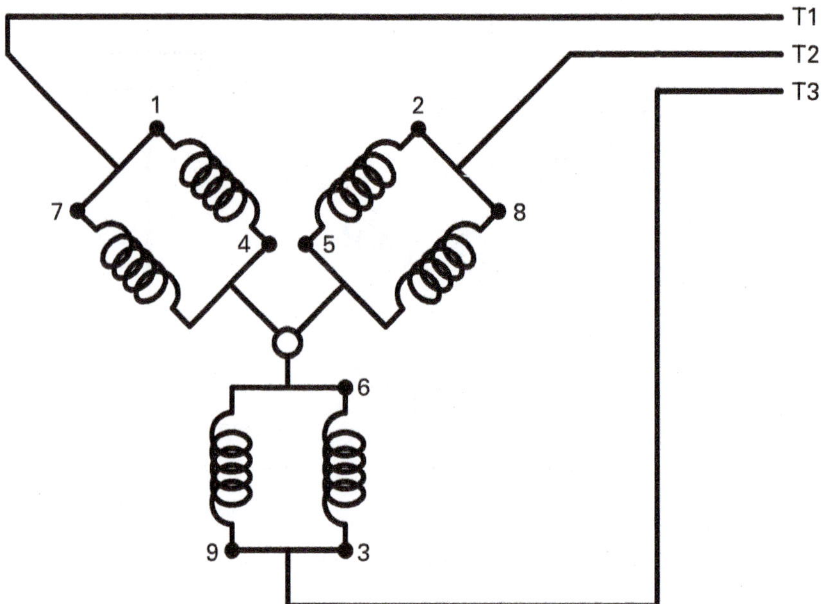

Figure 5 Equivalent parallel circuit.

The same standard numbering system is used for delta-connected motors, and many delta-wound motors also have nine leads, as shown in *Figure 6*. However, there are only three circuits of three leads each. The high- and low-voltage connections for a three-phase, delta-wound, dual-voltage motor are shown in *Figure 7*.

In some instances, a dual-voltage motor connected in a delta configuration will have 12 leads instead of 9. *Figure 8* shows the high-voltage and low-voltage connections for a dual-voltage, 12-lead, delta-wound motor.

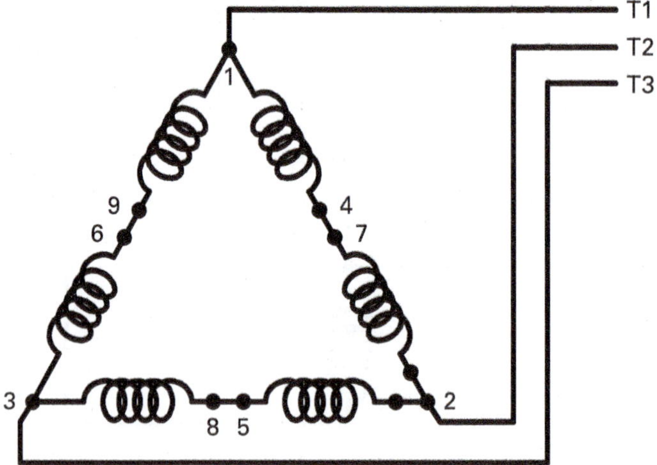

Figure 6 Arrangement of leads in a nine-lead, delta-wound, dual-voltage motor.

1.2.1 Dual-Voltage Connections

When a dual-voltage motor is operated at 240V, the operating current of the motor is double that of the operating current when powered by a 480V power source. For example, if a motor draws 10A of current when connected to 240V, it will draw only 5A when connected to 480V. The reason for this is the difference in impedance in the windings between a 240V and 480V connection. Remember that the low-voltage windings are always connected in parallel, while the high-voltage windings are connected in series.

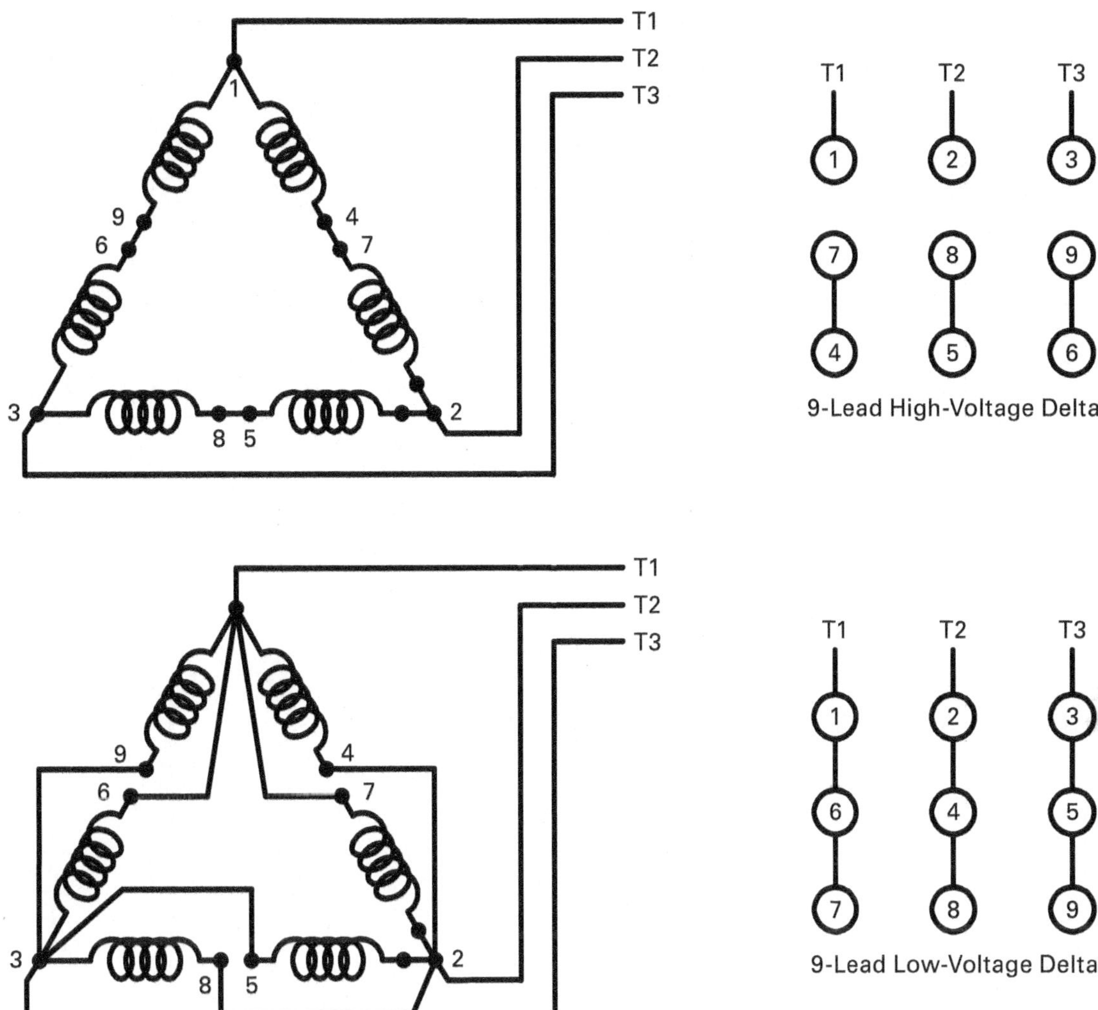

Figure 7 Load connections for a three-phase, dual-voltage, delta-wound motor.

For instance, assume that the stator windings of a motor (*R1* and *R2*) each have a resistance of 48Ω. If the stator windings are connected in parallel, the total resistance (R_T) may be found as follows:

$$R_T = (R1 \times R2) \div (R1 + R2)$$
$$R_T = (48\Omega \times 48\Omega) \div (48\Omega + 48\Omega)$$
$$R_T = 2{,}304\Omega \div 96\Omega$$
$$R_T = 24\Omega$$

Therefore, the total resistance of the motor winding connected in parallel is 24Ω, and if a voltage (*E*) of 240V is applied to this connection, the following current (*I*) will flow:

$$I = E \div R$$
$$I = 240\text{V} \div 24\Omega$$
$$I = 10\text{A}$$

If the windings are connected in series for operation on 480V, the total resistance of the winding is determined as follows:

$$R_T = R1 + R2$$
$$R_T = 48\Omega + 48\Omega$$
$$R_T = 96\Omega$$

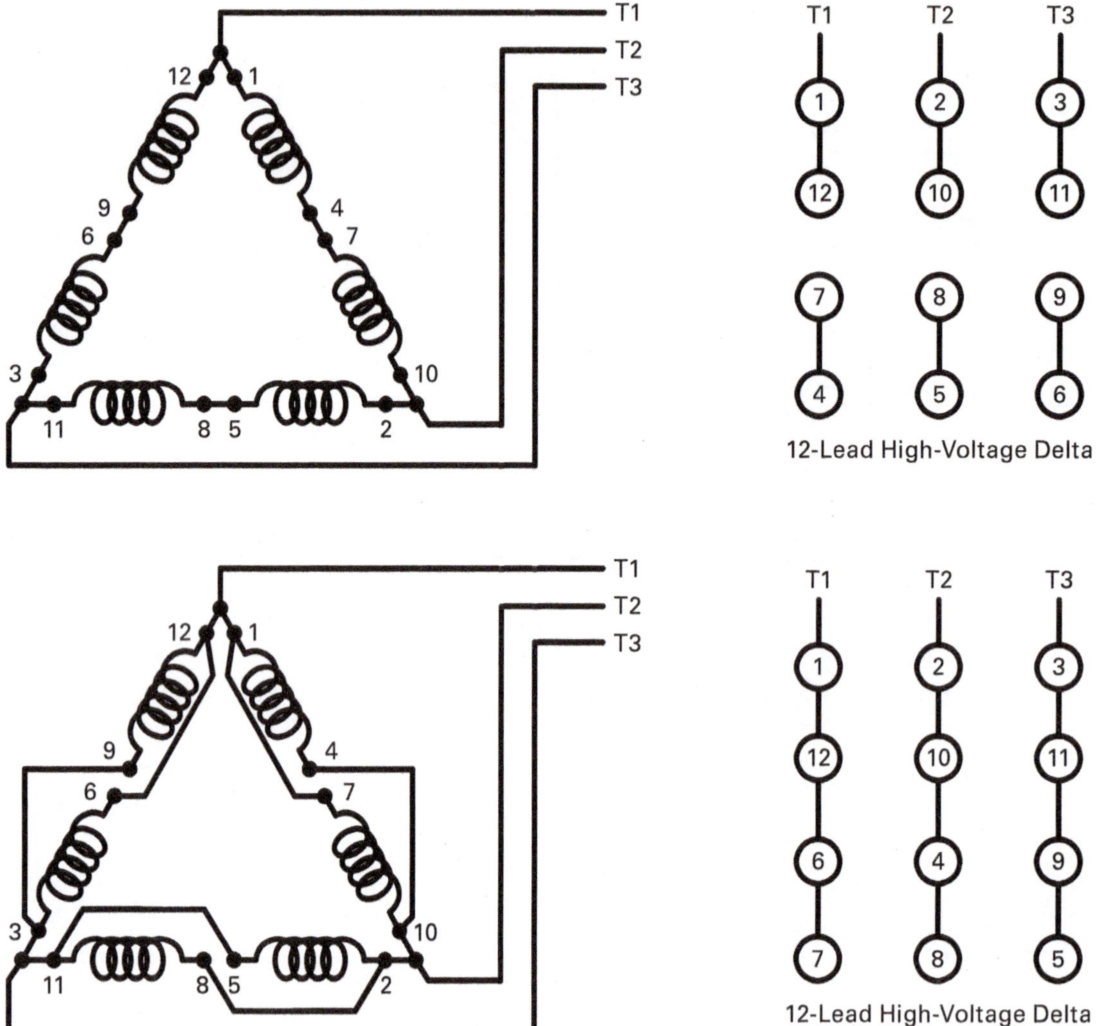

Figure 8 Lead connections for a 12-lead, dual-voltage, delta-wound motor.

Consequently, if 480V is applied to this winding, the following current will flow:

$$I = E \div R$$

$$I = 480\text{V} \div 96\Omega$$

$$I = 5\text{A}$$

This proves that twice the voltage means half the current flow and vice versa.

1.2.2 Special Connections

Some three-phase motors designed for operation at voltages greater than 600V may have more than 12 leads. Motors with 15 or 18 leads are common in high-voltage installations. A 15-lead motor has three coils per phase, as shown in *Figure 9*. Notice that the leads are numbered in the same spiral sequence as a 9-lead, wye-wound motor.

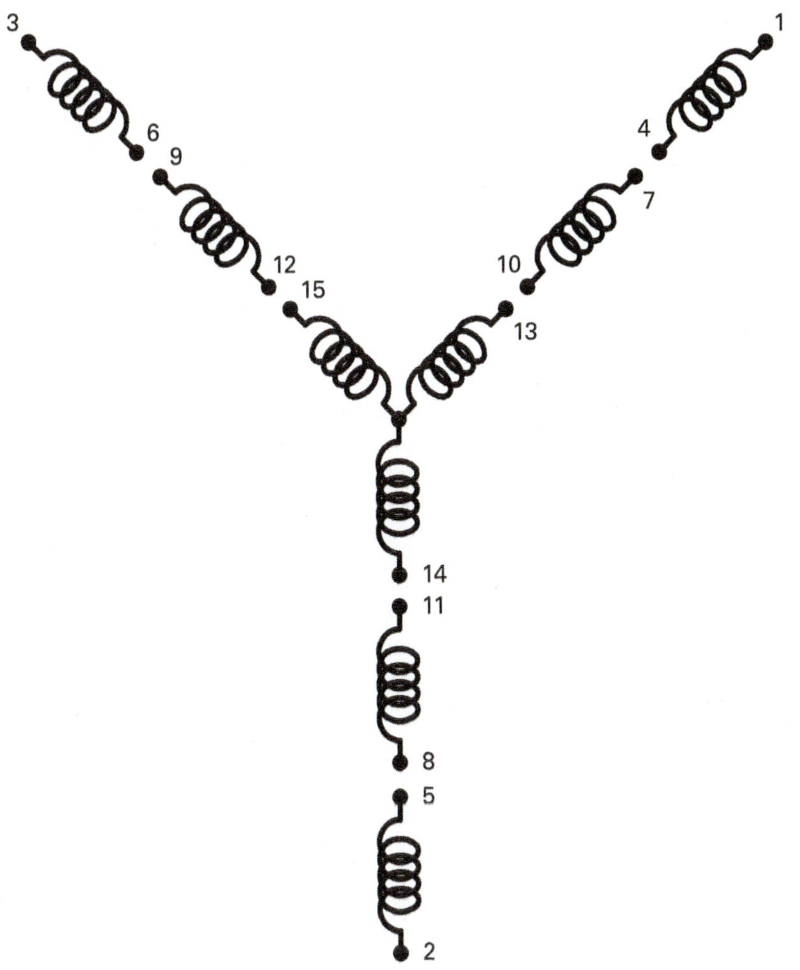

Figure 9 15-lead motor.

1.0.0 Section Review

1. The synchronous speed of a three-phase, 60 Hz motor with six poles is _____.
   a. 900 rpm
   b. 1,200 rpm
   c. 1,800 rpm
   d. 3,600 rpm

2. When a stator is connected for both starting and running, the running connection is a _____.
   a. wye configuration
   b. delta configuration
   c. spiral configuration
   d. star configuration

2.0.0 Motor Circuit Conductors

Performance Tasks

There are no Performance Tasks in this section.

Objective

Size motor circuit conductors.

a. Calculate conductor ampacities for a typical motor control center.

b. Calculate conductor ampacities for other motor applications.

The basic elements of a motor circuit are shown in *Figure 10*. Although these elements are shown separately in the illustration, there are certain cases in which the *NEC®* permits a single device to serve more than one function. For example, in some cases, one switch can serve as both a disconnecting means and a motor starter. In other cases, short circuit protection and overload protection can be combined in a single circuit breaker or set of fuses.

Know the Code

General
NEC Section 430.21

Know the Code

Over 1000 Volts, Nominal
NEC Article 430, Part XI

WARNING!

The motor circuit calculations reflected in this module and described in ***NEC Article 430, Part II*** apply only to motor circuits rated at a nominal 1,000V or less, as stated in ***NEC Section 430.21***. Calculations for motor circuits over 1,000V are not covered in this module. ***NEC Article 430, Part XI*** amends or adds to the other provisions in ***NEC Article 430*** for motor circuits over 1,000V.

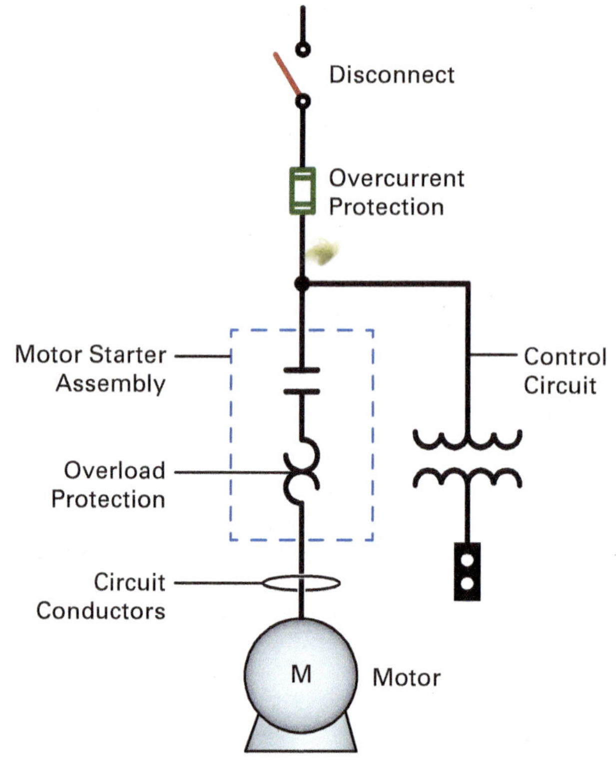

Figure 10 Basic elements of a typical motor circuit.

2.1.0 Conductor Ampacities for a Typical Motor Control Center

Know the Code

Full-Load Current, Three-Phase Alternating-Current Motors
NEC Table 430.250

A typical MCC (MCC) and branch circuits feeding four different motors are shown in *Figure 11*. This circuit contains both squirrel-cage and wound-rotor motors. The primary full-load current ratings for wound-rotor motors are listed in *NEC Table 430.250* and are the same for squirrel-cage motors.

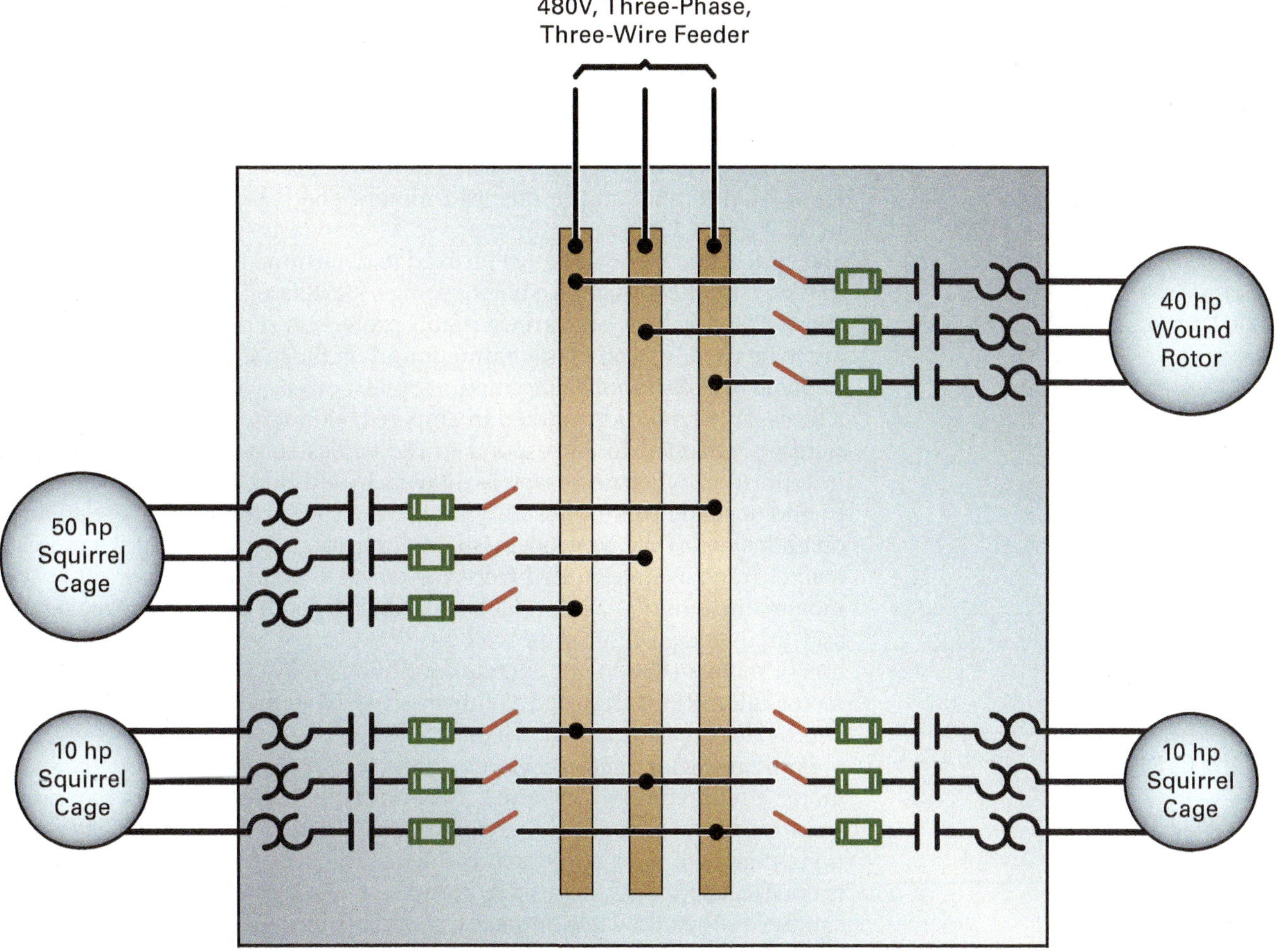

Figure 11 Typical MCC.

Conductors connecting the secondary leads of wound-rotor induction motors to their controllers must have an ampacity greater than or equal to 125% of the motor's full-load secondary current if the motor is used for continuous duty. If the motor is used for noncontinuous duty, the conductors must have an ampacity greater than or equal to the percentage of its full-load nameplate current given in *NEC Table 430.22(E)*. Conductors from the controller of a wound-rotor induction motor to its starting resistors must have an ampacity that meets the values found in *NEC Table 430.23(C)*.

NEC Section 430.99 requires the following information to be documented and made available:

- The available fault current at the MCC
- The date the calculation was performed

The following steps outline a general procedure for sizing feeder and branch circuit conductors for motors:

Step 1 Refer to *NEC Tables 430.247 through 430.250* for the full-load current of each motor.

Step 2 Determine the full-load current of the largest motor in the group.

Step 3 Calculate the sum of the full-load current ratings for the remaining motors in the group.

Step 4 Multiply the full-load current of the largest motor by 1.25 (125%) and then add the sum of the remaining motors to the result (*NEC Section 430.24*). The combined total will provide the minimum feeder ampacity.

Know the Code

Available Fault Current
NEC Section 430.99

NOTE

NEC Section 430.6(A)(1) specifies that, for general motor applications (excluding applications of torque motors and hermetic compressor motors), the values given in *NEC Tables 430.247, 430.248, 430.249, and 430.250* should be used instead of the actual current rating marked on the motor nameplate to size conductors, switches, and overcurrent protective devices. Overload protection, however, is based on the motor nameplate data.

Know the Code

Several Motors or a Motor(s) and Other Load(s)
NEC Section 430.24

Voltage Ratings

The motor voltage ratings used in the *NEC®* can be somewhat confusing. Motors are built and tested at various voltages. For example, the 460V referenced in ***NEC Tables 430.249 and 430.250*** is based on rated motor voltages. However, to standardize calculations and maintain minimum safety standards, the *NEC®* uses a higher value of 480V for motor calculations. Refer to ***NEC Section 220.5(A).***

Know the Code

Single Motor
NEC Section 430.22

Know the Code

Temperature Limitations
NEC Section 110.14(C)

NOTE

The motor circuit calculation worksheet shown in *Appendix 26309-23A* can be used for the types of motor circuits covered in this module. It is designed for use in motor circuit calculations for one or more standard single-speed, 3ϕ, 60 Hz continuous duty squirrel-cage, synchronous, and wound-rotor motors (without external resistors), all rated at 600V or less and with separate overcurrent protective devices and controllers.

Think About It

***NEC®* Motor Sizes**

How is the largest motor in a group determined by the *NEC®*? Is it by frame size, horsepower, weight, or the motor's full-load amps (FLA)?

When sizing feeder or branch circuit conductors for motors, be aware the previous and following procedures will give the minimum conductor ampacity rating based on the *NEC®* minimum only. Consequently, it is often necessary to increase the size of conductors to compensate for voltage drop and power loss in the circuit.

NEC Section 430.6(A)(1) states that, with the exception of low-speed (<1,200 rpm), high-torque, and multispeed motors, the full-load current values given in *NEC Table 430.250* for three-phase AC standard squirrel-cage, wound-rotor, and synchronous motors must be used to determine the ampacity of feeder and branch circuit conductors, switch ratings, starters or controllers, branch circuit short circuit values, and ground fault protection. The full-load current values are to be used instead of any rating found on the motor nameplate.

Some noncontinuous duty motor applications are allowed exceptions. Where a three-phase motor is marked in amperes but not horsepower, the horsepower rating is assumed to correspond to the values shown in *NEC Table 430.250.* Interpolation, when necessary, is allowed. In addition, *NEC Section 430.22* states that for a single continuous duty motor, the conductors must have an ampacity on both sides of the controller equal to or greater than 125% of the motor's full-load current, as determined from the tables. Considerations for other types of motors, including wye-start/delta-run, multispeed, high-torque, part-winding, and dual-voltage motors as well as wound-rotor motors with external resistors or motors that do not operate continuously, are not covered in the example motor calculations reflected in this module. Note that *NEC Informative Annex D* provides motor calculation examples.

Another factor that can affect conductor sizing for standard motors is the motor circuit component rating. *NEC Section 110.14(C)* states that conductor sizes must be selected so that the lowest temperature rating of any terminations, conductors, or additional connected devices (excluding the motor itself) is not exceeded. This means that regardless of a higher temperature rating for a conductor, the allowable ampacity selected from *NEC Tables 430.247 through 430.250* must not exceed the lowest temperature rating of the terminations, conductors, and additional devices.

If the temperature rating of any of the conductors and devices is unknown, the conductor sizes must be selected from the *60°C (140°F)* ampacity column of the referenced *NEC®* tables. In the case of termination provisions for equipment rated at 100A or less or marked for No. 14 AWG through No. 1 AWG, conductor sizes shall be selected from the *60°C (140°F)* ampacity columns. However, for motors marked with NEMA design letter B, C, or D, conductors rated at 167°F (75°C) or higher may be used if the ampacity required is 100A or less. In the case of termination provisions of equipment rated over 100A or marked for No. 1 AWG or larger, conductor sizes must be selected from the *75°C (167°F)* column of the tables.

Now we will complete the conductor calculations for the motor circuits shown in *Figure 11.* Assume all conductor terminations are rated at 167°F (75°C) and all the motors are NEMA design letter C.

Referring to *NEC Table 430.250*, the motor horsepower is shown in the far left-hand column. Follow across the appropriate row until you come to the column titled 460V, which is the center voltage (440V to 480V) of the motor circuits in *Figure 11.* The ampere ratings for the motors in question are as follows:

- 50 hp = 65A
- 40 hp = 52A
- 10 hp = 14A

The largest motor in this group is the 50 hp squirrel-cage motor, which has a full-load amperage (FLA) of 65A. Multiply the FLA of the largest motor by 1.25 (125%) and then add the total current of the remaining motors:

$$(1.25 \times 65A) + 52A + 14A + 14A = 161.25A$$

Therefore, the feeders for the 460V, three-phase, three-wire MCC must have a minimum ampacity of 161.25A. Referring to *NEC Table 310.16* under the column headed *75°C (167°F)*, the closest conductor size is 2/0 copper (rated at 175A) or 4/0 aluminum (rated at 180A).

The branch circuit conductors feeding the individual motors are calculated somewhat differently. *NEC Section 430.22* requires that the ampacity of branch circuit conductors supplying a single continuous duty motor must be greater than or equal to 125% of the motor FLA. Therefore, the ampacity of the branch circuit conductors feeding the four motors in question is calculated as follows:

50 hp motor = 65A × 1.25 = 81.25A

40 hp motor = 52A × 1.25 = 65A

10 hp motor = 14A × 1.25 = 17.5A

Referring to *NEC Table 310.16*, the closest size 167°F (75°C) THWN copper conductors permitted to be used on these various branch circuits are as follows:

- A 50 hp motor at 81.25A requires No. 4 AWG THWN conductors.
- A 40 hp motor at 65A requires No. 6 AWG THWN conductors.
- A 10 hp motor at 17.5A requires No. 14 AWG THWN conductors per *NEC Section 110.14(C)(1)(a)(3)* and *NEC Table 310.16*.

Refer to *Figure 12* for a summary of the conductors used to power the example MCC, along with the branch circuits supplying the individual motors.

If voltage drop and/or power loss must be considered, refer to the related information in NCCER Module 26302-23, *Conductor Selection and Calculations*.

Squirrel-Cage Motor Fuse Selection

Most squirrel-cage motors manufactured prior to 1996 are not marked with a design letter B, C, or D on their nameplates. However, unless otherwise marked, most common squirrel-cage motors fall into the design code B classification. They are considered design code B motors for the purpose of selecting overcurrent protection and conductor temperature.

Think About It

Feeder Size

If the 40 hp wound-rotor motor in *Figure 12* was replaced with a 50 hp squirrel-cage motor, the No. 6 THWN conductors powering that motor must be replaced with No. 4 THWN conductors. What size THWN, 167°F (75°C) feeders are also needed to accommodate the change?

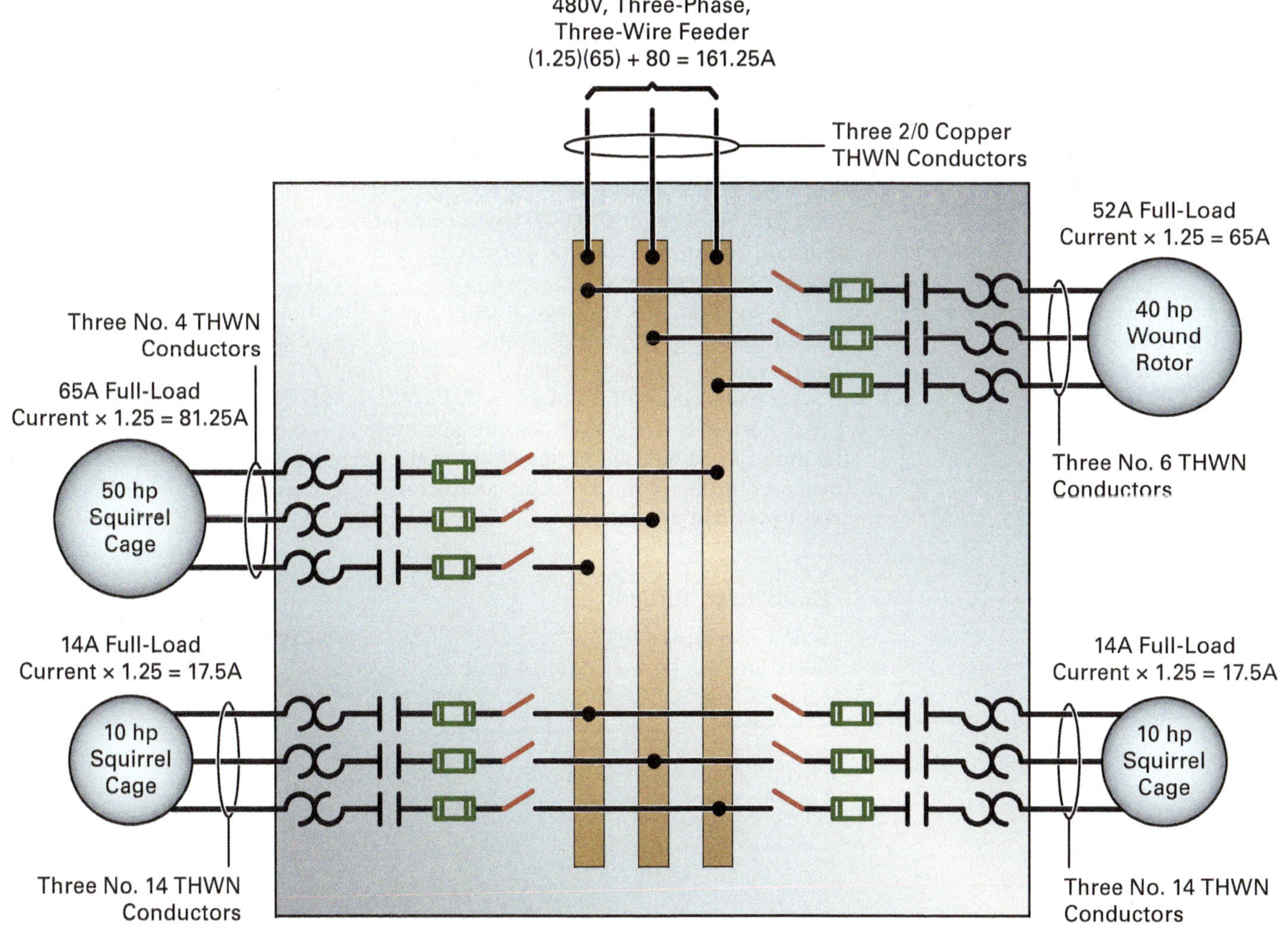

Figure 12 Selected motor branch circuit conductors.

For motors with other voltages (up to 2,300V) or for synchronous motors, refer to *NEC Table 430.250*.

2.2.0 Conductor Ampacities for Other Motor Applications

Some motor applications have specific sizing procedures and *NEC®* requirements. Some of these applications include intermittent duty motors, DC motors, and a variety of miscellaneous motor types. Always refer to the applicable section of the *NEC®* when calculating conductor ampacities for motor circuits.

2.2.1 Intermittent Duty Motors

In accordance with *NEC Section 430.22(E)*, branch circuit conductors serving motors used for short-time, intermittent, or other varying duty must have an ampacity equal to or greater than the percentage of the motor nameplate FLA shown in *NEC Table 430.22(E)*. However, to qualify as a short-time, intermittent motor, the driven equipment must be controlled so the motor cannot operate continuously with a load under any condition of use. Otherwise, the motor is considered continuous duty. Consequently, most motors encountered in the electrical craft must be rated for continuous duty, and the branch circuit conductors sized accordingly.

Duty Ratings

Not all motors are rated for continuous duty. Some are rated for 5-, 15-, 30-, or 60-minute operation. The *NEC®* makes special allowances for these duty ratings. Refer to ***NEC Section 430.22(E)***.

2.2.2 DC Motors

NEC Sections 430.22(A) and 430.29 cover the rules governing the sizing of conductors from a power source to a DC motor controller and from the controller to separate resistors for power accelerating and dynamic braking. *NEC Section 430.29*, with its table of conductor ampacity percentages, assures proper application of DC constant-potential motor controls and power resistors. However, when selecting overload protection, the actual motor FLA must be used.

Know the Code

Constant Voltage Direct-Current Motors — Power Resistors
NEC Section 430.29

Know the Code

Conductor Ampacity and Motor Rating Determination
NEC Section 430.6

Know the Code

Rating or Setting for Individual Motor Circuit
NEC Section 430.52

2.2.3 Miscellaneous Motor Types

Refer to *NEC Section 430.6(B)* when calculating conductor ampacity for torque motors, shaded-pole motors, permanent split-capacitor (PSC) motors, and AC adjustable-voltage motors.

NEC Section 430.6(B) specifically states that the motor's nameplate FLA is used to size ground fault protection for a torque motor. However, both the branch circuit conductors and the overcurrent protection are sized by the provisions listed in *NEC Article 430, Part II, and NEC Section 430.52*.

For hermetic refrigerant compressor motors, the actual nameplate FLA of the motor must be used in determining the current rating of the disconnecting means, controller, branch circuit conductor, overcurrent protective devices, and motor overload protection.

Breakdown Torque

Certain listed appliances are now rated by the *breakdown torque*. This is the force required to stop the motor from rotating at full speed. This allows the use of the FLA directly from the appliance's nameplate rather than having to refer to ***NEC Tables 430.248 and 430.250***.

2.0.0 Section Review

1. You are connecting a 10 hp motor rated at 14A and a 50 hp motor rated at 65A to a 480V MCC. The *minimum* feeder ampacity is _____.
   a. 65A
   b. 81.25A
   c. 95.25A
   d. 98.75A

2. Information on calculating conductor ampacity for shaded-pole motors can be found in _____.
   a. *NEC Section 427.45*
   b. *NEC Section 430.6(B)*
   c. *NEC Section 502.15*
   d. *NEC Section 511.4*

3.0.0 Motor Protective Devices

Objective

Size motor protective devices.

a. Size fuses for short circuit protection.
b. Size overcurrent protection for oversized motors.
c. Size motor overload protection.
d. Size short circuit protection for use with motor starters.
e. Size protective devices for multimotor branch circuits.
f. Size equipment grounding conductors.
g. Explain how to apply capacitors for power factor correction.

Performance Tasks

There are no Performance Tasks in this section.

The branch circuit protection for motors must also protect the circuit conductors, control apparatus, and the motor itself against overcurrent due to short circuits or ground faults.

Motors and motor circuits have unique operating characteristics and circuit components. Therefore, these circuits must be managed carefully. Generally, two levels of overcurrent protection are required for motor branch circuits:

- *Overload protection* — Motor overload protection is intended to protect the system components and motor from damaging overload currents. The overload relay thermal elements used to sense current are found between the motor starter contacts and the motor. When an overload condition is sensed, the overload relay opens and de-energizes the motor.
- *Short circuit protection (includes ground fault protection)* — Short circuit protection is intended to protect motor circuit components, such as the conductors, switches, controllers, overload relays, etc., against short circuit current or grounds. This level of protection is commonly referred to as *motor branch circuit protection*. Dual-element fuses are designed to provide this protection as long as they are sized correctly.

3.1.0 Fuses for Short Circuit Protection

The ampere rating of a fuse selected for motor protection depends on whether the fuse is of the dual-element, time-delay type, or the nontime-delay type.

NEC Table 430.52(C)(1) specifies that short circuit and ground fault protection nontime-delay fuses can be sized at 300% of the motor FLA for ordinary motors. Fuses for wound-rotor and DC motors may be sized at 150% of the motor FLA. The sizes of nontime-delay fuses for the four motors previously mentioned are listed in *Figure 13*.

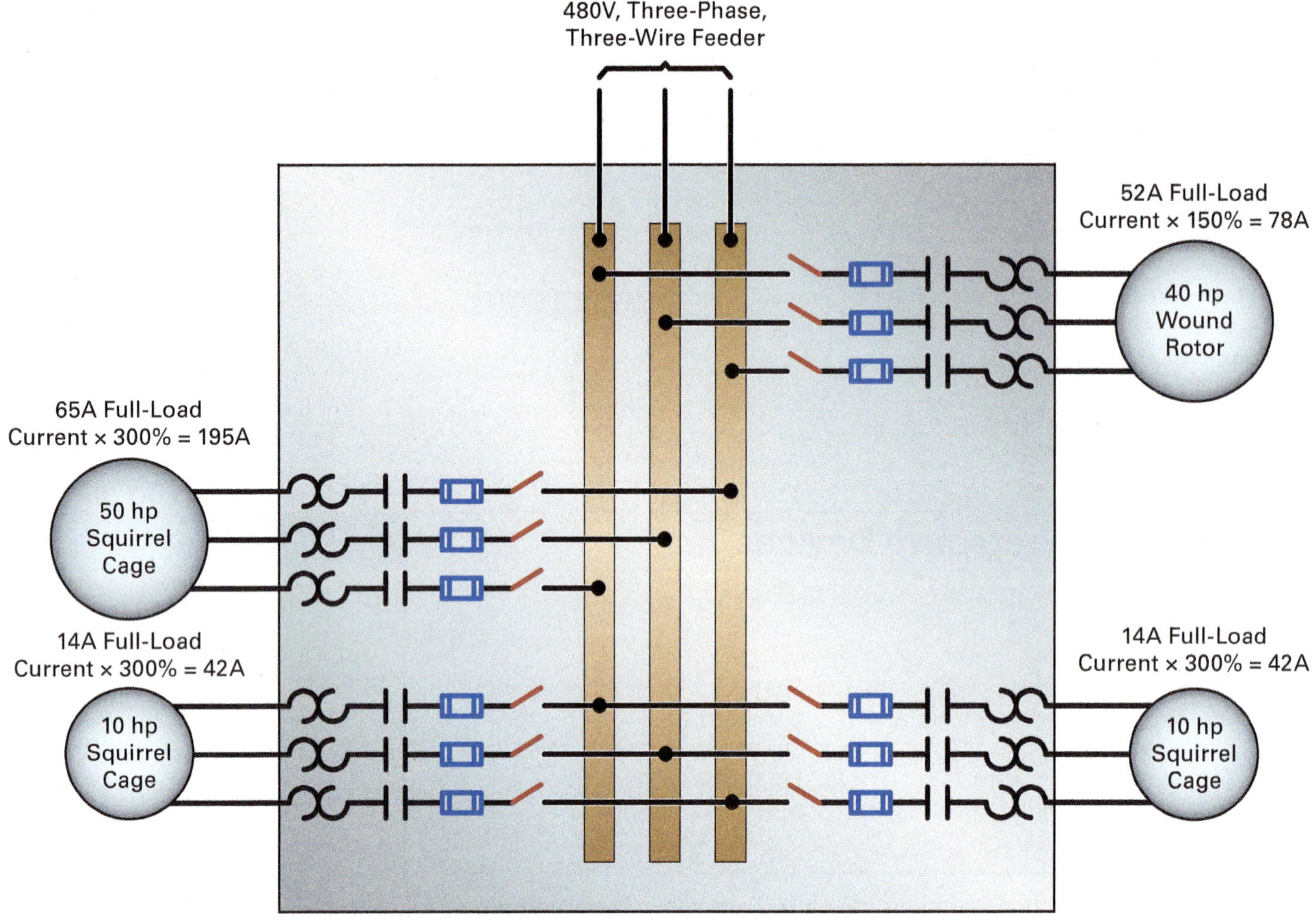

Figure 13 Ratings of nontime-delay fuses for typical motor circuits.

Know the Code

Rating or Setting
NEC Section 430.52(C)

Because none of these are standard sizes, *NEC Section 430.52(C)(1)(a)* permits the rating of the overcurrent protective device to be increased to the next available standard size. Also, where necessary to manage the inrush current of the starting motor, the size of the overcurrent protective device may be further increased but must never be more than 400% of the FLA, per *NEC Section 430.52(C)(1)(b)*.

Most electricians would choose a 200A nontime-delay fuse for the 50 hp motor, a 175A fuse for the 40 hp motor, and 45A fuses for the 10 hp motors. If any of these fuses do not allow the motor to start without blowing, the fuses may be increased to a maximum of 400% of the full-load current, which are 360A for the 50 hp motor, 208A for the 40 hp motor, and 56A for the 10 hp motors.

Per *NEC Table 430.52(C)(1)*, dual-element, time-delay fuses can withstand normal motor starting current and be sized closer to the actual motor rating than nontime-delay fuses. If necessary for proper motor operation, dual-element, time-delay fuses may be sized up to 175% of the motor's FLA for all standard motors except for wound-rotor and DC motors. These motors must not have fuses sized for more than 150% of the motor's FLA. Where necessary for proper operation, the rating of dual-element, time-delay fuses may be increased, but must never be more than 225% of the motor FLA, per *NEC Section 430.52(C)(1)(b)(2)*.

To size dual-element fuses for the four motors in *Figure 13*, find 175% of the FLA:

50 hp motor = 65A × 1.75 = 113.75A

40 hp motor = 52A × 1.75 = 91A

Each 10 hp motor = 14A × 1.75 = 24.5A

Figure 14 shows general fusing guidelines for motor branch circuits (*NEC Article 430, Part IV*). Remember that, in many cases, the maximum fuse size depends on the motor design letter, motor type, and starting method.

Type of Motor	Dual-Element, Time-Delay Fuses			Nontime-Delay Fuses
	Motor Overload and Short Circuit	**Backup Overload and Short Circuit**	**Short Circuit Only (Based on *NEC Tables 430.247 through 430.250* current ratings)**	**Short Circuit Only (Based on *NEC Tables 430.247 through 430.250* current ratings)**
Service Factor 1.15 or Greater or 40°C Temp. Rise or Less	125% or less of motor nameplate current	125% or next standard size (not to exceed 140% of motor nameplate current)	150% to 175%	150% to 300%
Service Factor Less Than 1.15 or Greater Than 40°C Temp. Rise	115% or less of motor nameplate current	115% or next standard size (not to exceed 130% of motor nameplate current)	150% to 175%	150% to 300%
	Fuses provide overload and short circuit protection.	Overload relay provides overload protection and fuses provide backup overload protection.	Overload relay provides overload protection and fuses provide only short circuit protection.	Overload relay provides overload protection and fuses provide only short circuit protection.

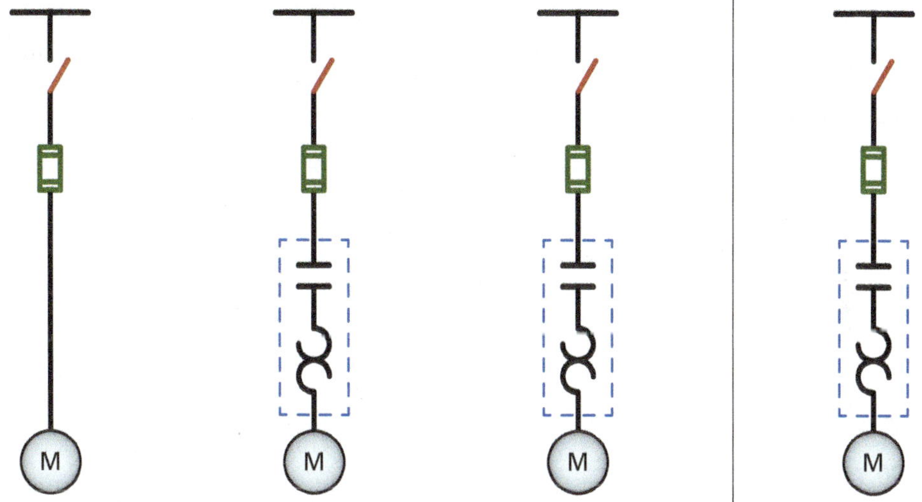

Figure 14 Fuse application guidelines for motor branch circuits.

Did You Know?

Fuses in MCCs

Fuses are commonly used in MCCs because they are less expensive than circuit breakers. Fuses are also more easily removed and replaced with a different size if the motor specifications change or if problems are encountered with the original fuse choice during startup. Fuses also have no moving parts and are very reliable.

3.2.0 Overcurrent Protection for Oversized Motors

Motors are often oversized for the applied load. For instance, a 5 hp pump motor may be installed when the load demand is only 3.5 hp, but there is no smaller motor that can manage the load. In these cases, a higher degree of overload protection can be obtained by sizing the overload relay elements and/or dual-element, time-delay fuses based on the actual FLA rather than the nameplate value.

The advantage to this is realized when a lightly loaded motor, especially larger motors, experiences a single-phase condition. Even though the relays and fuses may be sized correctly based on the motor FLA from the nameplate, circulating currents within the motor may cause damage before the protective devices can react.

In existing installations, the procedure for determining the actual running current and selecting appropriate overcurrent protection for oversized motors is as follows:

Step 1 With a clamp-on ammeter, determine the running rms current when the motor is fully loaded, as shown in *Figure 15*. It is best to measure the current on each of the three leads and average the results. Ensure the measured current does not exceed the nameplate FLA.

Step 2 Size the overload relay elements and/or overcurrent protection based on the average current.

Step 3 Use a labeling system to mark the type and ampere rating of the fuse that should be in the fuse clips. This simple system makes it easy to run spot checks for proper fuse replacements.

Individual motor disconnect switches must have a current rating of at least 115% of the motor FLA per *NEC Section 430.110(A)* or be sized as specified in *NEC Section 430.109*. The next larger switch size, equipped with fuse reducers, may sometimes be required.

Some installations require larger dual-element fuses when the following conditions exist:

- The motor uses dual-element fuses in high ambient temperature environments.
- The motor is started frequently, or the rotation is rapidly reversed.
- The motor is directly connected to a machine that cannot be brought up to full speed quickly. This is common for equipment with large flywheels, such as punch presses and other types of manufacturing and fabrication equipment.
- The motor is a NEMA Design B energy-efficient motor with full-voltage start.

NOTE

When installing fuses, it is often best to leave spare fuses on top of the disconnect or starter enclosure or in a cabinet adjacent to the MCC. This ensures that if the fuses open, replacement fuses are readily available.

Know the Code

Current Rating and Interrupting Capacity
NEC Section 430.110

Know the Code

Type
NEC Section 430.109

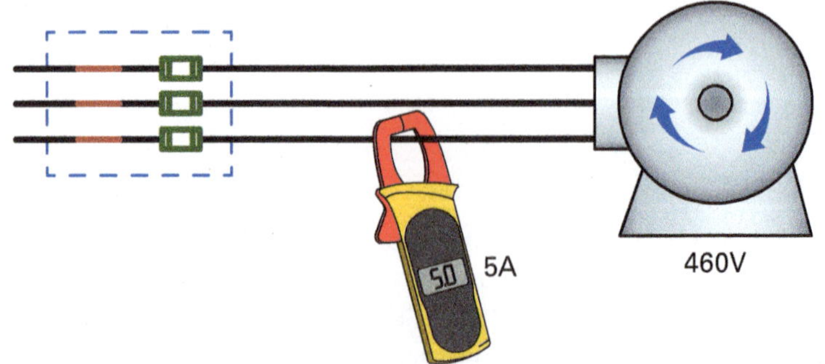

Figure 15 Determining running current with an ammeter.

3.3.0 Motor Overload Protection

A good electric motor that is properly cooled and protected against overloads can have a long service life. The goal of proper motor protection is to prolong motor life and postpone the failure that ultimately takes place. Good electrical protection provides both proper overload protection and current-limiting short circuit protection.

Remember that AC motors require protective devices with special characteristics due to the inrush current that occurs at startup. The FLA of a motor is substantially less than the current consumed as it begins to rotate. This is illustrated by the typical motor starting current curve shown in *Figure 16*.

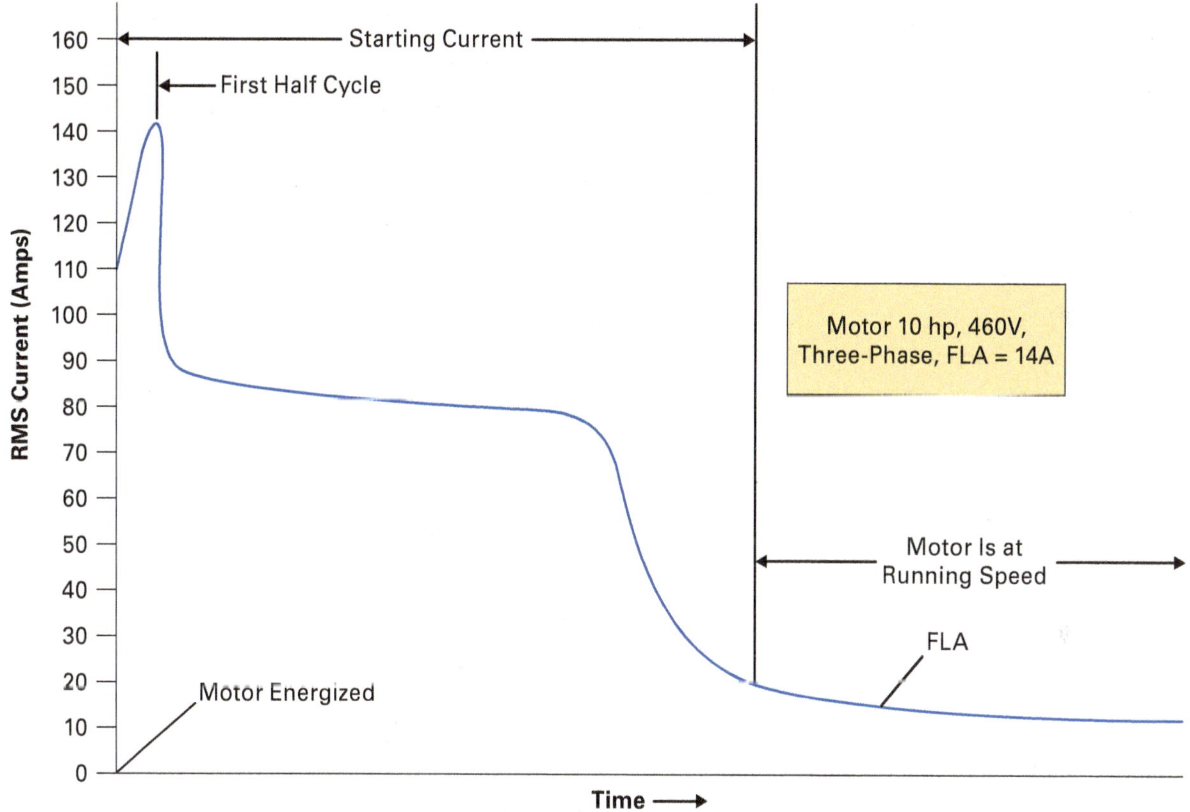

Figure 16 Motor starting current characteristics.

As the motor starts, the overcurrent protective devices in the motor circuit must be able to tolerate the substantial inrush current. For the first half cycle, the current can be 10 or more times the FLA. After this first half cycle, the current drops rapidly to about 6 times the FLA and often remains there for several seconds. This is called the *locked-rotor current* or *locked-rotor amperage*. As the motor approaches its normal rotational speed, the current drops to its normal running level. Note that motor starting currents can vary substantially, depending on the motor type, applied load, starting methods, and other factors.

The main types of devices used to provide overload protection include:

- Overload relays
- Fuses
- Circuit breakers

Overloads can occur for a variety of reasons. If the overload protective devices are properly selected and respond as they should, overloads can be interrupted before damage occurs. As shown in *Figure 17*, a curve representing the points where motor damage is likely to occur can be developed. To ensure proper protection, the overload thermal elements should have time-current characteristics shaped like the motor damage curve but slightly faster in their response.

Think About It

Overload Protection

For a 230V/460V dual-voltage motor such as the one shown here, which voltage requires the larger overload protective device?

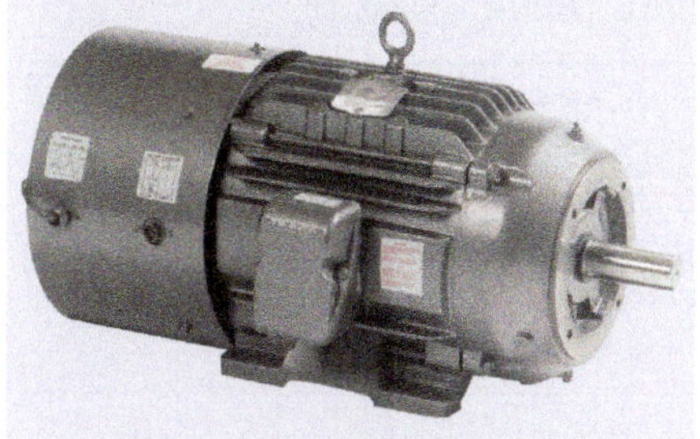

Source: Courtesy of Baldor Electric Company

As an example, we will consider a 10 hp motor and determine the proper circuit components that should be applied (*Figure 18*).

To begin, select the proper size overload relays. Typically, the overload relay is rated to trip at about 115% of the rated current. In this case, 115% of the 14A full-load current is 16.1A. The correct NEMA starter size for the motor is Size 1. The switch size that should be used is 30A. Switch sizes are based on *NEC*® requirements; dual-element, time-delay fuses allow the use of smaller switches.

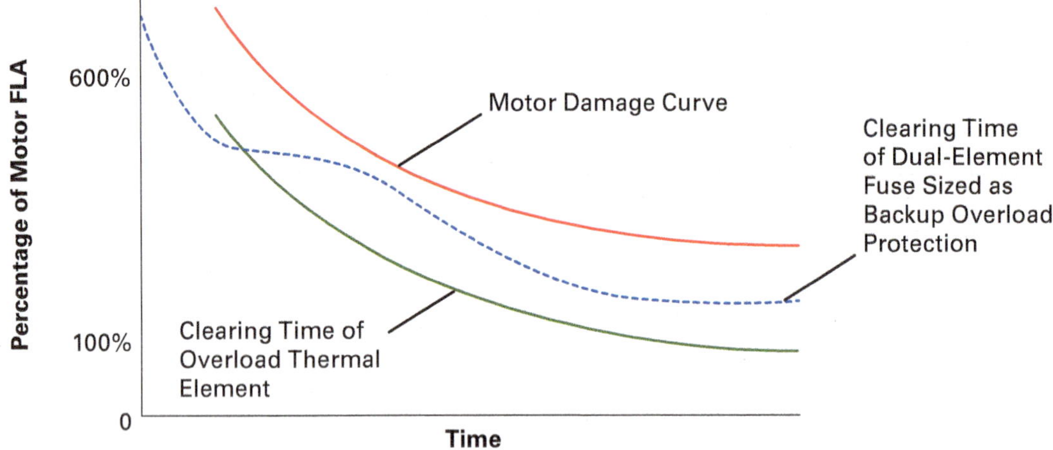

Figure 17 Time-current characteristics of dual-element fuses and overload heaters.

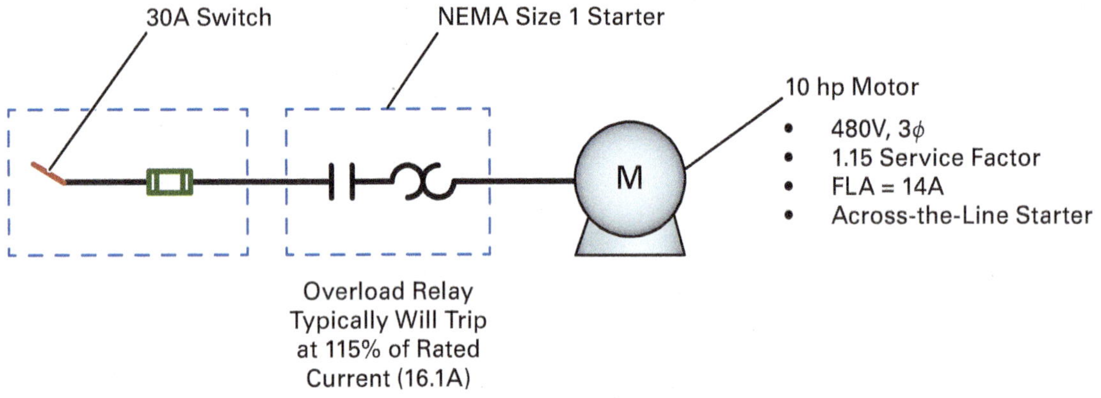

Figure 18 Circuit components of a typical 10 hp motor.

For short circuit protection on large motors with currents exceeding 600A, low-peak time-delay fuses are recommended. Most motors this size will have reduced voltage starters, and the inrush currents will not be as high. Low-peak fuses should be sized at approximately 150% to 175% of the motor full-load current.

Motor starters and controllers with overload relays commonly used on motor circuits provide motor overload protection. The overload relay selection or setting (for adjustable thermal elements) must comply with *NEC Section 430.32*. For backup protection, size dual-element fuses at the next current rating greater than the overload relay trip setting. This can typically be achieved by sizing dual-element fuses at 125% for motors with a 1.15 **service factor** and 115% for motors with a 1 service factor.

Know the Code

Continuous-Duty Motors
NEC Section 430.32

Service factor: A number by which the horsepower rating of a motor is multiplied to determine the maximum safe load a motor can be expected to carry continuously at its rated voltage and frequency.

3.4.0 Short Circuit Protection for Motor Starters

The *NEC*® recognizes the use of instantaneous trip circuit breakers (no time delay) for short circuit protection of motor branch circuits. These breakers are acceptable only if they are adjustable and used in combination motor starters. The starters must have coordinated overload, short circuit, and ground fault protection for each conductor and must be listed for the purpose per *NEC Section 430.52(C)(3)*.

This permits the use of smaller circuit breakers than would normally be allowed if a standard thermal-magnetic circuit breaker was used. In this case, smaller circuit breakers can offer faster operation for greater protection against grounds and short circuits. *Figure 19* shows a schematic diagram of a motor circuit protector (MCP) used in a combination motor starter.

MCPs are used only in combination with motor starters with integral overload protection. However, thermal elements in the overload relay protect the entire circuit and all equipment against overloads up to and including locked-rotor current. Adjustable thermal overload relays are commonly set to 115% to 125% of the motor FLA.

In these circuits, an adjustable circuit breaker can be set to take over the interrupting task at currents above the locked-rotor current and up to the short circuit duty of the supply system at the point of the installation. The magnetic trip in adjustable breakers can typically be adjusted from 3 to 13 times the breaker's current rating. For example, a 100A circuit breaker can be adjusted to trip anywhere between 300A and 1,300A. Consequently, the circuit breaker may serve as motor short circuit protection. However, instantaneous trip circuit breakers used in these installations cannot be adjusted to more than the value specified in *NEC Table 430.52(C)(1)*.

Know the Code

Instantaneous-Trip Circuit Breaker
NEC Section 430.52(C)(3)

Know the Code

Maximum Rating or Setting of Motor Branch-Circuit Short-Circuit and Ground-Fault Protective Devices
NEC Table 430.52(C)(1)

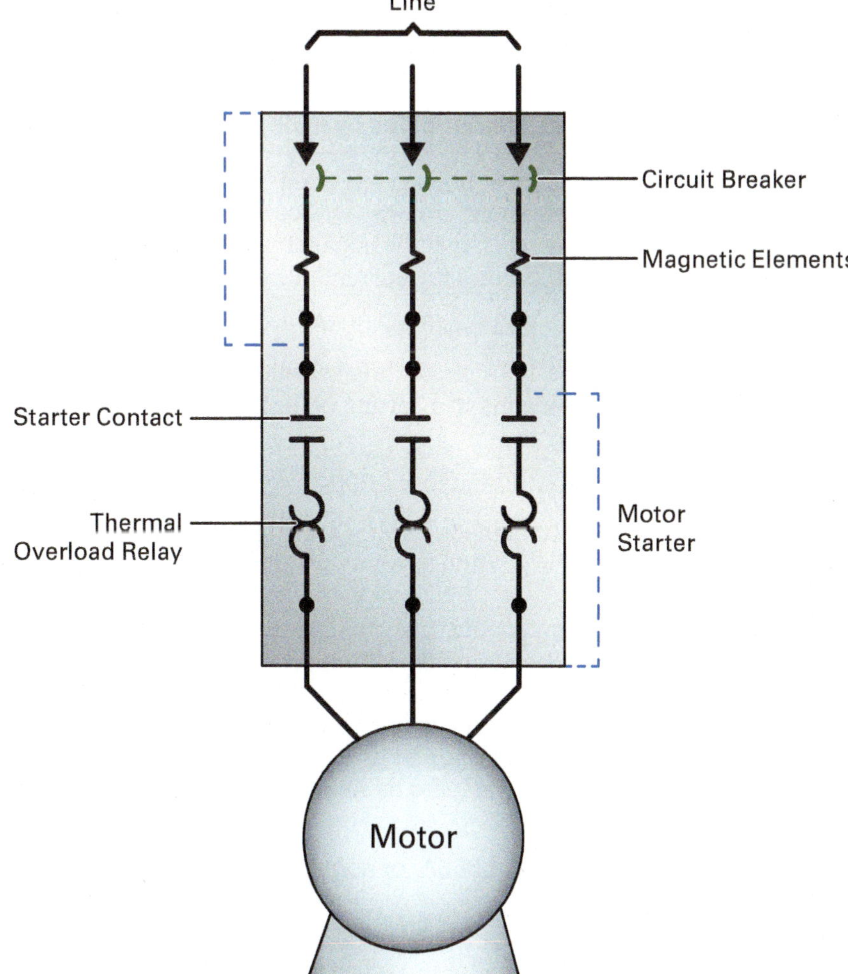

Figure 19 Combination motor starter with MCP.

3.4.1 Application of MCPs

We will compare the use of both thermal-magnetic and magnetic-only circuit breakers in the motor circuit shown in *Figure 20*. The task is to select a circuit breaker that will provide short circuit protection and qualify as the means of disconnect for the motor, as outlined in the following steps:

Step 1 Determine the motor full-load current from *NEC Table 430.250*. This is found to be 80A.

Step 2 A circuit breaker suitable for use as a motor disconnecting means must have a current rating of at least 115% of the motor FLA. Therefore, multiply the full-load current by 115%.

1.15 × 80A = 92A

NOTE

NEC Table 430.52 permits the use of an inverse-time circuit breaker rated at not more than 250% of the motor FLA. However, a circuit breaker could be rated as high as 400% of the motor FLA if necessary to manage the motor starting current, according to *NEC Section 430.52(C)(1)(b)(3)*.

Step 3 Assuming that a circuit breaker rated at 250% of the motor FLA will be used, multiply the full-load current by 250%.

2.5 × 80A = 200A

Step 4 Select a regular thermal-magnetic circuit breaker with a 225A frame that is set to trip at 200A.

Step 5 Refer to *Figure 20* and note that the motor is a NEMA Design B energy-efficient motor. Refer to *NEC Table 430.52(C)(1)* and note that an instantaneous breaker for a Design B motor should be no more than 1,100% of the motor FLA.

Step 6 Determine the circuit breaker rating by multiplying the FLA by 1,100%.

80A × 11 = 880A

The thermal-magnetic circuit breaker selected in *Step 4* will provide protection for grounds and short circuits without interfering with motor overload protection. Note, however, that the instantaneous trip setting of a 200A circuit breaker will be about 10 times the current rating, or:

$$200A \times 10 = 2{,}000A$$

Now consider the use of a 100A circuit breaker with thermal and adjustable magnetic trips. The instantaneous trip setting at 10 times the normal current rating would be:

$$100A \times 10 = 1{,}000A$$

Although this 1,000A instantaneous trip setting is above the locked-rotor current of the 30 hp motor, the starting current would probably cause the circuit breaker to trip.

If the setting of the instantaneous trip circuit breaker will not hold under the starting load in *NEC Section 430.52(C)(3)(b)(2)*, then the *NEC*®, under engineering evaluation, permits increasing the trip setting up to but not exceeding 1,700% for NEMA Design B energy-efficient motors.

Therefore, since it has been determined that this 30 hp motor has an FLA of 80A, the maximum trip current must not be set higher than 1,700% of the FLA, determined as follows:

$$80A \times 17 = 1{,}360A, \text{ or about } 1{,}300A$$

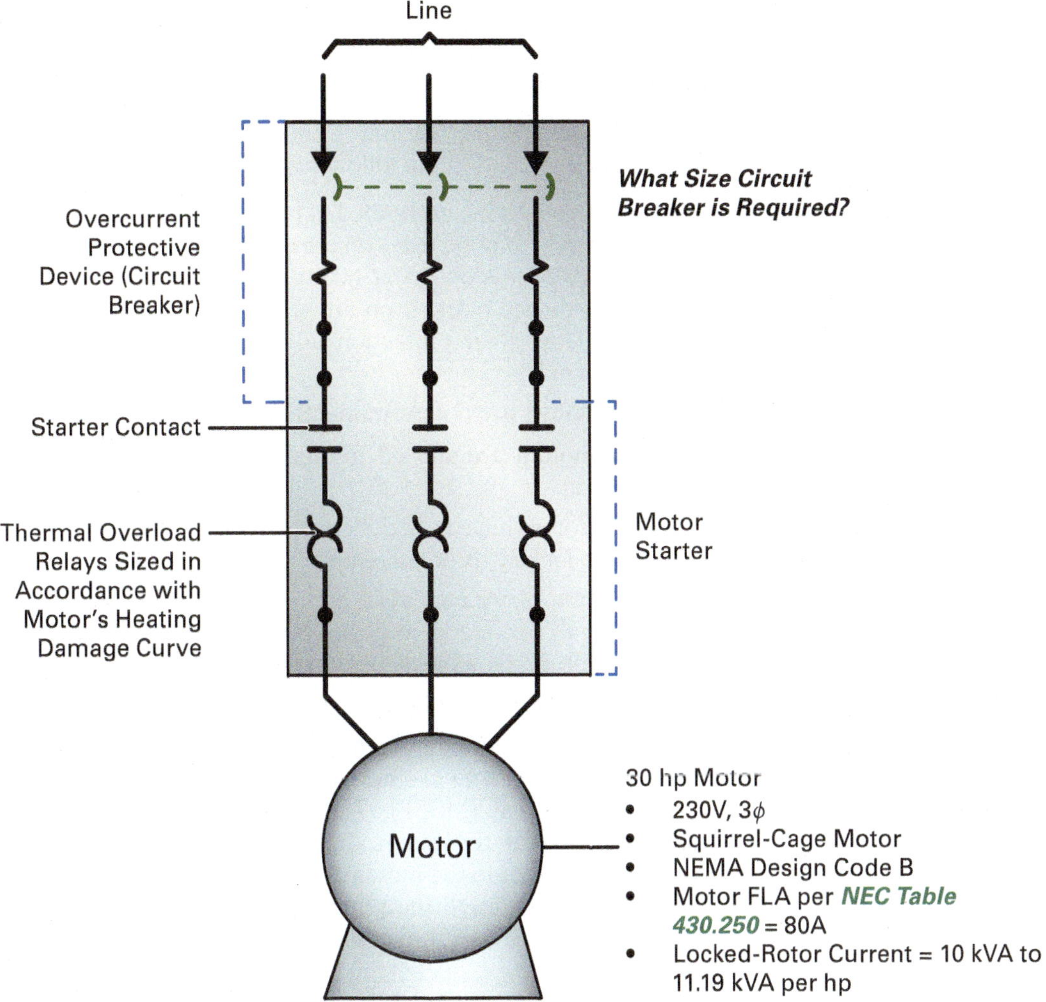

Figure 20 Typical 30 hp NEMA Design B energy-efficient motor circuit.

This means the circuit breaker determined previously would qualify as the circuit disconnect because it has a rating higher than 115% of the motor FLA. However, a magnetic-only circuit breaker does not protect against low-level grounds and short circuits in the branch circuit conductors on the line side of the motor starter overload relays. Such an application must be made only where the circuit breaker and motor starter are installed as a combination in a single enclosure.

3.4.2 Motor Short Circuit Protectors

Motor short circuit protectors (MSCPs) are fuse-like devices designed for use only in a special type of combination motor starter. The combination offers short circuit protection, overload protection, a disconnecting means, and motor control, all with assured coordination between the short circuit interrupter and the overload devices.

The *NEC*® recognizes MSCPs in *NEC Section 430.52(C)(7)*, provided the combination is identified for the purpose. This means it must be a combination motor starter equipped with an MSCP and be UL listed or listed by another nationally recognized third-party testing lab as a package called an *MSCP starter.*

Know the Code

Motor Short-Circuit Protector
NEC Section 430.52(C)(7)

3.5.0 Protective Devices for Multimotor Branch Circuits

Know the Code

Several Motors or Loads on One Branch Circuit
NEC Section 430.53

NEC Sections 430.53(A), (B), and (C) permit the use of more than one motor on a branch circuit, provided the following conditions are met:

- Two or more motors, each rated at not more than 1 hp and with an FLA not exceeding 6A, may be used on a branch circuit protected at not more than 20A at 120V or less or 15A at 1,000V or less. The rating of the branch circuit protective device marked on any of the controllers must not be exceeded. Individual overload protection is necessary unless one of the following conditions exists:
  - The motors are not permanently installed.
  - The motors are started manually and within sight of the controller location.
  - The motors have sufficient winding impedance to prevent overheating due to locked-rotor current.
  - The motors are part of an approved assembly that does not subject the motor to overloads, and the assembly also incorporates protection for the motor against locked-rotor current.
  - The motors cannot operate continuously under load.
- Two or more motors of any rating, each with individual overload protection, may be connected to a single branch circuit protected by a short circuit protective device (MSCP). The protective device must be selected in accordance with the maximum rating or setting that could protect an individual circuit to the motor of the smallest rating. This may be done only where it can be determined that the branch circuit device selected will not open under the maximum expected conditions of service.

These *NEC®* sections offer the wide application of more than one motor on a single circuit, particularly in the use of smaller motors installed on 208V, 240V, and 480V, three-phase commercial and industrial systems. Only these motors have an FLA low enough to permit more than one motor on a circuit fed from 15A protective devices.

Know the Code

Conductor Ampacity and Motor Rating Determination
NEC Section 430.6

Using these *NEC®* rules, we will consider a typical branch circuit with more than one motor connected (*Figure 21*) and see how the calculations are made.

The full-load current of each motor is taken from *NEC Table 430.250*, as required by *NEC Section 430.6(A)(1)*. A circuit breaker must be chosen that does not exceed the maximum 250% value of short circuit protection required by *NEC Section 430.52* and *NEC Table 430.52(C)(1)* for the smallest motor in the group. In this case, it is 1.5 hp. Since the listed FLA for the smallest motor here is 3A, the calculation is as follows:

$$3A \times 2.5 = 7.5A$$

The total load of the motor currents is calculated as follows:

$$4.8A + 3.4A + 3.0A = 11.2A$$

The total 11.2A full-load current for the three motors is well within the 15A circuit breaker rating, which has a sufficient time delay in its operation to permit starting of any one of these motors with the other two already operating. The torque characteristics of the loads on starting are not high. Therefore, the circuit breaker will not open under the most severe normal service. However, be certain that each motor is provided with the properly rated individual overload protection in their individual motor starters.

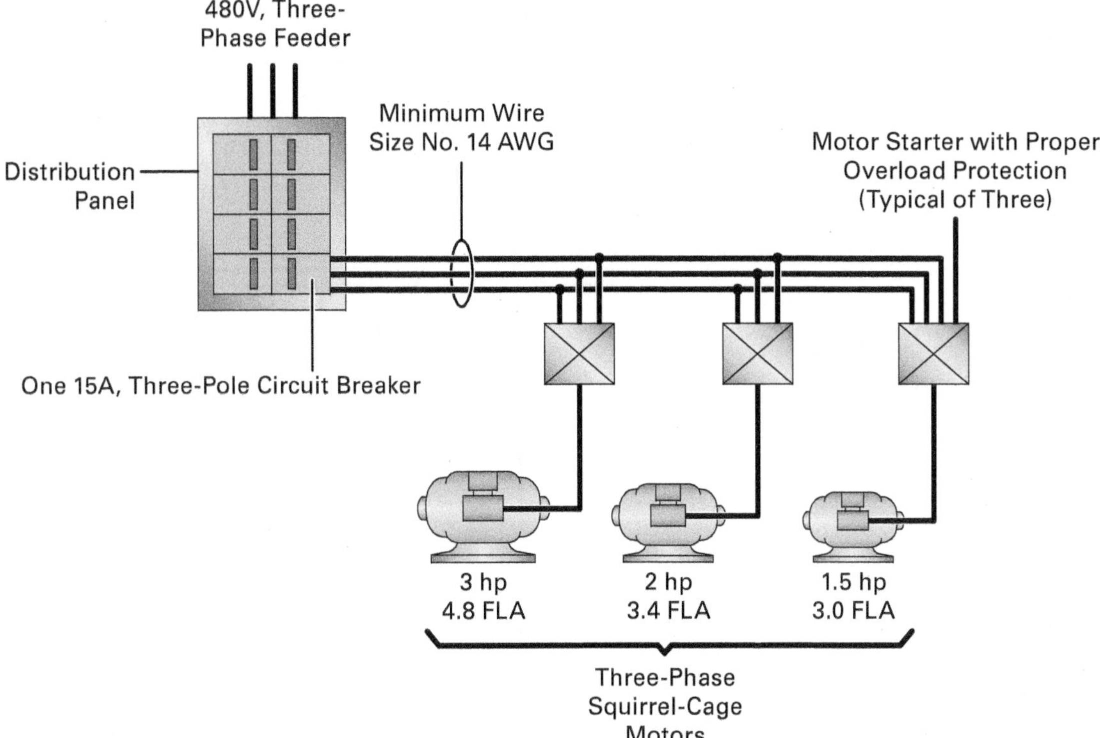

Figure 21 Several motors on one branch circuit.

Branch circuit conductors are sized in accordance with *NEC Section 430.24*. In this case, the following applies:

Conductor ampacity = 4.8A + 3.4A + 3.0A + 25% of largest motor

Conductor ampacity = 4.8A + 3.4A + 3.0A + (4.8A × 0.25)

Conductor ampacity = 4.8A + 3.4A + 3.0A + 1.2A = 12.4A

No. 14 AWG conductors rated at 167°F (75°C) will fully satisfy this application.

Another multimotor situation is shown in *Figure* 22. In this case, smaller motors are used. In general, *NEC Section 430.53(B)* requires branch circuit protection to be no greater than the maximum current permitted by *NEC Section 430.52* for the lowest-rated motor of the group. In this case, it is 1.1A for the 0.5 hp motors. With this in mind, we will size the circuit components.

From *NEC Section 430.52* and *NEC Table 430.52(C)(1)*, the maximum protection rating for a circuit breaker is 250% of the lowest-rated motor. Since this rating is 1.1A, the calculation is performed as follows:

2.5 × 1.1A = 2.75A

These two previous applications permit the use of several motors up to the circuit capacity. This is based on *NEC Sections 430.24 and 430.53(B)* and their starting torque characteristics, operating duty cycles of the motors, and the time delay of the circuit breaker. Such applications greatly reduce the number of circuit breakers and panels as well as the amount of wire used in the total system.

One limitation, however, is placed on this practice in *NEC Section 430.52(C)(2)*. It refers to the maximum branch short circuit and ground fault protective device ratings. When shown in the manufacturer's overload relay table for use with a motor controller or marked on the equipment, the ratings cannot be exceeded even if higher values are allowed.

Know the Code

Several Motors or a Motor(s) and Other Load(s)

NEC Section 430.24

Think About It

Motor Branch Circuit Requirements

If the motors in *Figure 22* were replaced with 5 hp, 3 hp, and 2 hp squirrel-cage motors, what size feeder and circuit breaker would be needed to accommodate the changes?

NOTE

Since 2.75A is not a standard rating for a circuit breaker, *NEC Section 430.52(C)(1)(A)* permits the use of the next higher rating. Because 10A is now the lowest standard rating of circuit breakers, it is the next higher device rating above 2.75A and satisfies *NEC®* rules governing the rating of the branch circuit protection.

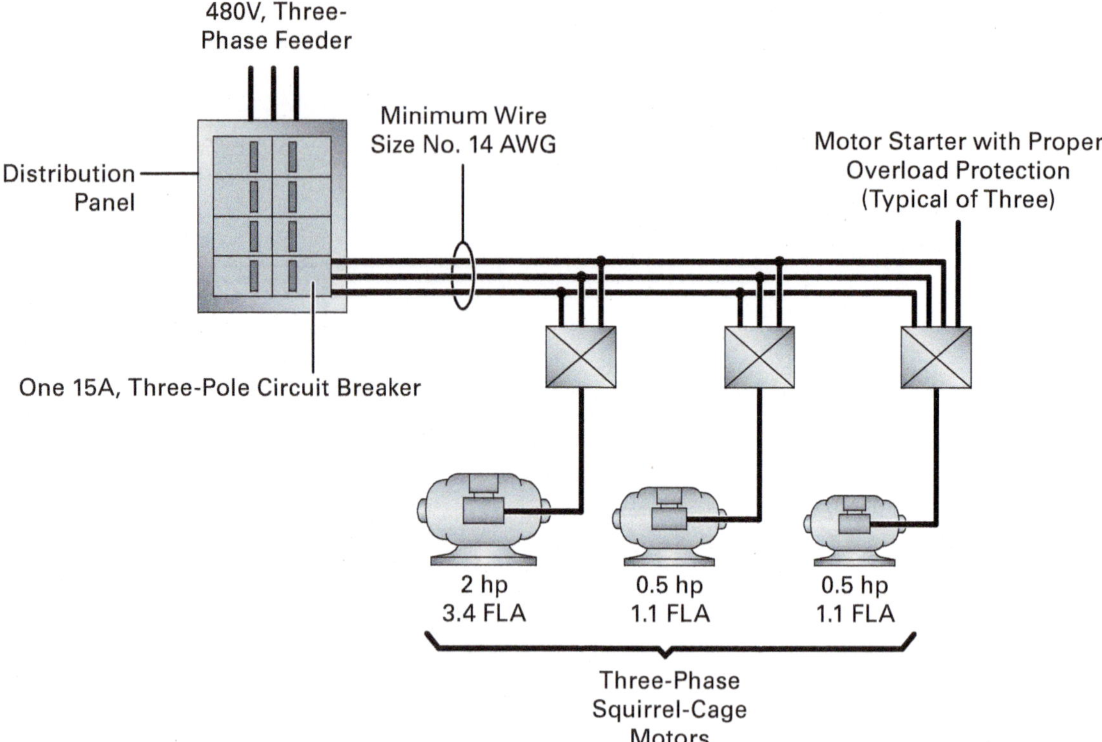

Figure 22 Several smaller motors supplied by one branch circuit.

3.6.0 Equipment Grounding Conductors

Know the Code

Size of Equipment Grounding Conductors
NEC Section 250.122

NEC Section 250.122(A) states that wire-based equipment grounding conductors must not be smaller than the size shown in *NEC Table 250.122*. For branch motor circuits serving continuous duty motors, the sizing is based on the overcurrent protective device, but it doesn't need to be larger than the circuit conductors supplying the equipment. Additionally, where the overcurrent protection is an instantaneous trip device, the equipment grounding conductor is permitted to be based on the rating of the overload protective device, but still not less than the size shown in *NEC Table 250.122* for that rating.

The equipment grounding conductor for a motor feeder circuit furnishing the branch circuits for continuous duty motors is determined by combining the rating of each branch circuit overcurrent protective device and, when applicable, the rating of each overload protective device.

3.7.0 Capacitors for Power Factor Correction

Generally, the most effective method of power factor correction is the installation of capacitors at the cause of the poor power factor—the induction motor. Doing so not only increases the power factor but also releases system capacity, improves voltage stability, and reduces power losses.

When power factor correction capacitors are used, the total corrective kilovolt-amperes reactive (kVAR) on the load side of the motor controller should not exceed the value required to raise the no-load power factor to unity. Corrective kVAR applied that exceeds this value may cause over-excitation, resulting in high transient voltages, currents, and torques that can create safety hazards and possibly damage the motor.

Potential locations for capacitors in the circuit are shown in *Figure 23*. If possible, capacitors should be located at position No. 2. In this position, they do not affect the current flowing through the motor overload protectors.

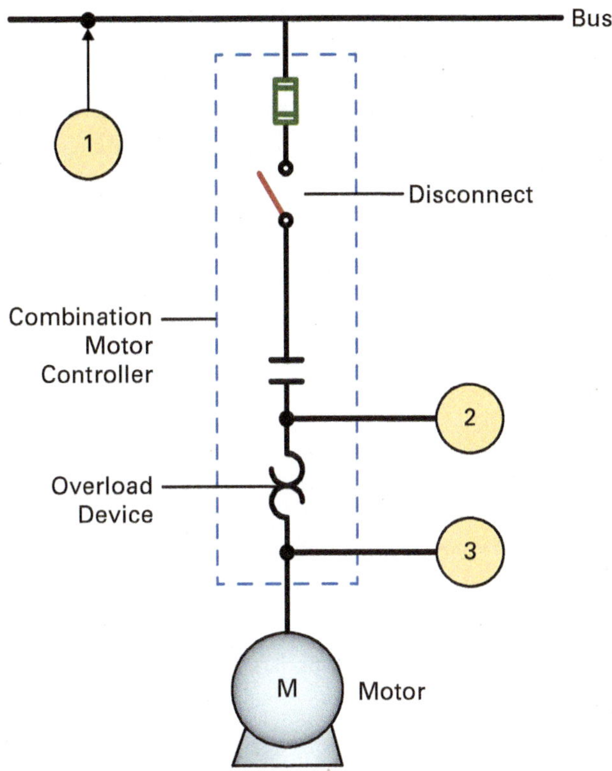

Figure 23 Placement of capacitors in motor circuit.

The connection of capacitors at position No. 3 requires a change of overload protectors since it is between the protective devices and the motor. Capacitors should be located at position No. 1 for any of the following applications:

- Elevator motors
- Multispeed motors
- Plugging or jogging applications
- Open transition, wye-delta, and autotransformer motor starters
- Some part-winding motors

Table 1A and *Table 1B* allow the determination of corrective kVAR required where capacitors are individually connected at motor leads. These values should be considered the maximum capacitor rating when the motor and capacitor are switched as a unit. The figures given are for three-phase, 60 Hz, NEMA Design B motors to raise the full-load power factor to 95%.

CAUTION

To avoid over-excitation, make sure that the bus power factor is not increased above 95% under all loading conditions. Do not connect power-factor correction capacitors at the motor terminals of the types listed above.

TABLE 1A Motor Power Factor Correction Table (3,600 rpm to 1,200 rpm)

Induction Motor (hp)	3,600 rpm		1,800 rpm		1,200 rpm	
	Capacitor Rating (kVAR)	Line Current % Reduction	Capacitor Rating (kVAR)	Line Current % Reduction	Capacitor Rating (kVAR)	Line Current % Reduction
3	1.5	14	1.5	15	1.5	20
5	2	12	2	13	2	17
7.5	2.5	11	2.5	12	3	15
10	3	10	3	11	3.5	14
15	4	9	4	10	5	13
20	5	9	5	10	6.5	12
25	6	9	6	10	7.5	11
30	7	8	7	9	9	11
40	9	8	9	9	11	10
50	12	8	11	9	13	10
60	14	8	14	8	15	10
75	17	8	16	8	18	10
100	22	8	21	8	25	9
125	27	8	26	8	30	9
150	32.5	8	30	8	35	9
200	40	8	37.5	8	42.5	9
250	50	8	45	7	52.5	8

TABLE 1B Motor Power Factor Correction Table (900 rpm to 600 rpm)

Induction Motor (hp)	900 rpm		720 rpm		600 rpm	
	Capacitor Rating (kVAR)	Line Current % Reduction	Capacitor Rating (kVAR)	Line Current % Reduction	Capacitor Rating (kVAR)	Line Current % Reduction
3	2	27	2.5	35	3.5	41
5	3	25	4	32	4.5	37
7.5	4	22	5.5	30	6	34
10	5	21	6.5	27	7.5	31
15	6.5	18	8	23	9.5	27
20	7.5	16	9	21	12	25
25	9	15	11	20	14	23
30	10	14	12	18	16	22
40	12	13	15	16	20	20
50	15	12	19	15	24	19
60	18	11	22	15	27	19
75	21	10	26	14	32.5	18
100	27	10	32.5	13	40	17
125	32.5	10	40	13	47.5	16
150	37.5	10	47.5	12	52.5	15
200	47.5	10	60	12	65	14
250	57.5	9	70	11	77.5	13

3.0.0 Section Review

1. The *maximum* size of a dual-element, time-delay fuse protecting a wound-rotor motor is _____.
   a. 100% of the motor's FLA
   b. 150% of the motor's FLA
   c. 175% of the motor's FLA
   d. 300% of the motor's FLA

2. It may be necessary to increase the size of a dual-element fuse for an oversized motor when the _____.
   a. motor operates for long periods without cycling on and off
   b. ambient temperature is 72°F (22°C)
   c. motor is connected to a machine that can be brought up to speed very quickly
   d. motor is frequently reversed

3. The locked-rotor current of a motor is typically _____.
   a. 2 times the normal running current
   b. 6 times the normal running current
   c. 12 times the normal running current
   d. 18 times the normal running current

4. The *minimum* circuit breaker size that can be used as the disconnecting means for a motor with an FLA of 20A is _____.
   a. 10A
   b. 23A
   c. 28A
   d. 30A

5. For two or more motors to be fed from the same branch circuit, each motor must be rated at *no* more than _____.
   a. 1 hp
   b. 3 hp
   c. 5 hp
   d. 10 hp

6. The *minimum* size equipment grounding conductor for a motor circuit can be found in _____.
   a. *NEC Table 250.122*
   b. *NEC Table 310.16*
   c. *NEC Table 430.250*
   d. *NEC Chapter 9, Table 8*

7. When calculating the power factor correction for a motor with a nominal speed of 1,200 rpm and a rating of 10 hp, the line current reduction should be _____.
   a. 20%
   b. 14%
   c. 11%
   d. 9%

Module 26309-23 Review Questions

1. Stator windings can be *best* described as _____.
   a. a structure of copper or aluminum coils or bars
   b. insulating fibers mounted on a shaft
   c. copper bars mounted on a spindle
   d. steel wires mounted on a shaft

2. Three-phase motor types are identified by their type of _____.
   a. stator
   b. rotor
   c. voltage
   d. rpm

3. The synchronous speed of a motor is the _____.
   a. speed at which the magnetic field rotates
   b. maximum motor speed
   c. speed at which the power factor is 0.7
   d. minimum motor speed

4. The synchronous speed of a 60 Hz, three-phase induction motor with two poles is _____.
   a. 3,600 rpm
   b. 1,800 rpm
   c. 1,200 rpm
   d. 900 rpm

5. The two most common three-phase motor configurations are _____.
   a. synchronous and rms
   b. box and star
   c. wye and delta
   d. star and wye

6. The *most* common number of motor leads found on a three-phase, wye-connected motor is _____.
   a. 3
   b. 6
   c. 9
   d. 12

7. Doubling the voltage on a dual-voltage motor _____.
   a. doubles the synchronous speed
   b. doubles the full-load current
   c. reduces the full-load current by half
   d. reduces the synchronous speed by half

8. If the stator windings of a dual-voltage motor both have an impedance of 10Ω and they are connected in parallel, the total impedance is _____.
   a. 5Ω
   b. 10Ω
   c. 15Ω
   d. 20Ω

9. With 240V applied to a total parallel resistance of 5Ω, the current is _____.
   a. 12A
   b. 24A
   c. 36A
   d. 48A

10. If two 10Ω windings are connected in series for operation on 480V, the total resistance is _____.
    a. 5Ω
    b. 10Ω
    c. 15Ω
    d. 20Ω

11. Assume that the stator windings of a dual-voltage motor each have a resistance of 96Ω. If the stator windings are connected in parallel for low-voltage operation, the total resistance (R_T) is _____.
    a. 48Ω
    b. 96Ω
    c. 192Ω
    d. 220Ω

12. If a dual-voltage squirrel-cage induction motor consumes 2A of current at 240V and is then connected to a 480V power source, the new current will be _____.
    a. 1A
    b. 2A
    c. 3A
    d. 4A

13. Of the following motors, the one *most* likely to have 15 or 18 motor leads is the _____.
    a. single-phase capacitor-start motor
    b. 120V shaded-pole motor
    c. 480V three-phase squirrel-cage motor
    d. 2,400V three-phase motor

14. The main purpose of motor protective devices that use thermal elements to sense current, which are found between the motor starter contacts and the motor itself, is to protect the motor from _____.
    a. short circuits
    b. ground faults
    c. damaging overload current
    d. locked-rotor current

15. A circuit breaker suitable for use as a disconnect switch must have a current rating of at *least* _____.
    a. 115% of the motor's FLA
    b. 175% of the motor's FLA
    c. 250% of the motor's FLA
    d. 300% of the motor's FLA

Answers to odd-numbered Module Review Questions are found in *Appendix A*.

Module 26309-23 Supplemental Exercises

1. Name the three basic types of three-phase motors.

   ____

   ____

2. What three factors cause the magnetic field to rotate?

   ____

   ____

3. What are the two factors that determine the synchronous speed of a three-phase motor?

   ____

   ____

4. When a motor is operated at 240V, the current consumed by the motor is ________ the current draw of a 480V connection.

5. If 480V is applied to 20Ω, what current will flow?

   ____

   ____

6. *NEC Section 430.22* states that for a single continuous duty motor, the conductors must have an ampacity greater than or equal to ________ of the motor's full-load current.

7. An MCC serves three motors, each with a rating of 10 hp and an FLA of 14A. What is the minimum feeder size?

   ____

   ____

8. The primary full-load current ratings for wound-rotor motors are listed in ________.

9. When sizing a dual-element, time-delay fuse for a DC motor, do *not* exceed ________ of the motor's FLA rating.

10. True or False? An MSCP is designed for use only in a special type of combination motor starter.

Answers to odd-numbered Supplemental Exercises are found in *Appendix B*.

Answers to Section Review Questions

Answer	Section Reference	Objective
Section 1.0.0		
1. b	1.1.0	1a
2. b	1.2.0	1b
Section 2.0.0		
1. c	2.1.0	2a
2. b	2.2.3	2b
Section 3.0.0		
1. b	3.1.0	3a
2. d	3.2.0	3b
3. b	3.3.0	3c
4. b	3.4.1	3d
5. a	3.5.0	3e
6. a	3.6.0	3f
7. b	3.7.0; *Table 1*	3g

Section Review Calculations

Section 2.0.0

Question 1

Find 125% of the current rating for the largest motor in the group and add the result to the current rating for the smaller motor.

$(1.25 \times 65A) + 14A = 95.25A$

The minimum feeder ampacity is **95.25A**.

Section 3.0.0

Question 4

Calculate 115% of the motor FLA to determine the minimum current rating.

$20A \times 1.15 = 23A$

The circuit breaker must have a current rating of at least **23A**.

User Update

Did you find an error? Submit a correction by visiting **https://www.nccer.org/olf** or by scanning the QR code using your mobile device.

MODULE 26310-23

Voice, Data, and Video

Source: iStock@kynny

Objectives

Successful completion of this module prepares you to do the following:

1. Recognize various types of structured cabling systems.
    a. Identify campus backbone subsystems.
    b. Identify equipment/telecom room subsystems.
    c. Identify riser subsystems.
    d. Identify horizontal subsystems.
    e. Identify work area subsystems.
    f. Identify distributed antenna systems.
2. Describe how to install and terminate various types of cable.
    a. Explain how to terminate UTP jacks and plugs.
    b. Explain how to terminate RG-6 F-type coaxial cable.
    c. Describe how to install fiber optic cable.
    d. Summarize how to ground and test VDV systems.

Performance Tasks

This is a knowledge-based module. There are no Performance Tasks.

Overview

A structured cabling system is a system in which the main components, which are divided into five subsystems, do not change. Electricians are often involved in the installation of structured voice, data, and video cabling systems in various commercial applications, with each subsystem having its own specific installation requirements. This module covers the installation, termination, and testing of these systems.

NOTE

NFPA 70®, *National Electrical Code®*, and *NEC®* are registered trademarks of the National Fire Protection Association, Quincy, MA.

Digital Resources for Electrical

Scan this code using the camera on your phone or mobile device to view the digital resources related to this craft.

NCCER Industry-Recognized Credentials

If you are training through an NCCER-accredited sponsor, you may be eligible for credentials from NCCER. The ID number for this module is 26310-23. Note that this module may have been used in other NCCER curricula and may apply to other level completions. Contact NCCER at 1.888.622.3720 or go to **www.nccer.org** for more information.

You can also show off your industry-recognized credentials online with NCCER's digital badges. Transform your knowledge, skills, and achievements into badges that you can share across social media platforms, send to your network, and add to your resume. For more information, visit **www.nccer.org**.

1.0.0 Structured Cabling Systems

Performance Tasks

There are no Performance Tasks in this section.

Objective

Recognize various types of structured cablins per second (Mbit/s)g systems.

a. Identify roughly campus backbone subsystems.
b. Identify equipment/telecom room subsystems.
c. Identify riser subsystems.
d. Identify horizontal subsystems.
e. Identify work area subsystems.
f. Identify distributed antenna systems.

As an electrician, you may be involved in the installation of structured voice, data, and video (VDV) cabling systems in various commercial applications. Structured cabling systems consist of cable, jacks, faceplates, patch or adapter panels (depending on cable type), cable trays, conduit systems, boxes, enclosures, and all associated support and suspension systems. Structured cabling systems may use unshielded twisted pair (UTP) cable, shielded twisted pair (STP) cable, **coaxial cable** (coax), or fiber optic cable, depending on the system design.

UTP cable has been used for many years for both voice and data transmission. It consists of anywhere from one twisted pair to 1,800 twisted pairs of solid copper conductors. The conductors range in size from No. 24 AWG to No. 22 AWG. In cables exceeding 600 pairs, an overall aluminum-steel shield is used to enclose the cable conductors. Each pair of conductors has a nominal impedance of 100Ω. UTP cable is available in the following categories with specified transmission rates that are recognized by the **American National Standards Institute/Telecommunications Industry Association (ANSI/TIA)** in the current version of the *ANSI/TIA 568* standard:

- *Category 3 cable* — Category 3 cable, also called *Cat 3 cable*, with transmission rates up to 16 MHz has historically been used for **Ethernet** networks with speeds of 10 **megabits per second (Mbit/s)**. It is not recommended for use in new installations except for analog data or voice-only applications.
- *Category 5e cable* — Category 5e cable (Cat 5e cable) is an enhanced version of Cat 5 cable with transmission rates up to and exceeding 100 MHz. Cat 5 cable is no longer recognized for use in new installations but exists in present residential and commercial installations. Cat 5e is used for Ethernet networks with speeds of 100 Mbit/s and sometimes Ethernet networks with speeds of a **gigabit per second (Gbit/s)** but is rapidly being replaced by newer cable types with higher transmission rates.

Coaxial cable: A type of cable that has a center conductor separated from the outer conductor by a dielectric. Coaxial cable is used to transmit very small currents or high-frequency signals and is commonly used in closed-circuit television (CCTV) systems.

American National Standards Institute/Telecommunications Industry Association (ANSI/TIA): ANSI oversees the development of voluntary consensus standards. TIA is a trade association representing the global information and communications technology industries. TIA is accredited by ANSI to develop voluntary industry standards that address telecommunications products, such as *ANSI/TIA 568*.

Megabits per second (Mbit/s): A rate of data transfer roughly equal to 1,000,000 bits per second.

Ethernet: A local area network (LAN) developed by Xerox Corporation that uses a bus or star topology. Ethernet is currently the most popular type of LAN.

Gigabit per second (Gbit/s): An extremely fast rate of data transfer roughly equal to 1,000,000,000 bits per second.

- *Category 6 cable* — Category 6 cable (Cat 6 cable) has transmission rates up to 250 MHz (typically considered a 1 Gbit/s data rate).
- *Category 6a cable* — Category 6a cable (Cat 6a cable) is an augmented version of Cat 6 cable. It has transmission rates up to 500 MHz (considered a 10-gigabit data rate). Cat 6a cable uses the same RJ-45 connector as Cat 6 but is thicker and bulkier.
- *Category 8 cable* — Category 8 cable (Cat 8 cable) offers transmission rates up to 2 GHz (2,000 MHz) and can support a speed of 25 Gbps (25GBase-T) or 40 Gbps (25Gbase-T). Cat 8 terminates in RJ-45 connections for Cat 8.1/Class I and non-R-J45 connections for Cat 8.2/Class II. Cat 8 is limited to a cable length of 98' (30 m). One of the benefits of Cat 8.1 is that its connector interfaces are backward compatible with the RJ-45 jacks (inserts) in existing cabling systems.

The video cables used in new installations are normally RG-6 coax types (RG means radio guide). RG-6 tri-shield and quad-shield are the most common, but dual-shield is also used. Dual-shield cable consists of a layer of foil usually bonded to the dielectric core surrounding the center conductor. The foil shield is covered with a braid shield followed by an insulation cover over the outside of the cable (foil/braid construction). Tri-shield cable includes a layer of foil shield over the braid shield of a dual-shield cable followed by the cable cover (foil/braid/foil construction). Quad-shield includes a braid shield over the foil of a tri-shield cable followed by the cable cover (foil/braid/foil/braid construction).

Structured cabling systems are divided into five subsystems, as shown in *Figure 1*. The subsystems include the campus **backbone**, equipment room, riser, horizontal cabling, and work area.

These subsystems are defined as follows:

- *Campus backbone subsystem* — The subsystem that provides **outside plant (OSP)** connectivity between buildings in a multibuilding campus environment. This includes the entrances for each building.
- *Equipment room subsystem* — The area where all subsystems for an individual building (or the entire campus) connect. This can be an equipment room (ER), telecommunications room (TR), or campus/building main **cross-connect** (MC).
- *Riser subsystem* — The blocks, cables, pathways, and hardware that provide connectivity between ERs, TRs, and MCs.
- *Horizontal cabling subsystem* — The patch panels, cables, support structures, and hardware that deliver voice/data connectivity to the end user outlet location.
- *Work area subsystem* — The telecommunications (telecom) outlets at the end user location and the associated patch cables providing final connectivity.

For each subsystem in a structured cabling system, there are specific requirements as to how the cable is to be run, the environmental protection that must be in place, standard support structures, expectations for workmanship, and standards for labeling. The following sections describe these requirements for each of the five subsystems.

NOTE

Cat 7 and Cat 7a cables are also available but are not currently recognized by the *ANSI/TIA 568* standard. Cat 7 cable offers transmission rates up to 600 MHz, while Cat 7a cable, an augmented version of Cat 7 cable, offers transmission rates up to 1,000 MHz. Both supply a data rate of 10 gigabits or more. Cat 7/7a cables can be terminated with GigaGate45 (GG-45) connectors, which are backward compatible with existing RJ-45 systems, or TERA connectors, which are not compatible with existing RJ-45 systems. The maximum cable length of Cat 7/7a cable is 328' (100 m).

NOTE

The type, style, size, and installation requirements for connectors and associated wiring vary widely by location. Always refer to the job specifications, as well as local codes and standards, when selecting and installing wiring and associated components.

NOTE

Any common pathway used for both telecom and power distribution, such as utility columns, must be equipped with a barrier and comply with applicable electrical codes. When a metallic barrier is used, it must be bonded to ground.

Backbone: A high-capacity network link between two sections of a network.

Outside plant (OSP): Refers to all the communications equipment (cable, conduit, cabinets, and hardware) located between switching facilities or between a switching facility and a customer.

Cross-connect: A facility for connecting cable runs, subsystems, and equipment using patch cords or jumper cable.

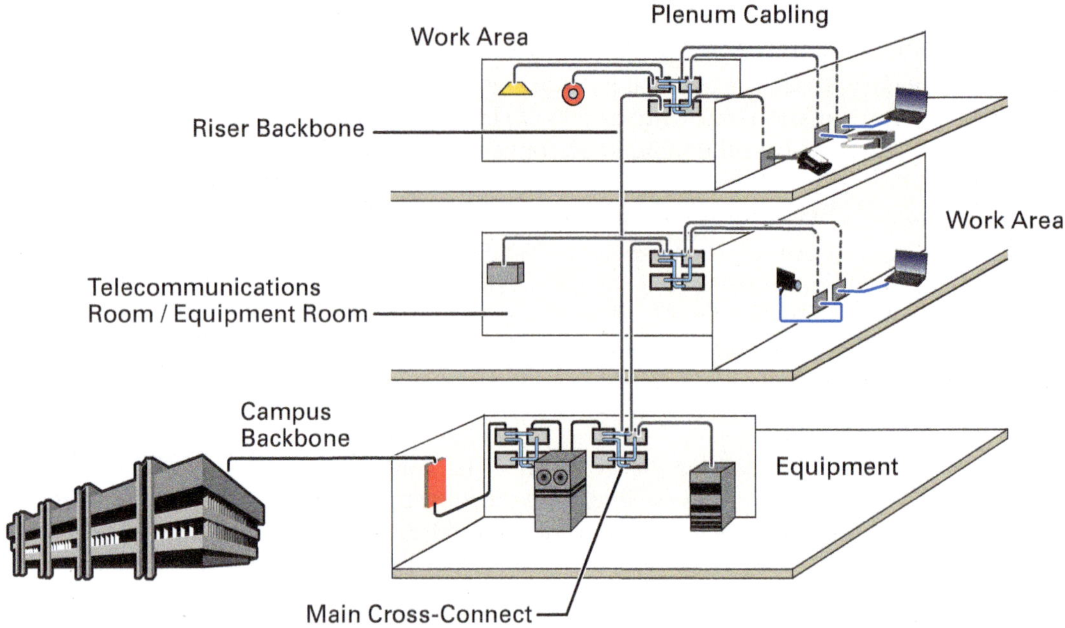

Figure 1 Diagram of a campus structured cabling system.

1.1.0 Campus Backbone Subsystems

The design of the backbone subsystem routes must reflect the most economical and flexible solution. Any hazards (roads, railroad tracks, ponds, seasonal waterways, etc.) that have the potential to cause an inconvenience when adding cables should be avoided. Pathways that have a greater potential for disturbance, such as those near sprinkler system manifolds or utility entrances (gas, water, electric, or community antenna television [CATV]), should be avoided. Any opportunity for others to disturb system pathways should be minimized.

When designing campus backbone conduit systems, install pull boxes (appropriately sized to the installed conduit) no more than 500' (152 m) apart on straight runs and after every 180° of bends (cumulative). Do not use standard 90° (Type L) conduit bodies (condulets) as they will make the pull difficult and may result in cable damage. If a 90° bend is unavoidable, special Smart conduit bodies can be used (*Figure 2*). Unlike traditional Type L conduit bodies that have a sharp 90° corner, Smart LB conduit bodies have a gentler bend radius that helps prevent cable damage and meets *ANSI/TIA 568* requirements for pulling optical fiber and twisted pair cable.

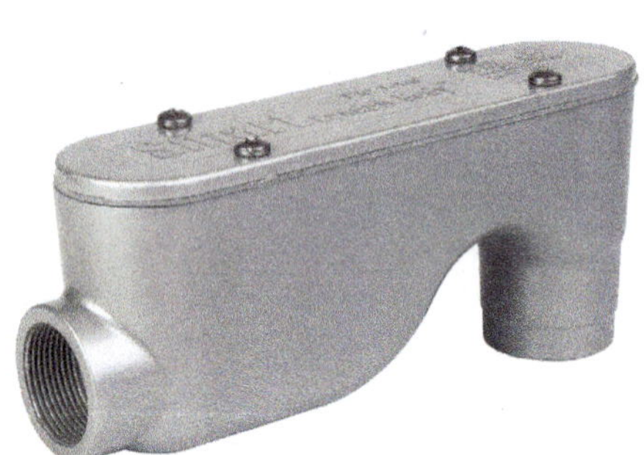

Figure 2 Smart Type L conduit body.
Source: Tri-City Electrical Contractors Inc

To minimize issues with water, maintain a negative slope away from the buildings. In addition, observe the following minimum distances for telecom conduit installation:

- Keep 3" (75 mm) away from power cables when encased in concrete, 4" (100 mm) away from power cables when surrounded by masonry, and 12" (300 mm) away from power cables when surrounded by well-tamped earth.
- Keep 6" (150 mm) away from pipes (gas, oil, water, etc.) when perpendicular and 12" (300 mm) away when parallel.

CAUTION

Refer to the authority having jurisdiction (AHJ) to determine the minimum burial depths for telecom conduit under roadways, railways, and other underground installations.

Aerial distribution systems must also be designed with safe clearances in mind. When designing aerial distribution systems, observe the following minimum attachment clearances for telecom cable:

- Keep 40" (1 m) away from all power cable.
- Keep 40" (1 m) away (+0.5" or 13 mm per kV over 8.7 kV) from open supply conductors.
- Keep 12" (300 mm) away from traffic signal and luminary drip loops.
- Keep 30" (750 mm) away from grounded supply equipment (transformers, etc.).

NOTE

Mid-span clearances must be 75% of the attachment clearances.

1.1.1 Environmental Protection

When installing nonmetallic (unarmored fiber) cabling underground, either direct buried or in nonmetallic pipe, add a metallic tape strip along the length of the cable to aid in locating the cable for future repair or maintenance. If metallic tape is not available, use copper wire (No. 18 AWG or larger) or two-pair or four-pair cable (No. 24 AWG).

All incoming conduit and **innerduct** should be plugged with water-blocking compounds or mechanical devices (jack plugs) at the last break (manhole, handhole, junction box, or weatherhead) before the building entrance, and at the first break (conduit stub, junction box, etc.) after the building entrance.

Innerduct: A type of high-density polyethylene (HDPE) conduit that is available in both corrugated and smooth types and is often placed inside a larger raceway to run communications cable.

Cables with a UV-resistant OSP sheath must be terminated and/or spliced to indoor-rated (riser or plenum) cables within 50' (15 m) of the building entrance or must access and run in conduit to its final termination point.

At the building entrance or the final termination point (whichever method was used), ground any metallic members in the sheath of the incoming cable to the building **telecommunications main grounding busbar (TMGB)**. Terminate individual metallic conductors on grounded lightning protection panels using solid-state protection fuses.

Telecommunications main grounding busbar (TMGB): A copper busbar typically located where the telecommunications cables enter the building. It serves as the grounding point for the telecommunications systems/equipment and is connected to the main electrical ground using an appropriately sized copper conductor.

1.1.2 Support Structure

OSP conduit must be buried a minimum of 24" (600 mm) below grade (or the frost line, depending on local climate). Live and dead load ratings must be calculated prior to installation. Soil type and backfill composition must be accounted for to ensure there is no conduit deformation. A soil bed may be required to provide a stable base to eliminate deformation.

When installing plastic or metallic OSP buried conduit (NEMA TC-6 and TC-8), all 90° bends must have a minimum sweep of 5' (1.5 m) and areas under vehicular traffic (roadways or railways) must be encased in concrete capable of withstanding 2,500 psi (17.2 megapascals or mPa).

When placing self-supporting cables aerially, use standardized tables to calculate span lengths according to the cable type (BHBS, BHAS, etc.), weight, and number of pairs. When supporting cables with metallic strand, never use 2.2M strand.

1.1.3 Workmanship

OSP pathways and cables for a campus backbone subsystem must be run in a parallel and perpendicular manner as much as possible. Avoid diagonal runs.

1.1.4 Labeling

All campus backbone subsystem cable, protection blocks, conduit, innerduct, handholes, and pedestals must be labeled. Labels must be machine-printed as large as reasonable, protected with a clear shield, and affixed to the following system components:

- Cable as close as possible to the point of termination
- Protection blocks at the top center of the front cover
- Conduit within 6" (150 mm) of the termination point
- Innerduct within 12" (300 mm) of the termination point
- Handhole cover(s)
- Pedestal at the top of the removable face or nearest the street-facing side

In addition, all OSP conduit must be labeled on each end and in each handhole, manhole, and pedestal. All innerduct within each conduit must be labeled on each end as well as any point where it exits the conduit, handhole, manhole, or pedestal.

Labeling

One of the most important aspects of cable installation is proper labeling. In fact, some jobs prohibit the use of handwritten labels. Labels are easy to produce using a handheld labeling machine, such as the one shown in the image.

1.2.0 Equipment/Telecom Room Subsystems

The design of the administrative subsystem routes requires a layout for all systems (fire alarm, A/V, security, access controls, etc.) to be housed in the equipment room subsystem area or MC. The first concern when laying out the MC is maintaining appropriate clearances from sources of electromagnetic interference (EMI). The following are general safe distance clearances for telecom cable:

- Keep 5" (125 mm) away from lighting.
- Keep 12" (300 mm) away from conduit and cables used for electrical distribution.
- Keep 48" (1.2 m) away from motors and transformers.

Lay out the MC backboard to minimize the number of times cross-connect (interconnect) cables must cross fields if using older technology punchdown blocks. Start left to right (preferable) or right to left and dedicate 16" (400 mm) vertically on the backboard for each carrier, switch in, switch out, all risers in order, and the MC field. If a wall-mounted switch is used, allocate 32" (800 mm) minimum between switch-in and switch-out fields.

Backboards should be constructed of heavy plywood painted on both sides and all four edges with two coats of fire-resistant paint. Minimize any opportunity for others to disturb system pathways. When designing administrative subsystem conduit systems, enter MC near the corners and keep the conduit at least 4" (100 mm) away from the nearest wall.

1.2.1 Environmental Protection

Equipment/telecom rooms contain sensitive electronic equipment that must be protected from temperature and humidity extremes. The room should be positively pressurized, and the heating, ventilation, and air conditioning (HVAC) system must be in service 24 hours a day, 365 days per year. Maintain a temperature of 64°F to 75°F (18°C to 24°C) with 30% to 55% relative humidity (noncondensing). Proper lighting is also essential. Paint the walls and ceiling white. Provide 50 footcandles (538 lux) of light when measured 36" (900 mm) above the floor.

All cables entering an MC from outdoors must be properly grounded. In addition, voice and data communication systems must include lightning protection. All incoming conduit and innerduct (non-OSP) should include **firestopping**. Firestopping must be UL Listed and installed in accordance with the job specifications.

Firestopping: A material or mechanical device used to block openings in walls, ceilings, or floors to prevent the passage of fire or smoke.

1.2.2 Support Structure

Whenever possible, do not use shared cable trays to distribute both telecom and power cables. If trays or wireways are shared, the power and telecom cables must be separated by a grounded metallic barrier.

Ladder racks should be mounted to allow for a minimum of 12" (300 mm) of clear vertical space above the tray. Ladder rack support systems include wall triangles or supports, pipe stands, auxiliary bar structures, equipment racks, independent threaded rod, or any combination of these. Support centers must be on 60" (1.5 m) centers (maximum span) unless designed for greater spans. A support must also be placed within 30" (750 mm) on each side of any connection to a fitting.

CAUTION

When installing supports, ensure that all fasteners seat securely in wall studs or solid anchor points. Failure to do so could result in collapse of the ladder rack system.

All metallic cable trays must be grounded but may also be used as grounding conductors. Clearly mark any cable tray that is used as an equipment grounding conductor.

1.2.3 Workmanship

Telecom cables must be routed and dressed in a neat and orderly manner to prevent misidentification, cable stress, and EMI. *Figure 3* shows cable trunk dressing. *Figure 4* shows proper cable termination. When using standard cross-connect wire (No. 24 AWG twisted pair), provide a two-finger slack at each block.

Figure 3 Cable trunk dressing.
Source: iStock@alacatr

Figure 4 Proper method of cable termination.
Source: Tri-City Electrical Contractors Inc

Install all cables with a service loop of up to 15' (4.6 m). The loop should be neatly coiled, labeled, and secured overhead. Cables on a ladder rack in an ER, TR, or MC should be neatly bundled using Velcro® straps. In the past, cable ties were commonly used but were often overtightened, resulting in changes to the internal wire twists and adversely affecting the performance of the cable. Many engineering firms now require the use of wide-format straps such as Velcro®. In addition, Velcro® is easy to undo if necessary. Cables must be dressed from the point of entry in the room to the termination point. Dressing means all cables are lying straight, next to each other, and with no knots, crossovers, or divers (a cable that crosses and disappears into the center of the bundle).

CAUTION

Excessive tightening of both cable ties and Velcro® straps can result in diminished cable performance. Tighten ties and straps only enough to provide neat and orderly cable dressing; do not overtighten.

Cables must be run on a support structure in the ER, TR, and MC. All routing must be parallel and perpendicular to horizontal, with no beelining or diagonal routing. Cables should be laid sequentially on the back of the patch panels. Field locations should also be laid out in a sequential manner whenever possible.

1.2.4 Labeling

All equipment/telecom room subsystem cable, conduit, innerduct, racks, and patch panels must be labeled. Labels must be machine-printed as large as reasonable, protected with a clear shield, and affixed to the following system components:

- Cable as close as possible to the point of termination
- Conduit within 6" (150 mm) of the termination point
- Innerduct within 12" (300 mm) of the termination point
- Rack top front cross member
- Patch panel at the top center of the front of the panel

In addition, all OSP conduit must be labeled on each end and in each handhole, manhole, and pedestal. All innerduct within each conduit must be labeled on each end as well as any point where it exits the conduit, handhole, manhole, pedestal, etc. Manholes must be labeled as to their function (i.e., communication, fiber optics, etc.) via an engraving on the lid.

1.3.0 Riser Subsystems

The routes for a riser subsystem should take into consideration the low-voltage services needed at each location (ER/TR/MC), such as 10/100/1000 for digital data. The connectivity requirements, including single-mode (SM) fiber, multimode (MM) fiber, coax, 100-pair UTP, and so on, must also be determined. Further, backbone pathway viability should be determined based on capacity requirements for all systems (fire alarm, A/V, security, access controls, etc.). Routes should be designed according to the following maximum distances:

- *Analog/digital voice* — 2,600' (792 m) on UTP
- *Digital data / voice over internet protocol (VOIP)* — 295' (90 m) on UTP
- *Digital data / VOIP* — 810' (247 m) on MM fiber
- **Closed-circuit television (CCTV)** — Distribution distance based on cable used

NOTE

These distances are rules of thumb and may vary by application.

Closed-circuit television (CCTV): The use of video cameras to transmit a signal to one or more television monitors.

Riser pathways should cross perpendicular to lighting and run no closer than 5" (125 mm). Keep telecom cables 12" (300 mm) away from electrical power cables or conduit and 48" (1.2 m) away from motors and transformers. Run riser pathways in the most direct route possible (through secure space vertically and secured in public space horizontally) and maintain the same vertical and/or horizontal plane as much as possible.

When installing riser sleeves, provide a minimum of three 4" (100 mm) sleeves between floors vertically and TRs horizontally. Allow a 4" (100 mm) offset from the nearest wall. Terminate conduit that protrudes through the structural floor 1" to 3" (25 mm to 75 mm) above the surface sides (this dimension may vary; always check the job specifications and manufacturer's requirements). Firestop all sleeves (newly installed or existing). Fill riser sleeves according to the firestop system specifications. Ream all sleeve ends and fit them with an insulated bushing.

When installing telecom cabling in a riser where the vertical run spans more than 30' (9 m) continuously, all cables must be secured in 30' (9 m) intervals (maximum span) using a split Kellems® grip or similar mechanical device. This may require installing vertical ladder racking to support high pair count copper riser cables. Use devices designed for this purpose, as improper riser support can degrade copper, fiber, and coax cable performance.

Riser conduit should be 48" (1.2 m) from motors or transformers, 12" (300 mm) from conduit and cables used for electrical power distribution, and 5" (125 mm) from lighting. All conduit should be reamed and bushed on each end and have no more than two 90° bends between pull points or pull boxes. As with campus backbone systems, conduit should contain no standard 90° (Type L) conduit bodies, as they will make the pull too difficult. If necessary, Smart Type L conduit bodies may be used. In addition, the conduit should contain no continuous sections longer than 100' (30 m) and be clean, dry, and unobstructed. Conduit must be bonded to common ground on one or both ends and terminated near the TR/ER corner of the room nearest the backboard. It should be capped/firestopped for protection upon installation, labeled for identification, and equipped with a plastic or nylon line.

See *Figure 5* for typical conduit fill information. Note that conduit fill may vary based on the brand and type of cable.

1.3.1 Environmental Protection

All cable runs in the riser subsystem should be rated for indoor use (CMR, CMP, etc.). If PVC or riser-rated cable is not specifically called for in the scope of work, assume that plenum-rated cables are required in the riser.

Firestopping must be installed in accordance with the job specifications. Any penetrations through walls or floors that are required for the riser pathways must be neatly sleeved and firestopped immediately upon completing the penetration. Size floor slots adequately but not extravagantly—remember that they will need to be firestopped. If a previously installed core hole / sleeve does not contain firestopping material, install it when the work is completed.

NOTE

The number of cables that can be installed horizontally is limited by the maximum allowable pulling tension on the cable. The number of cables that can be installed vertically is determined by the firestopping system requirements.

CAUTION

Do not run riser pathways over or adjacent to boilers, incinerators, hot water lines, or steam lines.

Figure 5 Conduit fill guide.

CAUTION

If trays or wireways must be shared, the power and telecom cables must be separated by a grounded metallic barrier.

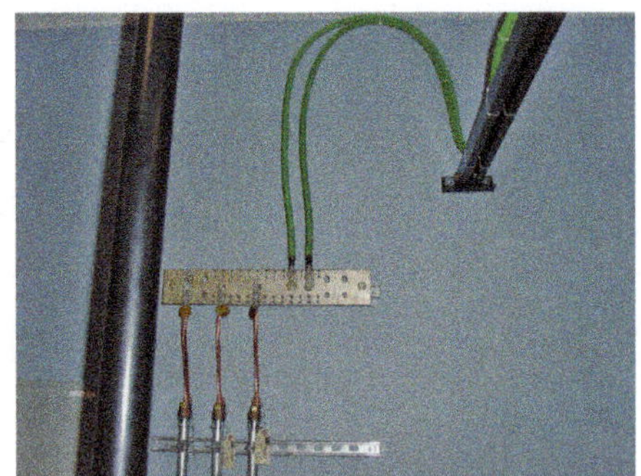

Figure 6 TMGB with conductors direct from the building ground ring.
Source: Tri-City Electrical Contractors Inc

Figure 7 Cable bundles secured with Velcro® straps.
Source: Tim Dean

CAUTION

When fiber optic cable is installed in innerduct without conduit, attach fiber optic warning tags to the innerduct every 7' (2.1 m).

1.3.2 Support Structure

Ladder racks or trays used in riser subsystems should be mounted at a minimum of 4" (100 mm) above the ceiling grid with a 12" (300 mm) vertical clearance above the side rails of the tray. Whenever possible, avoid using shared cable trays to distribute both telecom and power cables.

Riser cable support systems are not required to be continuous, but they must support the live loads exerted when placing cable. When installing telecom cabling in a riser where the vertical run spans more than 30' (9 m) continuously, all cables must be secured in 30' (9 m) intervals (maximum span) using a split Kellems® grip or similar mechanical device. This may require installing vertical ladder racking to support copper riser cables with a high pair count. Use devices designed for this purpose, as improper riser support can degrade copper, fiber, and coax cable performance.

When installing riser cables horizontally, the live loads exerted during cable placement require attention to lateral sway support. Field conditions, such as threaded rod diameter and length, will determine when and where this is required. Tray support centers must be on 60" (1.5 m) centers (maximum span) unless designed for greater spans. A support must also be placed within 30" (750 mm) on each side of any connection to a fitting. When using wall triangles or supports, ensure that fasteners anchor to wall studs.

All metallic cable trays must be grounded (*Figure 6*) but may also be used as ground conductors. Clearly mark any cable tray that is used as an equipment grounding conductor.

1.3.3 Workmanship

The riser support structure and cable must be straight, level, and plumb. Use sweeping 90° transitions with strict adherence to the cable manufacturer's bend radius. Cables should be dressed cable for cable and end to end as much as possible. Cables must be neatly bundled using Velcro® straps (*Figure 7*).

1.3.4 Labeling

All riser subsystem cable, conduit, and innerduct must be labeled. Labels must be machine-printed as large as reasonable, protected with a clear shield, and affixed to both ends of the following system components:

- Cable as close as possible to the point of termination (label must show pair count, cable start point, and cable termination point)
- Conduit within 6" (150 mm) of the termination point (label must note start point and conduit termination point)
- Innerduct within 12" (600 mm) of the termination point (label must list duct number, start point, and duct termination point)

1.4.0 Horizontal Subsystems

Routes for horizontal subsystems should be selected based on the best long-term route for the customer. Preferred routes will minimize interruption to the customer's daily operations. Typically, this means routing above hallways and common areas as opposed to directly overhead in offices or cubicle areas. Whenever possible, do not use shared cable trays to distribute telecom and power cables. If trays or wireways are shared, the power and telecom cables must be separated by a grounded metallic barrier.

Horizontal trunks should cross perpendicular to lighting and run no closer than 5" (125 mm). Keep telecom cables 12" (300 mm) away from electrical power cables or conduit and 48" (1.2 m) away from motors or transformers. Run conduit in the most direct route (over hallways as much as possible) and through nonsecure hallways and elsewhere when required by local code. Trunks must be a minimum of 4" (100 mm) above the ceiling grid and have a minimum 12" (300 mm) clear space directly above it. Remain on the same vertical or horizontal plane as much as possible. Minimize any opportunity for others to disturb system pathways. Provide segregation from other sources of EMI, such as low-voltage cables or systems.

When installing horizontal ER/TR/MC sleeves, provide a minimum of 3" (75 mm) between the ER, TR, or MC and the horizontal subsystem. Allow a minimum 4" (100 mm) offset above any office space ceiling grid. Also, allow a 4" (100 mm) offset from the nearest wall. Terminate sleeves that penetrate walls no less than 2" to 3" (50 mm to 75 mm) from the wall surface on each side. Adequately secure sleeves to prevent movement when cabling is installed. Firestop all sleeves (newly installed and existing) on both sides. Fill horizontal sleeves according to firestop system specifications. Ream all sleeve ends and fit them with an insulated bushing.

CAUTION

Never cut through a wall and run cable through the penetration without a sleeve.

Horizontal (field) conduit/stubs should be run in the most direct route possible (parallel and perpendicular to building lines). They should be located a minimum of 48" (1.2 m) away from motors or transformers, 12" (300 mm) from conduit and cable used for electrical power distribution, and 5" (125 mm) from lighting. In addition, they must be reamed and bushed on each end and have no more than two 90° bends between pull points or pull boxes. To ensure a smooth pull, they must be clean, dry, and unobstructed. They should contain no 90° conduit bodies and no continuous sections longer than 100' (30 m). They must be bonded to common ground on one or both ends (homerun conduit) and terminated near the tray/trunk. Terminations should be placed no closer than 6" (150 mm) on either side and vertically within 4" to 16" (100 mm to 400 mm) of the tray top, but not directly over it. The conduit must be bushed, capped, and firestopped upon finishing installation (*Figure 8*), labeled for identification, and equipped with a plastic or nylon pull cord with a minimum test rating of 200 lb (90 kg).

NOTE

Use flexible metal conduit only in situations where it is the only practical alternative. If flexible conduit must be used, increase the size by one trade size over the standard EMT size requirement.

When penetrating fire walls, consider using firestop products that allow for reentry and reuse. Allow a minimum 4" (100 mm) offset above any office space ceiling grid. Cut conduit that protrudes through the wall to leave 1" to 3" (25 mm to 75 mm) on both sides (this dimension may vary; always check the job specifications and manufacturer's requirements). Adequately secure the sleeve to prevent movement when cabling is installed. Use fire caulk to seal the rough opening around the conduit. Firestop all sleeves (newly installed and existing) according to the firestopping system and wall-rating specifications. Ream all conduit ends and fit them with an insulated bushing to eliminate sharp edges that can damage cables during installation or service. Where ladder or cable racking penetrates fire-rated walls, make the opening no bigger than necessary, and use fire bricks (also called *fire blocks*), firestop pillows (*Figure 9*), or other approved materials to stop the opening according to the firestop manufacturer's specifications.

When penetrating non-fire-rated walls, install sleeves but use standard caulk as opposed to fire caulk. When penetrating a ceiling (drop tile, gypsum, etc.), always secure conduit sleeves adequately to minimize movement when pulling cable.

1.4.1 Environmental Protection

All cables entering a plenum-rated horizontal subsystem must also be plenum-rated. Plenum subsystems require the use of plenum-rated Velcro® straps, supports, fasteners, and accessory equipment.

CAUTION

Do not run trunks over or adjacent to boilers, incinerators, hot water lines, or steam lines.

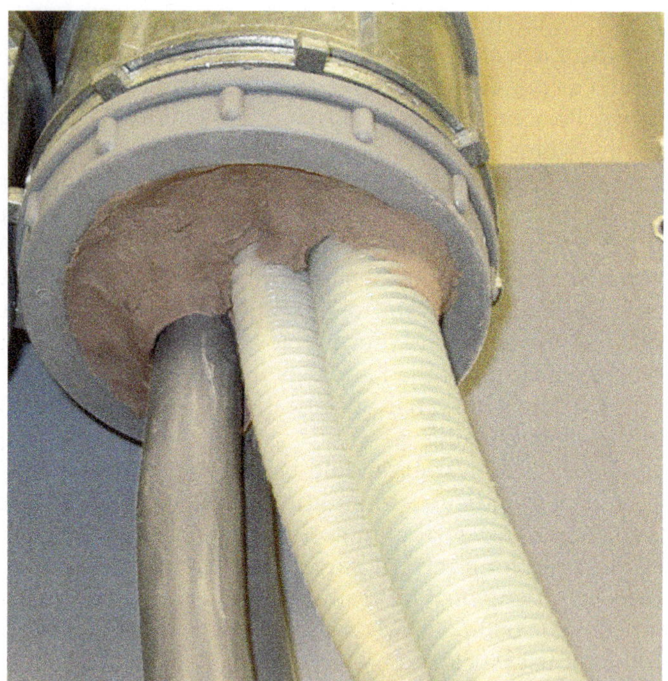

(A) Fire-Sealed Conduit

(B) Firestopping Sleeves

Figure 8 Firestopping.
Source: Tim Dean (8A); Hilti, Inc. (8B)

1.4.2 Support Structure

The support structure for a horizontal subsystem can consist of J-hooks, saddles, ladder racks, and cable trays. These items are explained as follows:

- *J-hooks and saddles* — When using J-hooks or saddles, locate them on 48" to 60" (1.2 m to 1.5 m) centers to adequately support and distribute the cable weight. These types of supports may typically hold up to fifty 0.25" (6.3 mm) diameter cables. Suspended cables must be installed with at least 3" (75 mm) of clear vertical space above the ceiling tiles and support channels (T-bars). For large quantities of cables (50 to 75) that convene at the telecom closet and other areas, provide cable trays or other special supports specifically designed to support the required cable weight and volume without adversely affecting cable performance. J-hooks must be wide to minimize strain on the cable.

Figure 9 Wall penetration protected by firestop pillows.
Source: Hilti, Inc.

- *Ladder racks and cable trays* — Ladder racks or cable trays should be mounted at least 4" (100 mm) above the top of ceiling tile grid. Allow for a minimum of 12" (300 mm) of clear vertical space above the tray. Ladder rack or tray support systems include independent threaded rod, wall triangles or supports, pipe stands, auxiliary bar structures, equipment racks, or any combination of these components.

All metallic cable trays must be grounded but may also be used as ground conductors. Clearly mark any cable tray that is used as an equipment grounding conductor. *Figure 10* shows a large-scale ladder rack / equipment rack grounding system. The ground cables are bonded using H-taps with covers, with approved cord securing the covers.

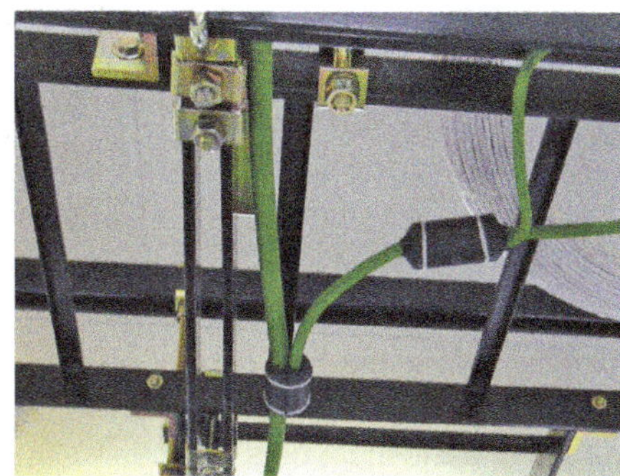

Figure 10 Ladder rack / equipment rack grounding system.
Source: Tim Dean

CAUTION

When installing supports, ensure that all fasteners seat securely in wall studs or solid anchor points. Failure to do so could result in collapse of the ladder rack system.

NOTE

Support stringers for horizontal cable support must be on 60" (1.5 m) centers (maximum span). A support must also be placed within 30" (750 mm) on each side of any connection to a ladder rack or tray splice.

1.4.3 Workmanship

The horizontal support structure and cable must be straight, level, and plumb. Use sweeping 90° transitions with strict adherence to the manufacturer's bend radius. Cables should be dressed cleanly with no visible kinks, knots, divers, holes, or wrapping. All cables must be installed with a service loop of up to 15' (4.6 m) on the ER, TR, and MC end, and 18" (450 mm) on the field end. The loop must be neatly coiled, labeled, and secured overhead. Cables in any horizontal support structure must be neatly bundled using Velcro® straps at 18" to 30" (450 mm to 750 mm) intervals. All routing must be parallel and perpendicular to horizontal, with no diagonal routing.

1.4.4 Labeling

All horizontal subsystem cable, conduit, innerduct, racks, and patch panels must be labeled. Labels must be machine-printed as large as reasonable, protected with a clear shield, and affixed to the following system components:

- Cable as close as possible to the point of termination
- Conduit within 6" (150 mm) of the termination point
- Innerduct within 12" (300 mm) of the termination point
- Rack top front cross member
- Patch panel at the top center of the front of the panel

NOTE

All innerduct within each conduit must be labeled on each end as well as at any point where it exits the conduit (pull boxes, junction boxes, etc.).

1.5.0 Work Area Subsystems

Determine the furniture layout before designing the work area subsystem. Locate outlet boxes within 36" (900 mm) of the electrical outlet for the intended device. Patch cables should be long enough to serve the intended device but no longer than necessary to minimize cable clutter.

The center line of outlets should be level with electrical outlets. The preferred configuration for a single wall outlet is ¾" EMT (MD 21) into a two-gang box with a single-gang adapter over the front. Wall phones should be mounted at the height indicated in the job specifications or, if not specified, in accordance with the Americans with Disabilities Act (ADA) requirements. Use stainless steel 8-pin modular faceplates.

When routing cables, maintain the following standard separations:

- 5" (125 mm) from lighting
- 12" (300 mm) from conduit and cables used for electrical distribution
- 48" (1.2 m) from motors and transformers

CAUTION

Do not share pathways with power. When installing telecom cables adjacent to power cables in modular furniture, there must be a grounded barrier between the two.

CAUTION

Never install surface-mounted boxes on furniture kick plates.

1.5.1 Environmental Protection

When installing boxes, leave no more than 6" to 12" (150 mm to 300 mm) of excess cable coiled within the box (two gang), with the service coil (where required) tied off above the conduit stub. When installing surface-mounted boxes under desktops, place split-wire loom over the exposed cables. When installing surface-mounted boxes, avoid positioning jacks so that the plug enters from directly above. This prevents dust from settling in the connection. In the case of floor outlets, make sure a spring-loaded cover is installed.

1.5.2 Support Structure

Wall boxes must be secured to the nearest structural member. If a new location must be cut in, box eliminators should be used. Stamped sheet metal with folding ears is not acceptable. Surface-mounted boxes and faceplates must be secured by mechanical means, such as screws, clips, snap-ins, etc. Double-sided tape is not acceptable. Utility columns that are used for both telecom and power distribution must be equipped with a barrier and comply with applicable electrical codes. When a metallic barrier is used, it must be bonded to ground.

1.5.3 Workmanship

All faceplates, surface-mounted boxes, etc., must be mounted flush, plumb, and level. Patch cables must be routed in such a manner as to limit cable clutter. All cables must be installed with a service loop. The loop must be neatly coiled, labeled, and secured overhead.

1.5.4 Labeling

All work area subsystem cables, faceplates, and jacks must be labeled. Labels must be machine-printed as large as reasonable, protected with a clear shield, and affixed to the cable as close as possible to the point of termination and in the windows of all faceplates and boxes (*Figure 11*).

1.6.0 Distributed Antenna Systems

Distributed antenna system (DAS): A network of antennas throughout a building that serves as a repeater to remedy areas of poor coverage inside the building.

A **distributed antenna system (DAS)** is a series of antennas in a building that disseminate a specific signal throughout the structure. In addition to providing enhanced coverage in large buildings, a DAS serves a critical safety function. NFPA 72®, *National Fire Alarm and Signaling Code®*, requires the use of a public safety DAS strictly for first responders during an emergency. The public safety DAS must have backup power in case of an outage, be on a different spectrum than the rest of the building's wireless network, and cover the entire building, including stairwells and basements.

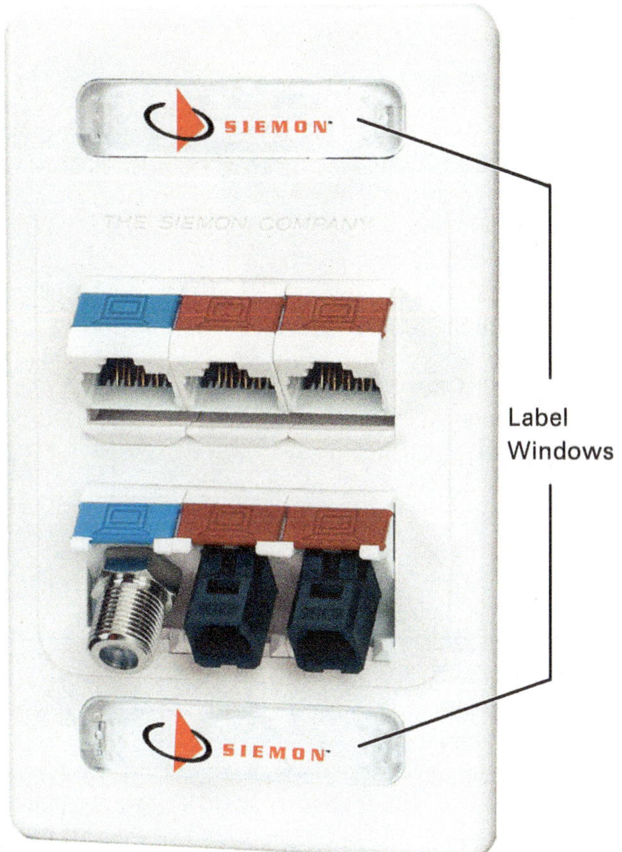

Figure 11 Label windows.
Source: Image courtesy of Siemon®

Distributed antenna systems are covered in NFPA 1221, *Standard for the Installation, Maintenance, and Use of Emergency Services Communications Systems.* DAS cable must be routed in conduit with a two-hour fire rating and with all passages firestopped with material that also has a minimum two-hour fire rating. A DAS replaces the fireman's jack in buildings to allow outside communications by building emergency services.

Some of the installation requirements for emergency services communications systems include the following:

- The DAS must cover 90% of the floor area for general building locations.
- The DAS must cover 99% of the floor area for critical locations (e.g., fire command centers, fire pump rooms, exit stairs, exit passageways, elevator lobbies, standpipe cabinets, sprinkler sectional valve locations, and other areas deemed critical by the AHJ).
- Antenna isolation must be provided to prevent a signal loop. The code sets the isolation at a minimum of 20 dB above the signal booster gain under all operating conditions.
- The signal must be strong enough to provide a minimum audio quality of 3.0, defined as speech understandable with a slight effort, and which requires occasional repetition because of noise and distortion.
- All active components (e.g., transmitters, receivers, repeaters, signal boosters, filters, and battery systems) must be housed in NEMA 4 or NEMA 4X enclosures.

Did You Know?

Bluetooth® Technology's "Hedy" Past

While the term *Bluetooth®* traces its name back to Harald "Bluetooth" Blatand, a tenth-century Danish King known for his ability to unify multiple countries, the technology itself is much younger. Jaap Haartsen and Sven Mattison are credited with perfecting the technology in the mid-1990s. What many do not know is that the technology's foundation was patented more than a half-century earlier by a Hollywood actress so beautiful that her face served as inspiration for Walt Disney's Snow White cartoon character. In the 1940s, Hedy Lamarr was considered by many to be one of the silver screen's most beautiful female actresses. Lamarr performed alongside famous actors like Clark Gable, Jimmy Stewart, and Spencer Tracey. Along with being an attractive actress, however, she was a pioneer in the wireless communications industry. Shortly after emigrating to the United States, Lamarr and co-inventor George Anthiel invented a technology intended to prevent the Nazis from jamming torpedo guidance signals. The design, which was patented in 1941, enabled a radio guidance transmitter and a torpedo's receiver to jump together from frequency to frequency. While their unique invention was initially rejected by the US Navy for being cumbersome, a refined version was ultimately installed on US ships patrolling the waters around Cuba during the Cuban Missile Crisis. Today, thanks to the work performed by Haartsen and Mattison, remnants of Lamarr's "frequency jumping" design are still included in billions of products using Bluetooth®, Wi-Fi®, GPS, and cellular technologies.

1.0.0 Section Review

1. When installing plastic or metallic OSP buried conduit (NEMA TC-6 and TC-8), all 90° bends must have a *minimum* sweep of _____.
   a. 2' (600 mm)
   b. 3' (900 mm)
   c. 4' (1.2 m)
   d. 5' (1.5 m)

2. In an ER or TR subsystem, the *minimum* clearance between telecom cables and motors is _____.
   a. 48" (1.2 m)
   b. 52" (1.3 m)
   c. 64" (1.6 m)
   d. 70" (1.8 m)

3. If trays or wireways must be shared, the power and telecom cables *must* be separated by a(n) _____.
   a. space of ½" (13 mm)
   b. clear plastic barrier
   c. grounded metallic barrier
   d. insulating pipe wrap

4. Innerduct on horizontal subsystems *must* be labeled within _____.
   a. 6" (150 mm) of the termination point
   b. 8" (200 mm) of the termination point
   c. 10" (250 mm) of the termination point
   d. 12" (300 mm) of the termination point

5. When installing a work area subsystem, outlet boxes should be located within _____.
   a. 36" (900 mm) of the electrical outlet for the intended device
   b. 44" (1.1 m) of the electrical outlet for the intended device
   c. 54" (1.4 m) of the electrical outlet for the intended device
   d. 60" (1.5 m) of the electrical outlet for the intended device

6. A DAS serves as a critical component of a _____.
   a. campus backbone system
   b. public safety communications system
   c. riser subsystem
   d. horizontal subsystem

2.0.0 Installation and Termination of Cable Systems

Objective

Describe how to install and terminate various types of cable.

a. Explain how to terminate UTP jacks and plugs.
b. Explain how to terminate RG-6 F-type coaxial cable.
c. Describe how to install fiber optic cable.
d. Summarize how to ground and test VDV systems.

Performance Tasks

There are no Performance Tasks in this section.

The type of termination required depends on the type and size of the conductor. UTP and coax cable terminations are made using jacks and plugs. F-type coax is terminated using a variety of crimp-on, twist-on, and other connectors. Fiber optic cable is terminated using mechanical connectors, fusion splicers, or field-installable epoxy connectors.

2.1.0 UTP Jacks and Plugs

UTP jacks and plugs for new structured cabling networks wired with Cat 5e, Cat 6, or Cat 6a UTP 4-pair cables must be wired in accordance with *ANSI/TIA 568A* and *568B* wiring schemes using Cat 5e, Cat 6, or Cat 6a RJ-45 modular jacks and mating RJ-45 plugs (RJ means registered jack) as required. The jacks are connected to horizontal homerun 4-pair UTP cables that originate from a patch panel and terminate in the work areas. The two wiring schemes, 568A and 568B, were originally included in *ANSI/TIA 568A* to accommodate installations that had been or were being installed to the 568B wiring scheme (formerly known as Western Electric Company [WECO] or AT&T Standard 258A). The preferred scheme for new installations was supposed to be the 568A wiring scheme; however, the most popular method used today is the 568B scheme because of the continued interface with equipment and systems that use the 568B scheme. This caused much confusion because, in the past, most patch panels and modular jacks were made using either the 568A or 568B scheme but were not labeled, and most suppliers carried only the 568B version. Today, however, almost all patch panels and modular jacks show diagrams for both versions (*Figure 12*). The only difference between the two is the interchanging of the pins used for the second and third pairs.

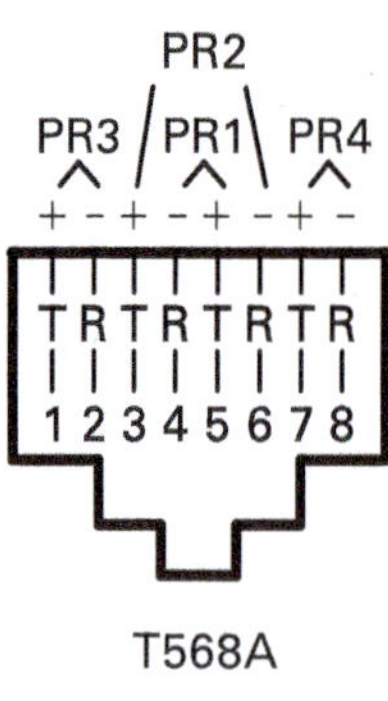

568A Wiring		
Pair #	**Wire**	**Pin #**
1 – White/Blue	White/Blue Blue/White	5 4
2 – White/Orange	White/Orange Orange/White	3 6
3 – White/Green	White/Green Green/White	1 2
4 – White/Brown	White/Brown Brown/White	7 8

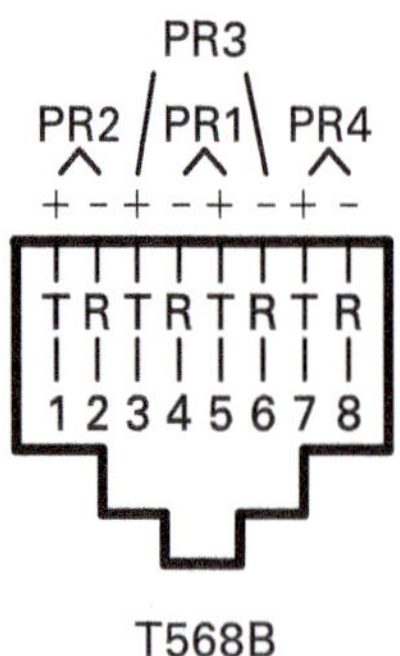

568B Wiring		
Pair #	**Wire**	**Pin #**
1 – White/Blue	White/Blue Blue/White	5 4
2 – White/Orange	White/Orange Orange/White	1 2
3 – White/Green	White/Green Green/White	3 6
4 – White/Brown	White/Brown Brown/White	7 8

Figure 12 *ANSI/TIA 568A* and *568B* wiring schemes.

For Cat 5e, Cat 6, and Cat 6a structured cable systems with 4-pair cables, typical 8-pin, 8-conductor (8P8C) RJ-45 modular jacks are used (*Figure 13*). Modular jacks are shaped so that they can accommodate 8-pin, 6-pin, or 4-pin plugs. Jacks may be crimped, punchdown types, or toolless (self-punching). For the toolless versions, a hinged cover is opened, and the wire pairs are inserted at the proper pins. When the cover is closed and latched, the wires are automatically punched down. To prevent system performance degradation, both types of jacks must be rated to match or exceed the category of UTP used for the system.

(A) Modular Connector Crimping Tools

(B) Punchdown with 110/66 Blade

Cat 6 Jacks Cat 6 Toolless Jack Cat 5e Toolless Jacks

(C) Jacks

Figure 13 RJ-45 modular jacks and tools.

For Cat 5e, Cat 6, and Cat 6a **patch cord** construction, appropriately rated 8P8C RJ-45 plugs are normally used. *Figure 14* shows typical instructions for constructing patch cords for Cat 6 cable.

Patch cord: A connecting cable used with cross-connect equipment and work area cables. It has three conductive portions separated by insulators.

For straight-through patch cords, it doesn't matter which color scheme is used for wiring the ends of the cord as long as the same scheme is used on both ends. Any scheme will pass the signal unchanged from one jack to another if both are wired to the same scheme. Today, most straight-through patch cords for Cat 5e, Cat 6, and Cat 6a cable are wired to the 568B scheme, including commercial patch cords, because most systems were previously wired using the 568B scheme.

However, when making a crossover patch cord, one end is wired to the 568A scheme and the other is wired to the 568B scheme. This allows signal transition from a modular jack wired to one scheme to another modular jack wired to the other scheme.

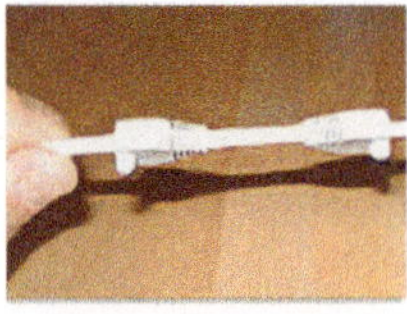

1. If you are planning to use boots, slide them onto the cable as shown. If you prefer not to use boots, start at Step 2.

Note: Making cables can be very labor intensive. Factory-made cables typically have better tolerances and consequently better quality than field-made cables.

2. Remove approximately 1½" of the cable jacket using a cable stripper.

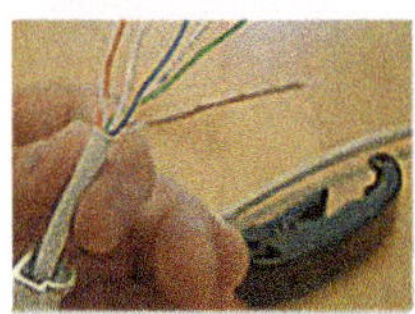

3. Partially untwist the pairs, leaving one twist remaining at the bottom. Be careful not to untwist into the cable jacket. Straighten and organize the conductors.

Note: Choose 568B (most common) or 568A wiring.

4. (Optional) Cut the ends of the conductors at an angle while holding them in the proper order. This will make it easier to install the load bar in the next step.

5. Slide the conductors into the load bar in their proper order with the hollow portion of the load bar facing the jacket. The holes in the load bar alternate up and down. For that reason, you may find it easier to insert the conductors one at a time. Double-check the color order.

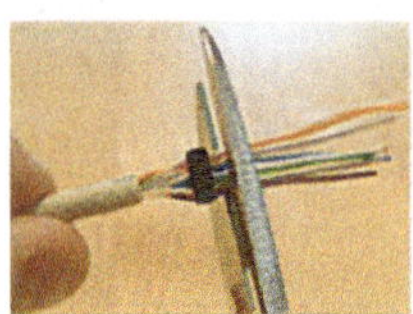

6. Make sure the pairs are twisted at least ¼" beyond the cable jacket. Push the load bar as far down as it will go. Then cut the conductors straight across approximately ⅛" from the front of the load bar. It is very important to get a straight and even cut. The use of a pair of electrician's scissors is highly recommended.

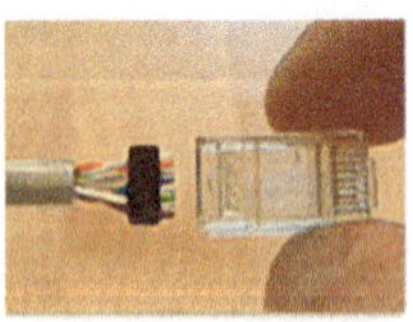

7. Pull the load bar back up near the cut end of the conductors. Then slide the wires and load bar into the connector body, holding it with the pins facing you. A very slight amount of jiggling may be needed to make the wires find their slots in the connector body.

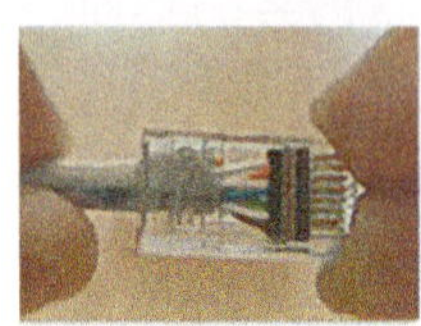

8. Once all of the wires have entered their slots, firmly push the connector body toward the cable. You will need to be sure that a) the wires have reached the end of the connector body, and b) the cable jacket is about halfway into the connector and past the first crimp point (the jacket crimp).

9. Crimp the connector using a high-quality crimp tool such as a ratchet-type RJ-45 tool.

Note: *Some RJ-45 connectors do not use loader bars.*

Figure 14 Typical Cat 6 patch cable construction.
Source: LANshack

2.2.0 RG-6 F-Type Coaxial Cable

A variety of coax F-type connector terminations are available, including crimp-on, push-on, twist-on, etc. Before a connection can be made, the cable must first be stripped using a special coax stripping tool (*Figure 15*). Select the proper tool based on the cable diameter and stripping requirements. In all cases, the first blade of a coax stripper is set to cut almost to the conductor at the proper distance from the end of the coax. A second blade, spaced behind the first at the proper distance, is adjusted to cut almost to the lowest layer of braid in the cable, depending on whether it is dual-, tri-, or quad-shield coax.

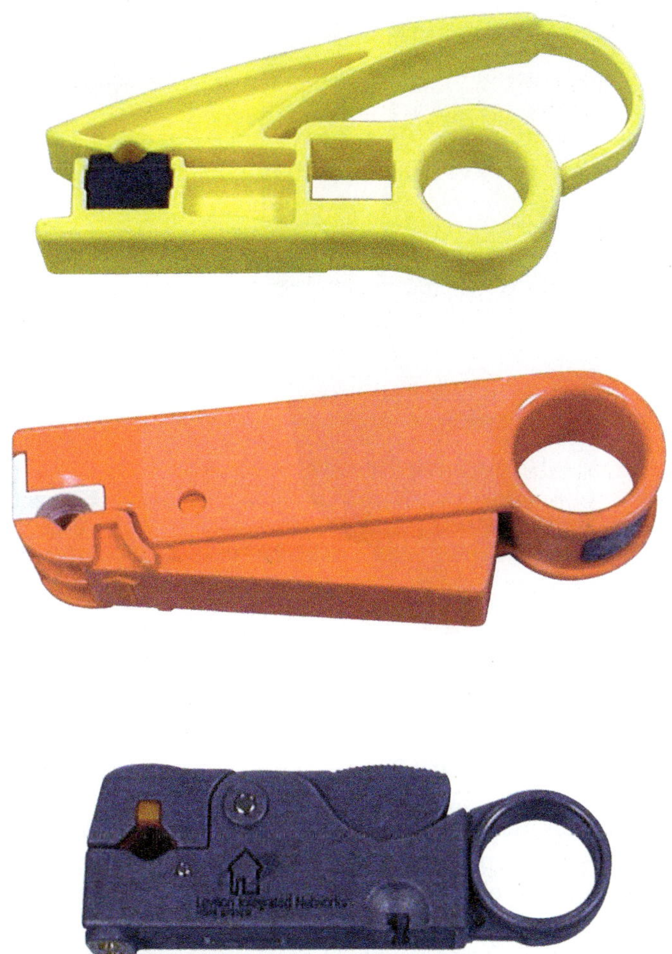

Figure 15 Coax stripping tools.

Most professional installers use compression connectors (*Figure 16*) because they are fast to install, the connection is more secure, and there is less chance of damaging the coax.

Some compression tools for F-type compression connectors are designed to fit only 19.6 mm or 20.3 mm connector barrels. Other compression tools are designed to handle both sizes. Proper tool selection and termination is essential. Many problems in cable systems can be traced to poor connections. Always follow the manufacturer's instructions for the tool in use.

Think About It

Connector Choice

What is the best type of connector to use for RG-6 coax?

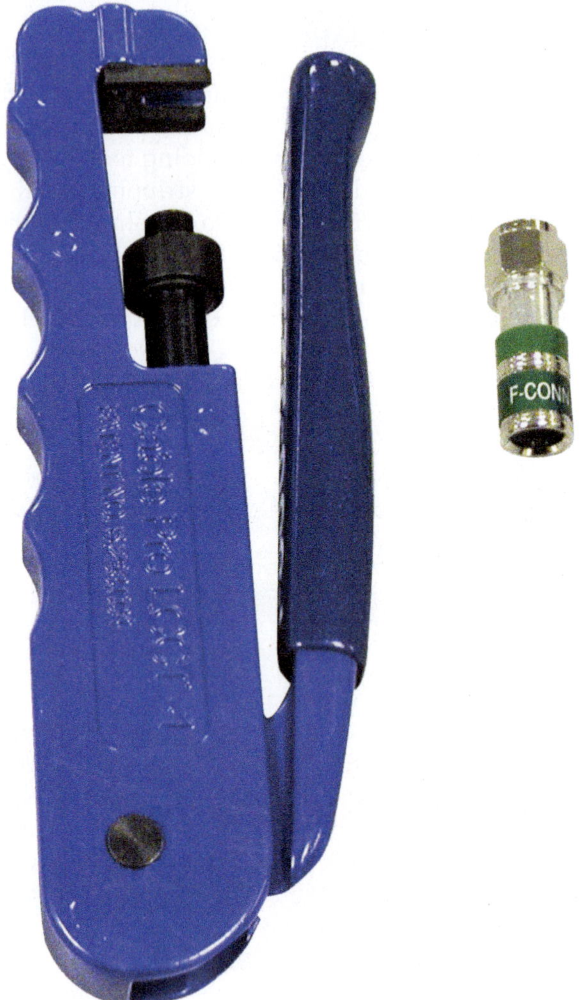

Figure 16 Coax compression tool and connector.

2.3.0 Fiber Optic Cable

Fiber optics is the transmission of light through glass fibers. Fiber optic transmission offers the benefits of high speed and low signal loss over long distances. It is also lightweight, immune to EMI and corrosion, and capable of carrying a much greater amount of information than copper or coaxial wire due to its extremely wide system **bandwidth**. In addition, it provides a very high level of safety and security because fiber optic cable presents no spark hazard, and it cannot be tapped without breaking the integrity of the system. A fiber optic system is ideal for making connections between two fixed points or between many fixed points, as in a computer network.

Bandwidth: The range of frequencies that will pass through a particular portion of a system.

Fiber optic cable is designated as simplex or duplex. Simplex cable sends data in one direction, while duplex cable can send data in both directions. Either type can consist of a single strand up to over a thousand strands. In addition, either type can be designated as SM or MM. SM fiber has a core size of 8.3 micrometers (µm) and is typically used to send data over long distances. MM fiber is available in core sizes of 50 µm and 62.5 µm and is often used to provide greater speed over shorter distances. However, the 62.5 µm size is typically found only in older systems. The 50 µm fiber is available in several configurations designated OM1 through OM5. OM3 and OM4 are the most common.

As with any other type of cable, fiber optic cable is not 100% efficient. The cable can experience signal loss known as **attenuation**. This loss is the result of several factors, the most crucial of which are proper installation and termination. The tools and techniques used with fiber optic cable differ greatly from those used for copper wire. Another area of difference is in the safety precautions you must take. Although there is no worry of being shocked by fibers, there is the possibility of eye damage if you are exposed to sufficient light energy.

Attenuation: The reduction of a signal due to system losses.

WARNING!

To avoid both retinal damage from light energy and physical damage from glass fiber entering the eyes, eye protection must be worn when repairing or troubleshooting an operating fiber optic system and when cutting fiber ends. In addition, the ends of stripped glass fibers are very hazardous to eyes and skin. The cleaved ends are very sharp and can easily penetrate the skin. Take care to avoid this hazard by keeping the work area cleaned of fiber bits. A piece of double-stick tape on the bench can be used to hold the fiber pieces for later disposal. In addition, never eat or drink in fiber handling areas.

Due to their light weight and flexibility, fiber optic cables are often more easily installed than copper conductors. Fiber optic cable can be more easily handled and pulled over greater distances. The minimum bend radius and maximum tensile rating are the critical specifications for any fiber optic cable during the installation process and over the installed life of the cable. Careful planning of the installation layout will ensure that the specifications are not exceeded. In addition, the installation process itself must be carefully planned.

The minimum bend radius and maximum tensile loading allowed on a cable differ during and after installation. An increase in tensile load first causes a reversible attenuation increase, then an irreversible attenuation increase, and finally, cracking of the fiber. The tensile loading allowed during installation is higher than that allowed after installation.

The minimum bend radius allowed during installation is larger than the bend radius allowed after installation. One reason for this is that the allowable bend radius increases with tensile loading. Because the fiber is under load during installation, the minimum bend radius must be larger. The allowable bend radius after installation depends on the tensile load.

Most indoor cables must be placed in conduit or trays. Because standard fiber optic cables are electrically nonconductive, they may be placed in the same ducts as high-voltage cables without the special insulation required by copper wire. However, this is not recommended. In addition, these cables cannot be placed inside air conditioning or ventilation ducts for the same reason as PVC-insulated wire: A fire inside these ducts could cause the outer jacket to burn and produce toxic gases that could be carried throughout the structure. Plenum cables, however, can be placed in any plenum area within a building without special restrictions. The material used in these cables does not produce toxic fumes when burned.

2.3.1 Installation in Trays and Ducts

The primary consideration in selecting a route for fiber optic cable through trays and ducts is to avoid potential cutting edges and sharp bends. Areas where particular caution must be taken are corners and exit slots on the sides of trays. In general, cables in trays and ducts are not subjected to tensile forces; however, in long vertical runs, the weight of the cable itself will create a tensile load of approximately 0.16 pounds per foot for simplex cable (0.07 dekanewtons or daN) and 0.27 pounds per foot (0.13 daN) for duplex cable. This tensile load must be considered when determining the minimum bend radius at the top of the vertical run.

Long vertical runs should be clamped at intermediate points—preferably every 3' to 6' (0.9 m to 1.8 m)—to prevent excessive tensile loading on the cable. The absolute maximum distance between clamping points is 330' (100 m) for duplex cable and 690' (210 m) for simplex cable. Clamping forces should be no more than is necessary to prevent the possibility of slippage. It is best determined experimentally because it is highly dependent on the type of clamping material used and the presence of surface contaminants on the clamp and jacket of the optical cable. Even so, the clamping force must not exceed 57 pounds per inch (0.11 newton-meter or Nm) and must be applied uniformly across the full width of the cable. The clamping force should be applied over the longest length of cable as is practical, and the clamping surfaces should be made of a soft material such as rubber or plastic. The tensile load during a vertical installation is reduced by beginning at the top and running the cable downward.

2.3.2 Installation in Conduit

Fibers are pulled through conduit by a wire or synthetic rope attached to the cable. Any pulling forces must be applied to the cable members and not to the fiber or jacket. For cables without connectors, pull wire can be tied to Kevlar® strength members or a pulling grip taped to the cable jacket or sheath (if approved by the cable manufacturer). Whenever possible, use an automated puller with tension control and a swivel pulling eye. Use extra caution when pulling cables with connectors installed.

The first factor that must be considered when determining the suitability of a conduit for a fiber optic cable is the clearance between the walls of the conduit and other cables that may be present. Sufficient clearance must be available to allow the fiber optic cable to be pulled through without excess friction or binding, as the maximum pulling force that can be used is 90 lb (400N). Because the minimum bend radius increases with increased pulling force, bends in the conduit itself and any fittings through which the cable must be pulled should not require the cable to make a bend with a radius of less than 5.9" (150 mm). Fittings, in particular, should be checked carefully to ensure they will not cause the cable to make sharp bends or press against corners. If the conduit must make a 90° turn, a turn fitting must be used to allow the cable to be pulled in a straight line and avoid sharp bends.

Pull boxes should be used on straight runs at intervals of 250' to 300' (61 m to 91 m) to reduce the length of cable that must be pulled at any one time and to reduce the pulling force. Also, pull boxes should be located in any area where the conduit makes several bends totaling more than 180°. To guarantee that the cable will not be bent too tightly while pulling the slack into the pull box, use a pull box with an opening length equal to at least four times the minimum bend radius. The tensile loading effect of vertical runs discussed previously is also applicable to conduit installations. Because it is more difficult to properly clamp fiber optic cables in a conduit than in a duct or tray, long vertical runs should be avoided if possible. If a clamp is required, the best type is a fitting that grips the cable in a rubber ring. Also, the tensile load caused by the weight of the cable must be considered along with the pulling force to determine the maximum total tensile load being applied to the cable.

2.3.3 System Components

In addition to the cable itself, fiber optic system components include enclosures, splice closures, adapter panels, patch cords, connectors, and outlet boxes. Equipment may vary depending on the type and size of the system being installed. Always read and follow the manufacturer's instructions when installing system components.

Within any fiber optic system, signals must be routed to their final destination. In large indoor applications, a main equipment room (MER) is used as a central distribution point. From this point, fibers can be routed to their destinations. A typical example is an outdoor multistrand cable brought into the MER, where each strand in the outdoor cable is terminated with connectors or spliced to patch cord assemblies. Such routing involves the distribution of signals and the fibers that carry them. There are various types of hardware that allow transitions from one fiber type to another or the distribution of fibers to different points, such as the following:

- *Fiber optic enclosures* — Fiber optic enclosures are rack- or wall-mounted cabinets designed to protect, manage, and organize fiber terminations or splices (*Figure 17*). When equipped with fiber adapter panels, the enclosures allow for cross-connections to other fiber cables or equipment.

  Rack enclosures are mounted in 19" (483 mm) and 23" (584 mm) equipment racks (*Figure 18*), which are standard sizes in the structured cabling and electronics industries in the United States.
- *Splice closures* — Fiber optic cable splices are protected mechanically and environmentally within a sealed closure. Closures have two main applications. The first is for straight splicing two cables when an installation span is longer than can be accommodated by a single cable. The second is to switch between cable types for various reasons (e.g., outdoor-rated to indoor-rated).

  In a typical application, the cable is removed to expose the strands at the point at which it enters the closure. The exposed length must be sufficient to loop it one or more times around the organizer within the closure. Such routing provides extra length for re-splicing or rearrangement of splices. Closures can hold one or more organizers to accommodate 12 to 144 splices.
- *Adapter panels* — Adapter panels provide a convenient way to rearrange fiber connections and circuits. One side of the panel is usually fixed, meaning the fibers are not intended to be disconnected. On the equipment side of the panel, fibers can be connected and disconnected to arrange the circuits as required using patch cords.

  Adapter panels are widely used in the structured cabling industry to connect circuits to transmission equipment (hubs, switches, routers, etc.). They can also be used in the MER of a building to rearrange connections. The splice organizer and adapter panel serve a similar function of distribution. The difference is that the organizer is intended for fixed, unchanged connections, whereas an adapter panel is used for changeable connections.
- *Patch cords* — A patch cord is used to connect the terminated fiber in an enclosure's adapter panel to equipment or other connectors. Connector ends match those in the adapter panel. Patch cords can be either duplex or simplex and are usually measured in meters.
- *Connectors* — Connectors are available in a variety of types, with local connector (LC), standard connector (SC), and straight tip connector (ST) types being the most common. Connectors can be either factory-installed or field-installed using epoxy, mechanical, or fusion-spliced connections.
- *Fiber outlet boxes* — Fiber optic cabling can be terminated into outlet boxes (*Figure 19*) mounted to electrical outlet boxes, allowing connection to cables carrying optical signals rather than current. A short simplex or duplex fiber patch cord (called a jumper) is used to connect the outlet box to the equipment being served.

Figure 17 Fiber optic enclosure.

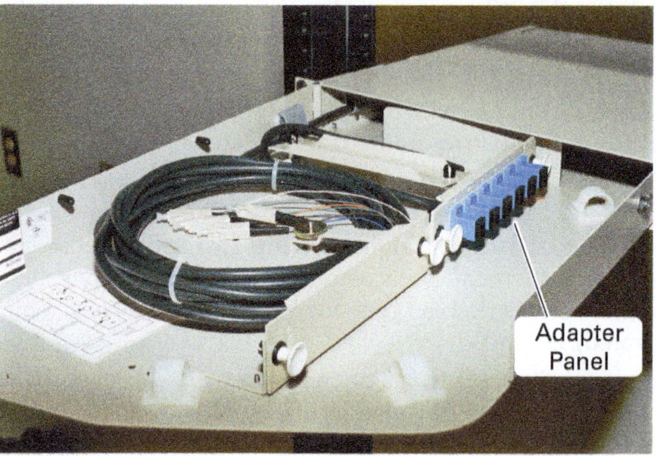

Figure 18 Open rack-mounted fiber termination box.
Source: Tri-City Electrical Contractors Inc

Fusion Splicers

Fusion splicers offer an advantage in that they provide a lower return loss than mechanical connections. A typical fusion splicer is shown in the image.

Source: Ridge Tool Company

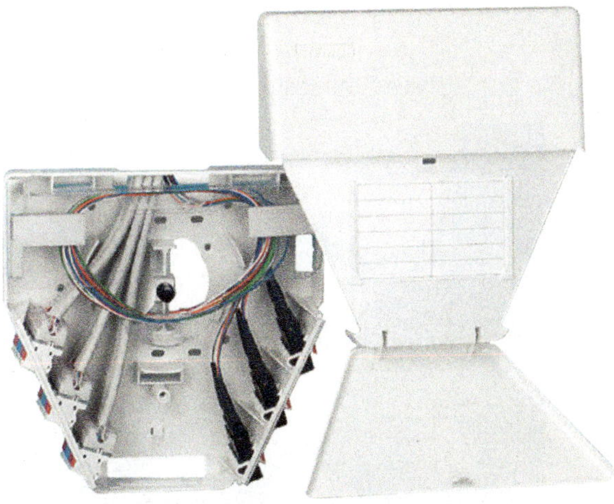

(A) Multiuser Outlet Assembly

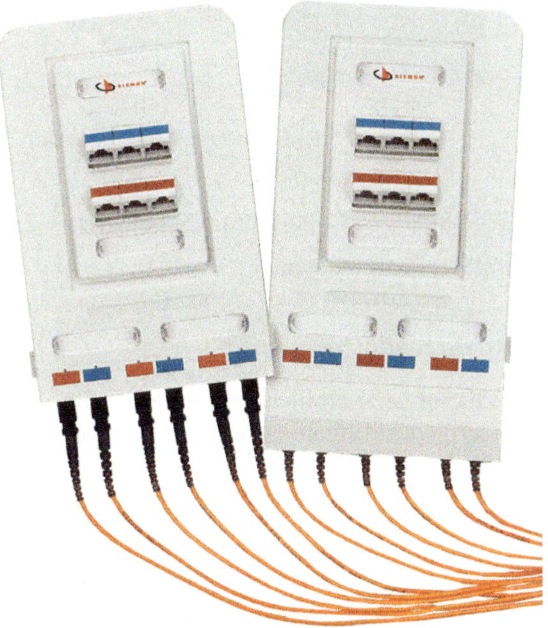

(B) Fiber Outlet Box

Figure 19 Fiber optic outlet boxes.
Source: Image courtesy of Siemon®

2.4.0 Grounding and Testing VDV Systems

All VDV systems must be grounded and bonded per *NEC®* requirements. They must also be tested after installation and periodically thereafter for troubleshooting and maintenance purposes.

2.4.1 Grounding and Bonding

All conductors for grounding must run in straight lines and be protected from physical damage. Each type of system has specific grounding requirements.

The minimum size grounding conductor for fiber optic cable is No. 14 AWG per *NEC Section 770.100(A)(3)*. The grounding means in buildings or structures shall be connected to the nearest accessible location on the following items per *NEC Section 770.100(A)(2)*:

- Building grounding electrode system
- Grounded interior metal water pipe within 5' (1.5 m) of the entrance to the building

Know the Code

Entrance Cable Bonding and Grounding
NEC Section 770.100

Fiber Optic Connections

Mechanical fiber optic connections require the use of cleaver and termination tools, such as those shown in these images.

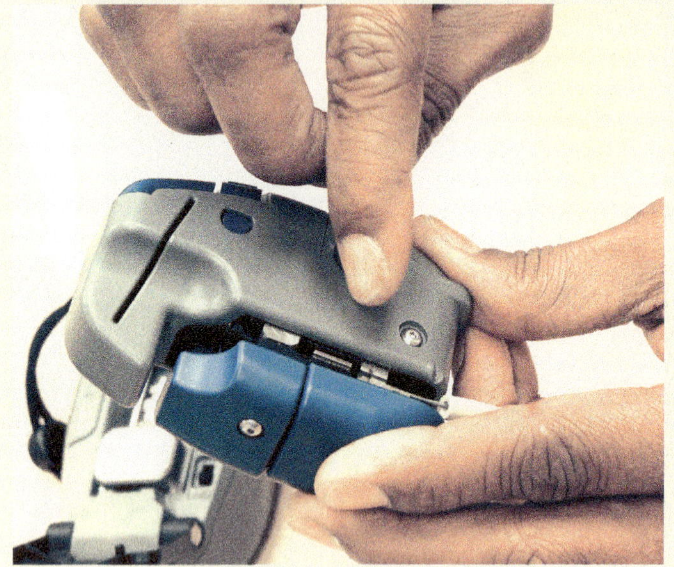

(A) Fiber Optic Cleaving Tool

(B) Fiber Optic Termination Tool

Network Analyzers

The network analyzer shown in the image can be used to replace multiple handheld devices to test a variety of data cables, including DSO-TIMS, up to 10G Ethernet, signaling cable, and more.

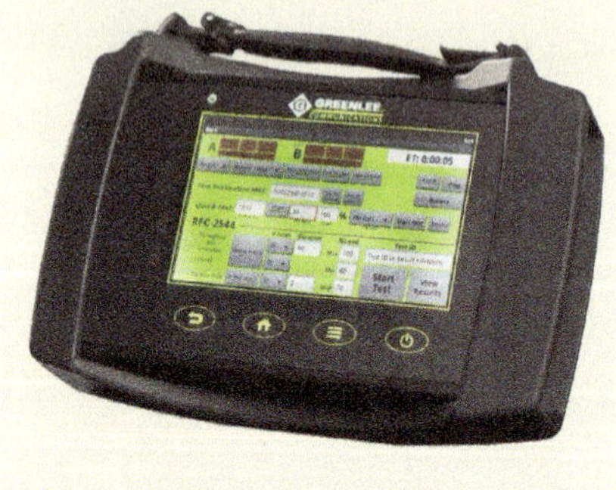

Source: Ridge Tool Company

Know the Code

Cable and Primary Protector Bonding and Grounding

NEC Section 800.100

Know the Code

Bonding Conductors and Grounding Electrode Conductors — Receiving Stations

NEC Section 810.21

- Power service accessible means external to enclosure
- Nonflexible metallic power raceway
- Service equipment enclosure
- Grounding electrode conductor

The bonding conductor size shall be no smaller than No. 6 AWG when connected to the power grounding electrode per *NEC Section 770.100(D)*. The conductors shall be copper, stranded, or solid.

The minimum size grounding electrode conductor for communications circuits (including CATV and radio systems, network-powered broadband systems, and premises-powered broadband systems) is No. 14 AWG and is not required to be larger than No. 6 AWG per *NEC Section 800.100(A)(3)*. It must have a current-carrying capacity of not less than the grounded metallic sheath member of the communications cable, the protected conductor of the communications cable, or the outer sheath of the coax cable.

Radio and television equipment (including satellite dishes) have slightly different requirements for grounding. The minimum size grounding electrode conductor is No. 10 AWG copper, No. 8 AWG aluminum, or No. 17 AWG copper-clad steel or bronze as listed in *NEC Section 810.21(H)*. The bonding jumper used to connect to the power grounding system shall not be smaller than No. 6 AWG per *NEC Section 810.21(J)*.

Other requirements for these systems can be found in the appropriate articles in the *NEC®*, some of which are listed in *Table 1*. The requirements in *NEC Article 800* apply to all communications circuits, CATV, radio, and broadband systems, unless modified by *NEC Articles 805, 820, 830, or 840*.

2.4.2 Testing

After the system installation is complete and the grounding conductors are in place, it must be tested. All testing can be done with cable analyzers such as the one shown in *Figure 20*. This cable analyzer can be used for copper certification, fiber optic loss testing, and to perform fiber end-face inspections. After testing, double-check the quality of the job before leaving the worksite.

TABLE 1 *NEC*® Requirements for VDV Systems

Description	Optical Fiber Cables	General Requirements for Communications Systems	Communications Circuits	CATV and Radio Distribution Systems	Network-Powered Broadband Communications Systems	Premises-Powered Broadband Communications Systems
Other Articles	*NEC Section 770.3*	*NEC Section 800.3*	—	*NEC Section 820.3*	—	—
Mechanical Execution of Work	*NEC Section 770.24*	*NEC Section 800.24*	—	—	—	—
Abandoned Cables	*NEC Section 770.25*	*NEC Section 800.25*	—	—	—	—
Spread of Fire	*NEC Section 770.26*	*NEC Section 800.26*	—	—	—	—
Overhead Wires	*NEC Section 770.44*	*NEC Section 800.44*	—	—	*NEC Section 830.44*	—
Protection	*NEC Section 770.93*	—	*NEC Section 805.90*	*NEC Section 820.93*	*NEC Section 830.90*	*NEC Section 840.90*
Grounding	*NEC Section 770.100*	*NEC Section 800.100*	*NEC Section 805.93*	*NEC Section 820.100*	*NEC Section 830.93*	*NEC Section 840.101*
Cable Installation	*NEC Section 770.133*	*NEC Section 800.133*	*NEC Section 805.133*	*NEC Section 820.133*	*NEC Section 830.133*	*NEC Section 840.133*
Cable Substitution Hierarchy	*NEC Table 770.154(b)*	*NEC Section 800.113*	*NEC Table 805.154*	*NEC Table 820.154*	*NEC Table 830.154*	—
Cable Applications	*NEC Table 770.154(a)*	*NEC Table 800.154(a), (b), and (c)*	—	—	—	—
Cable Marking	*NEC Table 770.179*	*NEC Section 800.179*	—	—	—	—

The following is a list of the tests that must be performed:

- *Cat 5e* — Tests include length, wire map, delay, **delay skew**, attenuation, **near end crosstalk (NEXT)**, return loss, **equal level far end crosstalk (ELFEXT)**, and power sum NEXT up to 100 MHz.
- *Cat 6* — Tests include length, wire map, delay, delay skew, attenuation, NEXT, return loss, ELFEXT, and power sum NEXT up to 250 MHz.
- *Cat 6a* — Tests include length, wire map, delay, delay skew, attenuation, NEXT, return loss, ELFEXT, and power sum NEXT up to 500 MHz.
- *Cat 8.1* — Tests include length, wire map, delay, delay skew, attenuation, NEXT, return loss, ELFEXT, and power sum NEXT up to 2,000 MHz.
- *Cat 8.2* — Tests include length, wire map, delay, delay skew, attenuation, NEXT, return loss, ELFEXT, and power sum NEXT up to 2,000 MHz.
- *MM fiber* — Tests include length, delay, loss/km @ 850 nanometers (nm), and loss/km @ 1,300 nm.
- *SM fiber* — Tests include length, delay, and loss/km @ 1,310 nm.
- *Coax* — Tests include length, open, shorts, and loss.

Delay skew: A change of timing or phase in a transmission signal.

Near end crosstalk (NEXT): A measurement of the interference (crosstalk) between two wire pairs.

Equal level far end crosstalk (ELFEXT): A value derived by subtracting the attenuation of the interfering pair from the far end crosstalk (FEXT) that it has caused in the interfered pair.

Cable Verifiers

This handheld cable tester is designed for use with both twisted pair and coax cable. It displays the cable length, wire map, cable identification, and distance to a fault.

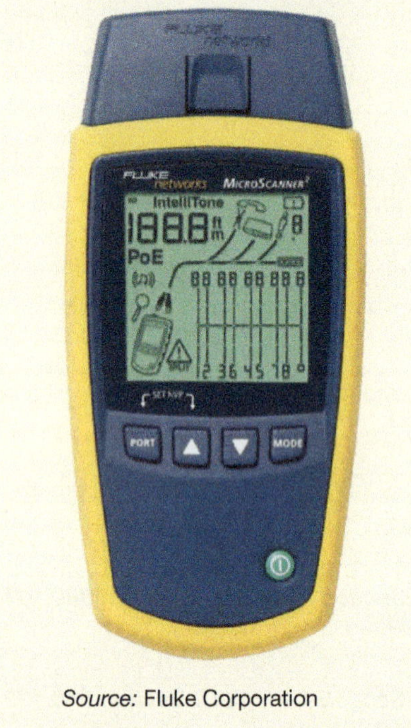

Source: Fluke Corporation

Figure 20 Cable analyzer/certification tester.
Source: Fluke Corporation

2.0.0 Section Review

1. In the term *RJ-45*, the letters RJ stand for _____.
   a. the manufacturer's initials
   b. registered jack
   c. ring joint
   d. radio joint

2. Coax cable is stripped using _____.
   a. a utility knife
   b. wire cutters
   c. standard wire strippers
   d. special coax strippers

3. The *maximum* pulling force of fiber optic cable is _____.
   a. 50 lb (222N)
   b. 90 lb (400N)
   c. 130 lb (578N)
   d. 170 lb (756N)

4. A type of cable that *must* be tested for NEXT is _____.
   a. MM fiber
   b. SM fiber
   c. coax
   d. Cat 5e

Module 26310-23 Review Questions

1. The video cable used in new installations is normally _____.
   a. Cat 5
   b. Cat 6
   c. RG-5
   d. RG-6

2. A structured cabling subsystem that connects buildings together in a multibuilding environment is called a(n) _____.
   a. riser subsystem
   b. equipment room subsystem
   c. campus backbone subsystem
   d. work area subsystem

3. Straight conduit runs used in campus backbone systems should have pull boxes installed *no* more than _____.
   a. 80' (24 m) apart
   b. 140' (43 m) apart
   c. 350' (107 m) apart
   d. 500' (152 m) apart

4. Label conduit for a campus backbone within _____.
   a. 6" (150 mm) of the termination point
   b. 12" (300 mm) of the termination point
   c. 18" (450 mm) of the termination point
   d. 24" (600 mm) of the termination point

5. Telecom cables in an ER or TR subsystem must be separated from electrical conduit or cable by at *least* _____.
   a. 5" (125 mm)
   b. 8" (200 mm)
   c. 10" (250 mm)
   d. 12" (300 mm)

6. Riser pathways should run perpendicular to lighting and be separated by a distance of at *least* _____.
   a. 5" (125 mm)
   b. 8" (200 mm)
   c. 10" (250 mm)
   d. 12" (300 mm)

7. Vertical cabling spans exceeding 30' (9 m) in riser support subsystems *must* be supported in _____.
   a. 15' (5 m) increments
   b. 20' (6 m) increments
   c. 30' (9 m) increments
   d. 40' (12 m) increments

8. When installing horizontal ER/TR/MC sleeves, the *minimum* distance between the ER/TR/MCs and the horizontal subsystem is _____.
   a. 3" (75 mm)
   b. 4" (100 mm)
   c. 5" (125 mm)
   d. 6" (150 mm)

9. RJ-45 connectors used for UTP 4-pair cables are identified as _____.
   a. 2P2C connectors
   b. 3P6C connectors
   c. 6P8C connectors
   d. 8P8C connectors

Answers to odd-numbered Module Review Questions are found in *Appendix A*.

10. The *minimum* size grounding conductor for fiber optic cable is _____.
    a. No. 10 AWG
    b. No. 12 AWG
    c. No. 14 AWG
    d. No. 16 AWG

Module 26310-23 Supplemental Exercises

1. Structured cabling systems may use UTP cable, STP cable, coax cable, or _________ cable.
2. Structured cabling systems are commonly divided into _________ subsystems.
3. Minimize water issues when designing campus backbone conduit systems by maintaining a(n) _________ slope away from the buildings.
4. True or False? OSP conduit must be buried a minimum of 24" (600 m) below grade.
5. Cable installation requires the use of large, machine-printed _________ for system components.
6. The first concern when laying out the MC is to maintain appropriate clearances from sources of _________ interference.
7. A cable in the bundle that crosses and disappears into the center of the bundle is known as a(n) _________.
8. The routes for a riser subsystem should consider what low-voltage services are needed at each location, such as _________, _________, or _________.
9. All cables entering a plenum-rated horizontal subsystem must be _________.
10. Most straight-through patch cords for Cat 5e, Cat 6, and Cat 6a cable are wired to the _________ scheme.
11. True or False? Fiber optic cable is 100% efficient.
12. True or False? Pulling forces should be applied to the fiber in fiber optic cable in order to avoid tearing the jacket.
13. All conductors for _________ must run in straight lines and be protected from physical damage.
14. The bonding conductor size shall be no smaller than _________ AWG when connected to the power grounding electrode per *NEC Section 770.100(D)*.
15. Cable installation is not complete without _________.

Answers to odd-numbered Supplemental Exercises are found in *Appendix B*.

Answers to Section Review Questions

Answer	Section Reference	Objective
Section 1.0.0		
1. d	1.1.2	1a
2. a	1.2.0	1b
3. c	1.3.2	1c
4. d	1.4.4	1d
5. a	1.5.0	1e
6. b	1.6.0	1f
Section 2.0.0		
1. b	2.1.0	2a
2. d	2.2.0	2b
3. b	2.3.2	2c
4. d	2.4.2	2d

User Update

Did you find an error? Submit a correction by visiting **https://www.nccer.org/olf** or by scanning the QR code using your mobile device.

MODULE 26311-23

Motor Controls

Source: iStock@Vithun Khamsong

Objectives

Successful completion of this module prepares you to do the following:

1. Identify relays, contactors, and motor starters.
   a. Identify control relays and their configurations.
   b. Identify magnetic contactors.
   c. Identify motor overload protective devices.
2. Explain how magnetic and manual motor starters are selected.
   a. Explain how NEMA contactors and motor starters are selected.
   b. Explain how IEC contactors and motor starters are selected.
   c. Identify various contactor and motor starter accessories.
3. Identify and describe control transformers and pilot devices.
   a. Identify and describe control transformers.
   b. Identify and describe various motor control and pilot devices.
4. State installation considerations for motor controls.
   a. Explain how to select contactor and motor starter enclosures.
   b. Interpret installation diagrams and their common symbols.
   c. Identify *NEC*® regulations related to motor control circuits.
   d. Explain how to connect motor controllers for specific applications.

Performance Task

Under supervision, you should be able to do the following:

1. Make all wiring connections for a magnetic motor starter controlled by two push-button stations, including a holding circuit interlock.

Overview

Motor starters and the many possible pilot devices associated with their control must be carefully selected and installed. This module introduces various relays, contactors, and motor controls, along with guidance in selecting and installing these devices.

NOTE

Digital Resources for Electrical

Scan this code using the camera on your phone or mobile device to view the digital resources related to this craft.

NCCER Industry-Recognized Credentials

If you are training through an NCCER-accredited sponsor, you may be eligible for credentials from NCCER. The ID number for this module is 26311-23. Note that this module may have been used in other NCCER curricula and may apply to other level completions. Contact NCCER at 1.888.622.3720 or go to **www.nccer.org** for more information.

You can also show off your industry-recognized credentials online with NCCER's digital badges. Transform your knowledge, skills, and achievements into badges that you can share across social media platforms, send to your network, and add to your resume. For more information, visit **www.nccer.org**.

1.0.0 Relays and Contactors

Performance Task

1. Make all wiring connections for a magnetic motor starter controlled by two push-button stations, including a holding circuit interlock.

Objective

Identify relays, contactors, and motor starters.

a. Identify control relays and their configurations.
b. Identify magnetic contactors.
c. Identify motor overload protective devices.

Relays, contactors, and motor starters are used to control electrical power to motors and major lighting circuits. Motor starters, also called *motor controllers*, are simply contactors with added motor overload protective devices selected to properly protect the motor. Conventional contactors, motor starters, and relays are electromagnetic mechanical devices that, when sized properly, minimize contact arcing during the opening (breaking) or closing (making) of the motor power circuit by operating very quickly.

Motor control circuit: A typically low-current electrical circuit that controls the operation of a contactor or motor starter but does not carry the current powering the motor.

Relays are generally used in a **motor control circuit** to control the contactor or starter. A low-current signal passing through a pilot device to another relay or directly to the contactor or starter is used to control the application of power to the motor.

A variety of relays, contactors, and motor starters are used for lighting, motor, and HVACR control circuits. Electromechanical relays, contactors, and motor starters are still common. *Electromechanical* simply means they are mechanical devices positioned by a coil that produces a magnetic field. Solid-state versions, though, with no moving parts or contacts, are growing in popularity but are not necessarily best for every application.

Contactors and motor starters are designed to handle higher current loads than relays. Without overload protection, contactors can be used to handle high-current, noninductive loads such as lighting or other resistance loads. Special solid-state, thermal, and magnetic relay devices are added for motor protection. Depending on the device, some or all of the following types of protection can be provided:

- Current overload
- Overvoltage and undervoltage
- Phase loss or unbalance
- Directional overcurrent
- Percentage of voltage or current differential

Contactors, motor starters, and their enclosures are available based on the National Electrical Manufacturers Association (NEMA) standards or standards developed by the International Electrotechnical Commission (IEC). The IEC is an international organization that produces performance and safety standards for electrical products. Comparing identical load ratings, IEC contactors are less expensive and of lighter-duty construction than NEMA contactors. IEC contactors are found in an increasing number of products, especially those that are internationally marketed. It is important to understand the differences between NEMA- and IEC-rated contactors, as you will encounter both in the field.

1.1.0 Relays

Relays are commonly used to manage low-current loads in control circuits. They vary in capacity from those rated only for milliamp loads to larger styles with maximum load ratings of 30A at 277VAC or 15A at 600VAC. Some relays can be plugged into a socket or printed circuit (PC) board. *Subminiature relays* are extremely small relays soldered onto PC boards. *Industrial relays* are physically much larger and may have several sets of contacts. *Reed relays* are delicate relays with low-current contacts sealed in a glass envelope.

1.1.1 General-Purpose Relay Configurations

General-purpose relays are available in various AC and DC voltage designs and current ratings. Coils can be specified in DC voltage ranges of 5V to 24V or AC ranges of 12V to 240V. DC coils can be activated with currents as little as 4 mA at 5VDC, making them compatible with certain integrated circuit logic gates.

Relays can have up to 12 poles with various combinations of *normally open* (NO) and *normally closed* (NC) contacts. NC and NO always refer to the position of the contacts when a relay is de-energized.

Figure 1 shows a single-phase, 25A, 240V, open-frame AC power relay with two poles (double pole, or DP). Each pole has an NC and NO contact with a common single-break (SB) moving contact. When a pole is equipped with both NC and NO contacts, it is designated as a *double-throw (DT)* arrangement. The power relay shown is thus defined as a *DPDT-SB relay*.

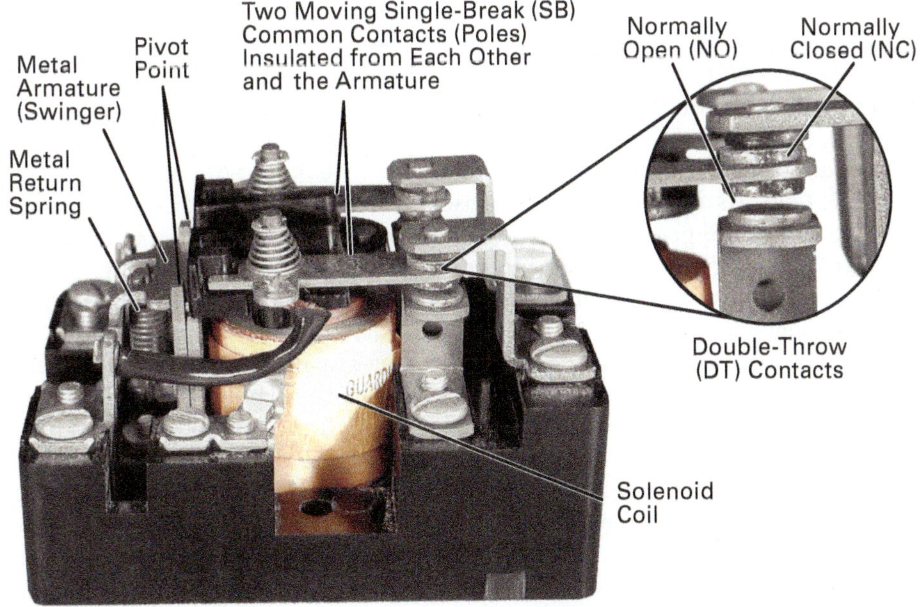

Figure 1 Typical double-pole, double-throw, single-break (DPDT-SB) open-frame power relay.

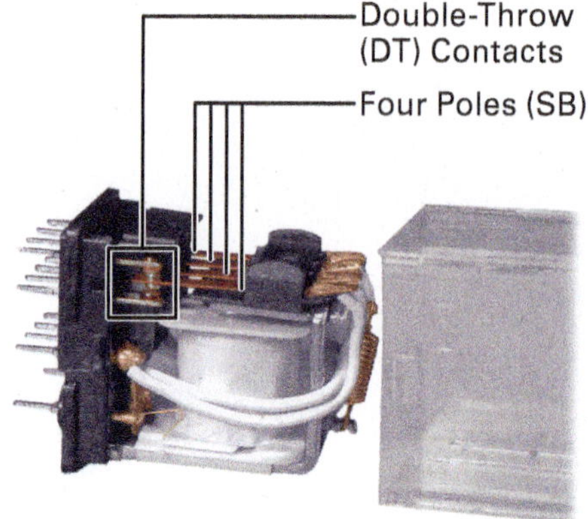

Figure 2 Miniature, four-pole, double-throw, single-break (4PDT-SB) plug-in relay.

Figure 2 shows a miniature, single-phase, 5A, 240VAC plug-in relay with four poles (4P). Each pole has an NC and NO contact (DT arrangement) with a common SB moving contact. This relay can be identified as a *4PDT-SB relay*. Note the pins on the base that make it suitable for soldering to a PC board. Some similar relays have more robust pins designed to plug into a socket.

As you can see, relays are identified by their number of poles, throws, and breaks. Relay pole and contact designations are shown in *Table 1*. Various relay contact configurations are shown in *Figure 3*. Some of these pole and contact designations also apply to contactors.

Manufacturers use a common code to simplify the contact sequence identification for relays. The code uses form letters to indicate the characteristics of each pole of a relay, as shown in *Figure 4*.

TABLE 1 Relay Contact and Pole Designations

Designation	Meaning
ST	Single Throw
DT	Double Throw
NO	Normally Open
NC	Normally Closed
SB	Single Break
DB	Double Break (industrial relays)
SP	Single Pole
DP	Double Pole
(#)P	# = numeric number of poles

1.1.2 Typical Operation

In the single-pole, single-break version of a relay shown in *Figure 5*, the relay operates when a control signal voltage is applied to its solenoid coil. The magnetic field created through the solenoid coil attracts the metal armature and draws it into contact with the coil core. The moving contact, attached to the armature (*Figure 5* and *Figure 6*), moves away from the NC contact and firmly connects with the NO contact. The input power at the common (COM) terminal is routed through the contacts to the NO terminal.

When voltage is removed from the solenoid coil, a return spring pulls the armature back to its de-energized position, thus causing the moving contact to break with the NO contact and connect with the NC contact. Any input power from the COM terminal is then switched to the NC contact terminal.

When the solenoid coil is powered by AC, a shading coil (aluminum or copper) at the top of the core is used to create a weak, out-of-phase, auxiliary magnetic field. As the main field collapses when the AC periodically drops to zero, the weak field generated by the shading coil is strong enough to keep the armature in contact with the core and prevent the relay from "chattering." If the shading coil is loose or missing, the relay will be noisy and experience abnormal wear and heat generation.

In a power relay of this size, the moving contact arm must be made of relatively thick metal to accommodate the current. Light-duty relays use flexible, thin-gauge copper spring stock for the moving contact arm. Because the armature is designed to swing over a slightly greater distance than necessary in both directions, the flexible moving arm bends slightly in the NO and NC positions. This is referred to as *armature overtravel*. This keeps the contacts firmly together. In relays and contactors that use an inflexible moving contact arm, a contact pressure spring is used instead.

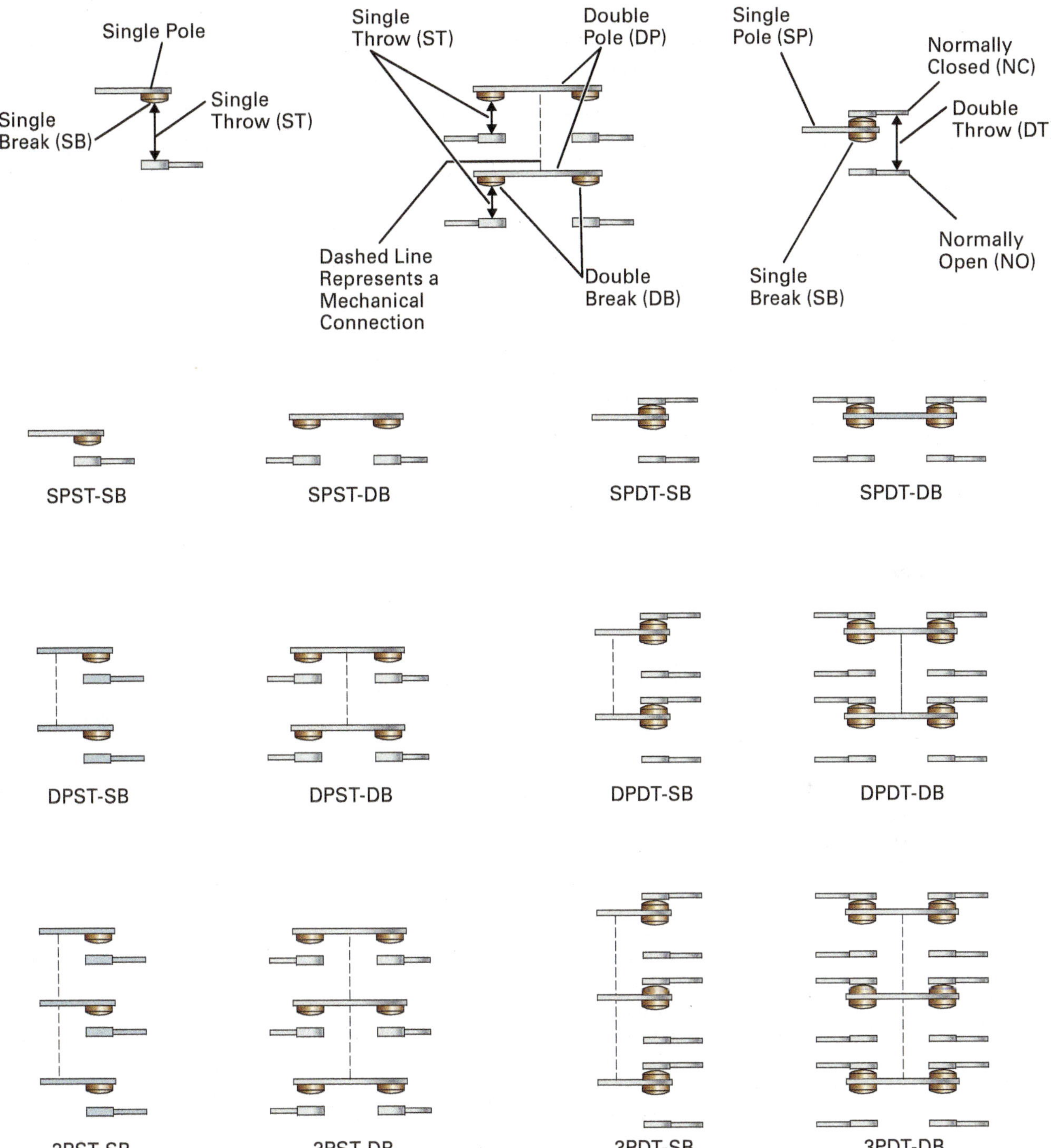

Figure 3 Examples of relay contact configurations.

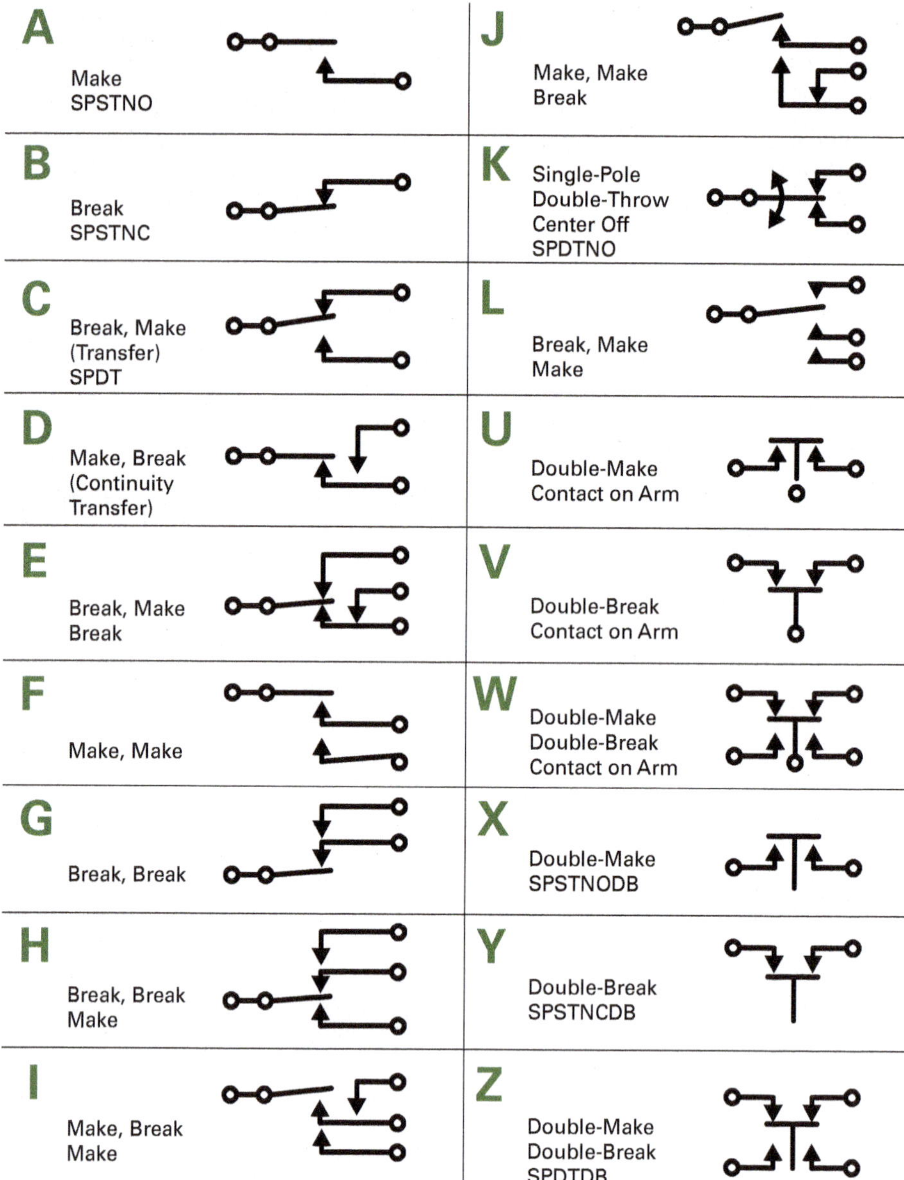

Figure 4 Relay contact form identification.

Contact Wiping

A degree of contact cleaning is accomplished by armature overtravel. As the contacts mate, the overtravel causes the contact surfaces to wipe across each other lightly. This action removes oxides from the mating surfaces of the contacts and helps keep contact resistance low.

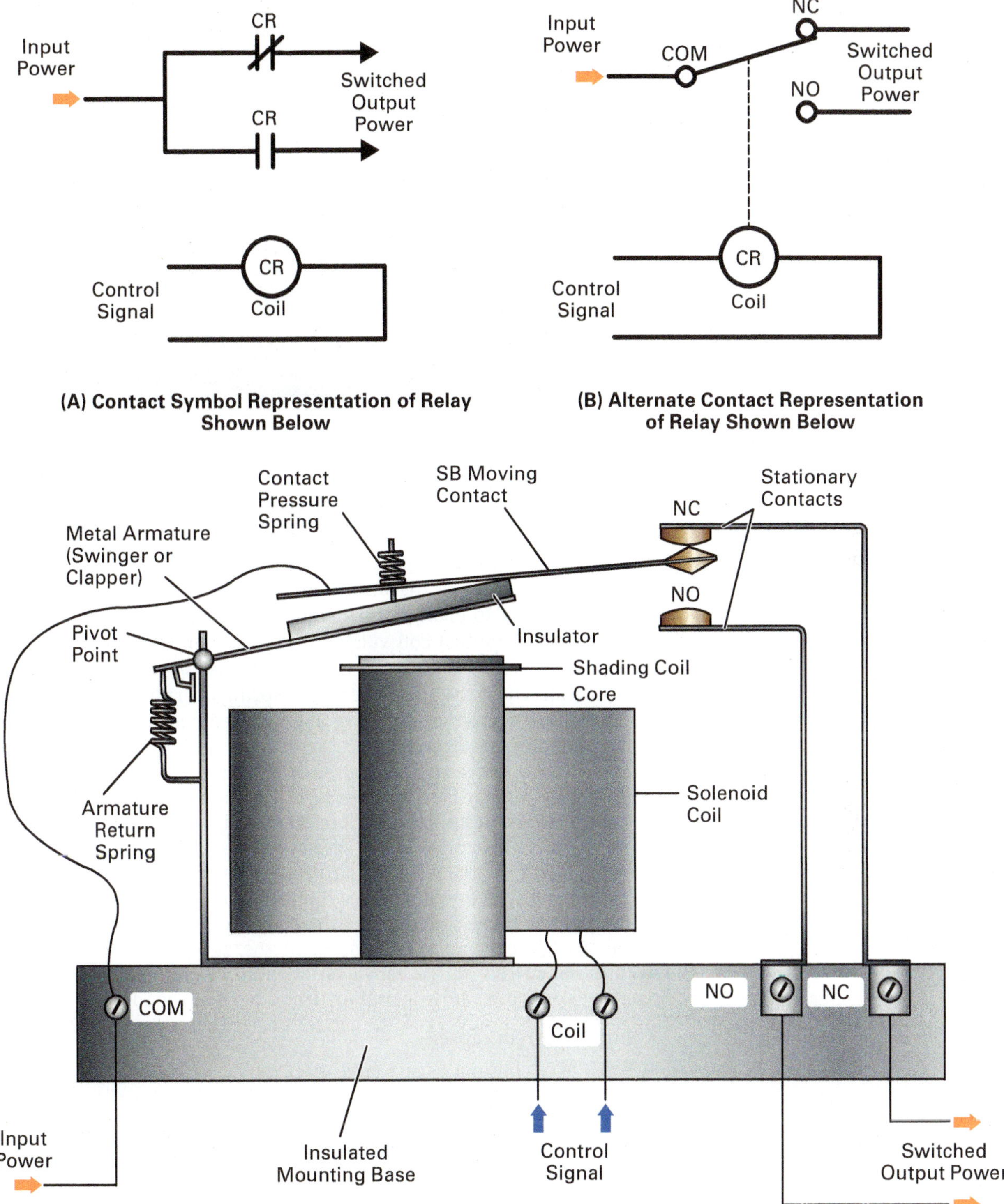

Figure 5 Typical SPDT-SB power relay and symbol representation.

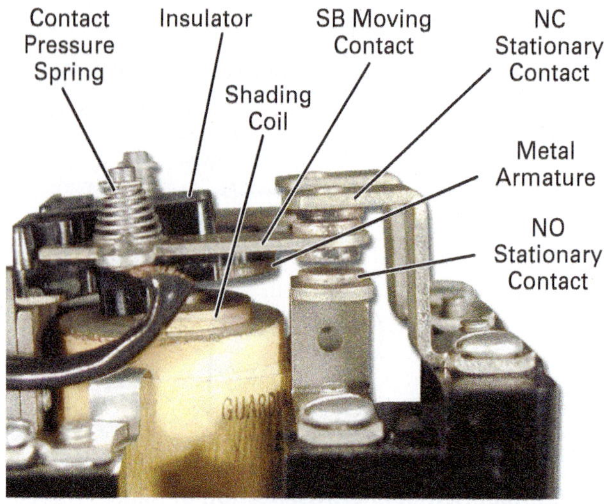

Figure 6 Detail view of a power relay.

1.1.3 Relay Selection Criteria

Relay selection criteria includes characteristics such as coil voltage, contact current and voltage ratings, and contact materials. To begin, we must define some terms related to coil voltage.

Pickup voltage: The minimum voltage applied to a relay or contactor coil that will allow it to energize. The device is not likely to operate below this voltage due to the weakness of the magnetic field.

Seal-in voltage: The minimum voltage applied to a relay or contactor coil that will maintain a sufficient magnetic field to keep the device energized. The seal-in voltage is usually less than the pickup voltage.

Dropout voltage: The voltage at which the armature return spring of a relay or contactor overpowers the operating coil's weakened magnetic field, allowing it to return to its de-energized state.

The **pickup voltage** is the minimum applied coil voltage that will cause a relay or contactor to change its position. Once energized, the **seal-in voltage** is the minimum applied coil voltage that will keep the device energized. It is usually a bit less than the pickup voltage. The **dropout voltage** is defined as the coil voltage at which the armature return spring overpowers the coil's magnetic field, and the contacts move to their de-energized position. The dropout voltage is lower than the seal-in voltage.

Most coils on relays and contactors are designed so that a dropout does not occur until the voltage is reduced to 85% of the nominal or specified coil voltage. They must also be designed to pick up or energize when the voltage rises above 85% of the specified coil voltage. These voltages are established by NEMA and are conservative. Most electromechanical devices manufactured today drop out and pick up at even lower voltages than specified. However, their operation is not guaranteed below the specified dropout and pickup voltages.

One very important consideration in the application selection process is the contacts' current rating. However, the ratings can be somewhat confusing. Sometimes only one rating is provided, but there can be three published ratings:

- Inrush current capacity
- Normal or continuous current capacity
- Breaking current capacity

For example, a typical industrial relay may have the following contact ratings at a given AC voltage:

- 15A resistive (noninductive) continuous load
- 8A inductive continuous load
- 75A inrush, 50A break (inductive or resistive)

Two common examples of resistive loads are heating elements and incandescent lights. Relay and contactor coils represent inductive loads. Extreme inrush current or sustained overloads can cause *contact welding*. This means a portion of the two contact surfaces melt and weld together. Therefore, for inductive loads that experience inrush current, the steady-state and inrush currents should be measured to determine the proper contact rating. Some typical loads and their approximate inrush currents are summarized in *Table 2*.

The contacts used in relays or contactors are made from various metals and alloys. Various materials have certain advantages and disadvantages, as shown in *Table 3*.

TABLE 2 Typical Loads and Their Inrush Currents

Type of Load	Approximate Inrush Current
Resistive heating	None; steady-state current experienced immediately
Sodium vapor lamps	1 to 3 times steady-state current
Mercury lamps	About 3 times steady-state current
Motors	5 to 10 times steady-state current
Transformers	5 to 15 times steady-state current
Incandescent lamps	10 to 15 times steady-state current
Solenoid coils	10 to 20 times steady-state current
Capacitive loads	20 to 40 times steady-state current

TABLE 3 Contact Material Characteristics

Material	Advantages	Disadvantages
Silver	Silver has the highest conductivity and thermal properties of any contact material.	Silver is subject to welding and sticking under arcing conditions. Rapid sulfidation, or *tarnishing*, creates a film that increases contact resistance. Normal, inherent contact wiping through armature overtravel usually removes the film. It is not used in intermittent or arcing applications.
Silver-cadmium alloy and silver-cadmium oxide alloy	These have good conductivity and thermal properties. Cadmium oxide alloy conducts even when oxide forms on surface of contacts.	These alloys resist arcing damage but are subject to some sulfidation. Normal contact wiping usually removes the film. When used in circuits drawing several amps at more than 12V, any accumulated sulfidation is burned off.
Gold-flashed silver	Gold-flashed silver has the same advantages as silver. Gold flashing protects against sulfidation. It is used in intermittent applications and is good for switching current of 1A or less.	Gold-flashed silver is not used in applications where arcing occurs because gold burns off too easily.
Tungsten and tungsten-carbide alloy	These are subject to minimal damage from arcing due to their high melting temperature. Both are good for high-voltage and repetitive switching applications.	These offer higher contact resistance than other materials.
Silver tungsten-carbide alloy	This alloy has the same advantages as tungsten or tungsten-carbide, but a lower contact resistance.	This alloy is subject to minor sulfidation; however, inherent wiping action and any arcing removes the film.

1.1.4 Contact Arc Suppression

The typical contact life of a relay ranges between 100,000 and 500,000 operations. The contact rating is based on the specified maximum current. When relays switch current that is below their full current rating, contact life is extended. If the current exceeds the contact rating or arcing occurs, the life of the contact is shortened due to overheating and burning.

In some applications, arc protection circuits can be added to the controlled circuit to reduce the arcing caused by inductive or high resistive loads. An arc protection circuit provides a nondestructive path for the voltage generated by the collapsing field of the inductive load when the relay contacts open. *Figure 7* shows three different protection circuits that can be used for arc suppression. Note that *Figure 7* (A) shows a DC circuit.

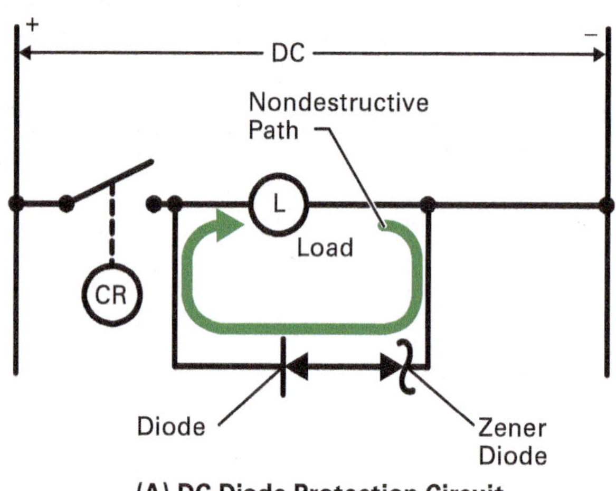

(A) DC Diode Protection Circuit

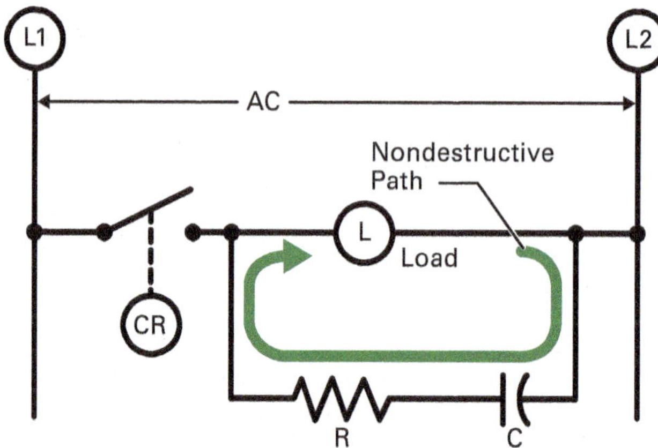

(B) RC Protection Circuit

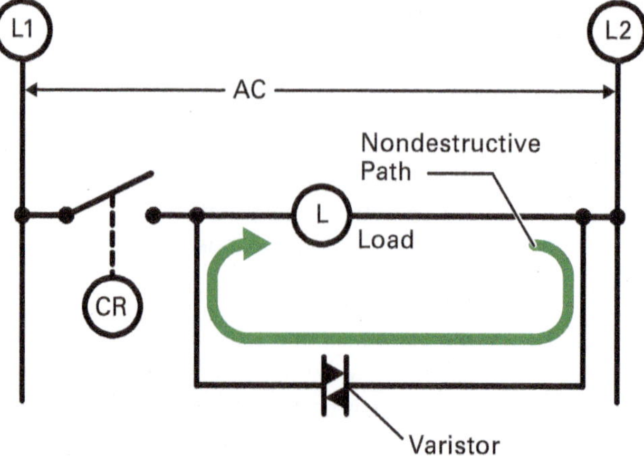

(C) Varistor Protection Circuit

Figure 7 Contact protection circuits.

In the DC diode protection circuit, a diode can be placed in parallel with the load to oppose any current flow when the relay contacts energize the load. When the relay contacts open, the reverse polarity voltage generated by the collapsing field is shorted out by conduction through the diode. By shorting out the generated voltage, the resulting current is dissipated across the load instead of across the relay contacts. The diode on-time and resulting dissipation time can be controlled by inserting a Zener diode. The cutoff voltage characteristic of the Zener diode halts the dissipation across the load quickly after the contacts have opened.

In an AC circuit, arc suppression is less of a problem because the arc is extinguished each time the current passes through zero. At a frequency of 60 Hz, this happens 60 times per second. This is why contacts of relays rated for AC use can tolerate a smaller separation at break than DC contacts. Arcing can still occur, however, and a resistor/capacitor (RC) circuit or a varistor can be used for contact protection to extend contact life.

In the RC protection circuit, the resistor and capacitor are selected to make the circuit effectively resistive, and its time constant is approximately equal to the time constant of the load. This provides a path through the load to dissipate the generated voltage and current when the relay contact is opened.

In the varistor protection circuit, the varistor is a solid-state device with a resistance inversely proportional to the voltage applied across it. When the relay contacts close and voltage appears across the load and the varistor, the varistor resistance is high. When the relay contacts open, a low-resistance path is provided through the varistor to dissipate the generated voltage and current across the load. The disadvantages of these circuits are that they slow down coil reaction time and cannot be used in electronic timer-controlled circuits.

Other circuits placed across the control relay contacts can be used for arc suppression, but they are not fail-safe. If a component shorts in a suppression circuit, the load may remain energized even though the relay contact opens.

Transient Suppression

Some manufacturers offer arc suppression devices as accessories for use across the coils of their relays or contactors. These devices are primarily designed for noise suppression caused by arcing.

Hermetically Sealed Coils

Many coils used today are hermetically sealed (encapsulated) in plastic to protect against corrosive environments. They can generally operate continuously at 110% of the nominal rated voltage without damage to the coil. However, to ensure reliable relay operation, the specified coil voltage should be maintained at the coil terminals when the coil is energized.

1.2.0 Magnetic Contactors

Magnetic contactors are control devices that energize or de-energize significant loads. In this respect, they are like relays. However, they can switch much larger currents. They are used instead of relays to repeatedly establish and interrupt high current loads. They are available from most manufacturers in four basic categories, summarized in *Table 4*.

TABLE 4 Contactor Categories and Applications

Category	Application
Lighting contactors	The current rating of lighting contactors is only for resistive loads. They do not have horsepower ratings for motor use. The current rating is for the maximum continuous current required by a resistive load. At the maximum current rating, the contactors are designed to withstand the large initial inrush currents of tungsten and ballast lamp loads, as well as other resistive loads, without contact welding. They can be locally or remotely controlled via AC or DC control circuits and may be magnetically or mechanically held or magnetically latched. The mechanically held or latched contactors are quiet (no AC hum) and remain closed during power interruptions so their loads will come back on when power is restored.
Definite-purpose (DP) contactors and motor starters	Low-duty cycle definite-purpose contactors are intended for use in applications where the control requirements are well-defined. These contactors carry dual ratings (ampere and horsepower) for either resistive or inductive loads found in applications such as refrigeration, air conditioning, resistance heating, and other Standard Industrial Classification (SIC) applications. Motor starters are usually equipped with bimetallic overload protective devices.
NEMA-rated contactors and motor starters	NEMA contactors are for general use and use in motor starters when combined with overload protective devices. They are identified in eleven overlapping ranges and rated in horsepower (for motor starters) and/or continuous current capacity. They are primarily designed for use with inductive motor loads and to withstand the interrupting current of locked rotor current. They are designed with reserve capacity to perform over a broad range of applications without the need for an assessment of life requirements. Motor starters can be manually or magnetically operated and are equipped with melting-alloy, bimetallic, or solid-state overload relays. Most of these contactors have replaceable contacts and encapsulated coils.
IEC-rated contactors and motor starters	IEC contactors and motor starters perform the same functions as NEMA-rated devices but are physically smaller than NEMA devices for the same horsepower or current ratings. As a result, they are very application sensitive and may require rating and life assessment analysis before application. US manufacturers normally provide tables like the NEMA tables to facilitate selection. The motor starters are usually equipped with bimetallic overload relays. Most IEC contactors do not have sealed coils and, except for the largest sizes, have non-serviceable contacts.

1.2.1 Construction and Operation

Like a relay, a magnetic contactor is actuated by a solenoid coil. *Figure 8* shows three types of mechanical actions used for the solenoid closure of contactors. In all cases, energizing the solenoid coil assembly causes a moving contact assembly with spring-loaded, double-break (DB) bridging contacts (one DB set of contacts for each pole) to mate with corresponding pairs of stationary contacts to close a circuit.

Contactors typically have between one and six poles. The bridging contacts and the materials used for the contacts allow higher current ratings for contactors. The angled contact structure allows a slight wiping action as the contacts open and close. Coil voltage and contact rating information is reported the same way as the information for relays.

Figure 9 is a typical NEMA-rated magnetic contactor. It has a three-phase configuration that is closed by a mechanical bell crank when the coil is energized. The energized indicator shows if the contactor is de-energized (flush) or energized (depressed). It can also be manually depressed to check if the contactor mechanism is free to move and the contacts close.

Many NEMA-rated contactors can be disassembled and their individual components replaced. Smaller NEMA contactors and many IEC contactors are not serviceable. The coil and contacts of the contactor shown in *Figure 9* can be replaced. Other types of contactors, including definite-purpose and lighting, are similarly constructed.

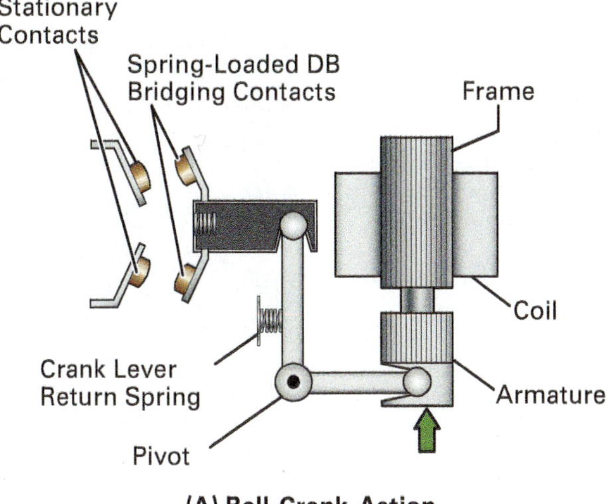

(A) Bell-Crank Action

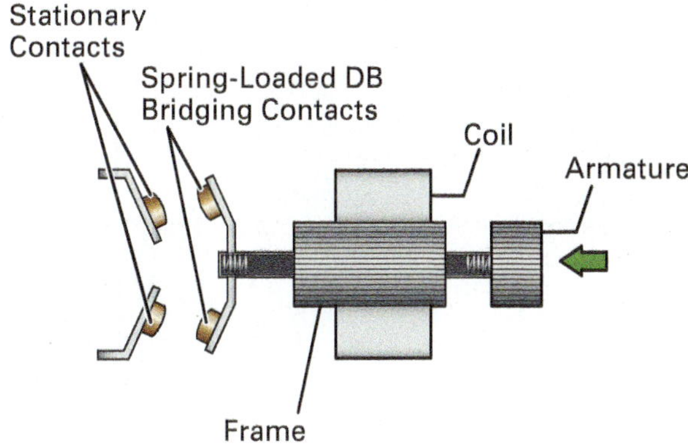

(B) Horizontal Action

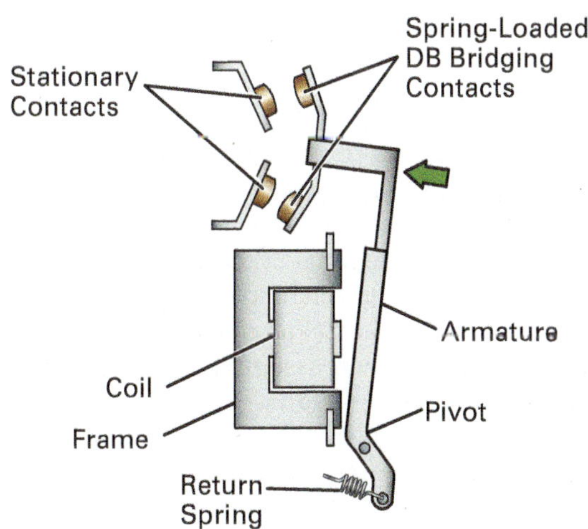

(C) Clapper Action

Figure 8 Mechanical actions used for magnetic contactors.

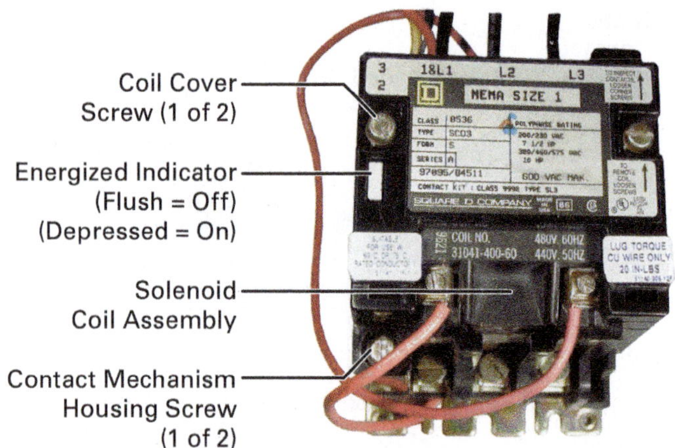

Figure 9 Typical NEMA bell-crank actuated, three-pole, magnetic contactor.

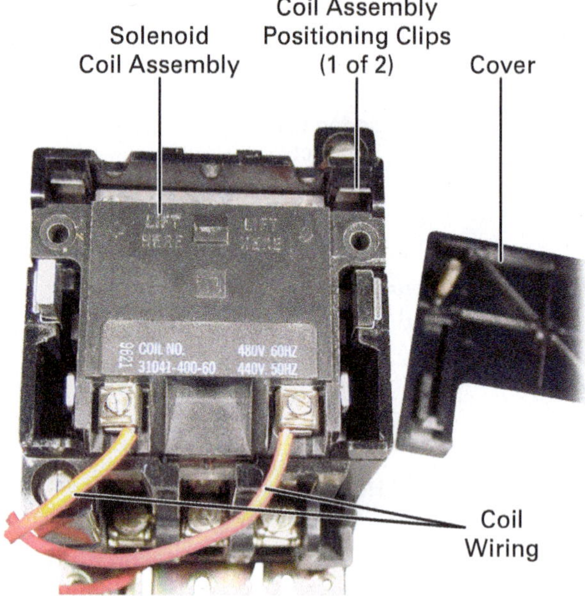

Figure 10 Contactor with cover removed.

For coil replacement, the two cover screws can be loosened or removed, and the cover removed (*Figure 10*). This exposes the solenoid coil assembly and its positioning clips that are released when the cover is removed. For the contactor in *Figure 10*, lugs on the cover are forced down between the clips and the contact mechanism housing to clamp the solenoid assembly in place horizontally.

After the wires to the solenoid coil assembly are disconnected, the assembly can be lifted out (*Figure 11*). When assembled, the lever hook of the armature fits over the spring-returned bell-crank lever. Energizing the coil causes the armature to pull in and move the bell crank so the moving contact assembly is pushed down to mate with the stationary contacts.

When the coil is energized, and the armature is pulled in, a slight air gap is maintained between the frame and armature for two reasons. The first reason is to prevent the steel armature from slamming against the steel frame, and the second is to prevent any residual magnetism in the frame from holding the armature in when the coil is de-energized. *Figure 12* shows the laminated frame and armature separated from the coil. Like a relay, the legs of the coil frame used with an AC coil are each fitted with shading coils to prevent chattering of the armature when the coil is energized.

To expose the contacts, the housing mounting screws can be loosened and the housing removed from the contactor base (*Figure 13*). However, the solenoid coil assembly does not have to be removed to gain access to the contacts.

Shading Coils

Missing or loose shading coils can cause excessive contactor noise and wear, along with coil damage due to overheating.

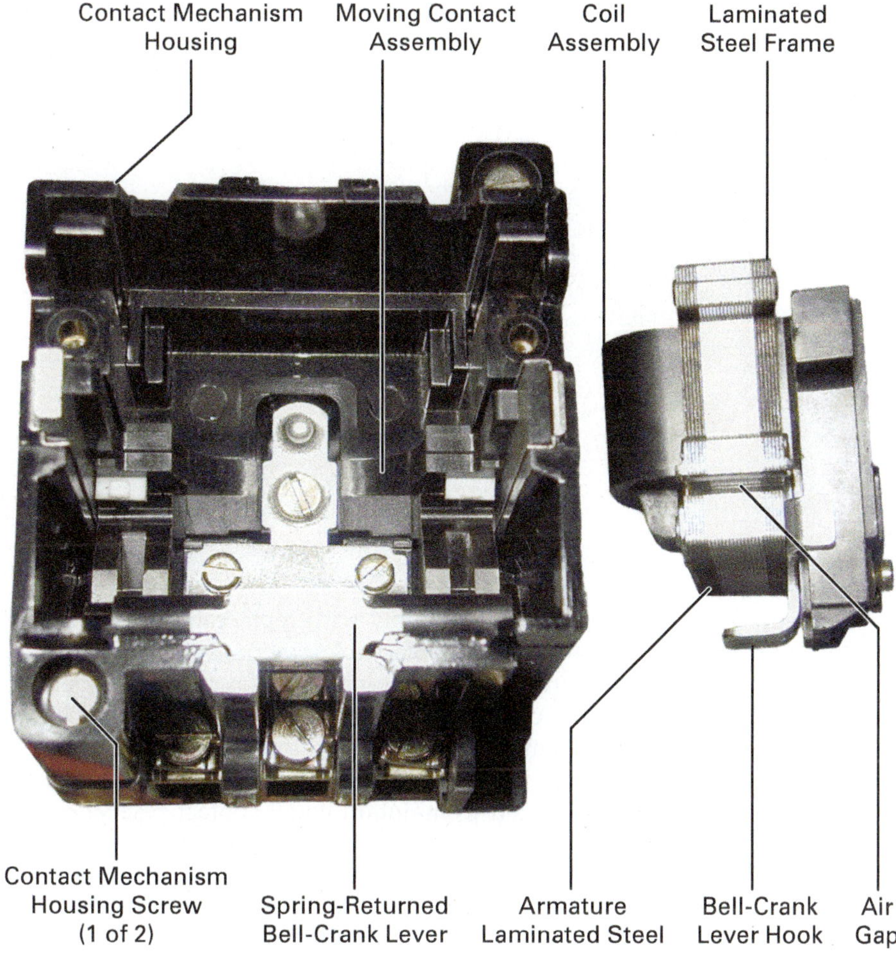

Figure 11 Contactor with solenoid coil assembly removed.

Figure 12 Disassembled solenoid coil assembly.

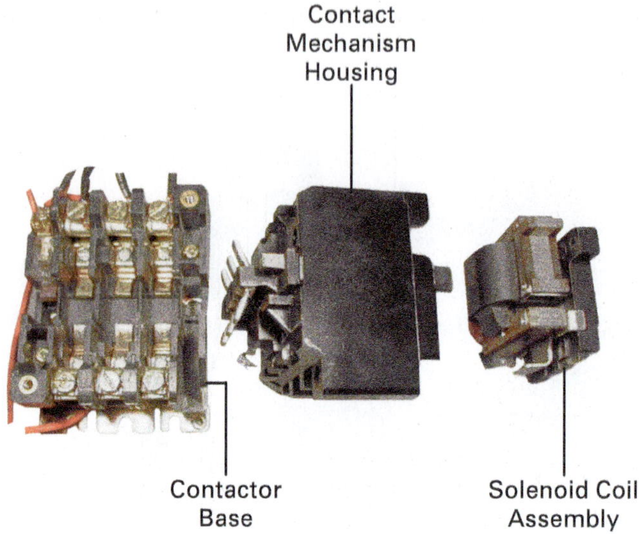

Figure 13 Disassembled contactor.

A close-up of the moving contact assembly with replaceable spring-loaded bridging contacts and corresponding replaceable stationary contacts in the contactor base is shown in *Figure 14*. Each bridging contact is removed by sliding it from under its spring-loaded clip in the moving contact assembly. Each pair of stationary contacts can be removed by removing the screws securing them to the contactor base. Note that if any contacts must be replaced, all contacts for each pole must also be replaced to maintain equal contact resistance and pressure.

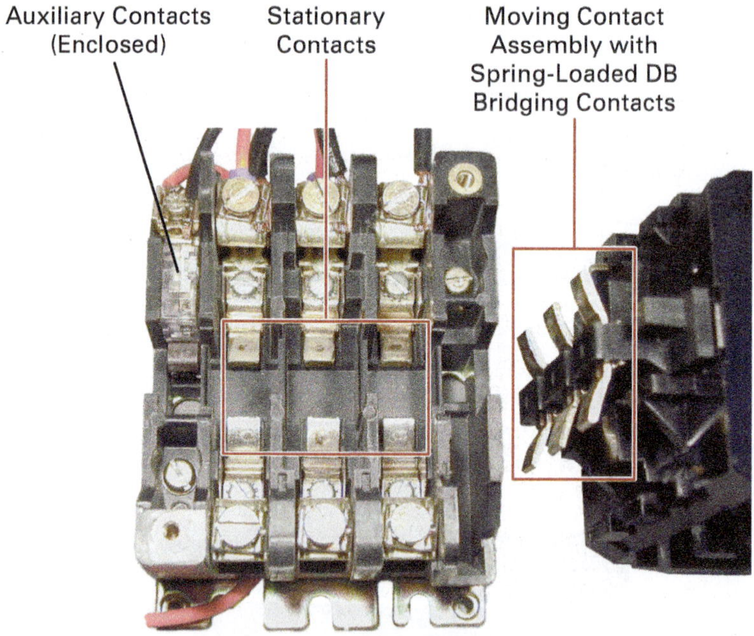

Figure 14 Close-up view of stationary and moving contacts.

1.2.2 Contactor Marking Conventions

Most contactors reflect various labeling conventions used by manufacturers. Many of these conventions exist because of NEMA and IEC standards, as well as requirements set by UL Solutions, formerly known as Underwriters Laboratories, Inc. Because some NEMA devices conform to IEC standards and vice versa, marking conventions used by manufacturers may include elements of both standards on the devices.

There are three categories of UL markings applied to contactors and motor starters:

- *UL Listed* — A UL Listed product complies with a governing UL Standard and qualifies for installation under specific provisions of the *NEC*®. Products that are UL Listed must have a UL Listing Mark (UL enclosed in a circle) on their nameplates adjacent to the ratings to which the marking applies (*Figure 15*). Next to the mark, the words "Listed Industrial Control Equipment" or "Listed Ind. Cont. Eq." must appear. UL Listed equipment is suitable for installation with general tools and for application at the ratings to which the listing mark is related. Other ratings and certifications, such as IEC hp or kW ratings or Canadian Certifications (CE or CSA), may also appear on the nameplate but are segregated from the ratings associated with the UL Listing Mark.
- *UL Recognized* — UL Recognized equipment is unsuitable for general use and must be combined with other items, under the stated conditions of acceptability, into another product that can be UL Listed. The conditions of acceptability are published in a required manufacturer's UL Component Recognition Report. A UL Recognized product carries the UL Recognized Component Mark called a backward UR—a printed, boldfaced, back-slanted letter U and R, joined and reversed as they would appear in a mirror. This mark must appear adjacent to the manufacturer's identification and catalog number.
- *UL Classified* — A UL Classified product carries the circled UL mark and the word "Classified" along with a notation that UL has evaluated the product for compliance with a specific characteristic, standard, or part of a standard. UL classification of a product for these items has no bearing on the product's ability to comply with the *NEC*®.

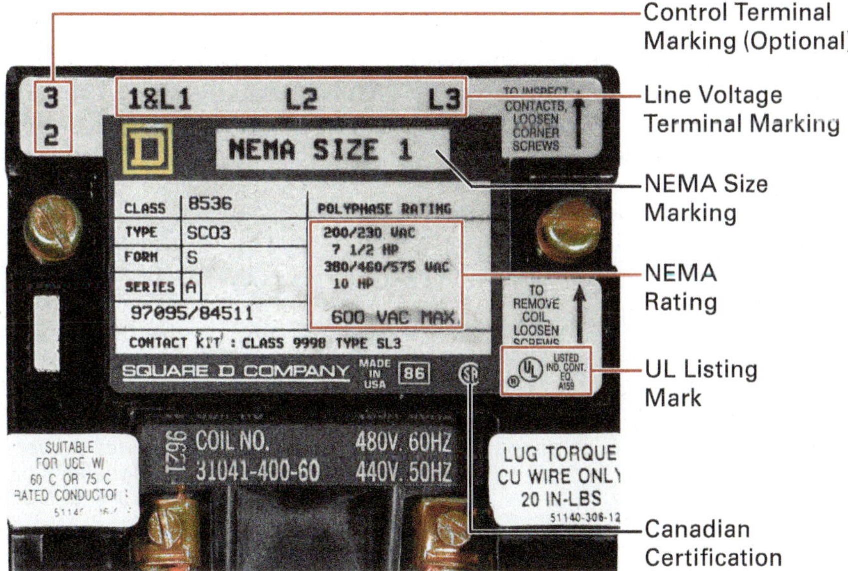

Figure 15 Typical NEMA contactor nameplate showing size, rating, and UL Listing Mark.

It is important to note that not all functions or ratings shown on a nameplate of a UL Listed, Recognized, or Classified product are qualified for use under all articles of the *NEC*® or meet other UL Standards.

The following ratings are normally marked on NEMA-rated contactor or motor starter nameplates:

- *NEMA size* — The NEMA size designation is a standardized rating system for contactors and motor starters (*Figure 15*). As defined by NEMA standards, horsepower, voltage, frequency, and/or current ratings are assigned to each size.

- *Horsepower and voltage* — The maximum horsepower at various voltages that correspond to the designated NEMA size.
- *Continuous current* — In addition to or instead of a horsepower marking, some contactors are marked with the maximum current that an enclosed starter or contactor may carry continuously or switch without exceeding the temperature rise permitted by NEMA standards.

The following ratings are normally marked on IEC contactor or motor starter nameplates:

- *Horsepower and kilowatts* — The maximum rating for each operational voltage and utilization category marked on the nameplate. The most common ratings given are for Utilization Category AC-3.
- *Utilization category* — Describes the types of service for which the controller is rated, such as AC-1, AC-2, or AC-3.
- *Thermal current (I_{th})* — The maximum current a contactor or starter, outside an enclosure, can carry continuously without exceeding the temperature rise allowed by the IEC standard.
- *Rated operational current (I_e)* — The maximum full-load amperage (FLA) at which a motor starter or contactor can manage for a given combination of voltage, frequency, and utilization category. A device may have more than one operational current rating.
- *Rated insulation voltage (U_i)* — A parameter sometimes shown that defines the insulation properties of the controller. This parameter is not usually needed for selection or application purposes.
- *Rated operational voltage (U_e)* — The voltage required for each listed hp or kW rating.
- *Standard designation* — The specific IEC standard to which the product has been tested.

The symbols shown in *Figure 16* and *Figure 17* for NEMA are generally used for all North American and Canadian products, including lighting, general-purpose, and IEC contactors sold in these countries. Many IEC contactors retain the IEC marking as additional information. However, imported equipment may only reflect the IEC marking.

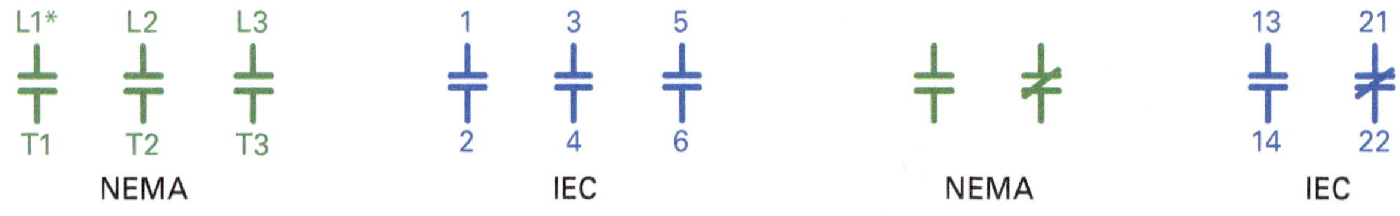

Alphanumeric corresponding to incoming line and motor terminal designations.

Single digit numeric. Odd numbers for line terminals and even numbers for load terminals.

No specific marking of component.**

Two digit numeric marking. First digit designates sequence. Second digit designates function as 1-2 for NC and 3-4 for NO.

(A) Power Terminals

(B) Control Terminals

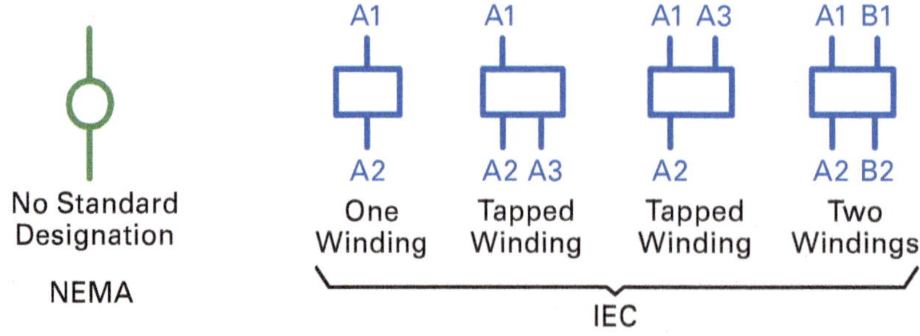

Figure 16 Conventional terminal markings.

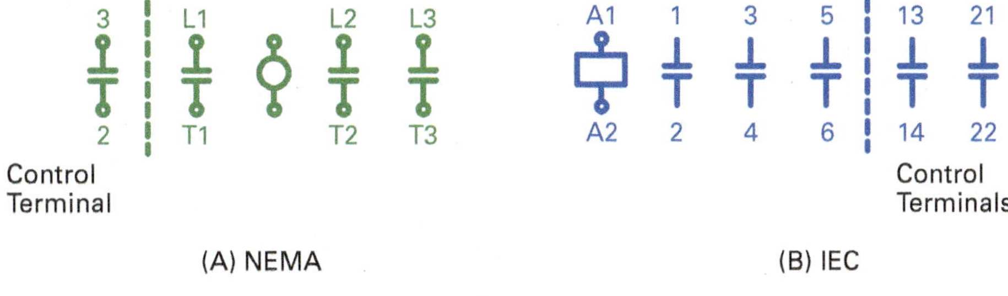

Figure 17 Typical contactor/motor starter marking and symbols.

The most common wiring practice for NEMA-marked products is to connect all the terminals with the same terminal markings. For example, in power circuits, the terminal marked T1 on a motor starter is connected to the terminal marked T1 on a motor, and in control circuits, terminals marked 1, 2, and 3 on the starter are connected to terminals 1, 2, and 3 on the control station. The IEC devices use power terminals marked 1, 3, 5, and 2, 4, 6 that correspond to L1, L2, L3, and T1, T2, T3 on NEMA devices. IEC device terminals carrying the same numbers are not typically connected. Control terminals on IEC contactors have two-digit markings. The first digit designates the location sequence, and the second is the control-device function. Most NEMA coils are single voltage, while IEC coils may be single or multivoltage.

1.2.3 Two-Wire and Three-Wire Contactor Control

As shown in *Figure 18*, two- or three-wire control using **pilot devices** can be used to energize an electrically held magnetic contactor. Pilot devices are switches or sensors that control a contactor.

Pilot devices: Devices such as electrical push buttons and switches that influence the operation of related relays and contactors.

Two-wire control of a contactor can be accomplished with a simple On/Off toggle switch. A latching Start/Stop switch, also called a *Start/Stop station*, often uses a holding contact since the push buttons only make contact momentarily. In the case of switches, a loss of control voltage causes the contactor to de-energize. When power is restored, the contactor will automatically re-energize if the pilot device remained closed.

Three-wire control with a Start/Stop switch is accomplished with a pair of momentary contact push-button switches (*Figure 19*). It is better than a switch when loss-of-voltage protection or personnel safety is at stake.

In the simplest Start/Stop configuration, a contactor is initially energized when the Start push button is momentarily depressed to connect L1 (via wire 1) through the closed Stop switch and the contactor coil to L2. To keep the contactor energized, a set of **auxiliary contacts** is connected in parallel with the Start switch via wires 2 and 3. These contacts hold or *seal in* the contactor coil after the Start button is released. The auxiliary contacts are normally mounted on the left side of NEMA contactors (*Figure 14*) and are mechanically closed by a plunger on the side of the main contact assembly when the contactor is energized. When factory-installed, the auxiliary contacts on NEMA contactors are sometimes marked as terminals 2 and 3. However, auxiliary contact sets can be field mounted to many NEMA contactors, and often on either side of the contactor.

Auxiliary contacts: An added set of contacts that are mechanically connected to a contactor, either opening or closing as the contactor itself moves. Auxiliary contacts can be normally open (NO) or normally closed (NC).

When closed by the moving contactor assembly, the auxiliary holding contact bypasses the Start push-button switch contacts so that, after the push button is released, control voltage will remain applied to the coil. If the source of the control voltage is momentarily interrupted or the Stop push button is momentarily depressed, the contactor de-energizes, the auxiliary contact opens, and the contactor remains de-energized until the Start push button is depressed again.

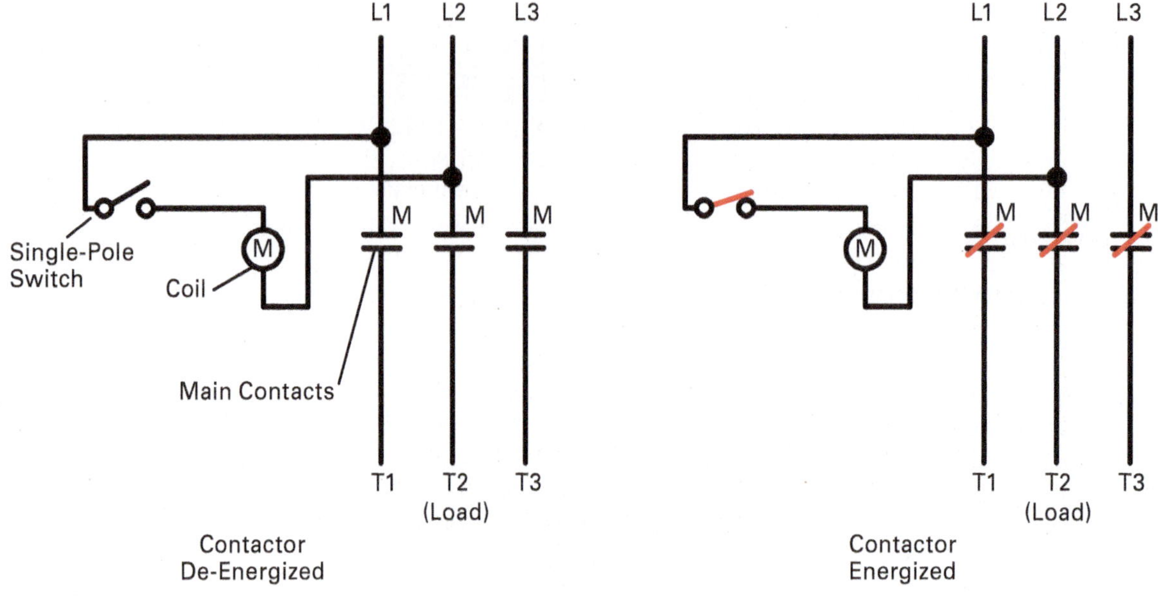

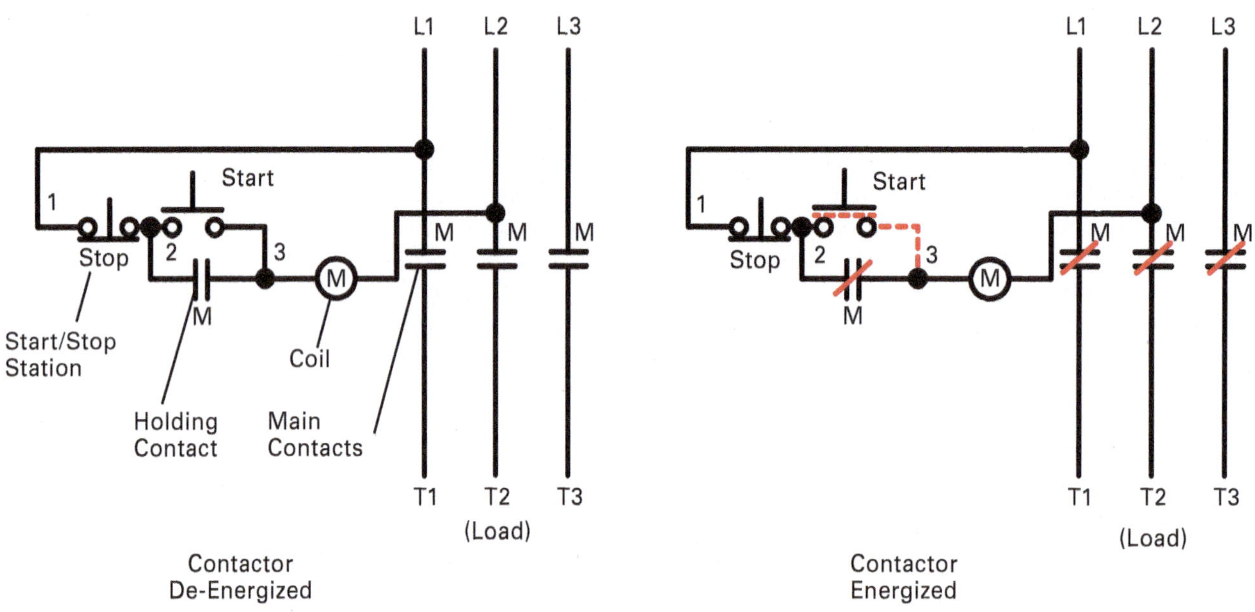

Figure 18 Two-wire and three-wire control schematics.

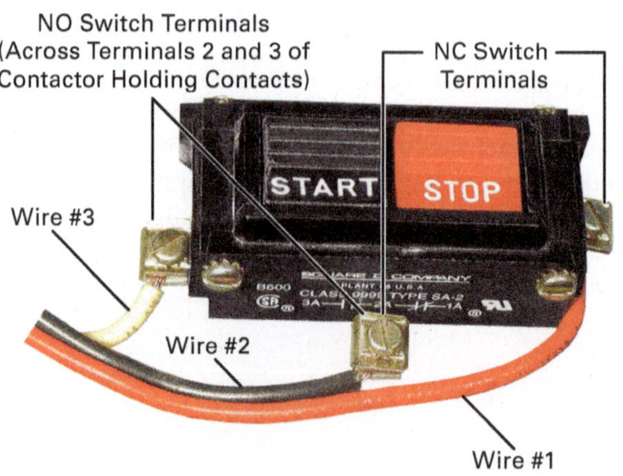

Figure 19 Start/Stop switch assembly.

1.3.0 Types of Motor Overload Protection

Overload protection, also known as *running protection,* for an electric motor is necessary to prevent burnout and ensure maximum operating life. If permitted, an electric motor will operate at an output of more than its rated capacity. Conditions of motor overload may be caused by an overload from driven machinery, a low line voltage, or an open line in a polyphase system that results in single-phase operation. Under any condition of overload, a motor draws excessive current that causes overheating. Because motor winding insulation deteriorates when subjected to overheating, there are established limits on motor operating temperatures. To protect a motor from overheating, overload relays are employed on motor controls to limit the amount of current drawn.

The ideal overload protection for a motor is a sensing element with current-sensing properties that are very similar to the heating curve of the motor (*Figure 20*). Through the overload relay, the element opens the control circuit when the motor full-load current is exceeded. The operation of the protective device should allow the motor to carry short overloads but must also react when an overload has persisted for too long.

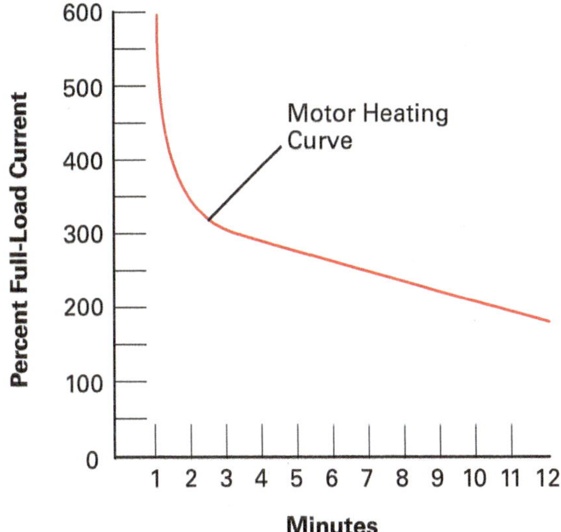

Figure 20 Typical motor heating curve.

Single-element, nontime-delay fuses are not designed to provide overload protection. Their basic function is overcurrent and short circuit protection. Motors draw a high inrush current when starting, and single-element, nontime-delay fuses cannot distinguish between a normal inrush current and a damaging overload. These fuses, if chosen based on motor full-load current, will blow every time the motor tries to start. On the other hand, if the chosen fuse is large enough to pass the starting or inrush current, it will not protect the motor against smaller, long-term, harmful overloads. Dual-element, time-delay fuses can provide a degree of motor overload protection but must be replaced when they open. Overload relays do a much better job of overload protection and are easily reset once the problem has been resolved.

The overload relay is the heart of motor protection. As stated previously, it normally has inverse trip-time characteristics, permitting short high-current draw as the motor starts yet protecting against overloads when the motor is running. Overload relays can withstand repeated trip and reset cycles without requiring replacement. However, unless equipped with an instantaneous overcurrent trip device, overload relays cannot replace proper fuses. Each method of protection is uniquely designed to serve a specific purpose.

The overload relay consists of a current-sensing unit connected in series with the motor windings plus a mechanism actuated by the sensing unit, which serves to break the circuit. In a manual starter, an overload trips a mechanical latch that causes the starter contacts to open and de-energize the motor. In a magnetic starter, the overload relay opens its contacts to break the control circuit. These contacts are wired in series with the contactor coil in the starter's control circuit. Breaking the control circuit de-energizes the coil, and the starter contacts open.

NOTE

Solid-state overload relays are presented in NCCER Module 26407-23, *Advanced Controls*.

Overload relays can be thermal, magnetic, or solid-state. As the name implies, thermal overload relays rely on rising temperatures caused by excessive current to trip the overload mechanism. They are also known as *inverse-time overload relays* because their response time varies inversely with the level of current flow. The ideal temperature curve of an overload relay corresponds to the motor's heating curve. However, magnetic overload relays react only to current excesses and do not correspond with the motor's heating curve.

Thermal overload relays can be further subdivided into melting-alloy and bimetallic types. Both melting-alloy and bimetallic types are usually available in three or four NEMA-designated trip time classes. Class 10 devices must trip within 10 seconds, Class 15 within 15 seconds, Class 20 within 20 seconds, and Class 30 within 30 seconds. The three most common classes available for NEMA devices are Class 10, Class 20 (considered standard), and Class 30. IEC starters usually use equivalent Class 10 devices.

Thermal units, also referred to as *heaters*, for both melting-alloy and bimetallic overload relays should always be selected based on the full-load current rating on the motor nameplate. The full-load current is more commonly referred to and shown on the nameplate as the FLA rating. If the nameplate is missing, the full-load current for a specific installation can be measured when the motor is operating under its maximum load conditions. As a last resort, manufacturers have charts that provide approximate FLA values for motors based on the horsepower ratings and operating voltages. However, any heaters chosen based on tables or measurements may have to be changed if nuisance trips occur. Using the motor data directly generally provides the most reliable results.

1.3.1 Melting-Alloy Thermal Overload Relays

Figure 21 shows a typical melting-alloy thermal overload relay used with a three-pole contactor. It comprises a housing with three changeable melting-alloy heater assemblies, an internal spring-loaded trip mechanism that operates an NO switch, and a reset button. The alloy, called a *eutectic alloy*, is a mixture of low-melting point metals that melt at fixed, predictable temperatures and rapidly changes from a solid to liquid. The alloy is unaffected by repeated melting and hardening cycles.

Figure 22 shows a simplified version of a melting-alloy overload relay along with its symbol representation and a typical application. When the alloy in the solder pot is solid, the shaft of the ratchet wheel is locked in place, and the trip lever holds the overload contacts closed. The closed overload contact is in series with the coil of the starter contactor. Motor current flows through the closed contactor contacts, through the overload heaters, and out to the motor connected to T1, T2, and T3. Excessive current passing through any one of the heater elements eventually raises the temperature enough to cause the alloy in the solder pot to melt. The ratchet wheel shaft then turns in the molten alloy due to the upward pressure of the spring-loaded trip lever. Through mechanical linkage, this action causes the overload contact to open, de-energizing the contactor and stopping the motor.

Figure 23 shows one of the heaters removed from the overload relay shown in *Figure 21* and turned over to show the ratchet wheel and solder pot containing the alloy. Also visible is one of the three trip levers that, through common linkage, open the spring-loaded overload contact. A cooling-off period is required to allow the alloy in the solder pot to harden before the overload relay assembly can be manually reset and motor service restored.

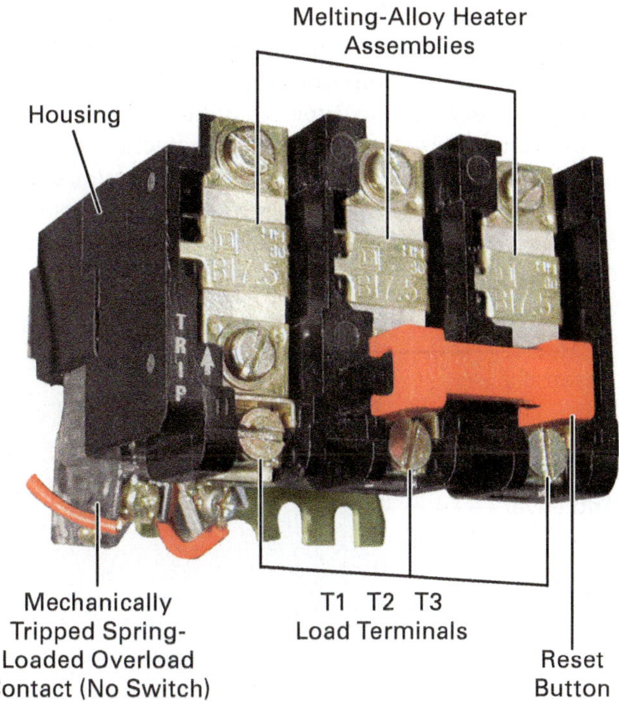

Figure 21 Typical melting-alloy overload relay.

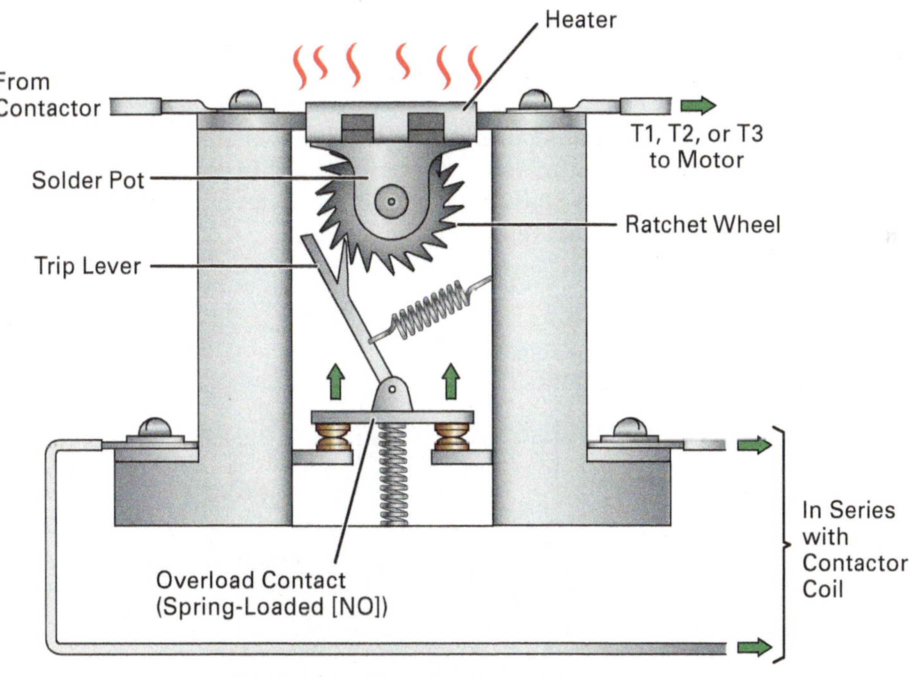

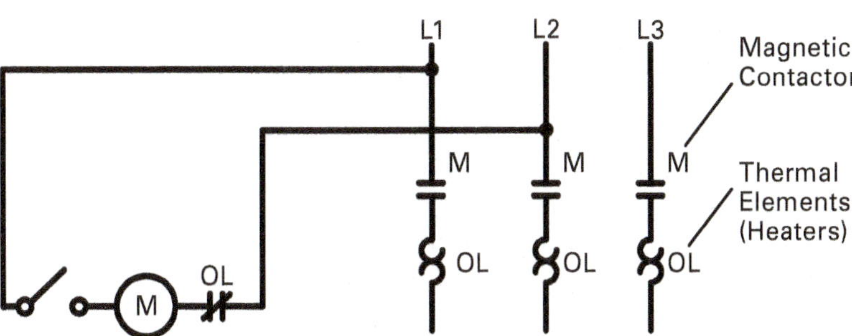

Figure 22 Thermal overload relay operation and wiring diagram.

Melting-alloy heater assemblies are factory sealed and interchangeable, ensuring a constant and reliable relationship between the heater element and solder pot. They are virtually tamperproof as a result. This important feature is not possible with all overload relay types. A wide selection of heater assemblies is available to provide precise motor overload protection for any motor. Special heaters are available for temperature compensation, which may be required when the motor and starter are in different environments. For example, the motor may be in a very hot room while its starter is in an air-conditioned equipment room.

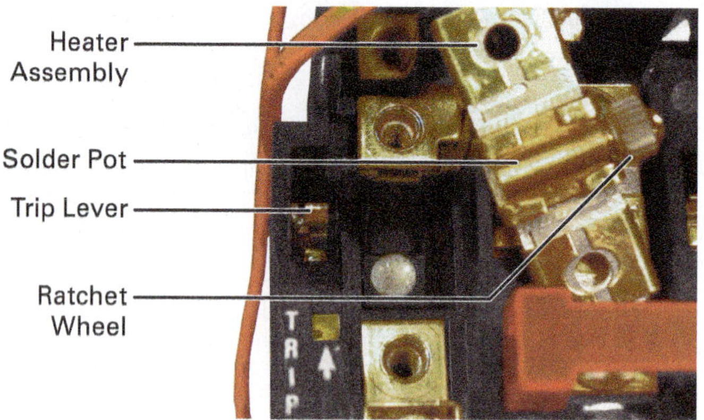

Figure 23 View of melting-alloy heater assembly and trip lever.

1.3.2 Bimetallic Overload Relays

Today, a great many overload relays are bimetallic (*Figure 24*). These include the ones often found in IEC motor starters. Except for the IEC devices, which have fixed heaters, bimetallic overload relays use interchangeable heaters to accommodate the current of various motors. If a different current range is required for an IEC starter with fixed heaters, or its heater must be replaced due to a failure, the entire overload relay must be replaced.

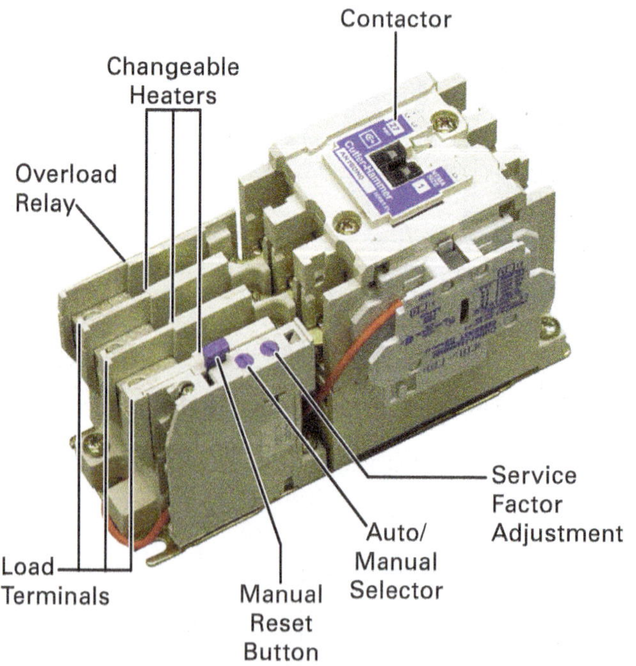

Figure 24 Typical three-pole, bimetallic overload relay.

A bimetallic overload relay has three advantages over a melting-alloy overload relay:

- The automatic reset feature is an advantage when the device is mounted in a location not easily accessible for manual operation or requires an automatic reset, such as in refrigeration units. Some bimetallic overload relays allow a choice of automatic reset or manual reset.
- Bimetallic overload relays usually have a motor service factor adjustment to adjust the trip point within a range of 85% to 115% of the nominal trip rating. This adjustment is useful when the recommended heater size might result in unnecessary tripping, while the next larger size would not provide adequate protection.
- Most bimetallic overload relays use automatic ambient temperature compensation so the overload relay can be at a different ambient temperature than the motor. The ambient compensation is accomplished by a second bimetal element that changes the trip point of the overload contacts. If no automatic compensation is included, special heaters may sometimes be available to partially correct the difference.

Figure 25 is a single-pole, auto-reset, bimetallic overload relay with the side cover removed, showing the location of the contacts, changeable heater, and heater bimetal element. As a motor overload develops, the increasing heat generated by the heater eventually causes the U-shaped element to straighten and press on the overload contact plunger. This action forces the ambient compensated overload contacts to open. The contacts are held open until the heater and bimetallic strip cool, allowing them to toggle back to their closed position.

Overload Relays

Although seldom identified, overload relays are also categorized by NEMA for their ability to consider the cumulative heating effects of motor operation. This characteristic is called *operating memory* and may be either volatile or nonvolatile, as shown in the list below.

All thermal overloads have a volatile operating memory (Category B) in that they must cool off before they can be reset. This mimics the heat retention of the motor to some extent. Solid-state overload relays may have volatile and/or nonvolatile memory (Category B and/or Category A), depending on whether the memory is stored by charging a capacitor or as a value in the nonvolatile microprocessor memory. Magnetic overload relays normally have no memory (Category C) and can be instantly reset even though the motor may still be hot. Therefore, the operating memory of each category is as follows:

- *Category A* — Nonvolatile
- *Category B* — Volatile
- *Category C* — No memory

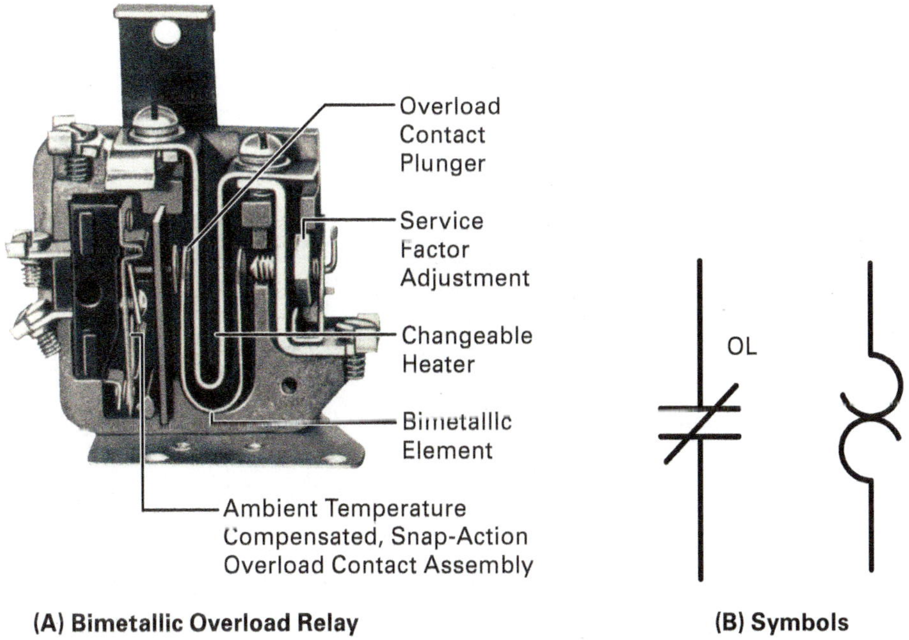

Overload Contact Plunger
Service Factor Adjustment
Changeable Heater
Bimetallic Element
Ambient Temperature Compensated, Snap-Action Overload Contact Assembly

(A) Bimetallic Overload Relay

OL

(B) Symbols

Figure 25 A single-pole, auto-reset, bimetallic overload relay.
Source: Schneider Electric

1.3.3 Magnetic Overload Relays

Magnetic overload relays are used for special applications in heavy industrial environments. In some cases, the specific motor current must be detected, or prolonged inrush currents may be encountered, such as when starting a high-inertia load. Under either of these conditions, the time/current curves of thermal overload relays would be unsuitable. Magnetic overload relays may offer settings for either instantaneous or inverse-time trips along with adjustable time or current settings.

Magnetic overload relays pass the load current through a heavy coil that functions as a solenoid in each relay pole. The position of a moving core within the solenoid coil is adjustable, so the core is sensitive to current flow. When equipped or set for inverse-time trip purposes, a fluid-filled adjustable **dashpot** (*Figure 26*) is attached to the moving core so the time delay of the relay can be adjusted over a wide range for various currents. When activated, the core's instantaneous or time-delayed total travel trips an overload contact, which must be manually reset.

Dashpot: A mechanical device that resists movement to cushion or dampen motion, such as the action of a shock absorber. A typical dashpot slows and controls mechanical movement by forcing air or liquid through an orifice with a fixed or adjustable diameter.

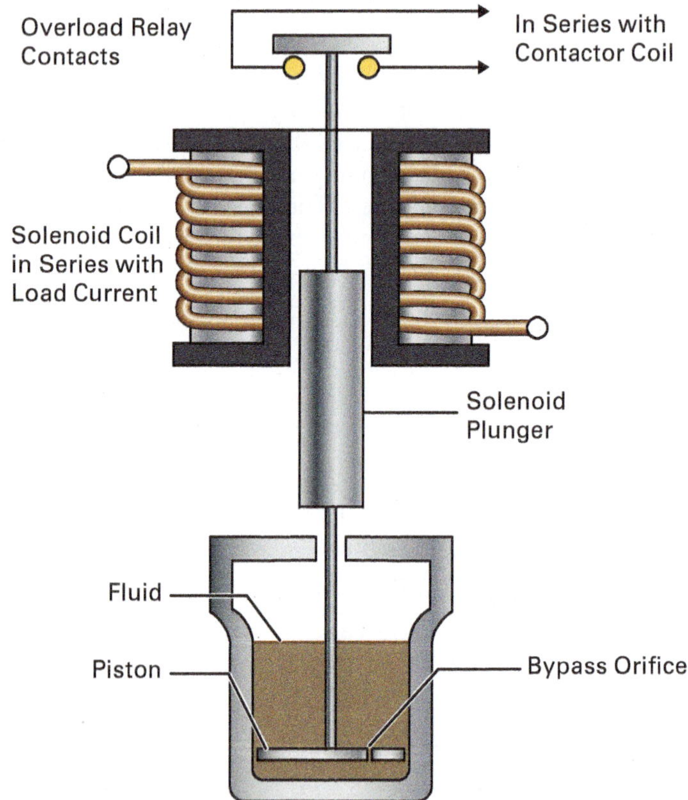

Figure 26 Magnetic overload relay dashpot.

Automatic Reset

NEC Section 430.43 prohibits the use of automatic overload reset devices in applications where an automatic restart could endanger personnel. Although an automatic reset feature sounds convenient, it can also be very dangerous and destructive.

1.0.0 Section Review

1. The abbreviation DB stands for _____.
   a. dual break
   b. double break
   c. delayed broadcast
   d. dual benefit

2. Equipment marked as UL Recognized has been examined by UL Solutions and determined to be _____.
   a. the same as equipment marked UL Listed
   b. suitable for general use
   c. acceptable for use in another product that can be UL Listed
   d. better than all similar products

3. Most IEC motor starters use _____.
   a. melting-alloy thermal overload relays with interchangeable heaters
   b. overload relays with fixed heaters
   c. thermistor overload relays with interchangeable heaters
   d. electronic overload relays with fixed heaters

2.0.0 Magnetic and Manual Motor Starters

Objective

Explain how magnetic and manual motor starters are selected.

a. Explain how NEMA contactors and motor starters are selected.
b. Explain how IEC contactors and motor starters are selected.
c. Identify various contactor and motor starter accessories.

Performance Task

1. Make all wiring connections for a magnetic motor starter controlled by two push-button stations, including a holding circuit interlock.

A NEMA-rated motor starter consists of an enclosure housing one or more NEMA-rated contactors with one or more associated overload relays on a common base plate. All wiring confined to the enclosure is typically done at the factory. The installer must connect any remote-control circuit wiring as well as the line voltage and load connections.

In contrast, IEC-rated magnetic motor starters are not usually factory wired and must be assembled in the field. However, several North American manufacturers also offer IEC magnetic motor starters as factory-assembled, internally wired units, like NEMA units. Field assembly, when required, includes assembling and mounting the contactor(s) and overload relay(s), as well as furnishing and connecting all the internal and external wiring.

Instantaneous Overload Relays

Instantaneous overload relays, including magnetic or solid-state versions, are sometimes referred to as *jam relays*. They are used in applications where the driven load is subject to jamming, such as a conveyor belt or stone crusher. In some applications, a magnetic overload relay may be combined with a thermal overload relay to provide both load jamming and standard running overload protection.

TABLE 5 IEC and NEMA Starter and Contactor Comparison

Subject	IEC	NEMA
Starter size	Physically smaller	Physically larger
Contactor performance	Electrical life typically 1,000,000 cycles for AC-3 (nonplugging/nonjogging) operations and 30,000 cycles for AC-4 (plugging/jogging) operations	Electrical life typically 2.5 to 4 times more than equivalently-rated IEC devices
Contactor application	Application sensitive—more knowledge and care in selection required	Application selection easier and less critical, with fewer parameters to consider
Overload relay reset characteristics	Class 10 (fast) typical	Class 20 (standard) typical
Overload relay adjustability	Fixed, noninterchangeable heaters; adjustable to suit different motors at the same horsepower	Field-changeable heaters allow adjustment for motors of different horsepower and greater flexibility
Overload relay reset mechanism	Manual/Auto typical; some use Reset/Stop dual function mechanism	Manual/Auto or Manual Only typical
Short circuit current rating	Typically designed for use with fast-acting, current-limiting fuses	Designed for use with common domestic current-rated fuses and circuit breakers

Table 5 summarizes the basic differences between IEC and NEMA devices.

The most common NEMA or IEC magnetic motor starters are non-reversing and reversing types for wye, delta, and reconfigurable wye/delta motors. *Figure 27* and *Figure 28* show typical non-reversing and reversing NEMA motor starters, respectively, along with their symbol representations. IEC motor starters are similar except for their terminal markings, which have been described previously.

As shown in the starters' symbol representations in *Figure 27* and *Figure 28*, the normal approach to wiring a non-reversing starter for forward motor rotation is to connect L1 to T1, L2 to T2, and L3 to T3. In the case of the reversing motor starter, the same wiring convention is observed for the forward contactor (F). The wiring convention for the reverse contactor (R) is to interchange L1 and L3 so that L1 is tied to T3, L2 is tied to T2, and L3 is tied to L1. This reverses the direction of the motor's rotation. Note that a mechanical interlock is sometimes used to prevent both contactors from being closed at the same time. If pilot devices are used that are not mechanically interlocked, a backup electrical interlock system can be used for protection against accidentally energizing both contactors simultaneously. As shown, the electrical interlock uses a set of NC auxiliary contacts (F and R) that are wired in series with their opposite contactor coils. The NO auxiliary contacts can be used as holding contacts for start switches or other pilot devices where needed.

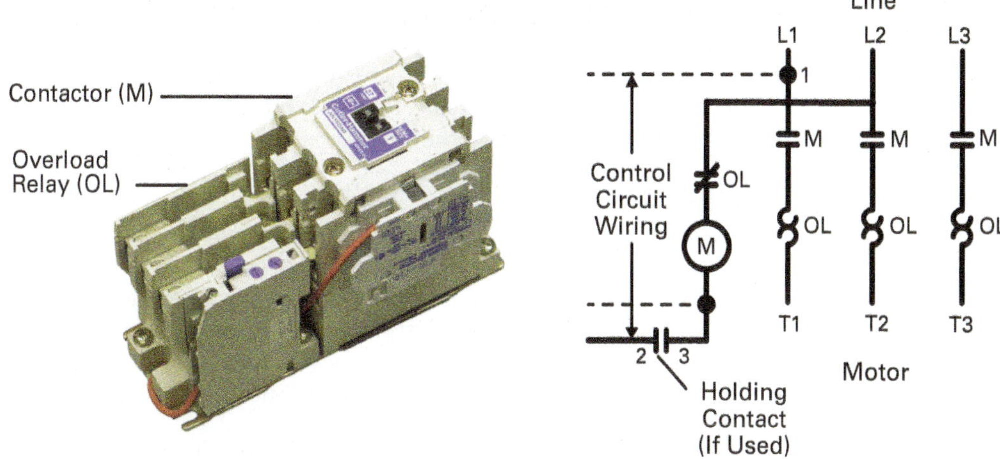

Figure 27 Typical NEMA non-reversing magnetic motor starter and schematic representation.

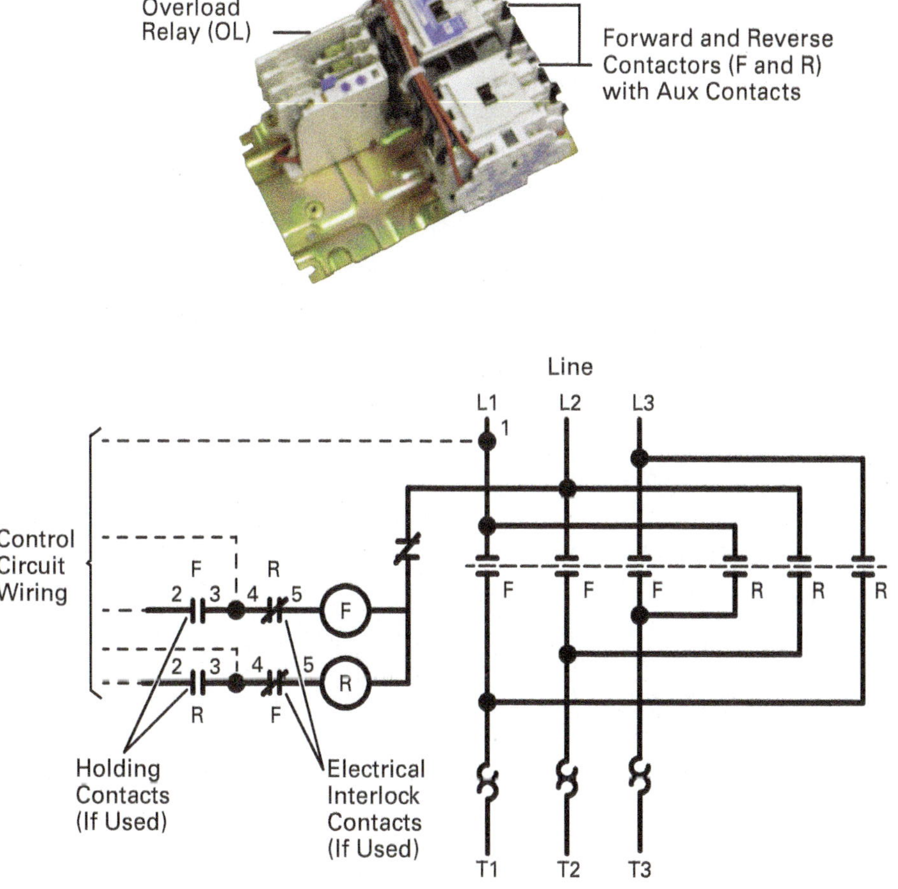

Figure 28 Typical NEMA reversing magnetic motor starter and schematic representation.

2.1.0 NEMA Contactors and Motor Starters

NEMA magnetic contactors and motor starters are selected based on NEMA sizes and motor nameplate data, including the NEMA design letter, voltage, horsepower, FLA, and other data.

2.1.1 Ratings

NEMA contactors and motor starters (*Figure 29*) are both designed to meet the size rating specified in the NEMA standards. The standards are used to provide electrical interchangeability among manufacturers for a given NEMA size. *Table 6A* and *Table 6B* list the NEMA sizes for contactors and motor starters and provide the maximum horsepower ratings at various voltages and unique motor applications, such as **plugging** and **jogging** operations. Jogging is also referred to as *inching*.

The table also lists the continuous current, kilowatts (kW), kilovolt-amperes (kVA), and kilovolt-amperes reactive (kVAR) ratings for other applications. Because NEMA contactors and motor starters must be able to safely interrupt the locked-rotor current of a motor in nonplugging or nonjogging applications, the maximum horsepower ratings shown for a particular NEMA size are based on the locked-rotor current for motors at the listed horsepower. For plugging and jogging, the maximum horsepower within any NEMA size is derated, as reflected in *Table 6A* and *Table 6B*.

Plugging: Braking a three-phase motor by reversing the phase sequence of the power supply to the motor to achieve both rapid stops and quick reversal of rotation.

Jogging: Intentional, repeated closure of a circuit used to start and stop a motor to initiate small, incremental movements in the driven load; also referred to as *inching*.

NOTE

When using a starter for plugging or jogging, the starter should be increased by one size to compensate for excess heating during locked-rotor conditions. *Table 6A* and *Table 6B* show this increase.

Figure 29 Typical NEMA Size 1 motor starter wired with a three-wire Stop/Start switch.

2.1.2 Motor Nameplate Data

The information on a motor nameplate is essential in selecting a compatible contactor or motor starter. The nameplate data must be used instead of actual current measurements or manufacturers' tables whenever possible. On the motor nameplate shown in *Figure 30*, the following parameters must be noted during the selection process:

- *Voltage and frequency* — 230V at 60 Hz AC
- *FLA* — 13.0A
- *Phase* — Three-phase
- *NEMA design letter* — B
- *Horsepower* — 5 hp
- *Service factor (SF)* — 1.0
- *Power factor* — 85.7 kW

NOTE

The power factor is only necessary if any capacitive correction is connected on the load side of the overload relays. If so, the overload relay manufacturer must be consulted to determine current elements for the relays.

Assuming the motor is not used for plugging or jogging, *Table 6A* and *Table 6B* shows that a 230V, polyphase, 60 Hz, 5 hp motor can be controlled with a NEMA Size 1 starter. The overload relay would be selected and equipped for an FLA of 13A in accordance with the manufacturer's motor starter overload relay selection tables and adjusted (if allowed) for a service factor of 1.0. The service factor rating is the amount of extra power demand that can be placed on the motor for a short period without damage to the motor. Common motor service factors range from 1.0 to 1.25, indicating that the motor can intermittently be required to produce up to 25% extra power over its nominal rating. If the service factor is unknown or not found on the nameplate, a value of 1.0 should be assumed.

The total excessive current required for intermittent power demands can be approximated by multiplying the motor FLA by the motor SF. For the motor SF value of 1.0, shown in *Figure 30*, the allowed extra current is 0%. This means any service factor adjustment would be equal to the value of 13A FLA. If the SF had been 1.25, then the adjustment would equal 16.25A (13A × 1.25 = 16.25A).

Any SF adjustment determined that is greater than the FLA permits the overload relay to allow intermittent, excessive power demands without nuisance tripping. For overload relays that are not ambient temperature compensated and where the starter is in a different temperature environment than the motor, an appropriate overload relay and heating elements must be selected as specified in special tables available from the starter manufacturer.

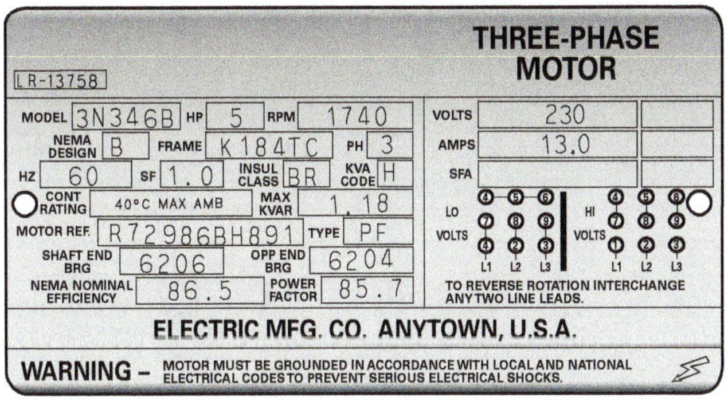

Figure 30 Typical motor nameplate.

Contactors Used for Plugging or Jogging Applications

Contactors used for plugging and jogging are derated. *Plugging* is the momentary stopping and/or reversing of a running motor. *Jogging* is the repeated momentary starting of a stopped motor. Both plugging and jogging are used in industrial machine operations that can subject the contactors to high inrush currents many times over a short period.

TABLE 6A NEMA Sizes for Contactors and Motor Starters (Reprinted by Permission of National Electrical Manufacturers Association) (1 of 2)

NEMA Size	Volts	Maximum hp Rating (Nonplugging and Nonjogging Duty[1])		Maximum hp Rating (Plugging and Jogging Duty[1])		Continuous Current Rating (Amperes) 600V Maximum[2]
		Single Phase	Polyphase	Single Phase	Polyphase	
00	115	$\frac{1}{3}$	—	—	—	9
	200	—	$1\frac{1}{2}$	—	—	9
	230	1	$1\frac{1}{2}$	—	—	9
	380 (50 Hz)	—	$1\frac{1}{2}$	—	—	9
	460	—	2	—	—	9
	575	—	2	—	—	9
0	115	1	—	$\frac{1}{2}$	—	18
	200	—	3	—	$1\frac{1}{2}$	18
	230	2	3	1	$1\frac{1}{2}$	18
	380 (50 Hz)	—	5	—	$1\frac{1}{2}$	18
	460	—	5	—	2	18
	575	—	5	—	2	18
1	115	2	—		—	27
	200	—	$7\frac{1}{2}$	—	3	27
	230	3	$7\frac{1}{2}$		3	27
	380 (50 Hz)	—	10	—	5	27
	460	—	10	—	5	27
	575	—	10	—	5	27
2	115	3	—	2	—	45
	200	—	10	—	$7\frac{1}{2}$	45
	230	$7\frac{1}{2}$	15	5	10	45
	380 (50 Hz)	—	25	—	15	45
	460	—	25	—	15	45
	575	—	25	—	15	45
3	115	$7\frac{1}{2}$	—	—	—	90
	200	—	25	—	15	90
	230	15	30	—	20	90
	380 (50 Hz)	—	50	—	30	90
	460	—	50	—	30	90
	575	—	50	—	30	90
4	200	—	40	—	25	135
	230	—	50	—	30	135
	380 (50 Hz)	—	75	—	50	135
	460	—	100	—	60	135
	575	—	100	—	60	135
5	200	—	75	—	60	270
	230	—	100	—	75	270
	380 (50 Hz)	—	150	—	125	270
	460	—	200	—	150	270
	575	—	200	—	150	270
6	200	—	150	—	125	540
	230	—	200	—	150	540
	380 (50 Hz)	—	300	—	250	540
	460	—	400	—	300	540
	575	—	400	—	300	540
7	230	—	300	—	—	810
	460	—	600	—	—	810
	575	—	600	—	—	810
8	230	—	450	—	—	1215
	460	—	900	—	—	1215
	575	—	900	—	—	1215
9	230	—	800	—	—	2250
	460	—	1600	—	—	2250
	575	—	1600	—	—	2250

Notes:

(1) These horsepower ratings are based on the locked-rotor current ratings given in NEMA Standard ICS 2. For motors with high locked-rotor currents, a controller should be used so the locked-rotor current rating is not exceeded.

(2) The continuous current ratings represent the maximum rms current, in amperes, which the controller shall be permitted to carry continuously without exceeding the temperature rises permitted by NEMA Standard ICS 1.

TABLE 6B NEMA Sizes for Contactors and Motor Starters (Reprinted by Permission of National Electrical Manufacturers Association) (2 of 2)

Service-Limit Current Rating (Amperes)[3]	Tungsten and Infrared Lamp Load (Amperes) 250V Maximum[2]	Resistance Heating Loads (kW), Other than Infrared Lamp Loads		KVA Rating for Switching Transformer Primaries at 50 or 60 Cycles		3ϕ Rating for Switching Capacitors (kVAR)
		Single Phase	Polyphase	Single Phase	Polyphase	
11	5	—	—	—	—	—
11	5	—	—	—	—	—
11	5	—	—	—	—	—
11	—	—	—	—	—	—
11	—	—	—	—	—	—
11	—	—	—	—	—	—
21	10	—	—	0.9	1.2	—
21	10	—	—	—	1.4	—
21	10	—	—	1.4	1.7	—
21	—	—	—	—	2.0	—
21	—	—	—	1.9	2.5	—
21	—	—	—	1.9	2.5	—
32	15	3	5	1.4	1.7	—
32	15	—	9.1	—	3.5	—
32	15	6	10	1.9	4.1	—
32	—	—	16.5	—	4.3	—
32	—	12	20	3	5.3	—
32	—	15	25	3	5.3	—
52	30	5	8.5	1.0	4.1	—
52	30	—	15.4	—	6.6	11.3
52	30	10	17	4.6	7.6	13
52	—	—	28	—	9.9	21
52	—	20	34	5.7	12	26
52	—	25	43	5.7	12	33
104	60	—	—	0.9	7.6	—
104	60	—	—	—	13	23.4
104	60	—	—	1.4	15	27
104	—	—	—	—	19	43.7
104	—	—	—	1.9	23	53
104	—	—	—	1.9	23	67
156	120	—	45	—	20	34
156	120	30	52	11	23	40
156	—	—	86.7	—	38	66
156	—	60	105	22	46	80
156	—	75	130	22	46	100
311	240	—	91	—	40	69
311	240	60	105	28	46	80
311	—	—	173	—	75	132
311	—	120	210	40	91	160
311	—	150	260	40	91	200
621	480	—	182		79	139
621	480	120	210	57	91	160
621	—	—	342	—	148	264
621	—	240	415	86	180	320
621	—	240	515	86	180	400
932	720	180	315	—	—	240
932	—	360	625	—	—	480
932	—	450	775	—	—	600
1400	1080	—	—	—	—	360
1400	—	—	—	—	—	720
1400	—	—	—	—	—	900
8590	—	—	—	—	—	—
8590	—	—	—	—	—	—
8590	—	—	—	—	—	—

Notes (continued):

(3) The service-limit current ratings represent the maximum rms current, in amperes, which the controller shall be permitted to carry for protracted periods in normal service. At service-limit current, temperature rises may exceed those obtained by testing the controller at its continuous current rating.

2.2.0 IEC Contactors and Motor Starters

The selection of IEC magnetic contactors and motor starters differs from that of NEMA starters. IEC-rated contactors and motor starters are covered by *IEC Standard 60947-4-1, Low-Voltage Switchgear and Controlgear – Part 4-1: Contactors and Motor-Starters – Electromechanical Contactors and Motor-Starters.*

2.2.1 Ratings

IEC Standard 60947-4-1 does not define any standard contactor or starter sizes like NEMA. An IEC-rated contactor or motor starter (*Figure 31*) indicates that the device has been evaluated by the manufacturer or a laboratory to meet the requirements of defined applications called *utilization categories.* There are many utilization categories covering 600VAC or less switchgear and control gear:

- AC-1 to AC-8
- AC-11 to AC-15
- AC-20 to AC-23

There are also utilization categories for similar DC equipment.

Of the AC categories, those used for motor or lighting control purposes are summarized in *Table 7. Table 8* lists and expands the application definitions of the most common categories used for motor control. Category AC-1 is included because most AC-3 and AC-4 devices have Category AC-1 included on their nameplate. Category AC-2 is omitted because these motors are uncommon.

TABLE 7 IEC Categories for Lighting and Motor Control

Category	Typical Application
AC-1	Noninductive or slightly inductive loads
AC-2	Slip-ring motor starting and switching off
AC-3	Squirrel-cage motor starting and switching off
AC-4	Squirrel-cage motor starting, plugging, and jogging
AC-5a	Switching electric discharge lighting loads
AC-5b	Switching large incandescent lighting loads
AC-6a	Switching transformer loads
AC-6b	Switching capacitor loads
AC-11	Control of AC electromagnetic circuits (auxiliary contacts)

TABLE 8 Common IEC Categories for Motor Control Devices

Category	Application
AC-1	AC-1 devices are used with noninductive or mildly inductive loads, such as resistive furnaces, fluorescent lights, and incandescent lights.
AC-3	AC-3 devices are used with squirrel-cage motors for starting and switching off while running at rated speed. They include a contact-make capability for locked-rotor current and break at full-load current. They are occasionally used for jogging and plugging for limited times, such as machine setup. During such periods, the number of operations should not exceed 5 per minute nor more than 10 in a 10-minute period.
AC-4	AC-4 devices are used with squirrel-cage motors for starting and switching off while running at less than rated speed. They provide jogging, plugging stop, and plugging reverse and include contact make-and-break capability for locked-rotor current. Very few applications in the industry are totally Category AC-4.

2.2.2 Selection

There may be several ratings for a given contactor or motor starter based on application and voltage. A designer can choose a preferred device for an application based on the device's ability to meet or exceed the required horsepower, voltage rating, and other factors, including performance. This technical data is available within the manufacturer's specifications. IEC contactors used in the United States are usually marked with voltage and horsepower ratings for maximum AC-3-rated operational current.

Like NEMA motor starters, the same motor nameplate data is required to select an IEC motor starter with an appropriate overload relay, as well as to set the adjustments on the overload relay.

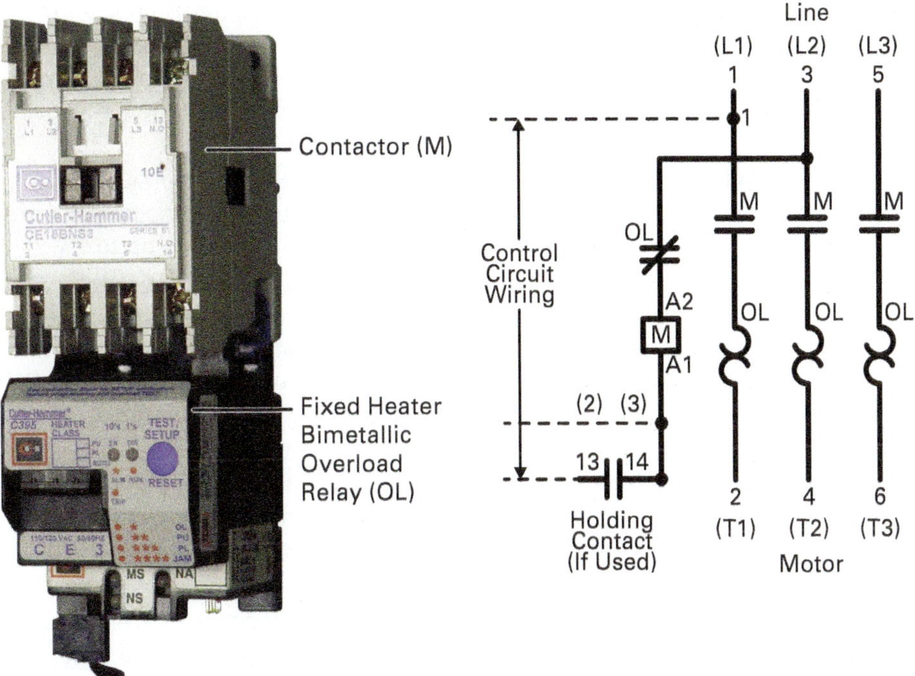

Figure 31 Typical IEC-rated, non-reversing motor starter with schematic.

Most US manufacturers provide tables that classify IEC contactors and motor starters in some sort of size ratings for horsepower, similar to NEMA, based on the maximum rated current in a range of standard line voltages. IEC contactor and motor starter selection is based on the percent that jogging and plugging (AC-4) is of the nonjogging and nonplugging (AC-3) condition in the duty cycle and the desired contact electrical life. US manufacturers often supply AC-3 derating tables for plugging and jogging duty.

In general, any time the duty cycle includes significant jogging or plugging, a larger size IEC contactor or motor starter is selected than would be needed for pure AC-3 applications. Alternatively, an appropriate AC-4-rated device is selected.

Manual motor starters (*Figure 32*) are used primarily on small machine tools, fans, pumps, conveyors, and similar equipment. They are available as single-, double-, or three-pole devices. Instead of magnetically closed contacts, NEMA-rated and/or UL-approved motor starters are operated by a mechanically linked and latched toggle handle or push button. The operating mechanisms are quick make-and-break toggle switches that cannot be teased into a partially open or closed position. They generally come equipped with melting-alloy or bimetallic overload relays.

Figure 32 Typical NEMA manual motor starter.

2.3.0 Contactor and Motor Starter Accessories

Manufacturers offer a variety of accessories that can be added to various contactors and motor starters to customize them. Technically, most of the accessories are directly related to the contactor that is part of a complete motor starter assembly. *Figure 33* shows some of these accessories. They are described further as follows:

- *Power-pole adder kits* — These kits can be used to increase the number of high-current power contacts actuated by the contactor. Depending on the manufacturer, some power-pole adder kits are mounted on the side or top of the contactor. In some cases, the internal springs or coil must be changed to accommodate the mechanical loads imposed by the extra poles.

- *Timer attachment (not shown)* — Mechanical or solid-state timers with an adjustable on-delay or off-delay are available for side mounting on the contactor. Some devices have a field-selectable on- or off-delay. They are equipped with one or more single- or double-throw, snap-switch control contacts.
- *Fuse kit* — When control circuit power tapped from the line inputs of a motor starter must be fused, side-mounted single- or double-fuse holder kits are available. Rated at 600V, they usually accept $^{13}/_{32}$" × $1^1/_2$" fuses up to 6A.
- *Transient suppression module* — Side-mounted transient suppression modules are available to reduce transient voltages and contact arcing in control circuits when the circuits are open. This eliminates electrical noise that interferes with nearby electronic circuits. Most modules consist of an RC circuit that is designed to suppress coil voltage transients to approximately 200% of peak coil supply voltage.
- *Internal auxiliary contacts* — Some contactors can have one or more internal NO or NC auxiliary contacts added for additional low-current circuits, including status feedback, or for electrical interlocking purposes. The devices have SB contacts.
- *External auxiliary contacts* — Side-mounted auxiliary contacts are available with single or double NO or NC contacts that are field convertible from NO to NC and vice versa. Some are available as DT contacts. The devices have SB contacts.

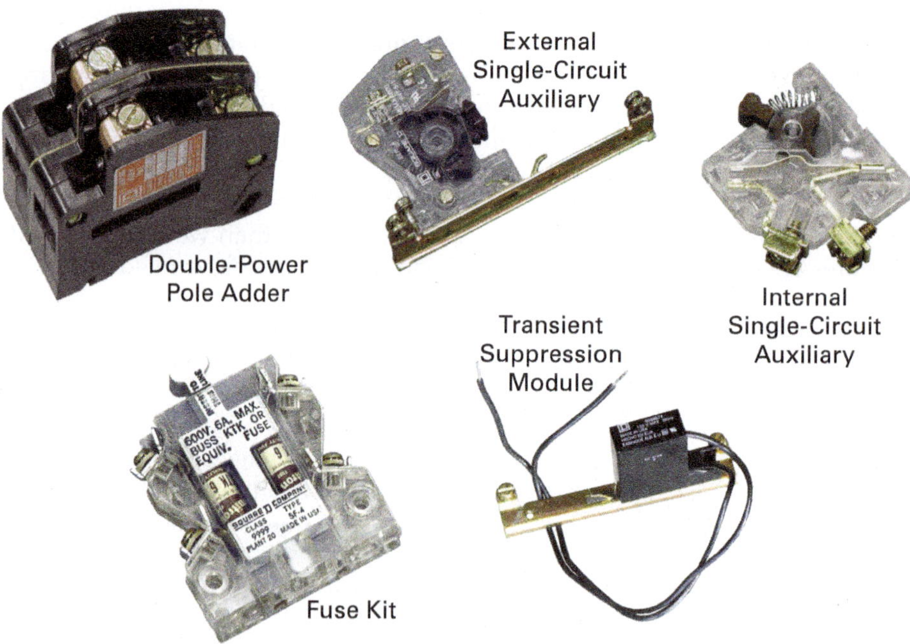

Figure 33 Typical contactor and motor starter accessories.

Manual Starters with Melting-Alloy Overloads

In versions of manual starters that use melting-alloy overloads, the latched contacts are tripped mechanically by the overload relay and must be manually reset with a separate button or by setting the starter to Off (stop) and then back to On (start). In bimetallic overload versions, the reset function may be selectable as automatic or manual, depending on the manufacturer.

2.0.0 Section Review

1. The NEMA-rated motor starter that must be used with a 460V, polyphase, 60 Hz, 90 hp motor *not* used for plugging or jogging applications is the _____.
   a. NEMA Size 2
   b. NEMA Size 3
   c. NEMA Size 4
   d. NEMA Size 5

2. The IEC lighting control used to switch a transformer load is a category _____.
   a. AC-1
   b. AC-5a
   c. AC-6a
   d. AC-11

3. Electrical noise from a motor starter that may interfere with nearby electronic circuits can be reduced using a(n) _____.
   a. adder kit
   b. transient suppression module
   c. status feedback module
   d. electrical interlock

3.0.0 Control Transformers and Pilot Devices

Objective

Identify and describe control transformers and pilot devices.

a. Identify and describe control transformers.
b. Identify and describe various motor control and pilot devices.

Performance Task

1. Make all wiring connections for a magnetic motor starter controlled by two push-button stations, including a holding circuit interlock.

This section presents control transformers and several types of pilot devices widely used in motor control circuits. The opening, closing, or transfer of pilot device contacts can govern the operation of other relays, contactors, and similar devices. Pilot devices are used to provide sequencing and automatic operation within certain parameters. Some commonly used types of pilot devices covered here include the following:

- Push-button switches
- Selector switches
- Pilot lights
- Temperature switches
- Pressure switches
- Limit switches
- Flow switches
- Float switches
- Foot switches
- Proximity switches and sensors
- Photoelectric switches and sensors
- Drum switches

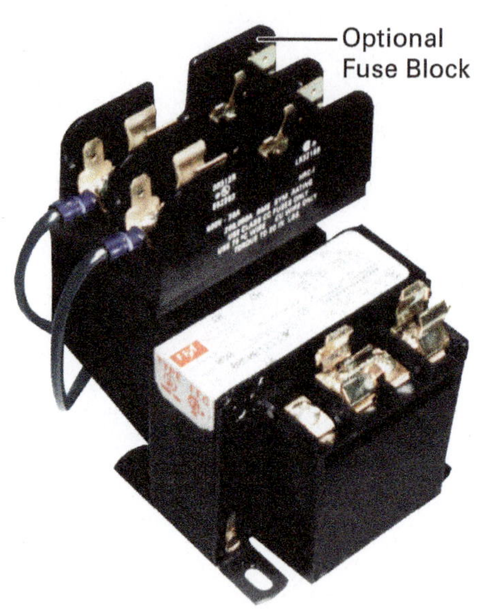

Figure 34 Industrial control transformer.
Source: Electro-Mechanical Corp./Line Power/Federal Pacific

3.1.0 Control Transformers

Control transformers (*Figure 34*) provide the operating voltage for motor control circuits and their components. Control transformers are step-down transformers that reduce the line voltage applied to the equipment to the operating voltage needed for the equipment's control circuits. They are designed to provide good voltage regulation where high inrush currents caused by contactors and relays occur.

Secondary fuse protection kits are often used with control transformers to protect the secondary winding and control circuit. The kits consist of fuse blocks and either cartridge or glass fuses of the proper size. Stepping down a higher line voltage to a lower voltage for use in control circuits has three advantages:

- Worker safety is increased at the control stations and other pilot devices.
- The reduced voltage decreases the chance of a fault occurring between lines of the control circuit wiring and ground.
- Use of components designed to operate at lower voltages lowers the cost for manufacturing the equipment.

Control transformers are made to accommodate many different primary and secondary voltage requirements. Control transformers in most residential HVAC equipment and similar systems normally step down an applied primary voltage of 120VAC or 240VAC to 24VAC. Control transformers used for industrial applications typically operate to produce a 24V or 120V secondary voltage. A typical industrial control transformer contains two primary windings and one or more secondary windings.

Figure 35 shows a control transformer with two primary windings and one secondary winding. The primary winding can be connected to either a 240VAC or 480VAC power supply. The secondary produces 24VAC. Terminals of most industrial control transformers are identified using an industry-standard method. As shown, the terminals of the 240V primary winding are marked H1 and H3, while the 480V terminals are marked H1 and H4. The terminals for the secondary winding are marked X1 and X2. The primary windings of most control transformers have the H2 and H3 terminals crossed, as shown in *Figure 35*. This is done to make the physical connection between the terminals of the primary windings easier.

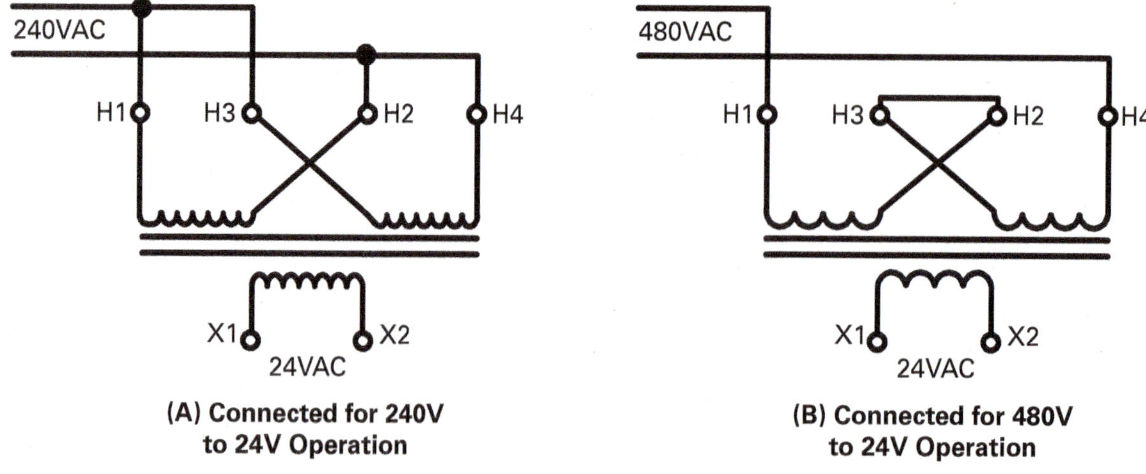

Figure 35 Control transformer schematic shown connected for 240V/480V primary and 24V secondary operation.

The correct connections for achieving the desired voltage output are normally shown on the transformer nameplate and in the manufacturer's datasheet. Metal links are normally used to make the jumper connections between the terminals of the primary and/or secondary windings to configure the transformer for the desired voltage(s). In *Figure 35,* note that a jumper is required between H2 and H3 when 480V is applied.

If this transformer must be connected to step down a primary voltage of 240V to produce 24V, the two primary windings are connected in parallel, as shown in *Figure 35* (A). Because the two primary windings are connected in parallel, each will receive the same voltage. This will produce a turns ratio of 10:1 between the primary and secondary windings. When 240V is connected to the primary of a 10:1 ratio transformer, the secondary voltage produced is 24V.

If this same transformer is connected to step down a primary voltage of 480V to 24V, the primary windings are connected in series, as shown in *Figure 35* (B). Terminal H2 of one primary winding is connected to H3 of the other primary winding with a jumper. This series connection of the two primary windings produces a turns ratio of 20:1.

Selecting a control transformer for an application requires that you know the following factors about the transformer and its related control circuit:

- *Inrush VA* — The inrush VA is the product of the load voltage times the current required during circuit startup. It is determined by adding the inrush VA requirements for all the circuit load devices, such as contactors, timers, relays, or pilot lights, that will be energized at the same time.
- *Sealed (steady-state) VA* — The sealed VA is the product of the load voltage times the current required to operate the circuit after the initial startup and under normal operating conditions. It is determined by adding the sealed VA requirements of all the electrical components that will be energized at any given time. The sealed VA requirements for each component can be obtained from the component manufacturer's datasheets.
- *Primary voltage* — The primary voltage is the voltage applied to the primary winding of the transformer.
- *Secondary voltage* — The secondary voltage is the voltage required for the control circuit, produced by the secondary winding.

The proper transformer can be selected by following the procedure often provided in the manufacturer's catalog or application bulletin. A typical procedure is described as follows:

Step 1 Determine the application inrush VA for the control transformer.

$$\text{Inrush VA} = \sqrt{(\text{inrush VA})^2 + (\text{sealed VA})^2}$$

- For example, assume there is an inrush VA of 1,000VA and a sealed VA of 100VA. Using the formula, the application inrush VA is calculated.

$$\text{Inrush VA} = \sqrt{(1000)^2 + (100)^2}$$

$$\text{Inrush VA} = \sqrt{1{,}010{,}000}$$

$$\text{Inrush VA} = 1{,}005\text{VA (rounded)}$$

Step 2 Using *Table 9,* select a proper secondary voltage column to use. If the primary voltage is stable and does not vary more than 5% from nominal, use the 90% secondary voltage column. If the primary voltage varies between 5% and 10% of nominal, use the 95% secondary voltage column. Note that to comply with NEMA standards, which require all magnetic devices to operate properly at 85% of rated voltage, the 90% secondary column is most often used in selecting a transformer.

Step 3 After determining the proper secondary voltage column to use, read down the column in *Table 9* until you find a value equal to or greater than the application inrush VA calculated in *Step 1*. For our example, an application inrush VA of 1,150 is closest to the 1,005VA. In no case should a value less than the application inrush VA be used! Now read left to the Transformer VA Rating column to find the proper transformer VA for your application. For this example, the transformer should have a VA rating equal to or greater than 200VA. As a final check, ensure the transformer VA rating is equal to or greater than the total sealed requirements. If not (rare), select the next size up with a VA rating equal to or greater than the total sealed VA.

Step 4 Using the transformer VA rating found in *Table 9*, refer to the transformer manufacturer's catalog or bulletin to identify the model and part number for the transformer.

TABLE 9 Regulation Data Chart (Inrush VA at 20% Power Factor)

Transformer VA Rating	95% Secondary Voltage	90% Secondary Voltage	85% Secondary Voltage
25	100	130	150
50	170	200	240
75	310	410	540
100	370	540	730
150	780	930	1,150
200	810	1,150	1,450
250	1,400	1,900	2,300
300	1,900	2,700	3,850
350	3,100	3,650	4,800
500	4,000	5,300	7,000
750	8,300	11,000	14,000

3.2.0 Switches, Sensors, and Pilot Lights

There are many types of switches, sensors, pilot lights, and other devices used in motor control systems. The following sections provide basic information on many of these devices. Always refer to the manufacturer's instructions for proper application, installation, and maintenance of these devices.

3.2.1 Push-Button Switches

Manually operated push-button and selector switches (*Figure 36*) are widely used pilot devices. They are made in standard- and heavy-duty versions. Heavy-duty switches can carry higher continuous and make-break currents. Different types of push-button and lever switches are needed to serve a wide variety of industrial motor applications. For this reason, switch manufacturers designed many of their industrial push-button and lever switches so that they can be assembled using different parts that can be interchanged according to the customer's specifications to meet the requirements of a specific application. This modular approach to building a switch also lowers the manufacturer's costs by reducing the number and type of switches and switch components that must be manufactured and maintained in inventory.

Figure 36 Typical push-button and selector switches.

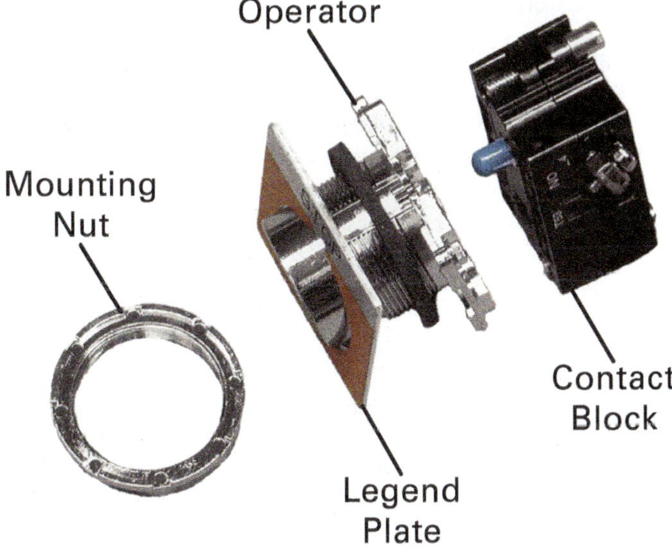

Figure 37 Typical parts of a push-button switch.

The parts used to assemble a typical push-button switch (*Figure 37*) include the **operator**, legend plate, one or more contact blocks, and related mounting adapters and hardware. The operator is part of the switch that is pressed or pulled by the person operating the switch. Operators are made in many different shapes, sizes, and colors. Red is typically used to indicate a Stop or Off function. Operators can also be non-illuminated or illuminated to indicate when they are active. The legend is a plate that displays the function of the push button, such as On, Off, or Inch. The contact block is part of the switch that secures the contacts.

Operator: A switch component that is pressed, pulled, or rotated by the person operating the circuit to activate the switch contacts.

Pushing the button initiates movement of the electrical contacts. The switch contact block can have NO, NC, or both NO and NC sets of contacts. Multiple contact blocks are often stacked together to form a switch with a significant number of contact sets. NO contacts are generally used to close the circuit when the operator is pressed to initiate a Start or On function. NC contacts generally initiate a Stop or Off function.

There are two types of push buttons: momentary and maintained. An NO momentary push button closes the contacts only while the button is held down. An NC momentary push button opens the contacts only while the button is held down. A maintained push button latches in place when pressed.

Figure 38 shows the schematic symbols for push-button switches. *Figure 39* shows a simple line diagram of Start and Stop push-button switches used in a basic contactor control circuit.

Push-button switches can also be of the *push-pull* type. A push-pull switch is typically used to replace two separate push buttons, such as Start and Stop buttons. There are three types of push-pull switches:

- *Maintained* — This is a two-position switch that remains in the pulled or pushed position until manually actuated to the opposite position.
- *Momentary* — This is a three-position switch. A spring returns the switch to an intermediate position when pulled or pushed and released.
- *Momentary pull, maintained push* — This is a three-position switch. A spring returns the switch to an intermediate position when pulled. In the push position, it maintains its position until manually returned to the intermediate position.

Push-button stations (*Figure 40*) are enclosures that house one or more push buttons, selector switches, and/or pilot lights to protect them from the environment. They can be bought unassembled or with the push button and/or lever switches installed. Cast metal, polyester, and stainless steel push-button stations are available in various NEMA enclosures and sizes. Always use the correct switch components and enclosure for the environment where they will be used.

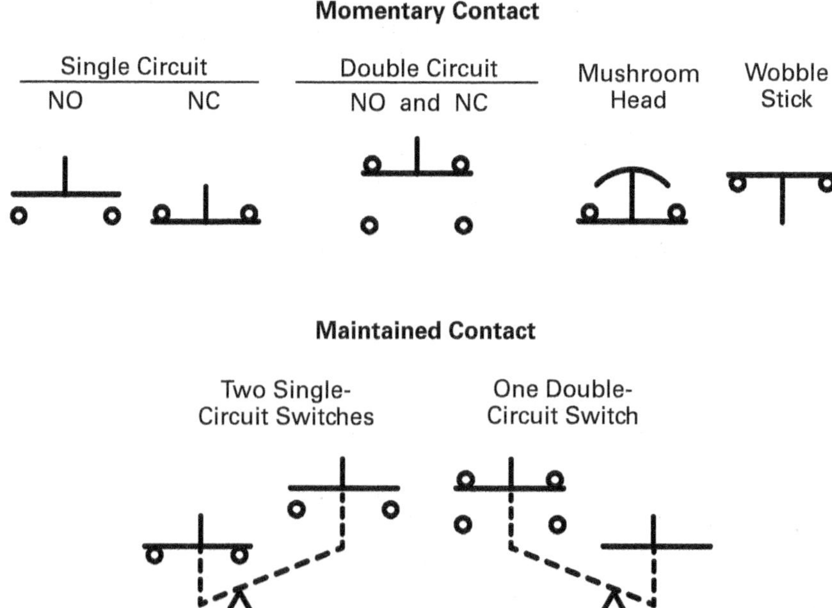

Figure 38 Schematic symbols for push-button switches.

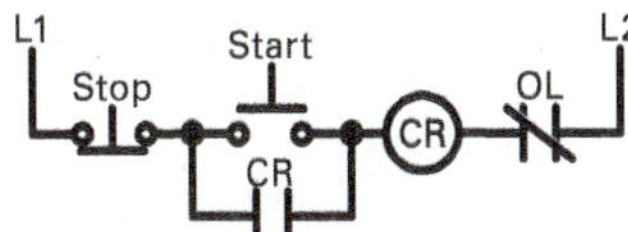

Figure 39 Stop and Start push-button switches in a basic control circuit.

Figure 40 Example push-button station.

Selecting the components for a typical push-button or lever switch and its related switch station involves using the switch manufacturer's catalog or datasheets to identify the proper components. You must make sure the electrical characteristics of the selected devices are compatible with the requirements of your application. Switch manufacturers describe the electrical characteristics of their push-button switches, lever switches, and contacts using codes defined in *NEMA Standard ICS 5* for categories of utilization and contact switching ratings. For example, category entries such as AC-14 or DC-13 define typical applications of the device. Contact rating designations such as A600 or B600 define the continuous current-carrying thermal capability of the device or contact. The codes and related operating characteristics are provided for review in *Table 10A*, *Table 10B*, and *Table 11*.

TABLE 10A Rating Codes for AC Control Circuit Contacts (Reprinted by Permission of National Electrical Manufacturers Association)

Rating Codes for AC Control Circuit Contacts at 50 Hz and 60 Hz											
Contact Rating Code	Thermal Continuous Test Current	120V		240V		480V		600V		Maximum VA	
		Make	Break	Make	Break	Make	Break	Make	Break	Make	Break
A150	10A	60A	6A	—	—	—	—	—	—	7200	720
A300	10A	60A	6A	30A	3A	—	—	—	—	7200	720
A600	10A	60A	6A	30A	3A	15A	1.5A	12A	1.2A	7200	720
B150	5A	30A	3A	—	—	—	—	—	—	3600	360
B300	5A	30A	3A	15A	1.5A	—	—	—	—	3600	360
B600	5A	30A	3A	15A	1.5A	7.5A	0.75A	6A	0.6A	3600	360
C150	2.5A	15A	1.5A	—	—	—	—	—	—	1800	180
C300	2.5A	15A	1.5A	7.5A	0.75A	—	—	—	—	1800	180
C600	2.5A	15A	1.5A	7.5A	0.75A	3.75A	0.375A	3A	0.3A	1800	180
D150	1A	3.6A	0.6A	—	—	—	—	—	—	432	72
D300	1A	3.6A	0.6A	1.8A	0.3A	—	—	—	—	432	72
E150	0.5A	1.8A	0.3A	—	—	—	—	—	—	216	36

TABLE 10B Rating Codes for DC Control Circuit Contacts (Reprinted by Permission of National Electrical Manufacturers Association)

Rating Codes for DC Control Circuit Contacts					
Contact Rating Code*	Thermal Continuous Test Current	Maximum Make or Break Current			Maximum Make or Break VA at 300V or Less
		125V	250V	301V to 600V	
N150	10A	2.2A	—	—	275
N300	10A	2.2A	1.1A	—	275
N600	10A	2.2A	1.1A	0.4A	275
P150	5A	1.1A	—	—	138
P300	5A	1.1A	0.55A	—	138
P600	5A	1.1A	0.55A	0.2A	138
Q150	2.5A	0.55A	—	—	69
Q300	2.5A	0.55A	0.27A	—	69
Q600	2.5A	0.55A	0.27A	0.1A	69
R150	1A	0.22A	—	—	28
R300	1A	0.22A	0.11A	—	28

*The numerical suffix designates the maximum voltage design values, which are to be 600V, 300V, and 150V, respectively. The test voltage is to be 600V, 250V, and 125V.

For maximum ratings at 300V or less, the maximum make-and-break ratings are to be obtained by dividing the volt-amperes rating by the application voltage, but these currents are not to exceed the thermal continuous test current.

TABLE 11 Utilization Categories for Control Circuit Switching Elements

Current Type	Category	Typical Applications
AC	AC-12	Control of resistive and solid-state loads with optical isolation
	AC-13	Control of solid-state loads with transformer isolation
	AC-14	Control of small electromagnetic loads (maximum of 72VA closed)
	AC-15	Control of electromagnetic loads (greater than 72VA closed)
DC	DC-12	Control of resistive and solid-state loads with optical isolation
	DC-13	Control of electromagnets
	DC-14	Control of electromagnet loads with an economy resistor in circuit

Installation of Push-Button Switches in a Station

When installing push-button switches in a push-button station, follow these guidelines:

- *Stop button location* — On all single-row push-button stations, the Stop button must be located below or to the right of the other buttons, including lights and selector switches.
- *Push buttons used with multispeed motors (mounted vertically)* — The lowest button in the station must be the Stop button. The lowest speed button must be above the Stop button. Buttons for consecutively higher speeds are positioned above the lowest speed button.
- *Push buttons used with multispeed motors (mounted horizontally)* — The button to the far right must be the Stop button. The lowest-speed button must then be directly to its left, followed by the buttons for consecutively higher speeds.

A general procedure for selecting a push-button or lever switch is provided here:

Step 1 Select the style of switch operator:

- Standard duty or heavy duty
- Chrome, plastic, or other material
- Oil-, water-, or dust-resistant; corrosion-resistant

Step 2 Select the type of switch operator:

- Push button or lever
- Nonilluminated or illuminated
- Color of push button

Step 3 Select the contact block(s) needed:

- NO or NC
- Standard or hazardous location
- Standard duty or heavy duty

Step 4 Select the appropriate operator identification or legend nameplate and color.

Step 5 Select the matching station enclosure for the type and number of operators involved.

If motors must be started from multiple locations, additional push-button stations will be needed. The Start buttons in these stations must be connected in parallel with the original Start button, and the Stop buttons must be connected in series with the original Stop button, as shown in *Figure 41*. The auxiliary contactor must also be connected in parallel with the Start buttons.

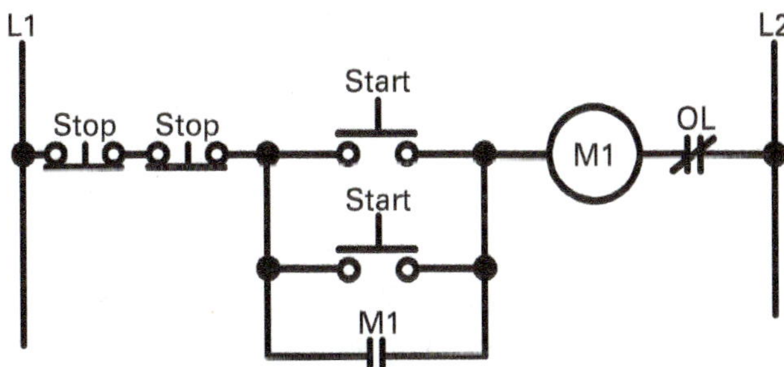

Figure 41 Multiple Stop and Start push buttons.

3.2.2 Selector Switches

Standard-duty selector switches have either two or three positions. Heavy-duty selector switches have two or more positions, with some having as many as twelve. The operator of a selector switch is rotated instead of pushed to activate the contacts.

Selector switches are assembled the same way as push-button switches, beginning with selecting suitable operator, legend, and contact block assemblies. *Figure 42* shows the schematic symbols for a two- and three-position selector switch. Schematic diagrams and switch manufacturers' product catalogs commonly show the contact positions for each selector switch position using a *truth table* placed near the switch. As shown in *Figure 42*, an X is placed in the truth table if a contact is closed in any position.

Figure 42 also shows a simple line diagram of a three-position selector switch being used to control a contactor. In the Off position, all the selector switch contacts are open, preventing the contactor coil from being energized. This is a safety position. In the Hand position, the upper contacts of the three-way switch are closed, enabling the contactor coil to be energized or de-energized by pressing the Start or Stop push buttons, respectively. In the Auto position, the lower contacts are closed, enabling the contactor coil to be energized by a two-wire pilot device. The pilot device can be a temperature switch, pressure switch, limit switch, or similar control.

Some selector switches have a joystick operator (*Figure 43*). Joystick selector switches can have two to eight positions. This type of switch is normally used in applications in which only one circuit is to be energized at one time, such as when operating a hoist or crane. In this application, movement of the operator for a five-position switch (four positions and center off) closes one circuit each in the Up, Down, Left, and Right positions, with all circuits open when the joystick is in the center position. Depending on the design, the joystick operator may only make contact while held, returning to the center position when released (spring return). Others are designed to remain in one of the four positions once moved (no spring return).

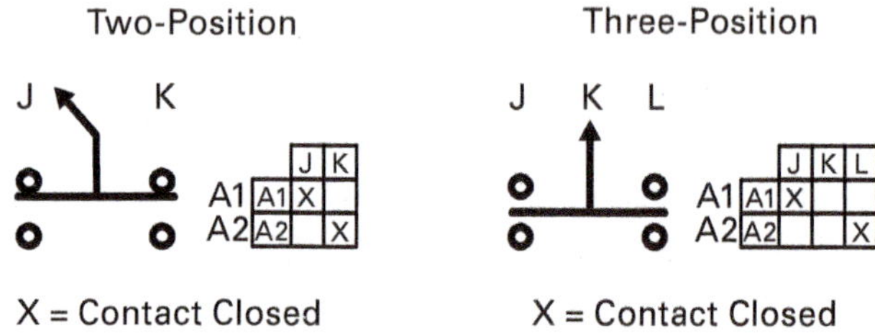

	J	K
A1	X	
A2		X

	J	K	L
A1	X		
A2			X

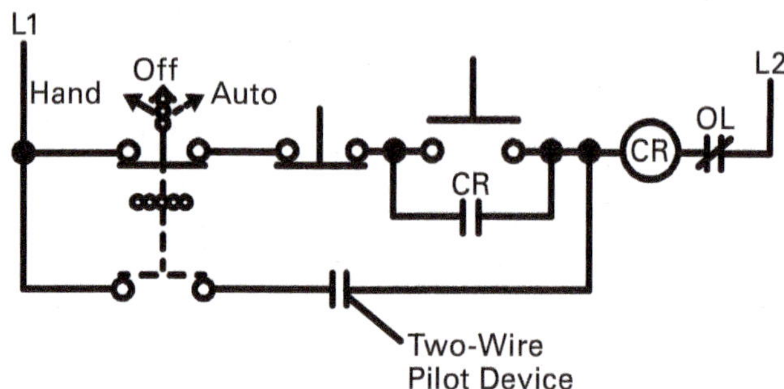

Figure 42 Selector switch schematic symbols and typical control circuit.

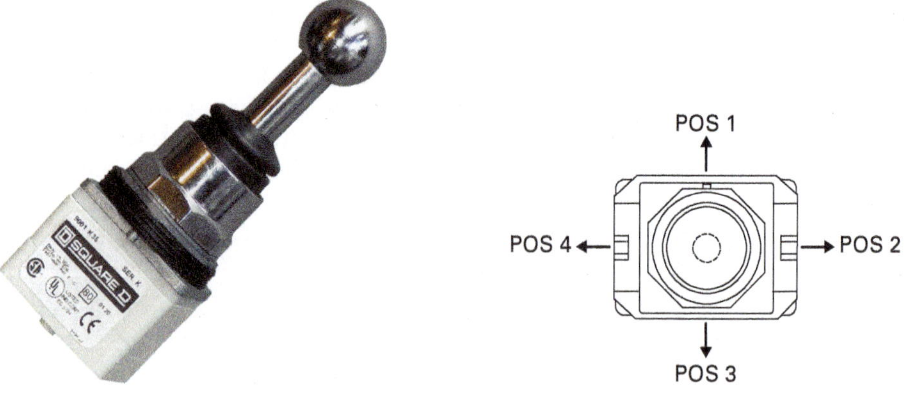

Figure 43 Joystick operated selector switch and diagram.

3.2.3 Pilot Lights

A pilot light is a small electric light used to visually show the status of a circuit. Pilot lights can be separate assemblies or built into a push-button switch assembly. Pilot light assemblies (*Figure 44*) typically have a polycarbonate or glass lens as a cap. Colored lenses are typically made in red, green, amber, blue, clear, white, or yellow.

Pilot lights often have a bayonet-style incandescent lamp that operates at full voltage or a reduced voltage via a transformer or resistor. Depending on the control circuit arrangement, pilot lamps operate at a voltage ranging from 6VAC or DC to 120VAC or DC. Some pilot lights use an LED instead of an incandescent lamp. Some are available with a local push-to-test capability, while others have a remote test capability used to test the operation of the indicator lamp.

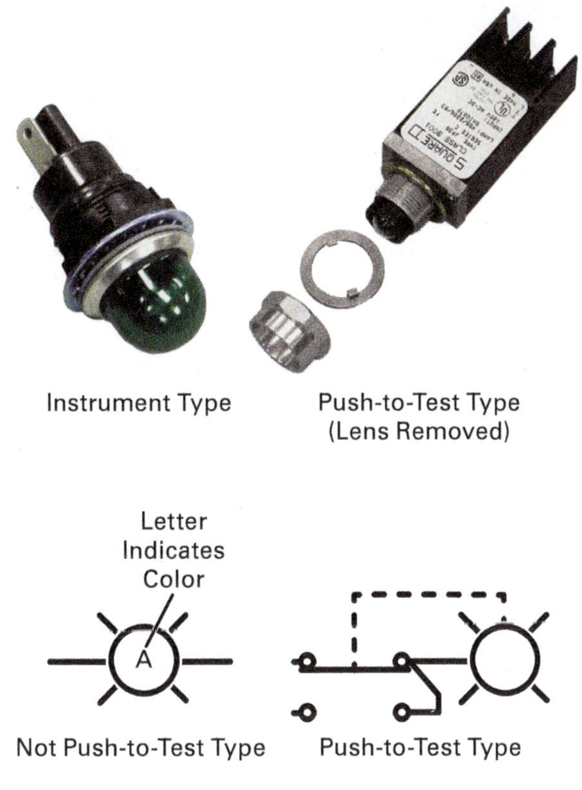

Figure 44 Typical pilot light assemblies with diagram.

3.2.4 Temperature Switches

Temperature switches (*Figure 45*) monitor temperatures in the control circuits for applications such as heating and cooling, damper, fire alarm, and process control systems. Temperature switch contacts open or close in response to a rise or fall in temperature. The temperature at which the contacts respond is normally determined by an operator-adjustable **setpoint**. The temperature-sensing elements used in mechanical temperature switches may be a bimetallic element or bulb and capillary tube element. There are also electronic temperature-sensing elements, but these tend to be used more often with temperature controllers rather than switches.

Setpoint: The desired value to which a switching device is set or adjusted to respond.

A bimetallic element is made of two different metals bonded together. One metal is usually copper or brass (*Figure 46*). The other, a special metal called *Invar steel*, contains 36% nickel. When heated, the copper or brass has a more rapid expansion rate than the steel, forcing a change in the element's shape. As it changes shape, it actuates the switch contacts. Although bimetal elements are constructed in various shapes, the coil-wound style is the most common.

Opens on Temperature Rise (High-Temperature Safety Switch; Disable Heating Mode)

Closes on Temperature Rise (Enable Cooling Equipment or a Pump)

(A) Schematic Symbols

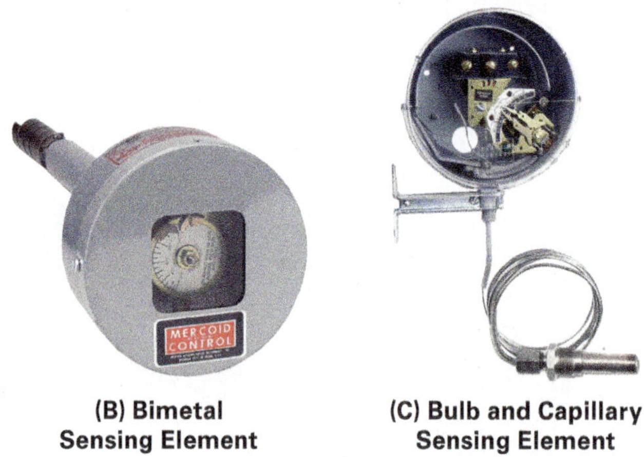

(B) Bimetal Sensing Element

(C) Bulb and Capillary Sensing Element

Figure 45 Typical Temperature switches with schematics.
Source: Dwyer Instruments, Inc. (45B and 45C)

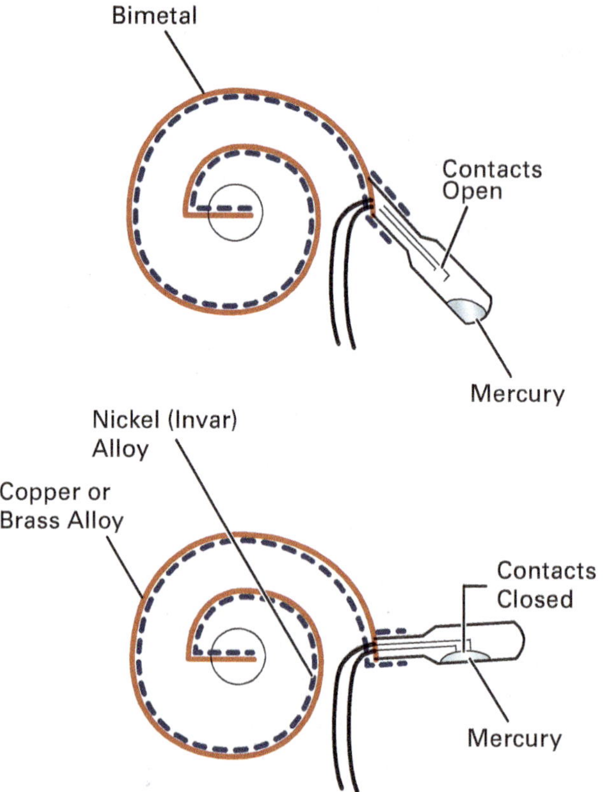

Figure 46 Bimetal strips used in temperature switches.

One widely used bimetal temperature switch uses a small pool of mercury as a conductor, as shown in *Figure 46*. When the bimetal strip causes the mercury tube to tip in one direction, the mercury moves across the switch contacts, closing the circuit. However, due to environmental concerns, most mercury-based switches are obsolete. However, they may still be found in industrial applications.

In many bimetal switches, the bimetal strip is connected directly to one of the switch contacts. This moving contact makes or breaks contact with a stationary switch contact in response to the expansion and contraction of the bimetal strip.

Temperature switches with a bulb and capillary tube are filled with a liquid/vapor mixture. When the temperature increases at the bulb, the fluid inside expands and vaporizes, causing an increase in internal pressure. The pressure is transmitted through the capillary tube and exerted on the mechanism that moves the switch contacts. The sensing bulb can be clamped to a pipe or duct, inserted in a heating or cooling coil, or placed in a tank.

Two types of switching mechanisms are commonly used: the Bourdon tube and bellows. *Figure 47* (A) shows a C-type Bourdon tube. It is a flattened metal tube open at one end and closed at the other. The tube straightens when a rise in pressure is applied to the open end of the tube and curls inward with a decrease in pressure. This movement is transmitted through mechanical linkage to the switch contacts, causing them to open or close. Spiral and helical Bourdon tubes are also encountered. Spiral and helical tubes produce more tip travel and torque at the free end than the C-type tube design.

In a bellows temperature switch, the pressure change in the bulb is transmitted through the capillary tube to a bellows inside the temperature switch, as shown in *Figure 47* (B). One end of the bellows is closed and the other is connected to the bulb pressure source via the capillary tube. The bellows is a round device with several folds, much like an accordion, that expands or contracts with changes in pressure. The expanding bellows acts on a calibrated spring mechanism that opens or closes the switch. The spring tension is adjustable so the switch setpoint can be adjusted.

Going Green

Environmental Hazard

Mercury is an environmental and health hazard. Never handle mercury with your bare hands. Do not dispose of thermostats containing mercury tubes in regular trash. Contact your local waste management or environmental authority for proper disposal and recycling instructions.

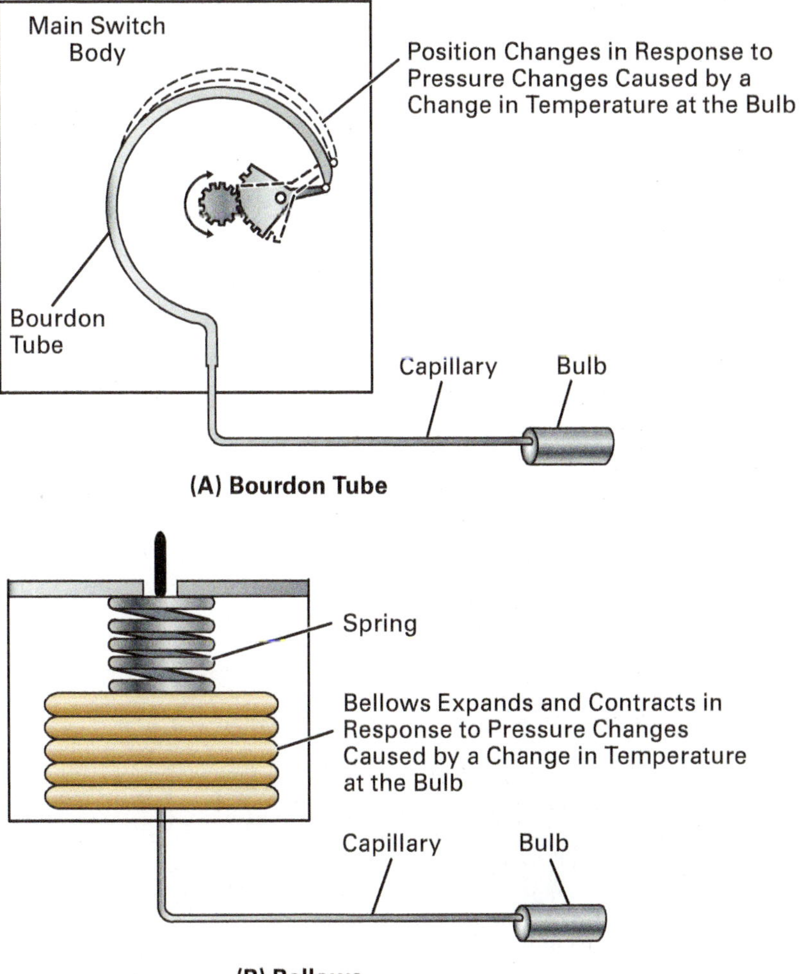

Figure 47 Bourdon tube and bellows temperature switch mechanisms.

To properly select temperature switches, you need to understand several terms used in the manufacturers' descriptions:

- *Allowable temperature limits* — The maximum and minimum temperatures to which a temperature switch may be exposed without altering its performance characteristics.
- *Cut-in temperature* — The temperature of the sensed medium at which a temperature switch is actuated.
- *Cut-out temperature* — The temperature of the sensed medium at which a temperature switch returns to its normal position.
- *Operating temperature differential* — The difference between the cut-in and cut-out temperatures. The operating temperature differential may also be called the **deadband**.
- *Operating temperature range* — The range between the maximum and minimum temperature settings at which a temperature switch will continue to operate within the manufacturer's specifications.
- *Remote temperature sensing* — The construction and installation of a temperature switch in which the sensing unit and the switch mechanism are thermally isolated so the temperature of the sensed medium does not affect the performance of the switch mechanism.

Deadband: The difference between the setpoint at which a switch activates as input increases and the point at which the switch resets to the former position when the input decreases.

3.2.5 Pressure Switches

Pressure switches (*Figure 48*) use mechanical motion in response to pressure changes to open or close contacts. Depending on the switch design and application, NO and NC contacts can be activated in response to positive, negative, or differential pressures.

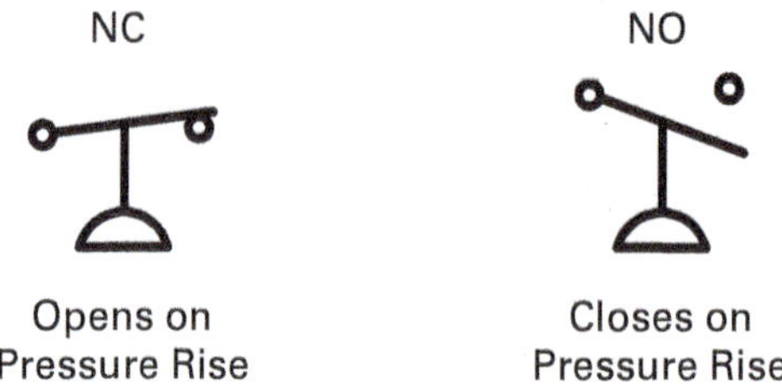

(A) Schematic Symbols

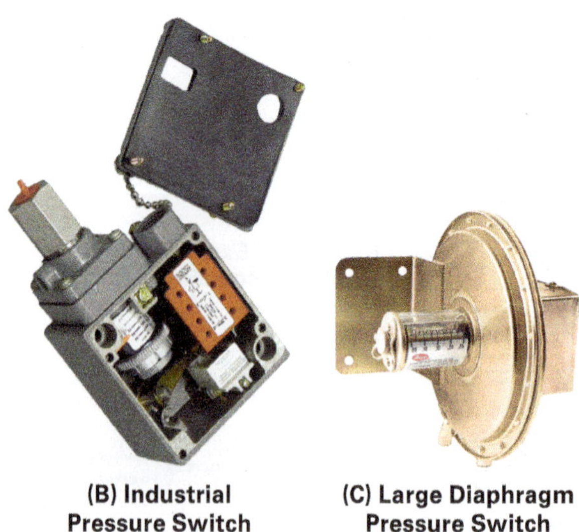

(B) Industrial Pressure Switch

(C) Large Diaphragm Pressure Switch

Figure 48 Typical pressure switches.
Source: Dwyer Instruments, Inc. (48C)

Single-stage switches are used in a wide variety of applications to protect the equipment from low or high pressures or initiate another aspect of a process. Dual-stage pressure switches contain two pressure switches. They are used in applications where it is practical to monitor two separate pressures using a single device. For example, in a refrigeration system, a dual-pressure switch is commonly used to monitor the compressor suction and discharge pressures for either a low-pressure or high-pressure condition, respectively. Differential pressure switches can signal that a predetermined pressure difference has developed between two separate pressure sources. One of those sources is often atmospheric pressure.

Both adjustable and nonadjustable types of pressure switches are used. An adjustable type provides far more flexibility. However, some machine processes and safety switches need to remain at a fixed point to avoid any unsafe conditions that an adjustment might cause. Fixed switches are used in these cases.

Different kinds of pressure-responsive elements are used in pressure switches. They can be a Bourdon tube, bellows, sealed piston, or diaphragm type. Bourdon tube and bellows switch sensors operate in the same way as described earlier for use in temperature switches.

In a sealed-piston pressure switch, the switch contacts are actuated by a piston that moves up and down in a cylinder in response to the pressure, as shown in *Figure 49* (A). The top of the piston moves against a calibrated spring, the tension of which actuates the switch contacts when the source pressure increases. The spring tension is typically adjustable so the setpoint can be changed.

In a diaphragm pressure switch, the edge of a disk-shaped diaphragm is firmly attached to the case of the switch, as shown in *Figure 49* (B). Pressure is applied against the underside of the diaphragm. In response, the center of the diaphragm opposite the pressure side moves toward or away from a calibrated spring. The switch **actuator** initiates movement of the contacts. Like temperature switches, some have a fixed pressure setting while others are adjustable.

Actuator: The mechanism of a switch or switch enclosure that moves to operate the contacts.

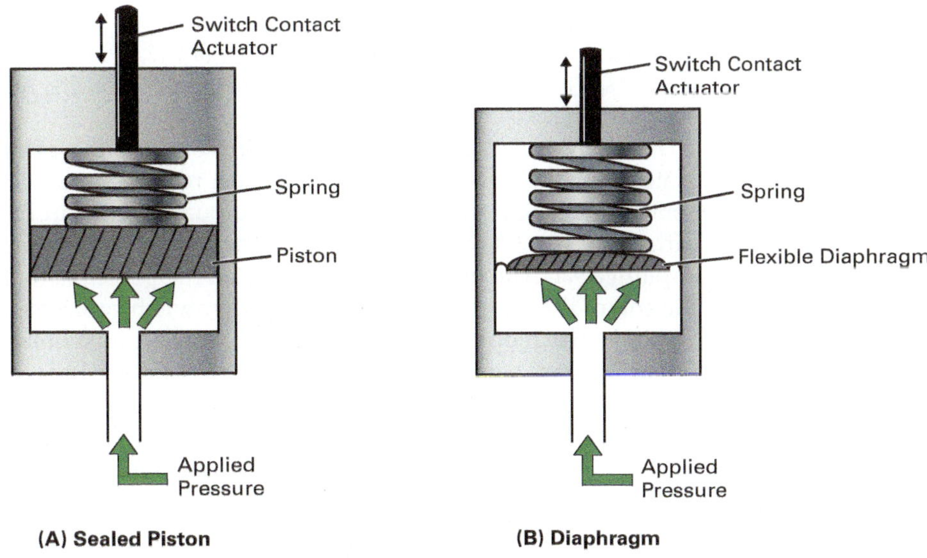

Figure 49 Sealed piston and diaphragm pressure switch mechanisms.

It is important to understand several key terms when selecting pressure switches:

- *Allowable pressure limits* — The maximum and minimum pressures to which a pressure switch can be exposed.
- *Cut-in pressure* — The pressure at which the switch moves the contacts from their normal position.
- *Cut-out pressure* — The pressure at which the switch returns the contacts to their normal position.
- *Deadband* — The pressure difference between the cut-in and cut-out pressures. The deadband may also be referred to as the *operating pressure differential*.
- *Drift* — An expected change in accuracy due to time or some number of cycles. Excessive drift is either resolved through setpoint adjustments, calibration, or switch replacement.
- *Maximum static pressure* — The continuous pressure the switch can sustain without damage. The maximum static pressure is sometimes called the *rated static pressure*.
- *Operating pressure range* — The maximum and minimum pressure settings the switch provides.
- *Proof pressure* — The amount of overpressure that can be applied to the switch without causing damage.
- *Pulsation snubber* — A device used to reduce the impact of pressure surges by slowing the rate a pressure change is transmitted to the switch.

3.2.6 Mechanical Limit Switches

Mechanical limit switches detect the position of an object by having the object make direct physical contact with the switch actuator. They are commonly used to limit the travel of machine tools, detect moving items on a conveyor belt, monitor an object's position, or detect that machinery safety guards are in place. Limit switches come in many shapes and sizes, with SPDT and DPDT versions being very common.

There are two types of limit switches: rotary lever-actuated (*Figure 50*) and plunger-actuated (*Figure 51*). Both consist of an actuator, switch body, and terminals. The switch actuator is the part of the switch that operates the switch contacts. The switch body houses the electrical contacts, and the terminals are the point of connection for the circuit wiring.

A rotary lever-actuated limit switch works on the principle that a **cam** or plate hits the end of the lever arm, which rotates a shaft to operate the switch contacts. In some rotary switches, the actuator arm can be interchanged with different styles (*Figure 52*). This allows the same switch to be used in many ways. A plunger-activated limit switch works on the principle that a cam or plate hits the end of the plunger, which operates the switch contacts.

Limit switch operating heads can be of two types: **momentary contact switch** or **maintained contact switch**. Momentary contacts, sometimes called *spring return contacts*, return from the actuated position to their normal position when the actuating force is removed. Maintained contacts remain in their placed position until another force moves them to another position. They are reset only by further movement of the operating head. For example, the contacts may be reset by shaft rotation in the opposite direction.

Cam: A machine component that applies a force to the actuator of a limit switch, causing the actuator to move.

Momentary contact switch: A switch with contacts that return from the operating position to their normal position when the actuating force is removed.

Maintained contact switch: A switch with contacts that remain in the operating position when the actuating force is removed, with other provisions to reset the switch to its normal position.

Pressure Switches

A pressure switch should not be used where it may be subjected to a pressure surge beyond its maximum allowable pressure. A pressure switch used in a system pressurized by a positive-displacement pump, for example, should be protected from surges by a snubber.

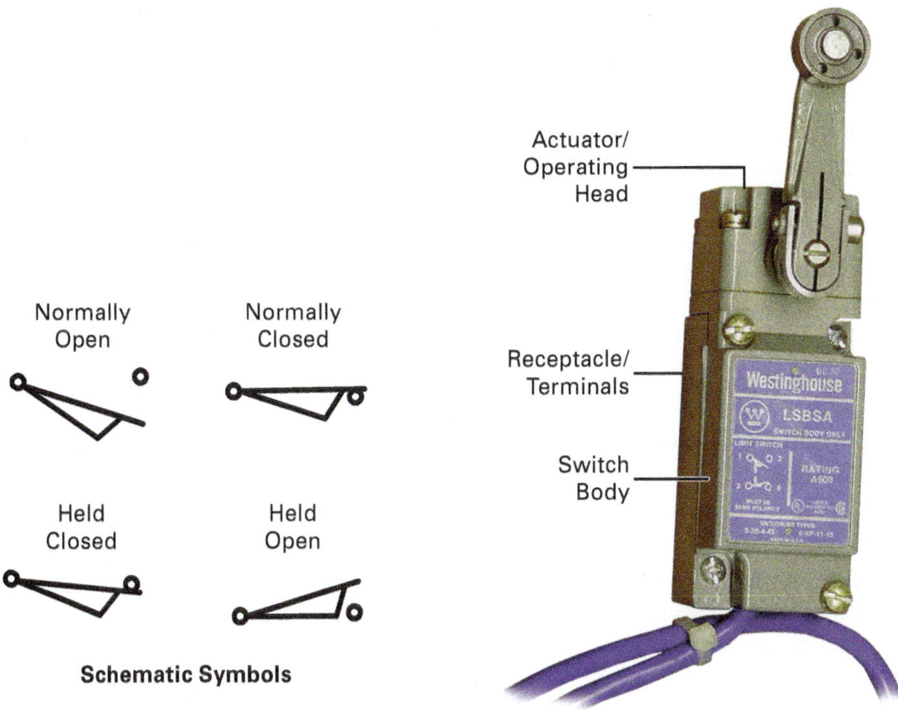

Figure 50 Rotary lever-actuated limit switch with schematic symbols.

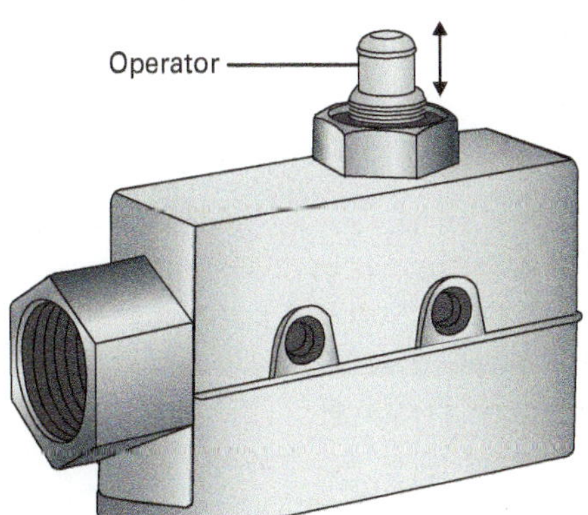

Figure 51 Plunger-actuated limit switch.

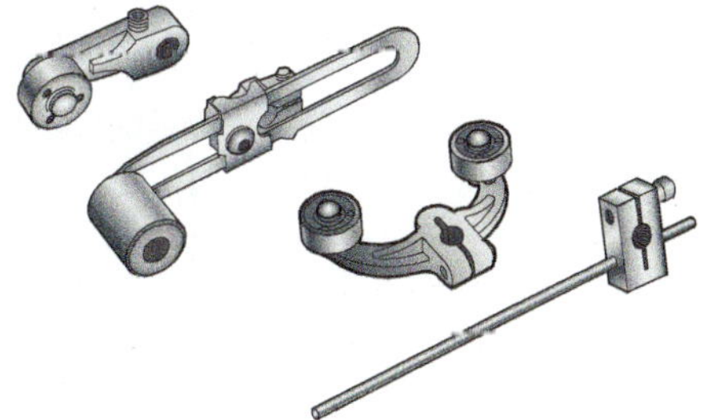

Figure 52 Rotary lever-actuated limit switch actuators.

Manufacturers of limit switches describe the position or motion of the switch actuator in specific terms. These terms are shown in *Figure 53* and described as follows:

- *Movement differential* — The distance or angle from the operating position to the releasing position.
- *Free (initial) position* — The switch actuator is positioned such that the limit switch contacts are in their normal, untriggered position.
- *Operating position or point* — The position of the actuator at which the switch contacts change from their normal position to their alternate, triggered position.
- *Operating torque* — The minimum force that must be applied to the actuator to cause the contacts to change from their normal position to the triggered state.
- *Overtravel* — In the case of the rotary limit switch, it is the distance or angle through which the actuator moves when traveling beyond the triggered position. In a plunger-actuated switch, the overtravel distance is the safety margin for the sensor to avoid breakage.
- *Pretravel* — The distance or angle through which the actuator moves from the free position to the position just before the contacts change state. The contacts are still in their normal position.
- *Release position or point* — The position of the actuator at which the contacts change from the triggered position back to the normal position.
- *Release torque* — The value to which the torque on the actuator must fall to allow the contacts to change from the triggered position to the normal position.

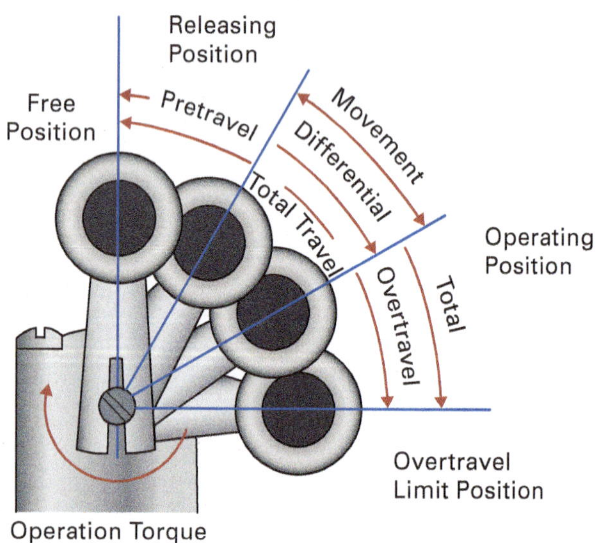

(A) Rotary Lever-Actuated Limit Switch

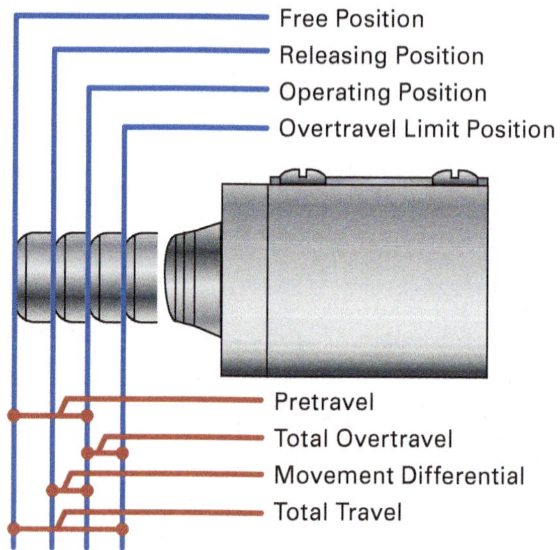

(B) Plunger-Actuated Limit Switch

Figure 53 Positions of limit switch actuators.

Depending on the limit switch design, the NO and NC contacts may or may not conduct current simultaneously during operation. SPDT limit switches can have make-before-break, break-before-make, and simultaneous make-and-break contacts. With make-before-break contacts, the NO contact closes before the NC contact opens. With break-before-make contacts, the NC contact opens before the NO contact closes. With simultaneous make-and-break contacts, the NC contact opens at the same time the NO contact closes. Manufacturers of limit switches often use contact function diagrams (*Figure 54*) to define the operation of their switches and show any overlap or nonoverlap of contact operation.

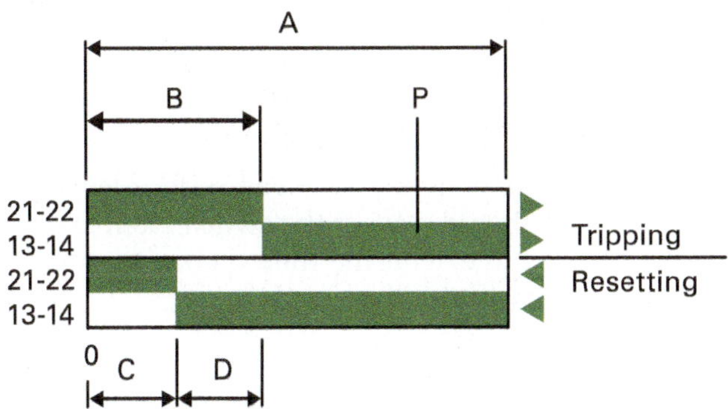

A = Maximum travel of the operator in mm or degrees
B = Tripping travel of the contact
C = Reset travel
D = B = C = Differential travel
P = Point from which positive opening is assured

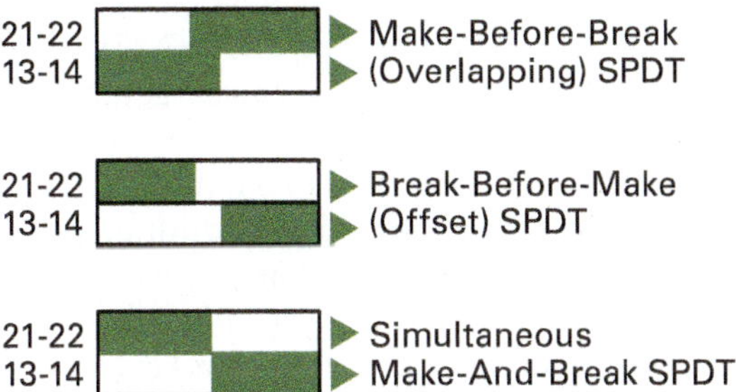

Figure 54 Example of a limit switch contact function diagram.

When selecting and installing limit switches, the following factors should be considered:

- A rotary lever-actuated limit switch is the best choice for most applications. It can be used in any application where the cam moves perpendicular to the lever's rotational shaft.
- A plunger-actuated limit switch is the best choice to monitor short, controlled machine movements or where space or mounting restrictions will not permit a lever-actuated switch.
- The switch contacts must be selected according to the proper voltage and current for the load. If the load current exceeds the switch contact rating, a relay, contactor, or motor starter must be used to interface the limit switch with the load.
- The switch contacts must be connected to the same polarity. Otherwise, arcing or welding of the contacts may occur. Do not connect voltages from different sources to the contacts of the same switch unless the switch is designed for such service.

Limit Switches

Most new industrial limit switches can be plugged into a matching receptacle, meaning the operating head, switch body, and receptacle are separate components. If the switch body becomes damaged or fails, it can easily be replaced in the field without replacing the entire assembly. Simply remove the switch body, and the wiring remains connected to the receptacle.

- Mount the switch so there is no possibility of unintentional actuation due to accidental movement. Mount it firmly in an easily accessible location with suitable clearance for service and replacement. Do not place the switch where machining chips or similar debris can accumulate, or the environment is unsuitable for the switch characteristics.
- Do not expose the switch to oils, coolants, or other liquids unless designed for that purpose. If liquid entry is possible, the switch should be mounted face down to prevent seepage through the seals on the operating head. All conduit connections must also be tightly sealed.
- For rotary lever-actuated limit switches, it is important that the correct lever be selected. The manufacturer's recommendations for selecting and applying limit switches and levers should always be followed. It is important that the selected lever be securely fastened to the shaft to prevent slipping.
- Where relatively fast motions are involved with a lever-actuated switch, the cam arrangement should protect the switch from a sharp impact. The cam should be tapered to extend the time it takes to engage the electrical contacts. This prevents wear on the switch.

3.2.7 Flow and Float Switches

Liquid levels are monitored using flow and float switches. Flow switches detect (prove) the presence of a liquid flowing through a pipe or airflow in a duct.

Mechanical flow switches (*Figure 55*) have a vane that extends into the pipe or duct. When the force of the fluid flow is sufficient to overcome the spring tension of the vane, the vane moves and actuates the switch contacts. When there is a loss or reduction in flow below the switch setpoint, the vane returns to the normal position.

The spring tension on the vane is typically adjustable, allowing it to actuate the switch contacts at the desired flow rate. The NO or NC switch contacts may be used to prove pump operation, open or close valves, actuate a warning lamp, or sound an alarm.

Float switches are used to monitor the level of a liquid in a tank or vessel. For example, when the level of a liquid reaches a predetermined point in a tank, the float switch can start or stop a pump or activate an alarm.

Float switch types include mechanical, electronic, ultrasonic, and optical designs. A basic mechanical float switch is shown in *Figure 56*. It consists of a float and a hermetically sealed, magnetically operated snap switch. As the float moves up and down with the level of the liquid in a tank, the NO or NC switch contacts are activated when the liquid level reaches a predetermined height. Two switches can be used to monitor and maintain minimum and maximum liquid levels.

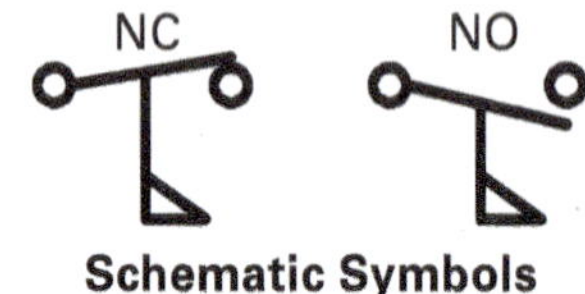

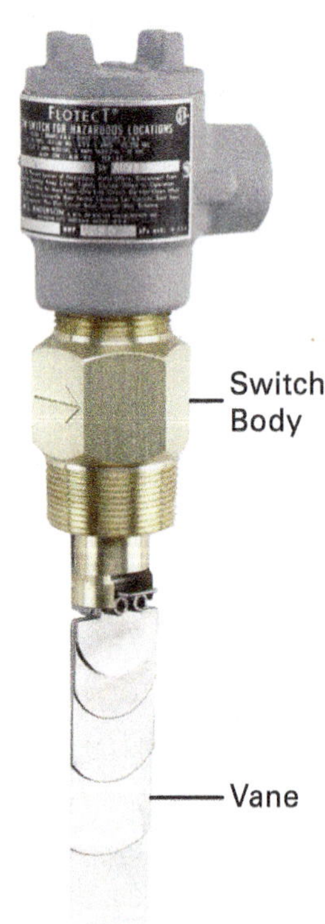

Figure 55 Liquid flow switch.
Source: Dwyer Instruments, Inc. (Bottom)

Microswitches

Various types of microswitches are used with limit switches. They are also used as equipment and instrument panel access and safety interlocks.

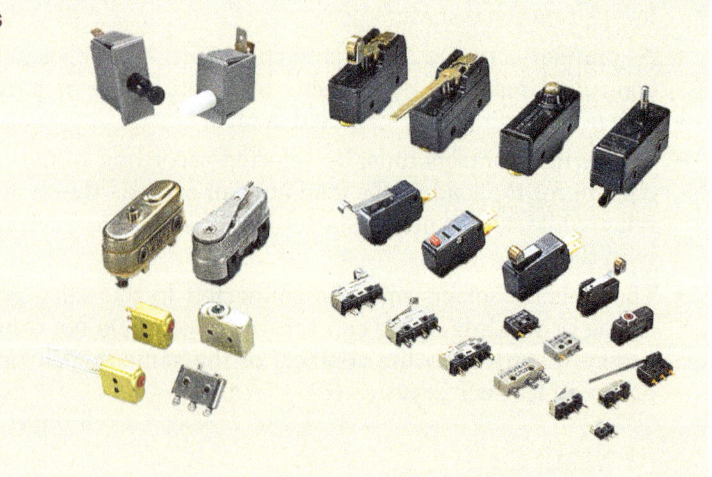

3.2.8 Foot Switches

Foot switches (*Figure 57*) are used where the task requires the operator to have both hands free. Foot switches usually have momentary contacts, but some are available with latches that enable them to be used as maintained-contact devices.

Foot switches typically have two positions: a toe-operated position and a spring-loaded Off position. Some have three positions: two positions that allow for toe and heel control and a spring-loaded Off position.

In applications where more than one foot switch is required, as when two or more persons are operating a machine, the foot switches are wired in series. Both foot switches must then be actuated before the machine can operate.

When selecting a foot switch, always refer to the applicable ANSI standards and OSHA regulations. Additional information can be found in the National Safety Council's Accident Prevention Manual. They must be applied wisely, with safety in mind.

3.2.9 Jogging and Plugging Switches

Many motor control applications require switches in the control circuit to initiate the jogging or plugging of a motor.

There are many ways that a push-button switch can be applied to facilitate jogging. *Figure 58* shows a line diagram of one basic control circuit where a push-button switch is used to jog a motor. In this circuit, the NC contacts of the Jog push-button switch are connected in series with the holding contact (M) of the motor starter. When the Jog push button is pressed, the NO contacts energize the starter. At the same time, the NC contacts disconnect the holding circuit. Therefore, when the Jog push button is released, the starter immediately opens to disconnect the motor from the line.

Plugging is typically done in machine tool applications when the tool must be stopped rapidly at some point in its cycle. Plugging can only be done if the driven machine and its load will not be damaged by the sudden reversal of the motor.

A plugging switch (*Figure 59*) is connected to the shaft of the motor or driven machinery. The rotation of the motor is transferred to the plugging switch contacts either by a centrifugal mechanism or by magnetic induction. The contacts of a plugging switch are designed to open and close as the shaft speed of the plugging switch varies.

There are several ways that a plugging switch can be connected. *Figure 60* shows a basic control circuit that uses a plugging switch to plug a motor to a stop from one direction only. In this circuit, the forward rotation of the motor closes the NO plugging switch contact. When the Stop button is pushed, the forward starter de-energizes. At the same time, the reverse starter is energized through the plugging switch and the NC forward interlock. This reverses the motor connections, and the motor is braked to a stop. When the motor is stopped, the plugging switch opens to disconnect the reverse contactor.

Some plugging switches have *anti-plugging protection*. This means the switch design prevents the application of any counter-torque until the motor speed is sufficiently reduced. A contact on the switch opens the control circuit of the reversing contactor, preventing it from closing until the motor speed has slowed to the setpoint. Then the other contactors can be energized.

Use of Foot Switches

When using a foot switch with equipment such as power presses, additional operator protection, such as point-of-operation guarding, must be provided. This is necessary because the operator's hands and other body parts are free to enter the pinch-point area, and serious injury can occur.

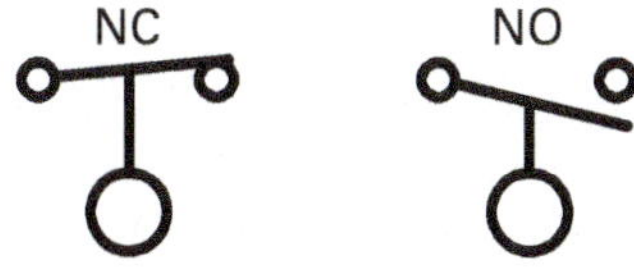

Schematic Symbols

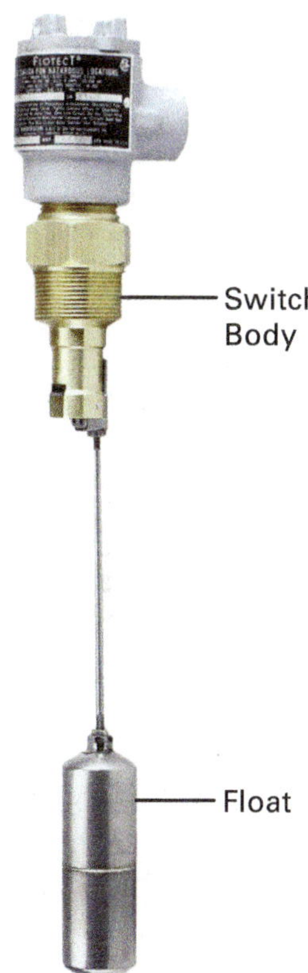

Figure 56 Liquid level float switch.
Source: Dwyer Instruments, Inc. (Bottom)

Schematic Symbols

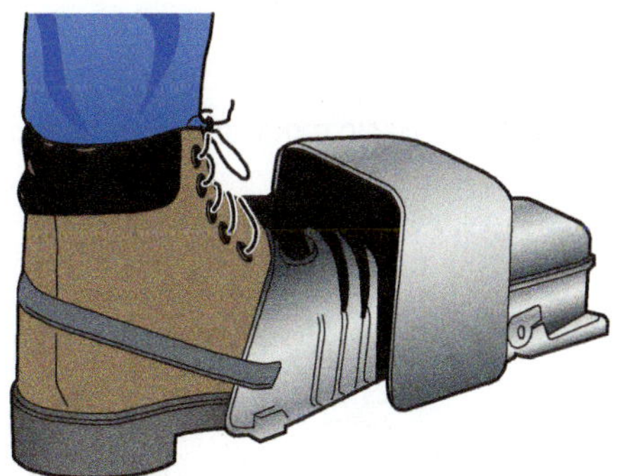

Figure 57 Foot-operated switch.

3.2.10 Proximity Switches/Sensors

Proximity switches, also called **proximity sensors**, are encapsulated solid-state devices that detect the presence or absence of targeted objects within a short range without physical contact. They incorporate both sensor and switch contacts that respond when a target is sensed. They can detect all sizes and types of objects, including very small ones.

Proximity sensors: Sensing switches that detect the presence or absence of an object using an electronic noncontact sensor.

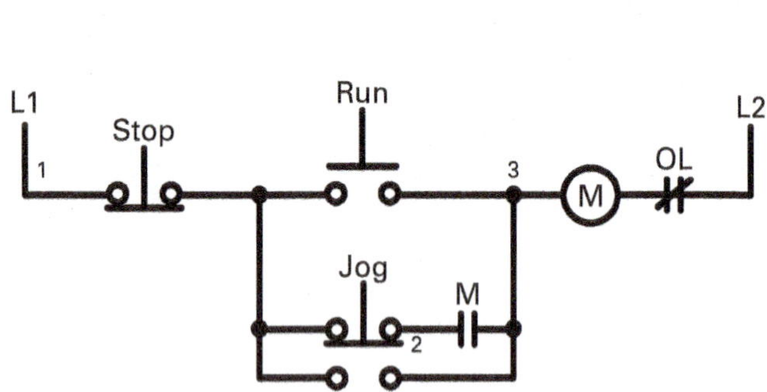

Figure 58 Basic motor jogging control circuit.

Figure 59 Plugging switch.

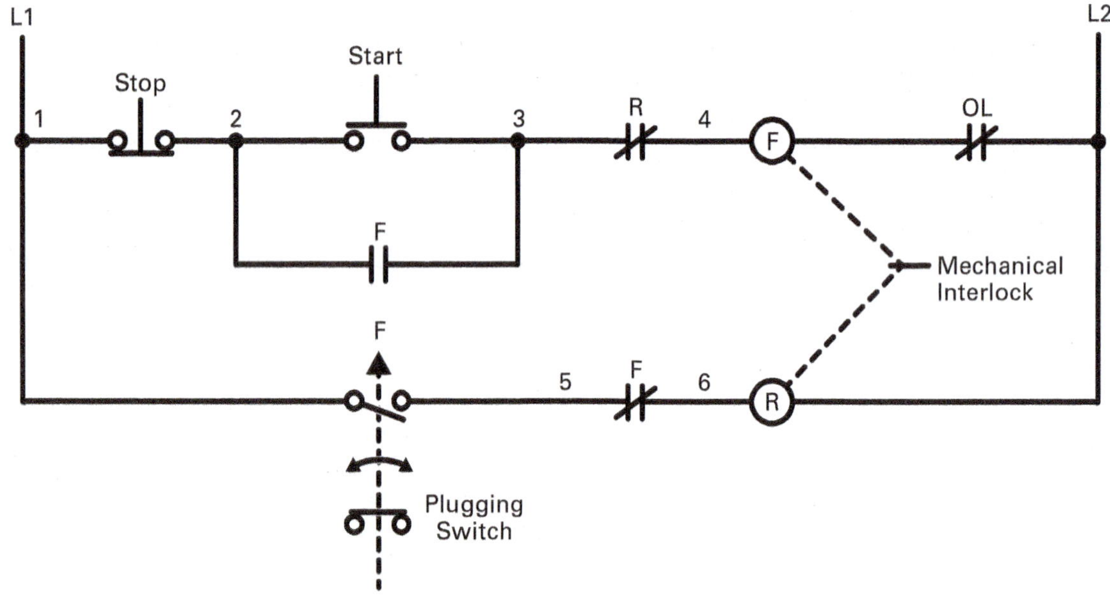

Figure 60 Basic motor plugging control circuit.

Plugging

Capacitor-start motors cannot be plugged to a stop because they cannot be reversed at full speed.

Proximity switches can be used in place of other types of switches in many applications. This is because they switch quickly and have no moving parts, making them more reliable and accurate. Being solid-state, they are compatible with electronic controllers and automated systems.

Proximity switches are intended for use in circuits with rated voltages not exceeding 250VAC or 300VDC. Two widely used proximity switches are the *inductive proximity sensor* and the *capacitive proximity sensor*. Visually, both types look the same. *Figure 61* shows two round proximity sensors. Block-shaped sensors are also available.

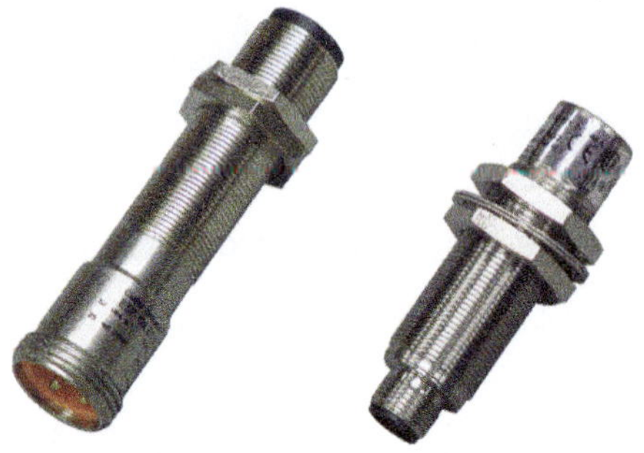

Figure 61 Examples of proximity sensors.

An inductive proximity sensor detects only metal targets within its range, typically 0.020" to 1.575" (0.5 mm to 40 mm). Its sensitivity depends on the size of the target and the metal from which the target is made. The internal circuitry consists of an oscillator circuit, output driver, and output switch (*Figure 62*). The oscillator circuit produces radio frequency (RF) oscillations. These oscillations are present at a coil located at the face of the sensor. This coil has two functions: it determines the tuning of the oscillator circuit and radiates an electromagnetic field created by the RF oscillations into the sensing field in front of the switch face.

When a target enters the sensing field, the electromagnetic field emitted by the proximity sensor causes **eddy currents** to be induced in the target. These eddy currents disrupt the electromagnetic field at the sensor face coil, causing a change in the tuning of the oscillator circuit. As a result, the oscillations cease. The output driver senses the loss of the oscillations and commands the output switch to operate, producing an On or Off output signal depending on the switch design. After the target passes out of the sensing field, the oscillator signal is again generated, causing the switch output signal to return to its normal inactive state.

Eddy currents: Currents induced in the body of a metallic material by an oscillating magnetic field.

The capacitive proximity sensor detects the absence or presence of both metallic and nonmetallic targets within its range, typically 0.012" to 0.59" (0.3 mm to 15 mm). Target detection is based on the principle that when a target moves into the sensing field, it changes the coupling capacitance of the sensor's sensing capacitor. Remember that a capacitor consists of two plates separated by an insulator (dielectric) and that any change in a capacitor's **dielectric constant** causes a change in the capacitor's capacitance value.

Dielectric constant: The ratio of the capacitance of a capacitor with a given dielectric to the capacitance of an identical capacitor using only air as a dielectric.

The internal circuitry of a solid-state capacitive proximity sensor consists of an oscillator, output driver, and output switch circuit (*Figure 63*). The sensing capacitor is part of the oscillator's tuning circuit. It is formed by two small plates separated by air and is located behind the front of the sensor face. In the absence of a target, the tuning of the oscillator causes it to be inactive. Note that this is the opposite of the inductive proximity sensor.

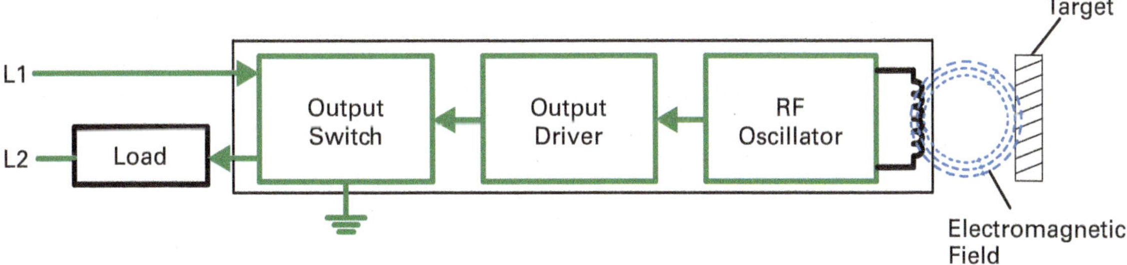

Figure 62 Block diagram of an inductive proximity sensor.

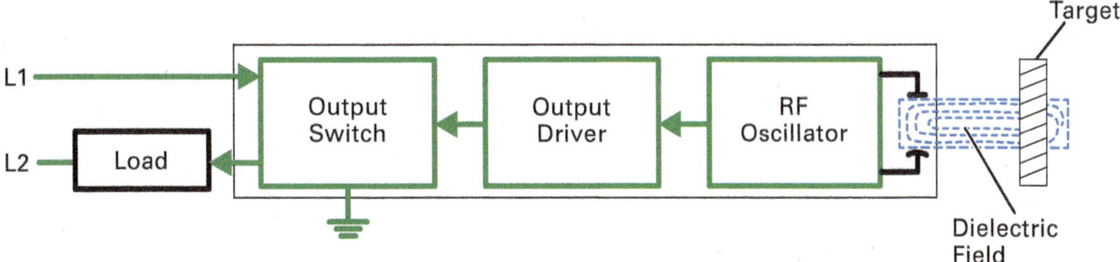

Figure 63 Block diagram of a capacitive proximity sensor.

Capacitive proximity sensors can detect any target with a dielectric constant slightly greater than air. Air has a dielectric constant of 1. When a target with a dielectric constant greater than 1 moves into range, its presence modifies the coupling capacitance. This change in capacitance tunes the oscillator circuit, so it begins to generate oscillations. The output driver senses these oscillations and commands the output switch to operate, producing an On or Off output signal depending on the switch design. After the target passes out of the sensing field, the oscillator signal ceases, causing the switch output signal to return to its normal inactive state.

Load-powered sensor: A two-wire sensor in series with the load that draws its operating current, referred to as *leakage* or *residual current*, through the load.

Line-powered sensor: A three-wire sensor that draws its operating current, referred to as *burden current*, directly from the line. The sensor's operating current does not flow through the load.

A proximity sensor can be either a **load-powered sensor** or a **line-powered sensor**. Load-powered sensors are two-wire sensors (excluding ground) connected in series with the controlled load (*Figure 64*). They draw their operating current, also commonly called *leakage* or *residual current*, through the load. When the sensor is in the open state (no target in range), it must draw a minimum operating current, sometimes called the *off-state leakage current*, through the de-energized load device to power the proximity sensor's electronics. However, this operating current must be low enough so it will not inadvertently energize the load.

When the sensor is in the closed state (target present), the current flowing in the control circuit is a combination of the proximity sensor operating current and the current drawn by the energized load. This means the proximity sensor must have a current rating high enough to carry the load current. Another factor to consider is that the current draw of the load must be high enough to ensure the sensor will operate when the load is energized. This current level is called the *minimum holding current*. If the current drawn by the load falls below the required minimum holding current, the sensor will not operate.

Two-wire proximity sensors can be connected in series or in parallel like other pilot devices. However, it is recommended that no more than three devices be connected in series or in parallel. This is because they can affect the load, as the sum of the individual voltage drops across each sensor can lower the voltage available to the load enough to render it inoperative. Typically, two-wire sensors have a voltage drop ranging from 3V to 8V. The lower the voltage drop, the more sensors can be connected in series.

Also, no more than three sensors should be connected in parallel because the sum of the operating currents flowing through the individual sensors can be enough to unintentionally energize the load. Additionally, when sensors are connected in parallel, the total amount of current the load draws when operating must be less than the maximum current rating of the lowest-rated sensor.

Line-powered proximity sensors are three-wire sensors (excluding ground) that derive power from the power lines, not the load (*Figure 65*). They have two wires for the power supply and one output wire. Some four-wire models are also available. These produce complementary NO and NC outputs. The operating current for a line-powered sensor drawn from the power lines is called the *burden current*.

Connecting line-powered sensors in series and in parallel can also affect the operation of the load. When connected in series, the first sensor in the line must carry the load current plus the operating currents for each downstream sensor. When connected in parallel, there is a danger that a nonconducting sensor will be damaged by reverse polarity. Note that this problem is usually eliminated by connecting a blocking diode in the output line of each sensor.

Capacitive Proximity Sensors

Because capacitive proximity sensors measure a dielectric gap, many have a sensitivity adjustment. This allows the sensor to be adjusted to compensate for target and application conditions. Do not handle a capacitive proximity sensor during adjustment—your hands have a dielectric constant greater than air, which can cause the sensor to lose its calibration.

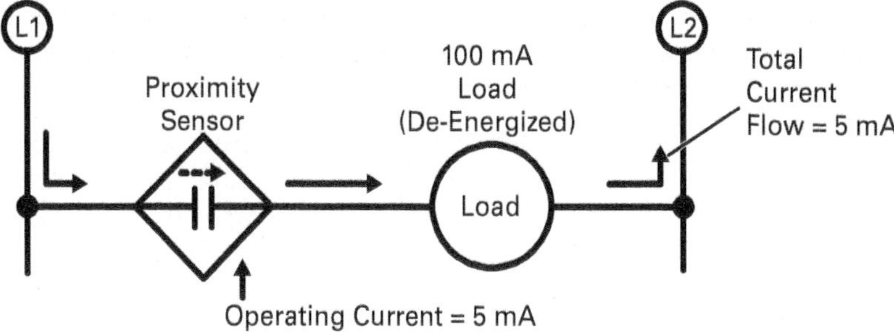

(A) Proximity Sensor Off (No Target Detected)

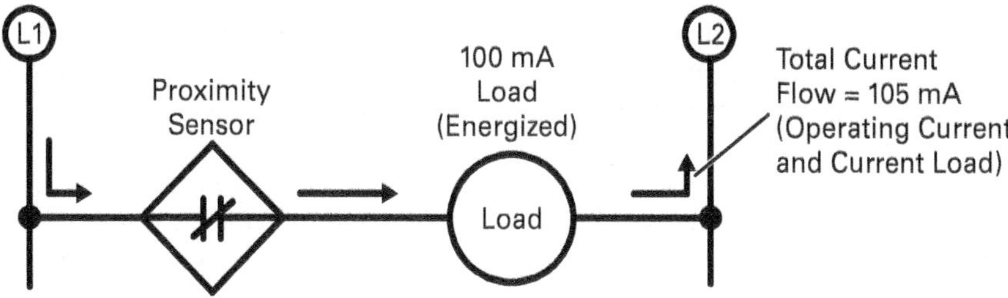

(B) Proximity Sensor On (Target Detected)

Figure 64 Load-powered proximity sensor circuit.

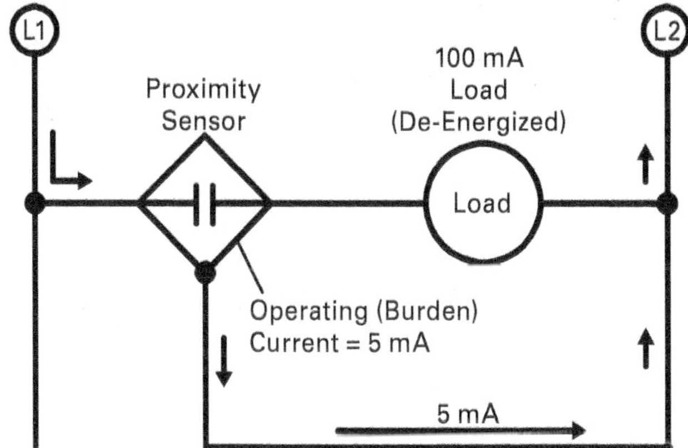

(A) Proximity Sensor Off (No Target Detected)

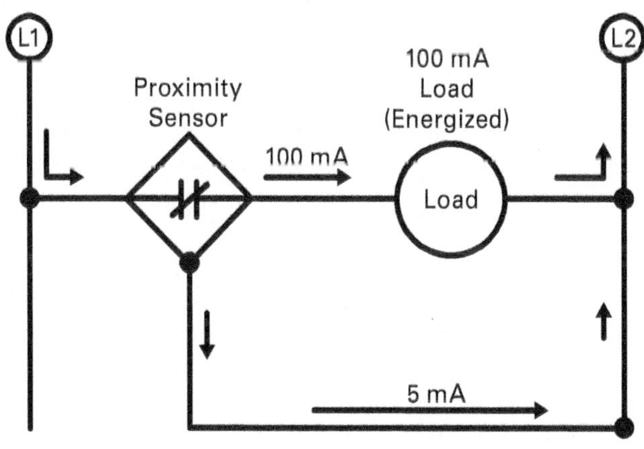

(B) Proximity Sensor On (Target Detected)

Figure 65 Line-powered proximity sensor circuit.

Proximity sensors should always be installed in accordance with the manufacturer's recommendations. Some guidelines for selecting and installing inductive and capacitive proximity sensors are provided here:

- The output switch in inductive and capacitive proximity sensors can use either a triac, silicon-controlled rectifier (SCR), or transistor as the switching device. Proximity sensors with triacs are used to switch AC loads. Those with SCRs and transistors are used to switch high-power DC loads and low-power AC loads, respectively. Some proximity sensors are designed for use in either AC or DC circuits. For DC and AC/DC versions, the switch can be connected to either positive (PNP) or negative (NPN) logic inputs.
- A shielded (flush-mounted) inductive sensor can be fully embedded in a metal mounting block without affecting the range. A non-shielded sensor needs a clearance around it called a *metal-free zone* that is determined by the sensing range. Otherwise, the sensor will sense the metal mounting and sense a target continuously.
- When mounting two or more sensors near each other, make sure to mount them in accordance with the manufacturer's instructions so their detection fields do not interfere with each other. A rule of thumb for flush-mounted sensors is to use a distance equal to or greater than twice the diameter of the sensors. When sensors of different diameters are used, base the separation distance on the largest diameter. For non-flush-mounted sensors, use a distance equal to or greater than three times the diameter of the sensors.
- When inductive and capacitive sensors are installed next to each other, their sensing fields can cause false readings in each other. For this reason, leave a space three times the diameter of the largest sensor between them. When installed opposite each other, leave six times the rated sensing distance between them.
- Where metal chips collect, the sensor should be mounted to prevent the chips from building up on the sensor face.
- Avoid installing inductive sensors in locations where the magnetic field from nearby electrical wiring can affect sensor operation.
- Avoid installing sensors in locations where electrical interference generated by nearby motors, solenoids, or relays can influence sensor operation.

3.2.11 Photoelectric Switches/Sensors

Figure 66 Photoelectric sensor.
Source: Schneider Electric

Photoelectric sensors (*Figure 66*), like proximity sensors, are solid-state devices used to detect the absence or presence of targets within their sensing range. Generally, they can detect objects more quickly and at greater distances than proximity switches.

A photoelectric sensor sends out a beam of light which is picked up by a receiver called a *photodetector*. When an object (target) moves into the path of the light beam, the beam is interrupted, indicating a target is detected.

Figure 67 shows a block diagram for a basic photoelectric switch. A photoelectric sensor consists of transmitter and receiver circuits and an output switch turned On or Off by the receiver output. Depending on the application, these circuits can be housed in the same encapsulated enclosure or separate enclosures.

Infrared (IR): A portion of the light spectrum that is not visible with a wavelength beyond 690 nanometers. IR radiation is emitted by certain LEDs.

The transmitter circuit generates **infrared (IR)** light emitted from an LED through a lens placed just behind the face of the sensor. The emitted infrared light beam is either an unmodulated beam or modulated beam. An unmodulated beam is emitted continuously, while a modulated beam is turned on and off at a very high frequency. Modulated-beam sensors tend to be more common because they do not respond to ambient light or other forms of light "noise." Unmodulated-beam sensors are typically used where the scanning range is very short and dirt, dust, and bright ambient light are not a problem.

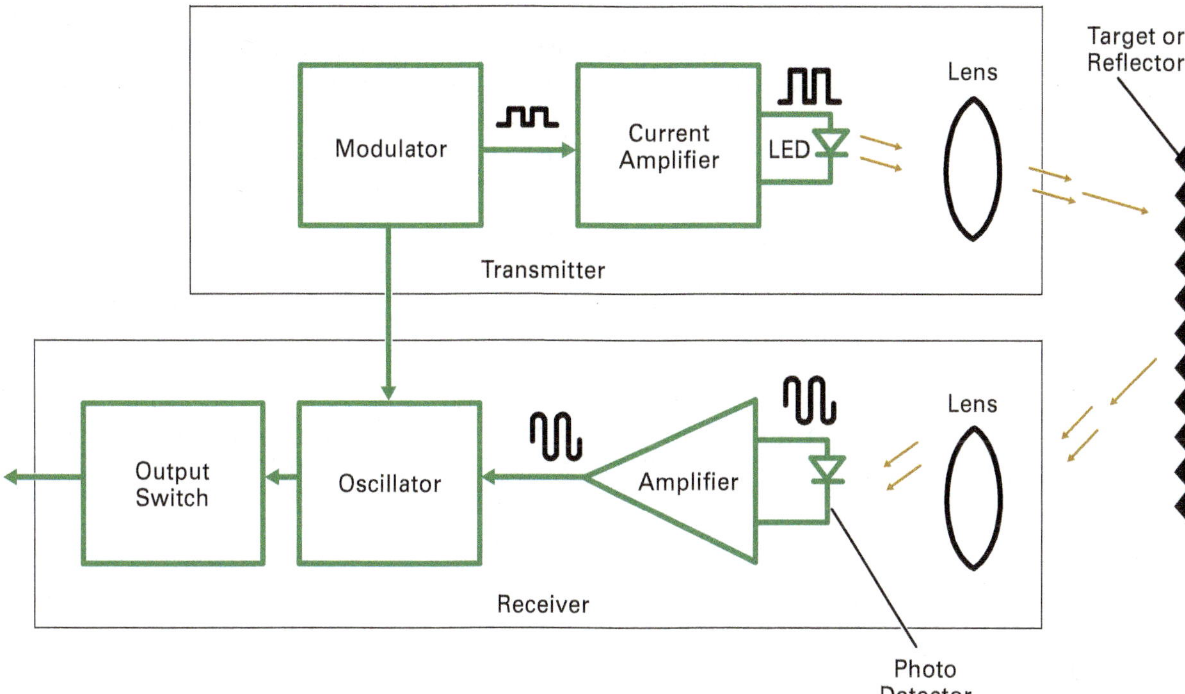

Figure 67 Block diagram of a modulated-beam photoelectric sensor.

A photodetector placed just behind a lens at the face of the sensor is the front end of the sensor's receiver circuit. Being light sensitive, the photodetector detects any received IR light and, in response, produces an output signal. This signal is further processed to amplify it to a usable level. In the receiver of a modulated-beam sensor, the amplified signal is also applied to a demodulator circuit. There, the characteristics of the signal are compared against those of the original modulated signal. This comparison helps to select the desired received signals and reject unwanted noise signals. The receiver output is sent to the output switch, commanding it On or Off depending on the application.

Photoelectric sensors generate an output any time an object is detected. If this occurs when the photodetector sees light, the sensor is classified as *working-in-the-light* mode. This means the sensor's output switch is energized when the target is missing. If the sensor generates an output when the photodetector does not see light, the sensor is classified as *working-in-the-dark* mode. This means the sensor's output switch is energized when a target breaks the beam of light.

Photoelectric sensors can be used in several different arrangements to meet the requirements of different applications. These include the following:

- *Direct scan (thru-beam)* — The transmitter and receiver are separate assemblies located opposite each other. The light beam from the transmitter is aimed directly at the receiver. The target passes between them, interrupting the light beam. This is the most basic approach.
- *Retroreflective scan* — The transmitter and receiver are housed in one assembly. The light beam from the transmitter is reflected to the receiver using a reflector located opposite the sensor. The target passes between the sensor and reflector, interrupting the light beam.
- *Polarized scan* — The transmitter and receiver are housed in one assembly. The light beam from the sensor is filtered by a special lens so it projects in one plane only. The receiver responds only to depolarized light reflected from a corner cube reflector or polarized tape.
- *Specular scan* — The transmitter and receiver are separate assemblies placed at equal angles from a highly reflective surface, requiring the receiver to be positioned precisely to receive reflected light. The transmitter sends the signal to the receiver by reflecting the signal off the surface.

- *Diffused scan* — The transmitter and receiver are housed in one assembly. The light beam from the transmitter is reflected to the receiver by the target.
- *Convergent scan* — The transmitter and receiver are housed in one assembly. The light beam from the transmitter is simultaneously focused and converged to a fixed point in front of the sensor. The detection point is positioned so that targets before or beyond the focal point are not detected.

Photoelectric sensors should be selected and installed in accordance with the manufacturer's recommendations. General guidelines for connecting and installing load- and line-powered photoelectric sensors are the same as those described for proximity sensors.

3.2.12 Drum Switches

Figure 68 Drum switch.

Drum switches are totally enclosed, multipole switches used in machine operation applications to start, stop, and reverse the direction of motor rotation. As shown in *Figure 68*, the switching elements are contained in a cylindrical housing that resembles a drum.

Drum switches are rated by maximum horsepower and vary in size and shape. They can have either maintained or momentary contact operation, with spring-assisted return to the center position. Drum switches do not contain any protective devices. When they control motor operation, circuit overload protection must be provided by a motor starter in line before the drum switch.

When the motor being controlled by a drum switch is not running in forward or reverse, the handle is in the Off (center) position. To reverse the direction of a running motor, the handle must first be moved to the Off position until the motor stops, then moved to the Reverse position. Motor reversal in single-phase motors is accomplished by the drum switch reversing the start winding connections. For a three-phase motor, the switch reverses two of the three motor leads. For a DC motor, the switch must reverse the current direction through the motor fields without changing the current direction through the armature or vice versa.

3.0.0 Section Review

1. While selecting a control transformer for a specific application, you have calculated its application inrush to be 3,500VA. If the primary input voltage is steady and not expected to vary more than 5% from nominal, the transformer VA rating should be _____.
   a. 150VA
   b. 250VA
   c. 300VA
   d. 350VA

2. The contacts of a switch are rated as A300. In a 240VAC application, the *maximum* break current for these contacts is _____.
   a. 1A
   b. 1.5A
   c. 3A
   d. 6A

4.0.0 Installation Considerations for Motor Circuits

Objective

State installation considerations for motor controls.

a. Explain how to select contactor and motor starter enclosures.
b. Interpret installation diagrams and their common symbols.
c. Identify *NEC*® regulations related to motor control circuits.
d. Explain how to connect motor controllers for specific applications.

Performance Task

1. Make all wiring connections for a magnetic motor starter controlled by two push-button stations, including a holding circuit interlock.

The accurate selection and installation of motor controls significantly contribute to the life span and reliability of both the motor and control devices. Always follow the system wiring diagrams, in addition to local codes and applicable *NEC*® requirements.

4.1.0 Selecting Motor Enclosures

To shield energized motor starters and pilot devices from accidental contact, an enclosure is necessary. A general-purpose sheet metal enclosure is usually sufficient. However, as specified by the *NEC*®, dust, moisture, or explosive gases make it necessary to use special enclosures for protection in some environments. It is essential to carefully consider the conditions under which the equipment must operate to ensure compliance with the *NEC*® and local codes. A general-purpose enclosure does not afford the required protection in many applications.

UL Solutions has defined the requirements for protective enclosures for various hazardous conditions, and NEMA has standardized enclosures to satisfy these requirements. Various NEMA enclosures (*Figure 69*) are described here:

- *General purpose (NEMA Type 1–IP40)* — A general-purpose enclosure is intended primarily to prevent accidental contact with the enclosed apparatus. It is suitable for indoor applications where it is not exposed to unusual service conditions. A NEMA Type 1 enclosure protects against dust, light, and indirect splashing but is not dust tight or raintight.
- *Dust tight, raintight (NEMA Type 3–IP52)* — This enclosure protects against rain and dust. It is suitable for outdoor applications.
- *Rainproof, sleet-resistant (NEMA Type 3R–IP52)* — This enclosure protects against rain and resists damage from exposure to sleet. It is designed with conduit hubs, external mounting, and drainage provisions.
- *Watertight (NEMA Type 4–IP65)* — This enclosure is designed to withstand high-pressure, hose-directed water and windblown dust and rain. It is constructed of steel with a gasketed cover. Pilot device holes are located on the enclosure cover and are provided with hole plugs.
- *Watertight, corrosion-resistant (NEMA Type 4X–IP65)* — These enclosures are similar to NEMA Type 4 enclosures except that they are made of a material that resists corrosion. They are used in locations such as meatpacking and chemical plants, where corrosive elements would destroy an ordinary steel enclosure.
- *Hazardous locations, Class I (NEMA Type 7C or D)* — Type 7C enclosures are designed to meet the application requirements of *NEC*® Class I hazardous locations where flammable gases or vapors may be present in quantities sufficient to produce explosive or ignitable mixtures. They must withstand the pressure generated by an internal explosion and be able to contain the explosion so that gases in the surrounding atmosphere do not ignite. Note that each letter following a NEMA type number indicates a particular group of hazardous locations as defined in the *NEC*®. The designation is incomplete without the suffix letter.

Enclosures

Most manufacturers provide a selection of engineered motor starter assemblies that include a NEMA enclosure as well as the fused disconnect device required by the *NEC®*.

- *Hazardous locations, Class II (NEMA 7 Type 9E, F, or G)* — These enclosures are designed to meet the application requirements of *NEC®* Class II locations where combustible dust is present. Again, each letter following the NEMA type number indicates the group of hazardous locations for which the enclosure is suited.
- *Industrial use (NEMA Type 12–IP62)* — This enclosure is designed to exclude dust, lint, fibers and other airborne materials, and oil or coolant seepage. It is mounted with flanges or mounting feet, as the enclosure has no conduit openings or knockouts. Pilot device openings are located on the enclosure cover and are provided with matching plugs.
- *Oiltight, dust tight (NEMA Type 13–IP62)* — These enclosures are generally cast iron and used in the same areas as NEMA Type 12 enclosures. The main difference is that, due to its cast housing, a conduit entry is provided, and blind holes are used for mounting. Pilot device openings with matching plugs are located on the enclosure cover.

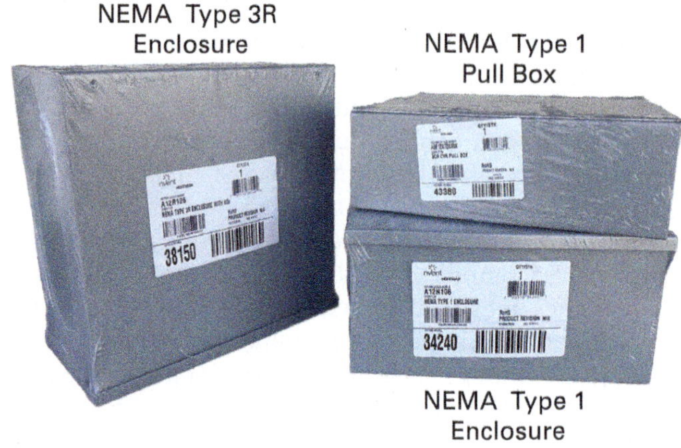

Figure 69 Typical NEMA enclosures.
Source: Courtesy of Honeywell International Inc.

Know the Code

Enclosure Types
NEC Section 110.28

IEC-compliant enclosures are identified by a two-digit IP protection rating system outlined in *Table 12*. Additional information for selecting enclosures for use in specific locations other than those considered hazardous can be found in *NEC Section 110.28* and *NEC Table 110.28*.

TABLE 12 IEC Enclosure IP Two-Digit Protection Code Definitions

First Digit	Description	Second Digit	Description
0	No protection	0	No protection
1	Protection against solid objects greater than 50 mm	1	Protection against vertically falling drops of water
2	Protection against solid objects greater than 12 mm	2	Protection against dripping water when tilted up to 15°
3	Protection against solid objects greater than 2.5 mm	3	Protection against spraying water
4	Protection against solid objects greater than 1 mm	4	Protection against splashing water
5	Total protection against dust; limited ingress (dust protected)	5	Protection against water jets
6	Total protection against dust (dust tight)	6	Protection against heavy seas
N/A	N/A	7	Protection against the effects of immersion
N/A	N/A	8	Protection against submersion

4.2.0 Interpreting Installation Diagrams

The information needed to install and troubleshoot motor control circuits is shown in several forms. These include wiring diagrams, circuit schedules, control ladder diagrams, and logic diagrams.

4.2.1 Wiring Diagrams

Connection diagrams show the actual point-to-point wiring connections for a motor and its control circuits. This type of diagram is normally found in the service literature for a specific piece of factory-wired equipment. Connection diagrams are useful when tracing and locating a specific wire connected between two components in a circuit, such as when locating an open or grounded circuit.

Figure 70 shows a typical connection diagram for a conveyor system. The format of connection diagrams varies depending on the type of equipment and the manufacturer. Some connection diagrams show power wiring as heavyweight lines and control circuit wiring as lighter-weight lines. Some will show the actual color of each wire in the point-to-point wiring scheme.

Where applicable, the terminal to which a specific wire is to be connected is shown. To help the technician interpret the diagram, a legend that describes any nonstandard symbols and abbreviations is provided.

4.2.2 Circuit Schedules

Circuit schedules, sometimes called *wire lists*, show actual wire or cable point-to-point connections for a motor and its control circuits. They are useful when installing a new system or modifying an existing one.

A circuit schedule is typically presented in tabular form. Each individual wire or cable is given a sequence or circuit identification number. The information normally includes the connection points and the type and size of the connecting wire. Some circuit schedules also include the color of the wire and termination hardware for each end of the wire. *Table 13* shows an example of a typical circuit schedule.

TABLE 13 Example of a Circuit Schedule

Circuit/Cable ID	Cable Type	Length	From	To
P108	3/c #10 w/ ground	140'	MCC 10 Section 4A	M108
C108	9/c #14	120'	MCC 10 Section 4A	PLC10-TB1
C108A	12/c #14	160'	MCC 10 Section 4A	JB108
C108B	5/c #14	40'	JB108	HS108
C108C	3/c #14	30'	JB108	HSS108A
C108D	3/c #14	10'	HSS108A	HSS108B
C108E	3/c #14	30'	JB108	AA108A
C108F	3/c #14	80'	JB108	AA108B
C108G	5/c #14	70'	JB108	SSL108
C108H	3/c #14	10'	JB108	ZS108A
C108J	3/c #14	10'	ZS108A	ZS108B
C108K	3/c #14	60'	JB108	ZS108C
C108L	3/c #14	10'	ZS108C	ZS108D

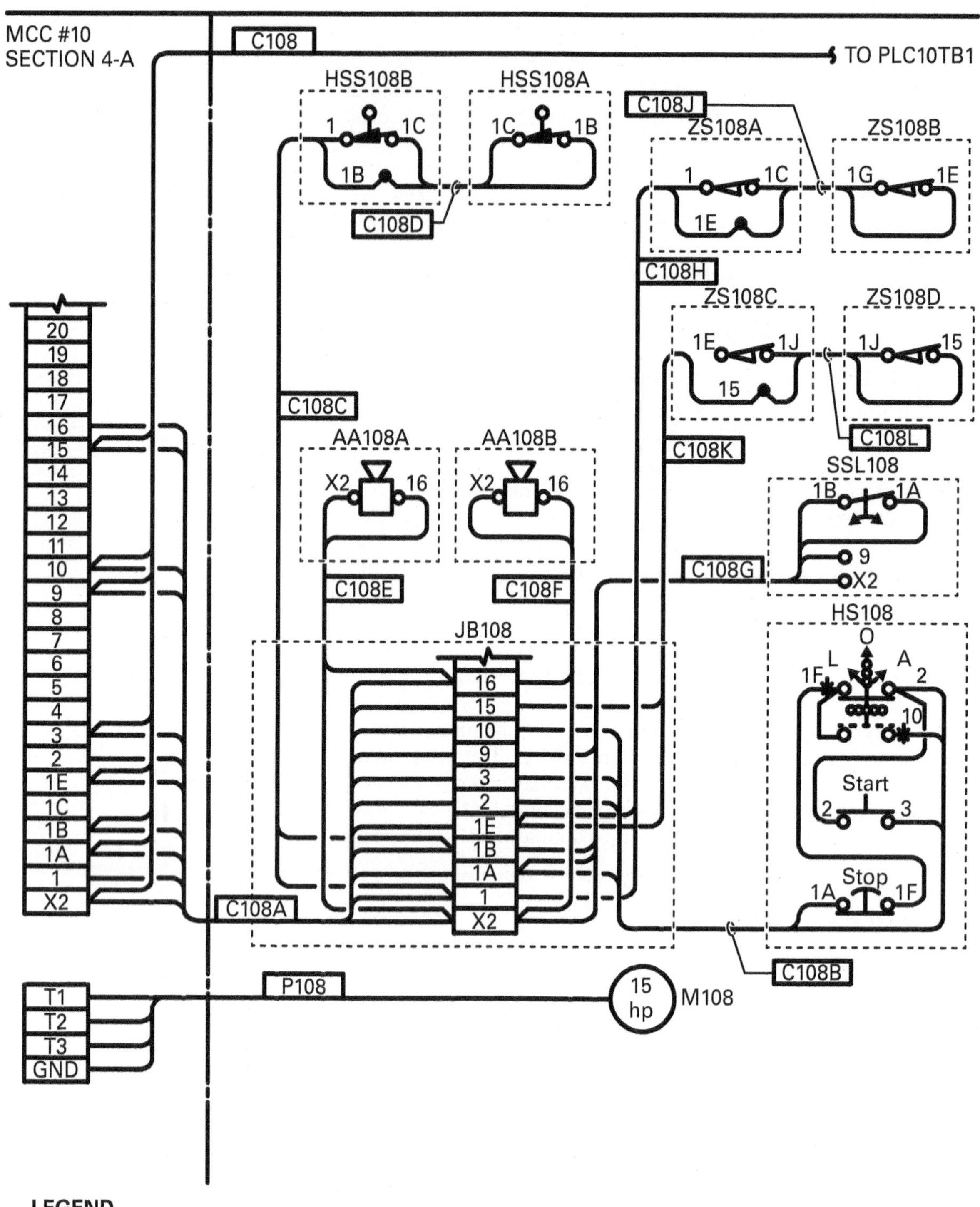

LEGEND

AA108A-D	Conveyor start warning horns
HS108	Local control station with stop mushroom maintained push button, start push button, and Local/Off/Auto selector switch
HSS108A-B	Emergency pull cord safety switches
PLC	Programmable logic controller
SSL108	Conveyor zero-speed switch
ZS108A-D	Belt runoff switches

Figure 70 Example connection diagram.

4.2.3 Ladder Diagrams

Control circuits that use various pilot devices and relays, as well as those that incorporate a microprocessor, are generally presented as ladder diagrams (*Figure 71*). A ladder diagram shows the operating sequence of a circuit using single lines and symbols.

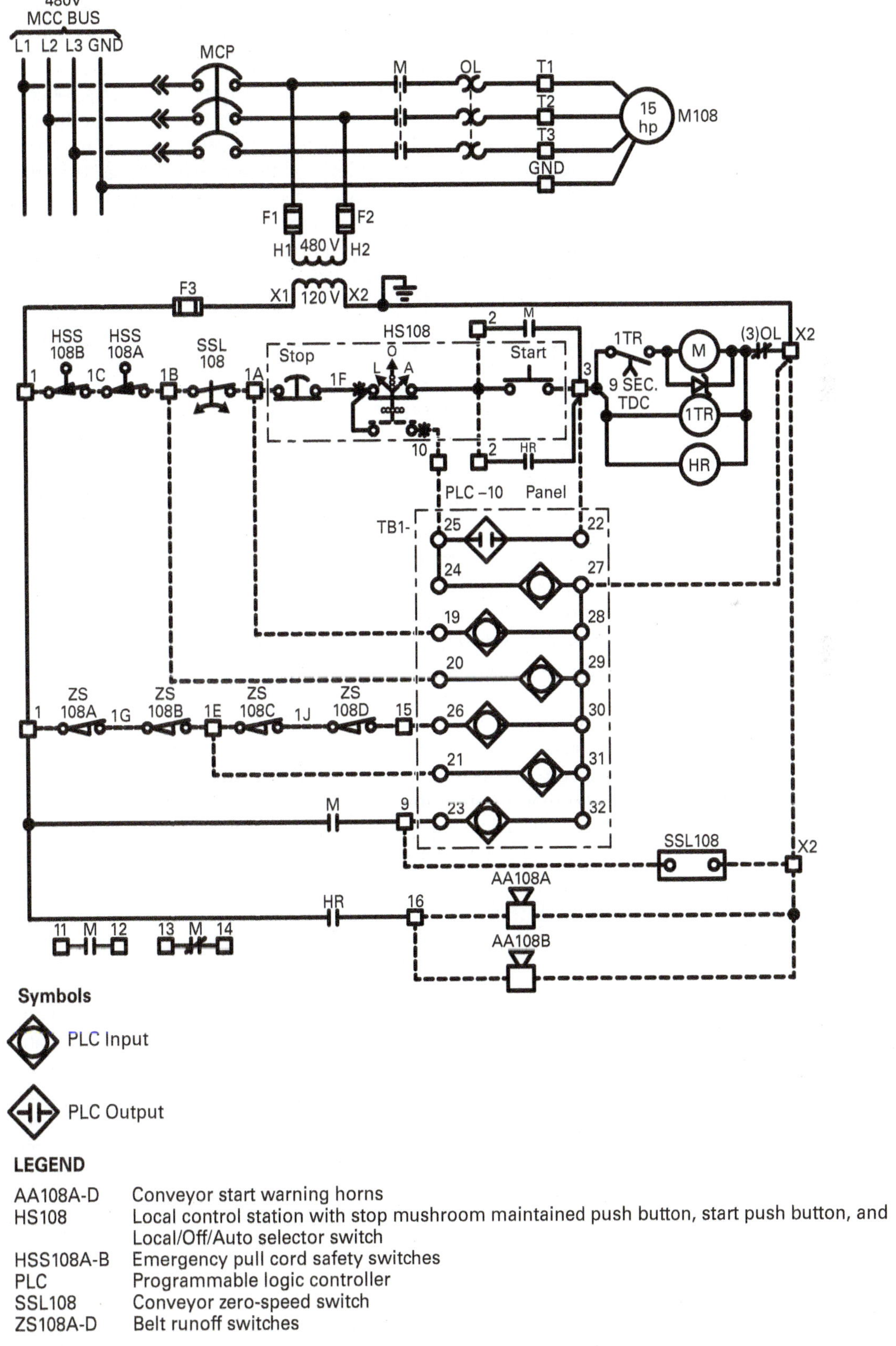

Figure 71 Example of a ladder diagram.

Ladder diagrams are usually presented in three sections between the two wires from the control voltage source, as shown in *Figure 72*. The three sections consist of a signal section, a decision section, and a direct/indirect action section. The signal and decision sections may be combined into a single section. These sections are described as follows:

- *Signal section* — Pilot devices are the source of the control signal. They can be manual switches, such as push buttons or foot switches; mechanical switches, such as limit switches; or automatic reset devices, like flow, pressure, or temperature switches.
- *Decision section* — The decision section adds, sorts, selects, and redirects the signals from the pilot devices to the load. The function of this area must be logical. In many cases, the pilot devices are part of the decision section. The way pilot device contacts are connected in the circuit reflects the circuit logic. There are three basic logic functions: AND, OR, and NOT. The electrical relay logic symbols for these logic functions, their related truth tables, and logic symbols are shown in *Figure 73*, where contacts A and B represent pilot device contacts. The status of each set of contacts is represented in the truth table by a zero (0) when the contacts are open and by a one (1) when they are closed. The output status, On (1) or Off (0), called Y, is shown for all the possible combinations of contact A and B positions.
- *Direct or indirect action section* — The action section usually energizes a relay or contactor coil. As shown in the truth tables, when Y is 1, the coil is energized; when it is 0, it is de-energized.

The logic symbols used in *Figure 73* for the AND, OR, NOT, NAND (NOT AND), and NOR (NOT OR) logic functions are a version of NEMA logic symbols. It is important to point out that the logic symbols used on a diagram differ depending on whether the application is electrical, electronic, hydraulic, or pneumatic. Even for similar types of equipment, the symbols used by different manufacturers often vary.

As shown in *Figure 73*, the electrical equivalent of an AND function is two NO sets of contacts connected in series. As indicated by its truth table, the AND function has an output that is on (1) only when both contacts A and B are closed (1). This is because the related coil CR is energized, causing its NO contacts to close. For all the other possible contact position combinations, the Y output remains off (0). This is because coil CR is de-energized, resulting in its NO contacts remaining open (0).

The electrical equivalent of an OR function is two NO sets of contacts connected in parallel. An OR function is so called because it will produce a 1 output (on) whenever contact A or contact B is closed (1). Output Y will be a 0 only when contacts A and B are open (0).

The third basic logic function is the NOT function, commonly called an *inverter*. Note that it has only one NO contact. The function of the NOT circuit is simple; it operates to produce an output always opposite the contact position. Thus, when contact A is closed (1), output Y is 0 (off); when contact A is open (0), output Y is 1 (on). For the circuit shown, the open NC contacts of coil CR accomplish the inversion.

A NOR function is the same as an inverted OR function. It consists of an OR function combined with an inverter (NOT) function. Comparison of the NOR truth table with the OR truth table shows that for the same contact A and B positions, the output Y for the NOR is always opposite that of the basic OR function.

A NAND function is the same as an inverted AND function. It consists of an AND function combined with an inverter (NOT) function. Comparison of the NAND truth table with the AND truth table shows that for the same contact A and B positions, the output Y for the NAND is always opposite the basic AND function.

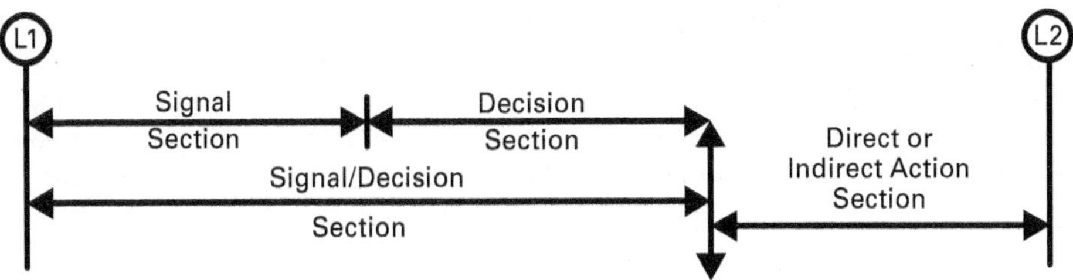

Figure 72 Typical ladder diagram sections.

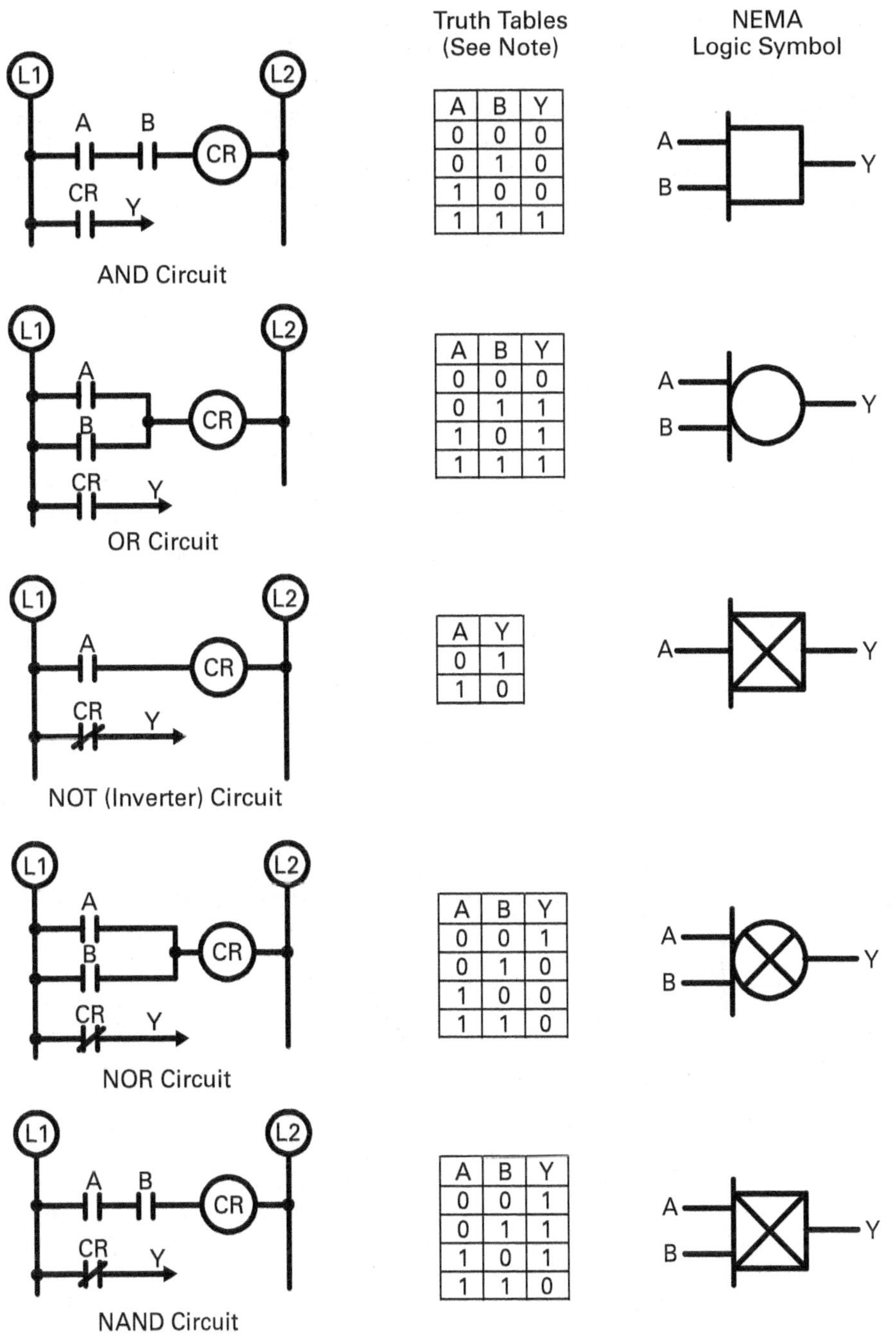

Truth Tables (See Note)

NEMA Logic Symbol

AND Circuit

A	B	Y
0	0	0
0	1	0
1	0	0
1	1	1

OR Circuit

A	B	Y
0	0	0
0	1	1
1	0	1
1	1	1

NOT (Inverter) Circuit

A	Y
0	1
1	0

NOR Circuit

A	B	Y
0	0	1
0	1	0
1	0	0
1	1	0

NAND Circuit

A	B	Y
0	0	1
0	1	1
1	0	1
1	1	0

NOTE:
When contacts A and/or B are shown as zero (0) in the truth table, it means that the contacts are open. If shown as one (1), it means that the contacts are closed.

Figure 73 Basic logic functions.

Except for the NOT function, any of the functions just described can have more than two contacts added in series or in parallel with the others in *Figure 73*, and the truth table can be expanded to include them. For example, if you have three series-connected contacts in an AND function, all three contacts must be closed to get a Y output of 1. Otherwise, the output is 0. Similarly, three parallel-connected contacts in an OR function would mean that the Y output is 1 if any one of the three contacts is closed.

Up to this point, relay coils and contacts have been used to describe the operation of the basic AND, OR, NOT, NOR, and NAND logic functions. These logic functions are commonly incorporated electronically into integrated chips, PLCs, and similar devices used in modern solid-state motor control equipment.

As a simple example, close examination of *Figure 74* reveals an AND function overlapped by an OR function. The AND function is the Stop/Start switch and MOL contact. The Stop contacts, Start contacts, and MOL contacts must be closed to energize the contactor coil M. When energized, an OR function consisting of either the Start switch or the M holding contacts keeps the coil energized.

Ladder diagrams normally reference the type of machine or process to which the diagram applies. Careful study of the ladder diagram can help a technician determine or confirm the correct sequence of operation for a particular piece of motor control equipment. As previously described, ladder diagrams are divided into sections horizontally and read as follows:

- *Left to right* — Control devices are connected between L1 and the operating coil (or load) from left to right. All control functions occur on the left side, or L1 side, of the coil. Only overload contacts are connected on the right side, or L2 side, of the coil. Stop and shutdown circuits occur before the start and run circuits. Where practical, external connections, such as buttons and switches, occur before internal connections, such as relay connections.
- *Top to bottom* — Ladder diagram horizontal lines are usually numbered from top to bottom, with the first function starting on line one and subsequent or dependent functions proceeding down the ladder until the last function occurs.

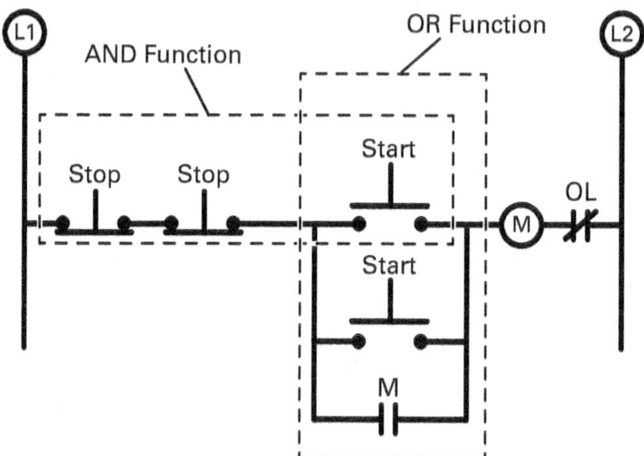

Figure 74 Example AND and OR functions in a motor starter control circuit.

Figure 75 shows common conventions used on a ladder diagram. However, diagrams furnished by manufacturers or designers vary widely, and not all the conventions pointed out in *Figure 75* may appear in all diagrams. The common conventions are as follows:

- *Component identification* — The diagram should identify the various control circuit devices and loads by reference (PB1, S1, CR1) and/or function (low-pressure switch, on-delay timer). If a control device is remotely located from the main equipment enclosure, the diagram should also identify the device's location.

- *Line fuse* — Many codes require line-voltage control circuits to be fused if they exist outside the enclosure for the relevant relays or contactors or if a step-down transformer is not power limited or is located outside the enclosure.
- *Step-down transformer and fuse* — In many cases, a low-voltage step-down transformer, along with a fuse, is required for control circuits. If not, these devices are omitted, and line voltage is used as the control voltage.
- *Component terminal numbers* — If a component has terminals marked with numbers or letters, they should be shown on the diagram.
- *Ladder line numbers* — The horizontal lines on the ladder should be numbered sequentially from top to bottom for contact cross-reference. This aids in locating all the contacts of a relay, contactor, or other device.
- *Mechanical linkage symbols* — If the items are located on adjacent lines, a dashed line may be used to indicate a mechanical link. An alternate approach is an arrowed box containing the ladder line reference to the location of the linked device. Appropriately numbered arrowed boxes are used at both locations of the linked item.
- *Wire numbers* — Unique wire numbers are assigned to each wire or group of wires that are electrically common. Labeling the wires in the actual equipment with these numbers supports the installation and future troubleshooting.
- *Line references for contacts* — Relays and most motor starters have one or more sets of control circuit contacts. At the right-hand side of the line containing the coil for the device, the location of any control contacts not shown on the same line is referenced by ladder line numbers, separated by commas. Underlined line numbers are for NC contacts, and nonunderlined numbers are for NO contacts.

4.2.4 Logic Diagrams

Logic diagrams show a pictorial representation of circuit elements in the form of logic symbols. They convey the logic behind system, machine, and/or network operation. They are used to communicate logic requirements to a control system's supplier or programmer. Engineers, technicians, and operators also use them to agree on the rules of system operation before the programming and circuit design begins.

Logic diagrams are useful for technicians because they show the equipment's decision-making processes regarding system operation and sequencing. These decision processes may not be obvious in a related ladder or schematic diagram because many actions are accomplished solely by the software.

The logic symbols used on logic diagrams do not represent the types of electronic components used, but they do represent their logical functions. As mentioned earlier, there are many symbol sets used on logic diagrams. Even for similar types of equipment, the symbols used by different manufacturers often vary. *Figure 76* shows a typical logic diagram of a motor control circuit. The symbols used on this diagram are the same NEMA logic symbols described earlier in this module.

The presentation of information on a logic diagram typically flows from the top of the page to the bottom. Logic functions associated with *permissive signals* are normally shown first. Permissive signals are typically generated by safety switches, stop switches, mode switches, or related software that enable or inhibit system operation. Logic functions associated with *running signals* are shown second. Running signals are signals that instruct the machine to operate. Logic functions associated with *miscellaneous signals*, such as alarm signals, are shown last.

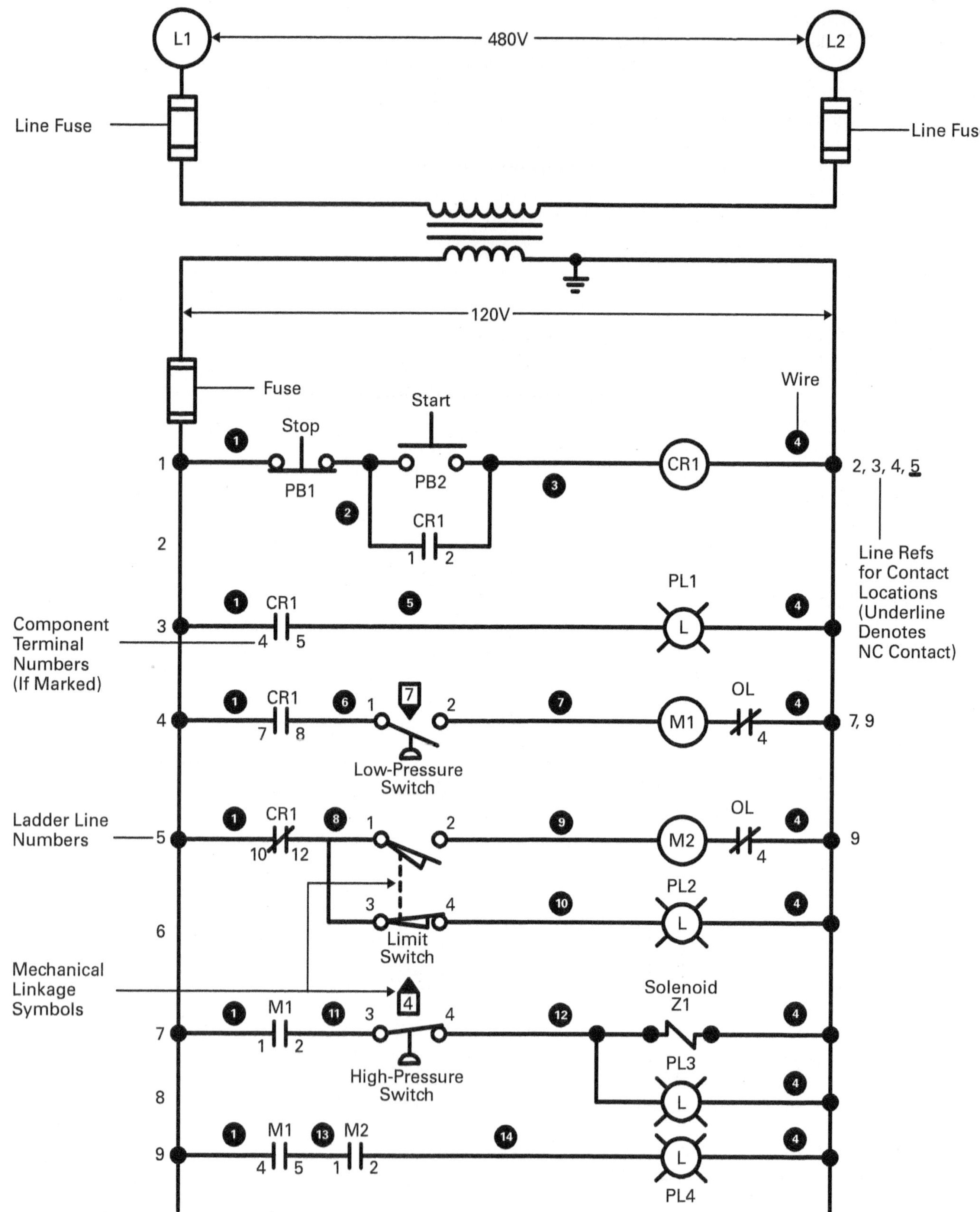

Figure 75 Typical ladder diagram conventions.

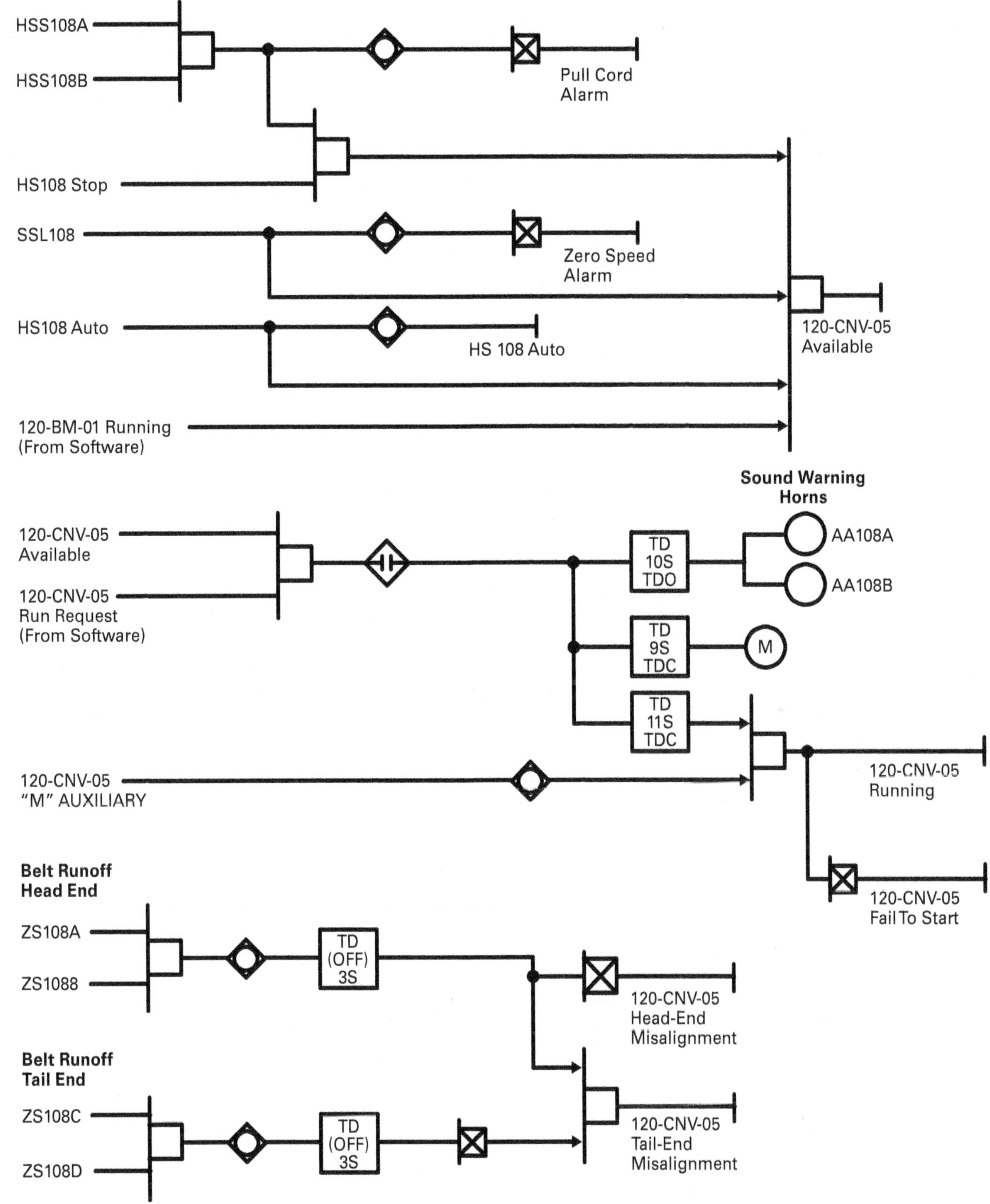

Figure 76 Example of a logic diagram.

The logic flow path shown on the diagram is from left to right. Sometimes the output line from the last logic element on the right-hand side of the drawing may loop back to the left side of the drawing to show the remainder of a logic function.

Labels shown on the left side of the diagram normally identify a device or signal name input to the logic element. Those on the right side show the output or action from a logic element. Note that the inputs and outputs shown on a logic diagram can be applied from or sent to hardware devices, while others may be applied from or sent to the system-controlling software.

Understanding the information shown on a logic diagram will help an electrician do the following:

- Understand how a programmable logic controller (PLC) controls the operation of a system without the need to sort through the specific PLC programming details.
- Understand the details of system operation that are not obvious on a ladder diagram.
- Generate a ladder diagram.
- Determine the status of devices shown on the drawings.
- Determine how and where to implement wiring modifications because of changes in operating requirements.
- Explain to others how the equipment works.

4.2.5 Relating Diagrams to Equipment Wiring and Operation

You have now been introduced to the different types of diagrams that can be used to show the wiring and operation of a system. The diagrams and circuit schedules used in this discussion are related to a mill conveyor system designated as 120-CNV-05. The conveyor has a 15 hp drive motor that transfers screened oversize materials to a **ball mill** designated as 120-BM-01.

Ball mill: A grinding unit used to grind and/or blend rocks and minerals for use in manufacturing or other processes.

Based on this description, you should examine the different diagrams and try to relate what they show to the equipment and its function. The connection diagram and circuit schedule (*Figure 70* and *Table 13*) show the wiring connections between the different components of the conveyor system. The ladder and logic diagrams (*Figure 71* and *Figure 76*) show the control sequence and design logic, respectively. Refer to the ladder diagram shown in *Figure 71* while following the description of the conveyor operation in the following paragraphs.

Placing the Local/Off/Auto switch (part of HS108) in the Auto position selects the normal operation mode for the conveyor. In this mode, the conveyor is controlled by the PLC (PLC10), which provides sequential interlocking for the 120-BM-01 ball mill.

Emergency pull cord safety switches (HSS108A and HSS108B) prevent the conveyor motor from starting if one or both are activated (contacts open). If activated when the conveyor motor is running, the motor will stop immediately. It will also cause the PLC to generate an alarm.

A conveyor zero-speed switch (SSL108) is mounted on the tail end of the belt pulley. It monitors the belt for slippage or seizure. This switch is energized when the conveyor motor is started. It contains an internal time-on-delay circuit whose contacts allow sufficient time for the conveyor belt to reach normal speed. If the conveyor belt is rotating at the correct speed by the end of the delay period, the contacts remain closed, allowing conveyor operation to continue. Should the belt slow down at any time, or if its speed isn't sufficient by the end of the delay, the switch contacts open, opening the control circuit and shutting the motor down. An alarm is then generated by the PLC.

Belt runoff switches (ZS108A through ZS108D) serve to detect belt misalignment at the head end (ZS108A and ZS108B) and tail end (ZS108C and ZS108D) of the conveyor. Should a misalignment be detected, the switch signals cause the PLC to generate an alarm. These switches do not shut down the motor.

Conveyor-start warning horns (AA108A and AA108B) provide a local area alarm through a horn relay (HR) that sounds for 10 seconds when a motor start command is initiated. This warns people that the conveyor is about to start. Motor startup is delayed for nine seconds by a time-delay relay (1TR) after a start command is issued.

Placing the Local/Off/Auto switch (part of HS108) in the Local position allows the conveyor to be operated for maintenance and testing independent of PLC control and its sequential interlocking. The safety features provided by the emergency pull-cord switches, zero-speed switch, and conveyor-start warning horns remain active even in the local mode. The functions of belt runoff switches are inactive during the local mode of operation.

Remember that the purpose of a logic diagram is to convey the design logic behind the operation of a system. It is important to point out that many of the functions shown on a logic diagram may be accomplished by the operating software rather than the hardware. For this reason, there is not always a direct correlation between what is shown on the logic diagram and what is shown on the related ladder diagram. To get the entire picture of how a system operates, always refer to both diagrams as well as any written sequence of operation.

Refer to the logic diagram shown in *Figure 76* for the conveyor system. The top part of the logic diagram shows the logic elements that develop the inputs to a four-input AND. As shown, the field inputs in this part of the diagram are applied from the system emergency pull cord safety switches (HSS108A and HSS108B), the Stop button (part of HS108), the conveyor zero-speed switch (SSL108), and the Auto switch (part of HS108). Also applied from the system software is an interlock input signaling that the 120-BM-01 ball mill is running. Each of these permissive signal inputs is applied through contacts that must be closed for the conveyor to run.

Tracing the emergency pull cord safety switch and the Stop button inputs through their respective two-input AND logic gates shows that when all three input signals are logical, the related input to the four-input AND is logical. The remaining three inputs applied to the four-input AND from the conveyor zero-speed switch, Auto switch, and the system software are all logical. This causes the output signal from the four-input AND to be logical. This signal (120-CVN-05 available) is applied as an input to the next group of decision logic. Should any of the input signals be a logical zero, the output from the four-input AND will be a logical zero, indicating the conveyor system is not available for use.

The middle of the diagram shows the development of the system running control signals. The first logical decision made here is whether the conveyor system is ready to run and whether the software or operator wants it to run. As shown, this decision is made by a two-input AND gate. The decision depends on the status of its input signals, 120-CVN-05 available and 120-CVN-05 run request applied from the software. If both inputs are logical, the output from the AND is logical. It causes a PLC contact closure that begins the conveyor start sequence by enabling a 10-second, timed open function that causes the conveyor start warning horns to sound for 10 seconds. A look at the ladder diagram shows that this function is accomplished in the conveyor system hardware by the 10-second horn relay (HR) and the conveyor-start warning horns (AA108A and AA108B). Simultaneously, a 9-second, timed closed function is also enabled that causes the conveyor motor contactor (M) to energize after a 9-second delay. In the hardware, this function is performed by a 9-second timing relay (1TR) and contactor (M).

The remainder of the decision logic shown in this section consists of an 11-second, time-closed, time-delay (TD) function; a PLC input; a two-input AND gate; and an inverter. The purpose of this logic is to verify that when the conveyor is commanded to run, it is indeed running. If the input to the 11-second TD function is logical for 11 seconds after conveyor start, the TD function outputs a logical one to the two-input AND. The purpose of the 11-second delay is to allow time for contactor M to pull in and the internal time delay of the conveyor zero-speed switch (SSL108) to expire. The second input to the two-input AND (input 120-CNV-05 M aux) via the PLC is the status of the contactor (M) auxiliary contact signal. With both inputs to the two-input AND being logical, it outputs a logical signal (120-CNV-05 running) to the software that verifies the conveyor is running. If either the TD function output or the 120-CVN-05 M aux signal input is a logical zero, the AND gate output (120-CNV-05 running) will be a logical zero, indicating the conveyor is not running. The logical zero output from the AND gate is applied through an inverter to generate a logical 120-CVN-05 fail-to-start signal for application to the software.

The bottom of the diagram shows the decision logic used to determine whether a belt runoff misalignment condition exists and, if it does, whether the problem is caused by head-end or tail-end misalignment. The individual status of belt runoff switches ZS108A and B (head end) and Z108C and D (tail end) are applied as inputs to the two-input AND gates. The outputs from the head-end and tail-end AND gates are applied via PLC inputs to three-second, timed-off, TD functions. The purpose of the three-second time delay is to eliminate the generation of nuisance alarms.

When no conveyor belt misalignment exists, the inputs applied to both of the three-second TD functions are logical, causing the outputs from both three-second TD functions to be logical. Should a head-end belt misalignment exist (or both a head-end and tail-end belt misalignment) for more than three seconds, the outputs from both three-second TD functions are a logical zero. Both are a logical zero in this case because the tail-end runoff switches (Z108C and D) are in series with the head-end runoff switches (ZS108A and B), as shown on the ladder diagram. Therefore, if the head-end input is open, the tail-end input will also be open. When only a tail-end belt misalignment occurs, the output of the head-end three-second TD function will be logical, and the output of the tail-end three-second TD function will be a logical zero. The statuses of the three-second TD function outputs for the various belt conditions are summarized in *Table 14*.

As shown, the output from the head-end three-second TD function is applied as an input to an inverter and one input of a two-input AND gate. The output from the tail-end three-second TD function is applied through an inverter as the second input to the two-input AND gate. The resultant logic status of the inverter output signal (120-CNV-05 head-end misalignment) and the AND gate output signal (120-CNV-05 tail-end misalignment) for the different belt conditions are shown in *Table 14*. These signals are used by the PLC software to determine the status of belt alignment.

TABLE 14 Three-Second TD Function Outputs

Belt Alignment Condition	Logical Signal Output from Head-End, Three-Second, Timed-Off Function	Logical Signal Output from Tail-End, Three-Second, Timed-Off Function	120-CNV-05 Head-End Misalignment Logical Signal Output	120-CNV-05 Tail-End Misalignment Logical Signal Output
No misalignment	1	1	0 (no alarm)	0 (no alarm)
Head-end misalignment or both head-end and tail-end misalignment	0	0	1 (alarm)	0 (no alarm)
Tail-end misalignment	1	0	0	1 (alarm)

4.3.0 NEC® Regulations for Motor Control Circuits

NEC Article 430 governs the installation of motors, motor circuits, and controllers. *NEC Article 440* applies to motors related to hermetic air conditioning and refrigeration compressors. Other articles that pertain to special motor circuits and hazardous locations are referenced in *NEC Table 430.5*.

Figure 77 shows a summary of the *NEC Article 430* requirements that apply to the installation of motors and their control systems. Detailed information can be found in the *NEC*® under the articles or sections referenced in the diagram.

Some *NEC Article 430* requirements related to motor controls are also outlined here:

- A magnetic starter cannot serve as a disconnecting means.
- Controllers must be able to start and stop the motors they control, including the ability to interrupt their locked-rotor current.
- A controller must have a horsepower rating no lower than the rating of the controlled motor.
- Branch circuit protective devices can be used as the controller for small stationary motors ($^1/_8$ hp or less) and portable motors ($^1/_3$ hp or less). A general-use switch rated greater than or equal to twice the FLA may be used with stationary motors up to 2 hp and operating at 300V or less.
- Where a controller does not serve as both the controller and disconnect switch and a fused switch provides the required disconnect, short circuit protection, and overload protection, the controller does not need to break all the conductors to the motor. It only needs to break enough conductors to stop and start the motor.
- The rating of TD fuses used as motor-running protective devices in a combination fuse holder and switch cannot exceed 125% or, in some cases, 115% of the FLA.
- *NEC Table 110.28* provides the selection criteria for enclosures suitable for nonhazardous locations.
- All conductors of a remote motor control circuit outside the controller, such as push-button stations, must be installed in a raceway or provided with some other form of protection.
- Where one side of the motor control circuit is grounded, the circuit must be wired so that an accidental ground in the controls or related wiring will not start the motor or bypass any safety shutdown devices.

Know the Code

Motors, Motor Circuits, and Controllers
NEC Article 430

Know the Code

Air-Conditioning and Refrigerating Equipment
NEC Article 440

Know the Code

Other Articles
NEC Table 430.5

Know the Code

Enclosure Selection
NEC Table 110.28

4.4.0 Connecting Motor Controllers for Specific Applications

This section gives a few examples of basic control circuit wiring for common applications such as the following:

- Air compressor motors
- Pump motors
- Multiple pump motors
- Multiple remote motor controls
- Three-phase motor reversal
- Conveyor system motors

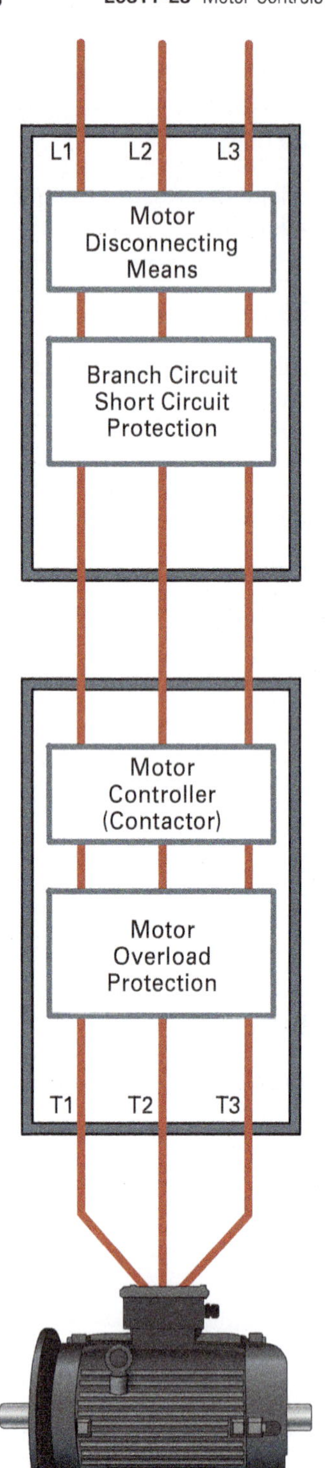

NEC Article 430, Part IX

Disconnects motor and controllers from circuit.

1. Continuous rating of 115% or more of motor FLA.
2. Disconnecting means shall be as listed in *NEC Section 430.109*.
3. Must be located in sight of motor location and driven machinery. The controller disconnecting means can serve as the disconnecting means if the controller disconnect is located in sight of the motor location and driven machinery.

NEC Article 430, Part IV

Protects branch circuit from short circuits or ground faults.

1. Must carry starting current of motor.
2. Rating must not exceed values in *NEC Table 430.52(C)(1)* unless not sufficient to carry starting current of motor.
3. The values of branch circuit protective devices must not exceed those shown in *NEC Section 430.52(C)(1)(b)*.

NEC Article 430, Part VII

Used to start and stop motors.

1. Controllers that are not inverse-time circuit breakers or molded-case switches must have horsepower ratings at the motor voltage equal to or greater than the motor horsepower per *NEC Section 430.83(A)(1)*.
2. Must be able to interrupt FLA per *NEC Section 430.82(A)*.
3. Must be rated as specified in *NEC Section 430.83*.

NEC Article 430, Part III

Protects motor and controller against excessive heat due to motor overload.

1. Must trip at following percent or less of motor FLA for continuous motors rated more than 1 hp.
   (a) 125% FLA for motors with a marked service factor of not less than 1.15 or a marked temperature rise of not over 40°C.
   (b) 115% FLA for all others. (See *NEC*® for other types of protection.)
2. Three thermal units required for any three-phase AC motor.
3. Must allow motor to start.
4. Select size from FLA on motor nameplate per *NEC Section 430.6(A)(2)*.

NEC Article 430, Part II

Specifies the sizes of conductors capable of carrying the motor current without overheating.

1. To determine the ampacity of conductors, switches, branch circuit overcurrent devices, etc., the full-load current values given in *NEC Tables 430.247 through 430.250* must be used instead of the motor FLA found on the nameplate per *NEC Section 430.6(A)(1)*.
2. According to *NEC Section 430.22*, branch circuit conductors supplying a single motor must have an ampacity equal to or greater than 125% of motor FLA, as directed by *NEC Section 430.6(A)(1)*, or not less than shown in *NEC Sections 430.22(A) through (G)*.

Figure 77 Summary of requirements for motors, motor circuits, and controllers.

4.4.1 Air Compressor Motors

Controlling an air compressor motor is done similarly to many other motors. However, an NC pressure switch used as a cycling control is added to cycle the compressor motor based on the air pressure in the receiver.

Figure 78 shows a line diagram for this circuit. In this circuit, setting the On/Off switch to On allows current to pass from L1 through the NC pressure switch contacts, the contactor coil (C), the NC overloads, and on to L2. This energizes the contactor coil, closing its NO power contacts (not shown). This starts the air compressor motor.

The motor will continue to run until the air pressure builds up to the setpoint of the pressure switch. When this occurs, the NC contacts of the pressure switch open, causing the contactor coil to de-energize and its related NO power contacts to open. It will remain off until the pressure in the system falls below setpoint, causing the NC contacts of the pressure switch to close again. This energizes the contactor coil again and starts the motor. This automatic sequence of cycling an air compressor motor in response to system pressure continues until the On/Off switch is set to Off.

Note that many circuits will also incorporate an NC high-pressure safety switch to shut the compressor down if the cycling control fails to do so.

4.4.2 Pump Motors

A control circuit controlling a pump that maintains the level in a tank or vessel often has one or more float switches. *Figure 79* shows the line diagram for a pump being controlled by a Hand/Off/Auto switch and a single float switch.

When the Hand/Off/Auto switch in the circuit in *Figure 79* is set to Off, no current can flow through the contactor coil to L2. The pump remains off. When the switch is placed in the Hand position, current from L1 can pass through the NO switch contacts, the contactor coil (C), the NC overloads, and on to L2. This energizes the contactor coil, closing its NO power contacts (not shown) and starting the pump motor. The pump motor will continue running until the Hand/Off/Auto switch is set to Off. This provides manual operation.

When the Hand/Off/Auto switch is set to Auto, current can flow from L1 through the closed switch contacts, the NC float switch contacts, the contactor coil, the NC overloads, and on to L2. This energizes the contactor coil, closing its NO power contacts and starting the pump motor. The pump will continue running until either the Hand/Off/Auto switch is set to Off or the liquid in the tank raises the float switch, causing its NC contacts to open. The contactor coil is then de-energized, and its NO power contacts open. The pump motor will remain off until the level in the tank allows the float switch to drop, causing its NC contacts to close. The contactor coil is energized again and starts the pump motor. This cycling sequence to maintain the tank level continues until the Hand/Off/Auto switch is set to Off.

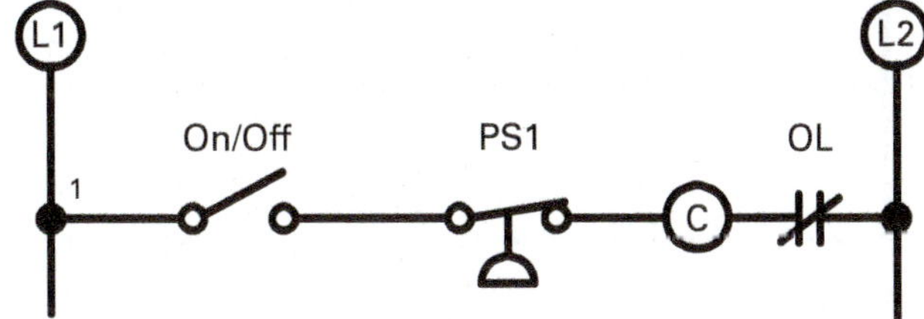

Figure 78 Control circuit for an air compressor.

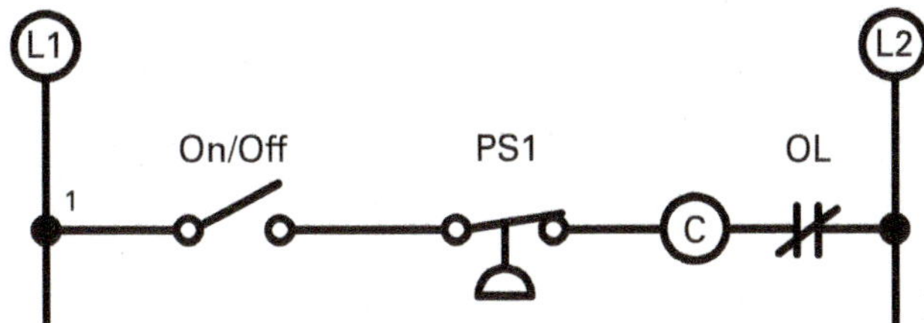

Figure 79 Control circuit for a pump motor.

4.4.3 Controlling Two Pumps

Assume that controlling two pumps is needed so they alternately operate to automatically fill and drain a tank based on the liquid level. *Figure 80* shows one way of connecting such a circuit.

Pressing the Start push button allows current to flow from L1 through the Stop switch NC contacts, the closed Start switch contacts, the control relay coil (CR), and on to L2. This energizes the control relay, causing its three sets of NO contacts to close. Assuming the tank is empty or partially full, coil M2 is de-energized. This allows current to flow from L1 through the closed control relay contacts, closed M2 auxiliary contacts, the NC float switch contacts, coil M1, the NC overloads, and on to L2. This energizes coil M1, closing its NO power contacts (not shown) to start the fill pump. It also causes its NC auxiliary contacts to open. This provides an electrical interlock in the control circuit that prevents coil M2 from being energized at the same time as coil M1.

The fill pump motor will continue to operate until the fluid in the tank raises the float switch, causing its NC contacts to open and its NO contacts to close. This de-energizes coil M1 and its related NO power contacts open, stopping the fill pump.

Activation of the float switch also allows current to flow from L1 through the closed control relay contacts, closed auxiliary M1 contacts, the closed NO float switch contacts, coil M2, the NC overloads, and on to L2. This energizes coil M2, closing its NO power contacts (not shown) to start the drain pump. It also causes its NC auxiliary contacts to open. This provides an electrical interlock in the control circuit that prevents coil M1 from being energized at the same time as coil M2.

The drain pump motor will continue to operate until the fluid in the tank is drained to a level that allows the float switch to drop, causing its NC contacts to close and its NO contacts to open. This causes coil M2 to de-energize and its related NO power contacts to open, stopping the drain pump. It also allows the fill pump motor to be started again, as described earlier.

This cycling of the pumps will continue until the Stop push button is pressed. Pressing the Stop push button opens its NC contacts, removing power from the control relay coil. This causes all its NO contacts to open, preventing both coils M1 and M2 from being energized.

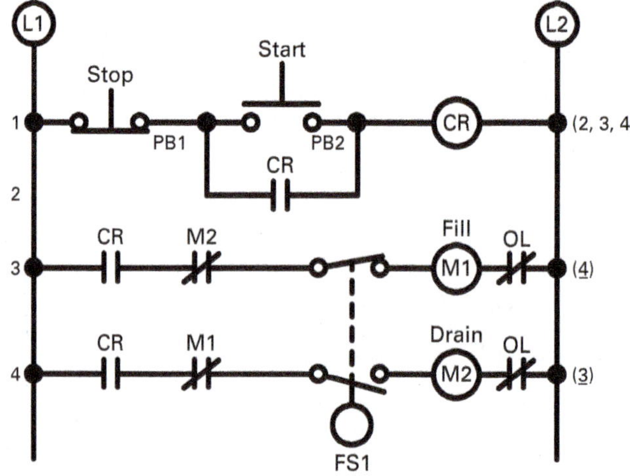

Figure 80 Control circuit for two pumps.

4.4.4 Multiple Remote Motor Controls

Figure 81 shows a line diagram for a control circuit where a magnetic starter can be energized from any of three locations. In this circuit, pressing any one of the Start push buttons causes coil M1 to energize. This causes the auxiliary contacts M1 to close until coil M1 is de-energized by pressing any one of the Stop buttons. The important thing to remember is that, regardless of the number of stations, the Start push buttons must all be connected in parallel, and the Stop push buttons must be connected in series.

4.4.5 Three-Phase Motor Reversal

Reversing three-phase motors is usually done by reversing the L1 and L3 power leads to the motor using a reversing starter and related control circuit (*Figure 82*). Technically, reversing any two of the leads provides the same result.

Circuit operation is started by pressing either the Forward or Reverse push button. Pressing the Forward push button allows current to flow from L1 through the NC contacts of the Stop switch, the NC contacts of the Reverse switch, the closed contacts of the Forward switch, the NC auxiliary contacts of the reverse coil, the forward coil (F), the NC closed overload contacts, and on to L2. This energizes coil F, closing its NO forward-power contacts (not shown), causing the motor to start in the forward direction. It also activates the NO and NC auxiliary contacts of coil F. The NO auxiliary contacts close, and the NC auxiliary contacts (located in the reverse coil (R) control circuit) open, providing an electrical interlock that prevents coil R from being energized simultaneously with coil F.

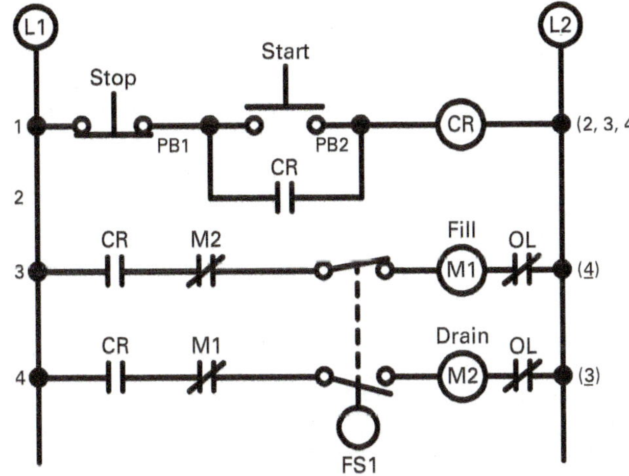

Figure 81 Control circuit for starting and stopping a single motor from three locations.

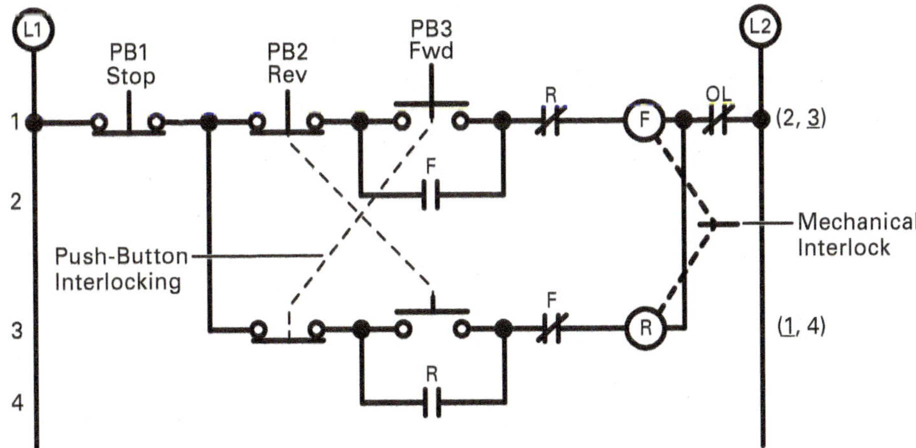

Figure 82 Control circuit for reversing a three-phase motor.

Pressing the Reverse push button allows current to flow from L1 through the NC contacts of the Stop switch, the NC contacts of the Forward switch, the closed contacts of the Reverse switch, the NC auxiliary contacts of the forward coil, the reverse coil (R), the NC closed overload contacts, and on to L2. This energizes coil R, closing its NO reverse-power contacts (not shown), causing the motor to run in the reverse direction. It also activates the NO and NC auxiliary contacts of coil R. The NO auxiliary contacts close, and the NC auxiliary contacts (located in the forward coil (F) control circuit) open, preventing coil F from being energized simultaneously with coil R. Pressing the Stop push button de-energizes the functioning F or R coil, stopping the motor.

This circuit uses three methods of interlocking. One method is a built-in mechanical interlock incorporated into the reversing starter. Another method is the push button interlocking of the Forward and Reverse push buttons. Push-button interlocking uses both NO and NC contacts mechanically connected on each push button. This ensures the forward and reverse coils cannot be energized simultaneously. As has already been pointed out, the third method involves the NC auxiliary contacts for the forward and reverse coils being used as safety interlocks to prevent the forward and reverse coils from being energized simultaneously.

4.4.6 Conveyor System Motors

Multiple conveyor belt systems must often be sequenced so that one conveyor system is started before another one can be started. *Figure 83* shows one basic method for wiring a control circuit to sequentially control three conveyor systems.

Pressing the Start push button energizes the conveyor 1 starter coil (M1) via current flow through the closed contacts of the Stop and Start switches and NC overload contacts. When coil M1 energizes, it closes its NO power contacts (not shown), causing the conveyor 1 motor to start. Energizing coil M1 also closes its three sets of NO auxiliary contacts. One set of closed NO auxiliary contacts provides a holding circuit for the Start switch. The second and third sets are in the control circuits for coils M2 and M3, respectively.

When the set of NO auxiliary contacts in coil M2's control circuit closes, it provides a current path from L1 through the closed NO contacts, coil M2, and the NC overload contacts to L2. This energizes coil M2. Energizing coil M2 closes its NO power contacts, causing the conveyor 2 motor to start. Energizing coil M2 also closes its NO set of auxiliary contacts located in coil M3's control circuit. With both the M1 and M2 sets of NO auxiliary contacts closed in coil M3's control circuit, coil M3 energizes. This closes the coil 3 NO power contacts, causing conveyor motor 3 to start.

Because there are no additional conveyors to energize, the NO auxiliary contacts for coil M3 are not used. Pressing the Stop push button de-energizes, as applicable, coils M1, M2, and M3, stopping all conveyor operations.

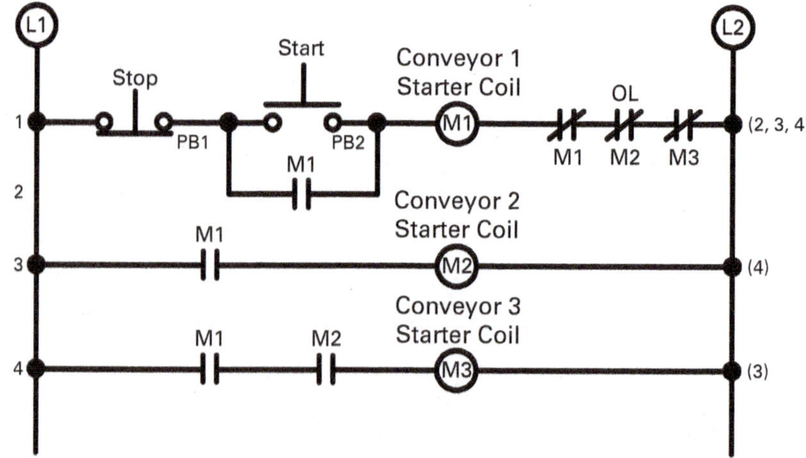

Figure 83 Control circuit for controlling three conveyor systems sequentially.

> **Think About It**
>
> **Overload Contacts**
>
> Why are all three overload contacts shown in *Figure 83* connected in series with the conveyor 1 starter coil?

4.0.0 Section Review

1. The IEC code for an enclosure that is dust tight and provides protection against splashing water is _____.
   a. IP36
   b. IP44
   c. IP54
   d. IP64

2. The three basic logic functions are _____.
   a. YES, NO, and GO
   b. AND, OR, and NOT
   c. STOP, START, and HOLD
   d. FORWARD, BACK, and STOP

3. A circuit breaker can be used as the controller for small stationary motors under _____.
   a. 1 hp
   b. ½ hp
   c. ⅓ hp
   d. ⅛ hp

4. Reversing a three-phase motor can be done by _____.
   a. reversing the L1 and L3 power leads
   b. reversing the L1 and L2 power leads and the L2 and L3 power leads
   c. adding a start capacitor
   d. spinning it in the desired direction before applying power

Module 26311-23 Review Questions

1. Contactors without overload protection may be used to control _____.
   a. fractional horsepower single-phase motors
   b. single-phase motors
   c. three-phase motors
   d. noninductive loads

2. Contacts that are defined as being NO are _____.
   a. open when the relay coil is energized
   b. left unused for that application
   c. open when the relay coil is de-energized
   d. closed when the relay coil is de-energized

3. The voltage at which the armature return spring of a power relay overpowers the coil's magnetic field, causing the contacts to move to their de-energized position, is called the _____.
   a. dropout voltage
   b. seal-in voltage
   c. pickup voltage
   d. transition voltage

4. IEC-rated contactors and motor starters perform the same functions as equivalent NEMA-rated contactors and motor starters, but they are usually _____.
   a. physically larger
   b. more application sensitive
   c. more expensive
   d. better designed

5. When replacing the contacts of a magnetic contactor, you *must* replace _____.
   a. all moving contacts for all poles
   b. all stationary contacts for all poles
   c. all moving and stationary contacts for all poles
   d. the damaged contacts only

6. Running overload protection for motors is provided by _____.
   a. circuit breakers
   b. general-purpose, current-limiting fuses
   c. overload relays
   d. backup current-limiting fuses

7. Which of the following is the NEMA-rated class for the fastest trip time of an overload relay?
   a. Class 30
   b. Class 20
   c. Class 15
   d. Class 10

8. The thermal units for overload relays should always be selected based on the motor _____.
   a. locked rotor amps
   b. full-load current from *NEC Tables 430.247 through 430.250*
   c. nameplate FLA
   d. horsepower

9. The *least* important consideration when selecting overload relays is the _____.
   a. motor NEMA design letter
   b. motor FLA rating
   c. motor horsepower
   d. motor manufacturer

10. The NEMA motor starter that must be used with a 230V, single-phase, 60 Hz, 10 hp motor *not* used for plugging or jogging applications is _____.
    a. NEMA Size 2
    b. NEMA Size 3
    c. NEMA Size 4
    d. NEMA Size 5

11. When selecting a control transformer for a specific application, you have calculated its application inrush to be 1,850VA. If the primary input VA rating is expected to vary between 5% and 10% from nominal, the transformer voltage should be _____.
    a. 150VA
    b. 250VA
    c. 300VA
    d. 350VA

12. A manufacturer states that the contacts of a particular switch are rated as A600. In a 120VAC application, the *maximum* break current for these contacts is _____.
    a. 1A
    b. 1.5A
    c. 3A
    d. 6A

13. A manufacturer lists the utilization category for a particular switching device as AC-15. The intended application for this device is the control of _____.
    a. electromagnetic loads greater than 72VA closed
    b. solid-state loads with transformer isolation
    c. small electromagnetic loads at a maximum of 72VA closed
    d. resistive loads and solid-state loads with optical isolation

14. A line-powered proximity sensor draws its operating current from _____.
    a. the load
    b. the power lines
    c. self-induction
    d. eddy currents

15. Motor and controller circuit overload protection is covered in *NEC Article 430,* _____.
    a. *Part III*
    b. *Part IV*
    c. *Part VII*
    d. *Part X*

Answers to odd-numbered Module Review Questions are found in *Appendix A.*

Module 26311-23 Supplemental Exercises

1. What is the purpose of a shading coil?

   ____

   ____

2. Explain how a melting-alloy thermal overload relay works to protect the motor.

   ____

   ____

3. List three advantages of a bimetallic overload relay when compared to a melting-alloy overload relay.

   ____

   ____

4. Internal auxiliary contacts are used for _________ or _________.

5. What is the deadband of a pressure switch?

   ____

   ____

6. An inherent change in accuracy for a given setting due to time or some number of cycles is known as _________.

7. Which type of switch would be used to detect the position of a mechanism or piece of equipment?

____

____

8. __________ is an operation in which a motor is brought to a rapid stop.

9. True or False? Capacitive proximity sensors can detect any target with a dielectric constant slightly greater than that of air.

10. Why should no more than three proximity sensors be connected in parallel?

____

____

11. What are the two categories of proximity sensors?

____

____

12. True or False? Drum switches contain protective overloads and therefore do *not* require additional circuit overload protection.

13. A typical NEMA enclosure that would be used outdoors to serve construction activities is a(n) __________.

14. Name three types of pilot devices you would typically find depicted on a ladder diagram.

____

____

15. Where in the *NEC*® would you find the requirements for most motor controllers?

____

____

Answers to odd-numbered Supplemental Exercises are found in *Appendix B.*

Answers to Section Review Questions

Answer	Section Reference	Objective
Section 1.0.0		
1. b	1.1.1; *Table 1*	1a
2. c	1.2.2	1b
3. b	1.3.2	1c
Section 2.0.0		
1. c	2.1.1; *Tables 6A and 6B*	2a
2. c	2.2.1; *Table 7*	2b
3. b	2.3.0	2c
Section 3.0.0		
1. d	3.1.0; *Table 9*	3a
2. c	3.2.1; *Table 10A*	3b
Section 4.0.0		
1. d	4.1.0; *Table 12*	4a
2. b	4.2.3	4b
3. d	4.3.0	4c
4. a	4.4.5	4d

User Update

Did you find an error? Submit a correction by visiting **https://www.nccer.org/olf** or by scanning the QR code using your mobile device.

APPENDIX A REVIEW QUESTION ANSWER KEYS

MODULE 26301-23

Answer	Section Reference
1. a	1.0.0
3. d	1.1.0
5. b	1.1.0
7. b	1.2.2
9. d	1.2.3
11. c	2.1.0
13. d	2.2.0
15. b	2.5.0
17. b	2.7.0
19. c	3.1.0

MODULE 26301-23 CALCULATIONS

Question 9

Use the following formula:

$$VD = \frac{\sqrt{3} \times L \times R \times I}{1000'}$$

Insert circuit values into formula based on the resistance per 1,000' found in ***NEC Chapter 9, Table 8***, *Conductor Properties*:

$$VD = \frac{\sqrt{3} \times 145' \times 0.778\Omega \times 26.5A}{1{,}000'}$$

$$VD = \frac{5{,}177.91}{1{,}000'}$$

$$VD = 5.18V$$

The voltage drop is **5.18V**.

Question 13

Multiply the amperage by the voltage to determine the capacity:

20A × 120V = 2,400VA

Divide the result by the rating per receptacle as specified in ***NEC Section 220.14(I)***:

2,400VA ÷ 180VA = 13.33, rounded down to 13

There would be **13 receptacles at 180V each**.

Question 15

Subtract 12 kW from the nameplate rating of the range to determine the number of kW exceeding 12:

16.75 kW – 12 kW = 4.75 kW, rounded to 5 kW

Multiply the result by 5 to determine the total percentage by which the demand must be increased:

5 kW × 5 = 25%

Multiply the maximum Udemand listed in Column C of ***NEC Table 220.55*** for one range (8 kW) by 125% to account for the 25% demand increase:

8 kW × 1.25 = 10 kW

Convert kW to VA:

10 kW × 1,000W/kW = 10,000VA

Divide by circuit voltage to find demand load in amps:

10,000VA ÷ 240V = 41.67A

The demand load in amps is **41.67A**.

Question 17

Determine the total load by adding the load of each unit.

1,000W + 1,000W + 1,000W+ 1,000W = 4,000W (VA)

OR

4 × 1,000W = 4,000W (VA)

Divide the total load by the circuit voltage to determine ampacity:

4,000VA ÷ 208V = 19.23A

Find 125% of the load since this is considered a continuous load:

19.23A × 1.25 = 24.04A

The minimum ampacity is **24.04A**.

Question 19

Since each foot of the multioutlet assembly is considered as one outlet of at least 180VA capacity, multiply each foot by 180VA:

6' × 180VA per outlet = 1,080VA

The load is **1,080VA**.

MODULE 26302-23

Answer	Section Reference
1. d	1.0.0
3. d	1.0.0
5. b	1.1.1
7. b	1.1.2
9. c	1.2.0

MODULE 26303-23

Answer	Section Reference
1. b	1.1.1
3. d	1.1.3
5. c	1.2.5
7. a	2.0.0
9. d	2.1.1
11. c	2.1.6
13. c	3.0.0
15. a	3.2.0

MODULE 26303-23 CALCULATIONS

Question 9

18' × 1.5 = **27'**

MODULE 26304-23

Answer	Section Reference
1. c	1.0.0
3. d	1.1.0
5. a	1.4.0; *Table 4*
7. a	2.2.2
9. d	2.2.3

MODULE 26305-23

Answer	Section Reference
1. b	1.2.0
3. a	2.1.1
5. d	2.3.0
7. b	2.3.1
9. d	2.3.2
11. d	2.3.5; *Table 1*
13. c	3.0.0
15. a	3.1.0
17. c	4.2.1; *Table 4*
19. c	4.2.1; *Table 5*

MODULE 26306-23

Answer	Section Reference
1. c	1.0.0
3. a	1.1.2
5. c	1.4.1
7. b	1.5.2
9. a	1.5.5
11. d	1.6.2
13. c	2.2.3

MODULE 26307-23

Answer	Section Reference
1. c	1.0.0
3. a	1.1.1
5. b	1.1.1
7. b	1.2.0
9. a	1.3.2
11. b	1.3.5
13. d	2.1.0
15. b	3.1.4

MODULE 26308-23

Answer	Section Reference
1. b	1.0.0
3. c	1.1.1
5. c	1.1.3
7. c	1.1.5
9. b	1.2.4

MODULE 26309-23

Answer	Section Reference
1. a	1.0.0
3. a	1.1.0
5. c	1.2.0
7. c	1.2.1
9. d	1.2.1
11. a	1.2.1
13. d	1.2.2
15. a	3.2.0

MODULE 26310-23

Answer	Section Reference
1. d	1.0.0
3. d	1.1.0
5. d	1.2.0
7. c	1.3.0
9. d	2.1.0

MODULE 26311-23

Answer	Section Reference
1. d	1.1.0
3. a	1.1.3
5. a	1.2.1
7. d	1.3.0
9. d	2.1.0
11. c	3.1.0; *Table 9*
13. a	3.2.1; *Table 11*
15. a	4.3.0; *Figure 77*

APPENDIX B SUPPLEMENTAL EXERCISE ANSWER KEYS

MODULE 26301-23

Question 1

Per *NEC Section 210.19(A)(1)*, branch circuit conductors are required to be sized with an ampacity rating that is no less than the maximum load to be served.

Section Reference 1.0.0

Question 3

20A × 240V = **4,800VA**

Section Reference 1.1.0

Question 5

A continuous load is a load where the maximum current is expected to continue for three hours or more.

Section Reference 1.1.0

Question 7

lowest

Section Reference 1.2.1

Question 9

nameplate rating of the equipment

Section Reference 2.5.0

Question 11

continuous load

Section Reference 2.6.0

Question 13

three; continuous; 125%; 80%

Section Reference 3.3.0

Question 15

primary current; duty cycle

Section Reference 3.7.0

MODULE 26302-23

Question 1

copper

Section Reference 1.0.0

Question 3

No. 14 AWG, No. 1 AWG, No. 1/0 AWG, No. 4/0 AWG

Section Reference 1.0.0

Question 5

overcurrent protective device, such as a fuse or circuit breaker

Section Reference 1.1.0

Question 7

branch circuit overcurrent device

Section Reference 1.2.2

Question 9

10A

Section Reference 1.1.2

Question 11

The ampacity of eight conductors in a raceway is reduced to 70%.

260A × 0.7 = **182A**.

Section Reference 1.2.0

Question 13

nonconductive colored tape; paint

Section Reference 1.2.2

Question 15

Use a larger conductor.

Increase the voltage.

Section Reference 2.3.1

MODULE 26303-23

Question 1

local

Section Reference 2.1.0

Question 3

False

Section Reference 1.2.3

Question 5

Open strip or channel

Section Reference 1.2.5

Question 7

High CRI

Section Reference 2.1.2

Question 9

indirect lighting

Section Reference 2.1.4

Question 11

25' × 1.5 = **37.5'**

Section Reference 2.1.1

Question 13

Type C

Section Reference 1.1.2

Question 15

False

Section Reference 3.1.0

MODULE 26304-23

Question 1

A hazardous location is any area in which the atmosphere or materials are subject to ignition from the arcs of operating electrical devices.

Section Reference 1.0.0

Question 3

Class II locations are those that are hazardous due to the presence of combustible dust.

Section Reference 1.2.0

Question 5

True

Section Reference 1.4.0

Question 7

Class I, Division I

Section Reference 1.4.1

Question 9

NEC Article 517

Section Reference 1.4.3

Question 11

gases; vapors; flames

Section Reference 2.2.3

Question 13

drain/seal

Section Reference 2.2.3

Question 15

200°F (~93°C)

Section Reference 2.2.3

MODULE 26305-23

Question 1

line-to-ground

Section Reference 1.2.0

Question 3

The protective device can be damaged or destroyed. The current can also cause severe damage to equipment and injure personnel.

Section Reference 1.2.0

Question 5

250V and 600V

Section Reference 2.1.1

Question 7

False

Section Reference 2.1.3

Question 9

10,000A

Section Reference 2.3.0

Question 11

True

Section Reference 2.3.3

Question 13

125% of 16A = 20A

Section Reference 4.1.1

Question 15

300%

Section Reference 4.1.5

MODULE 26306-23

Question 1

Air circuit breakers

Oil circuit breakers

Vacuum circuit breakers

Gas circuit breakers

Section Reference 1.1.2

Question 3

½" (13 mm)

Section Reference 1.2.3

Question 5

HVL and MVL switches can provide both switching and overcurrent protection.

Section Reference 1.3.0

Question 7

120V; 69.3V

Section Reference 1.5.5

Question 9

The interrupting rating is the maximum current a device is capable of interrupting without suffering damage.

Section Reference 2.1.1

Question 11

TDO

PS

XS

BLPB

Section Reference 2.2.0

Question 13

ground-return method; residual method; zero-sequence method

Section Reference 3.3.1

Question 15

$1\frac{1}{2}$

Section Reference 3.3.6

MODULE 26307-23

Question 1

change the output voltage

Section Reference 1.0.0

Question 3

two coils or windings formed on a single magnetic core

Section Reference 1.0.0

Question 5

10 ÷ 100 = 0.1

0.1 × 240V = **24V**

Section Reference 1.1.2

Question 7

butt lap, wound, and mitered

Section Reference 1.2.2

Question 9

Lower purchase price

Lower operating cost due to lower losses

Smaller size and easier to install

Better voltage regulation

Lower sound levels

Section Reference 1.3.4

Question 11

potential, small service, and control transformers

Section Reference 2.1.3

Question 13

The secondary circuit(s) does not produce a voltage.

Section Reference 3.1.1

Question 15

surface tracking and reduced insulation values

Section Reference 3.2.2

MODULE 26308-23

Question 1

conduit; service conductors; panelboard; overcurrent protective devices

Section Reference 1.0.0

Question 3

service drop

Section Reference 1.0.0

Question 5

safety

Section Reference 1.0.0

Question 7

True

Section Reference 1.1.1

Question 9

doughnut

Section Reference 1.1.3

Question 11

False

Section Reference 1.1.7

Question 13

service conductors

Section Reference 1.2.1

Question 15

outdoor

Section Reference 2.1.0

MODULE 26309-23

Question 1

Squirrel-cage induction motors; Wound-rotor induction motors; Synchronous motors

Section Reference 1.0.0

Question 3

Number of stator poles; Frequency of the AC line in Hz

Section Reference 1.1.0

Question 5

24A

$I = E \div R$

$I = 480V \div 20\Omega$

$I =$ **24A**

Section Reference 1.2.1

Question 7

45.5A

(1.25 × 14A) + 14A + 14A = **45.5A**

Section Reference 2.1.0

Question 9

150%

Section Reference 3.1.0

MODULE 26310-23

Question 1

fiber optic

Section Reference 1.0.0

Question 3

negative

Section Reference 1.1.0

Question 5

labels

Section Reference 1.1.4

Question 7

diver

Section Reference 1.2.3

Question 9

plenum-rated

Section Reference 1.4.1

Question 11

False

Section Reference 2.3.0

Question 13

grounding

Section Reference 2.4.1

Question 15

testing

Section Reference 2.4.2

MODULE 26311-23

Question 1

The shading coil prevents the relay from chattering.

Section Reference 1.1.2

Question 3

The automatic reset feature; Service factor adjustment; Automatic ambient temperature compensation

Section Reference 1.3.2

Question 5

The deadband is the pressure difference between the setpoint at which a switch activates when the pressure increases and the point at which the switch resets when the pressure drops.

Section Reference 3.2.5

Question 7

Limit switch

Section Reference 3.2.6

Question 9

True

Section Reference 3.2.10

Question 11

Load-powered sensors and line-powered sensors

Section Reference 3.2.10

Question 13

NEMA Type 3

Section Reference 4.1.0

Question 15

NEC Section 430

Section Reference 4.3.0

APPENDIX 26306-23A

IEEE Identification System

The devices in switching equipment are referred to by numbers with appropriate suffix letters when necessary, according to the functions they perform.

These numbers are based on a system adopted as standard for automatic switchgear by IEEE and incorporated in *American Standard C37.2-1970*. This system is used in connection diagrams, instruction books, and specifications.

Device Number	Definition and Function
1	**Master Element** – The initiating device, such as a control switch, voltage relay, float switch, that serves either directly, or through such permissive devices as protective and time-delay relays, to place equipment in or out of operation.
2	**Time-Delay Starting or Closing Relay** – A device that functions to give a desired amount of time delay before or after any point of operation in a switching sequence or protective relay system, except as specifically provided by device functions 48, 62, and 79 described later.
3	**Checking or Interlocking Relay** – A device that operates in response to the position of a number of other devices (or to a number of predetermined conditions) in equipment to allow an operating sequence to proceed, to stop, or to provide a check of the position of these devices or of these conditions for any purpose.
4	**Master Contactor** – A device, generally controlled by device No. 1 or equivalent, and the required permissive and protective devices, that serves to make and break the necessary control circuits to place equipment into operation under the desired conditions and to take it out of operation under other or abnormal conditions.
5	**Stopping Device** – A control device used primarily to shut down equipment and hold it out of operation. [This device may be manually or electrically actuated, but excludes the function of electrical lockout (see device function 86) on abnormal conditions.]
6	**Starting Circuit Breaker** – A device whose principal function is to connect a machine to its source of starting voltage.
7	**Anode Circuit Breaker** – A device used in the anode circuits of a power rectifier for the primary purpose of interrupting the rectifier circuit if an arc-back should occur.
8	**Control Power Disconnecting Device** – A disconnective device, such as a knife switch, circuit breaker, or pullout fuse block, that is used for the purpose of connecting and disconnecting the source of control power to and from the control bus or equipment. ***Note:*** Control power is considered to include auxiliary power which supplies such apparatus as small motors and heaters.
9	**Reversing Device** – A device used for the purpose of reversing a machine field or for performing any other reversing functions.
10	**Unit Sequence Switch** – A device used to change the sequence in which units may be placed in and out of service in multiple-unit equipment.
11	Reserved for future application.
12	**Over-Speed Device** – Usually a direct-connected speed switch that functions on machine over-speed.
13	**Synchronous-Speed Device** – A device such as a centrifugal-speed switch, slip-frequency relay, voltage relay, or undercurrent relay, that operates at approximately the synchronous speed of a machine.
14	**Under-Speed Device** – A device that functions when the speed of a machine falls below a predetermined value.
15	**Speed- or Frequency-Matching Device** – A device that functions to match and hold the speed or frequency of a machine or of a system equal to, or approximately equal to, that of another machine, source, or system.

Device Number	Definition and Function
16	Reserved for future application.
17	**Shunting or Discharge Switch** – A device that serves to open or close a shunting circuit around any piece of apparatus (except a resistor), such as machine field, machine armature, capacitor, or reactor. ***Note:*** This excludes devices that perform such shunting operations as may be necessary in the process of starting a machine by devices 6 or 42, or their equivalent, and also excludes the device 73 function, which serves for the switching of resistors.
18	**Accelerating or Decelerating Device** – A device used to close or to cause the closing of circuits that are used to increase or decrease the speed of a machine.
19	**Starting-to-Running Transition Contactor** – A device that operates to initiate or cause the automatic transfer of a machine from the starting to the running power connection.
20	**Electrically Operated Valve** – An electrically operated, controlled, or monitored valve in a fluid line. ***Note:*** The function of the valve may be indicated by the use of suffixes.
21	**Distance Relay** – A device that functions when the circuit admittance, impedance, or reactance increases or decreases beyond predetermined limits.
22	**Equalizer Circuit Breaker** – A breaker that serves to control or to make and break the equalizer or the current-balancing connections for a field, or for regulating equipment, in a multiple-unit installation.
23	**Temperature Control Device** – A device that functions to raise or lower the temperature of a machine or other apparatus, or of any medium, when its temperature falls below or rises above a predetermined value. ***Note:*** An example is a thermostat that switches on a space heater in a switchgear assembly when the temperature falls to a desired value as distinguished from a device that is used to provide automatic temperature regulation between close limits and would be designated as 90T.
24	Reserved for future application.
25	**Synchronizing or Synchronism-Check Device** – A device that operates when two AC circuits are within the desired limits of frequency, phase angle, or voltage to permit or to cause the paralleling of these two circuits.
26	**Apparatus Thermal Device** – A device that functions when the temperature of the shunt field or the armortisseur winding of a machine or that of a load limiting or load shifting resistor or of a liquid or other medium exceeds a predetermined value; it also functions if the temperature of the protected apparatus, such as a power rectifier, or of any medium decreases below a predetermined value.
27	**Undervoltage Relay** – A device that functions on a given value of undervoltage.
28	**Flame Detector** – A device that monitors the presence of the pilot or main flame in such apparatus as a gas turbine or steam boiler.
29	**Isolating Contactor** – A device used expressly for disconnecting one circuit from another for the purposes of emergency operation, maintenance, or testing.
30	**Annunciator Relay** – A nonautomatic reset device that gives a number of separate visual indications upon the functioning of protective devices and that may also be arranged to perform a lockout function.
31	**Separate Excitation Device** – A device that connects a circuit, such as the shunt field of a synchronous converter, to a source of separate excitation during the starting sequence or one that energizes the excitation and ignition circuits of a power rectifier.

Device Number	Definition and Function
32	**Directional Power Relay** – A device that functions on a desired value of power flow in a given direction or upon reverse power resulting from arc-back in the anode or cathode circuits of a power rectifier.
33	**Position Switch** – A device that makes or breaks its contacts when the main device or piece of apparatus that has no device function number reaches a given position.
34	**Master Sequence Device** – A device, such as a motor-operated multi-contact switch or the equivalent, or a programming device, such as a computer, that establishes or determines the operating sequence of the major devices in equipment during starting and stopping or during other sequential switching operations.
35	**Brush-Operating or Slip-Ring Short-Circuiting Device** – A device used for raising, lowering, or shifting the brushes of a machine; for short-circuiting its slip rings, or for engaging or disengaging the contacts of a mechanical rectifier.
36	**Polarity or Polarizing Voltage Device** – A device that operates or permits the operation of another device on a predetermined polarity only, or one that verifies the presence of a polarizing voltage in equipment.
37	**Undercurrent or Underpower Relay** – A device that functions when the current or power flow decreases below a predetermined value.
38	**Bearing Protective Device** – A device that functions on excessive bearing temperature or on other abnormal mechanical conditions, such as undue wear, that may eventually result in excessive bearing temperature.
39	**Mechanical Condition Monitor** – A device that functions upon the occurrence of an abnormal mechanical condition (except that associated with bearings as covered under device function 38), such as excessive vibration, eccentricity, expansion, shock, tilting, or seal failure.
40	**Field Relay** – A device that functions on a given or abnormally low value or failure of machine field current or on an excessive value of the reactive component of armature current in an AC machine indicating abnormally low field excitation.
41	**Field Circuit Breaker** – A device that functions to apply or remove the field excitation of a machine.
42	**Running Circuit Breaker** – A device whose principal function is to connect a machine to its source of running or operating voltage. This function may also be used for a device, such as a contactor, that is used in series with a circuit breaker or other fault protecting means, primarily for frequent opening and closing of the circuit.
43	**Manual Transfer or Selector Device** – A device that transfers the control circuits so as to modify the plan of operation of the switching equipment or of some of the devices.
44	**Unit Sequence Starting Relay** – A device that functions to start the next available unit in multiple-unit equipment on the failure or non-availability of the normally preceding unit.
45	**Atmospheric Condition Monitor** – A device that functions upon the occurrence of an abnormal atmospheric condition, such as damaging fumes, explosive mixtures, smoke, or fire.
46	**Reverse-Phase or Phase-Balance Current Relay** – A device that functions when the polyphase currents are of reverse-phase sequence or when the polyphase currents are unbalanced or contain negative phase-sequence components above a given amount.
47	**Phase-Sequence Voltage Relay** – A relay that functions upon a predetermined value of polyphase voltage in the desired phase sequence.
48	**Incomplete Sequence Relay** – A relay that generally returns the equipment to the normal or off position and locks it out if the normal starting, operating, or stopping sequence is not properly completed within a predetermined time. If the device is used for alarm purposes only, it should preferably be designated as 48A (alarm).

Device Number	Definition and Function
49	**Machine or Transformer Thermal Relay** – A relay that functions when the temperature of a machine armature, or other load-carrying winding or element of a machine or the temperature of a power rectifier or power transformer (including a power rectifier transformer) exceeds a predetermined value.
50	**Instantaneous Overcurrent or Rate-of-Rise Relay** – A relay that functions instantaneously on an excessive value of current or on an excessive rate of current rise, indicating a fault in the apparatus or circuit being protected.
51	**AC Time Overcurrent Relay** – A relay with either a definite or inverse time characteristic that functions when the current in an AC circuit exceeds a predetermined value.
52	**AC Circuit Breaker** – A device that is used to close and interrupt an AC power circuit under normal conditions or to interrupt this circuit under fault or emergency conditions.
53	**Exciter or DC Generator Relay** – A relay that forces the DC machine field excitation to build up during starting or which functions when the machine voltage has built up to a given value.
54	Reserved for future application.
55	**Power Factor Relay** – A relay that operates when the power factor in an AC circuit rises above or below a predetermined value.
56	**Field Application Relay** – A relay that automatically controls the application of the field excitation to an AC motor at some predetermined point in the slip cycle.
57	**Short-Circuiting or Grounding Device** – A primary circuit switching device that functions to short-circuit or ground a circuit in response to automatic or manual means.
58	**Rectification Failure Relay** – A device that functions if one or more anodes of a power rectifier fail to fire, to detect an arc-back, or on failure of a diode to conduct or block properly.
59	**Overvoltage Relay** – A relay that functions on a given value of overvoltage.
60	**Voltage or Current Balance Relay** – A relay that operates on a given difference in voltage or current input or output of two circuits.
61	Reserved for future application.
62	**Time-Delay Stopping or Opening Relay** – A time-delay relay that serves in conjunction with the device that initiates the shutdown, stopping, or opening operation in an automatic sequence.
63	**Pressure Switch** – A switch that operates on given values or on a given rate of change of pressure.
64	**Ground Protective Relay** – A relay that functions on failure of the insulation of a machine, transformer, or other apparatus to ground or on flashover of a DC machine to ground. ***Note:*** This function is assigned only to a relay that detects the flow of current from the frame of a machine or enclosing case or structure of a piece of apparatus to ground or one that detects a ground on a normally ungrounded winding or circuit. It is not applied to a device connected in the secondary circuit or secondary neutral of a current transformer connected in the power circuit of a normally grounded system.
65	**Governor** – The assembly of fluid, electrical, or mechanical control equipment used for regulating the flow of water, steam, or other medium to the prime mover for such purposes as starting, holding speed or load, or stopping.

Device Number	Definition and Function
66	**Notching or Jogging Device** – A device that functions to allow only a specified number of operations of a given device or equipment or a specified number of successive operations within a given time of each other. It also functions to energize a circuit periodically or for fractions of specified time intervals or that is used to permit intermittent acceleration or jogging of a machine at low speeds for mechanical positioning.
67	**AC Directional Overcurrent Relay** – A relay that functions on a desired value of AC overcurrent flowing in a predetermined direction.
68	**Blocking Relay** – A relay that initiates a pilot signal for blocking of tripping on external faults in a transmission line or in other apparatus under predetermined conditions, or a relay cooperates with other devices to block tripping or to block reclosing on an out-of-step condition or on power swings.
69	**Permissive Control Device** – Generally a two-position, manually operated switch that in one position permits the closing of a circuit breaker or the placing of equipment into operation and in the other position prevents the circuit breaker or the equipment from being operated.
70	**Rheostat** – A variable resistance device used in an electric circuit that is electrically operated or has other electrical accessories, such as auxiliary, position, or limit switches.
71	**Level Switch** – A switch that operates on given values or on a given rate of change of level.
72	**DC Circuit Breaker** – A circuit breaker used to close and interrupt a DC power circuit under normal conditions or to interrupt this circuit under fault or emergency conditions.
73	**Load-Resistor Contactor** – A contactor used to shunt or insert a step of load limiting, shifting, or indicating resistance in a power circuit, to switch a space heater in a circuit, or to switch a light or regenerative load resistor of a power rectifier or other machine in and out of a circuit.
74	**Alarm Relay** – A device other than an annunciator, as covered under device No. 30, that is used to operate or to operate in connection with a visual or audible alarm.
75	**Position Changing Mechanism** – A mechanism that is used for moving a main device from one position to another in equipment (for example, shifting a removable circuit breaker unit to and from the connected, disconnected, and test positions).
76	**DC Overcurrent Relay** – A relay that functions when the current in a DC circuit exceeds a given value.
77	**Pulse Transmitter** – A device used to generate and transmit pulses over a telemetering or pilot-wire circuit to remove the indicating or receiving device.
78	**Phase Angle Measuring or Out-of-Step Protective Relay** – A relay that functions at a predetermined phase angle between two voltages, between two currents, or between voltage and current.
79	**AC Reclosing Relay** – A relay that controls the automatic reclosing and locking out of an AC circuit interrupter.
80	**Flow Switch** – A switch that operates on given values, or a given rate of change of flow.
81	**Frequency Relay** – A relay that functions on a predetermined value of frequency, either under, over, or on normal system frequency or rate of change of frequency.
82	**DC Reclosing Relay** – A relay that controls the automatic closing and reclosing of a DC circuit interrupter, generally in response to load circuit conditions.
83	**Automatic Selective Control or Transfer Relay** – A relay that operates to select automatically between certain sources or conditions in equipment or that performs a transfer operation automatically.

Device Number	Definition and Function
84	**Operating Mechanism** –The complete electrical mechanism or servo-mechanism, including the operating motor, solenoids, position switches, and for a tap changer, induction regulator, or any similar piece of apparatus that has no device function number.
85	**Carrier or Pilot-Wire Receiver Relay** – A relay that is operated or restrained by a signal used in connection with carrier-current or DC pilot-wire fault directional relaying.
86	**Locking-Out Relay** – An electrically operated relay that functions to shut down and hold equipment out of service on the occurrence of abnormal conditions. It may be reset either manually or electrically.
87	**Differential Protective Relay** – A protective relay that functions on a percentage of phase angle or other quantitative difference of two currents or of some other electrical quantities.
88	**Auxiliary Motor or Motor Generator** – A device used for operating auxiliary equipment, such as pumps, blowers, exciters, and rotating magnetic amplifiers.
89	**Line Switch** – A switch used as a disconnecting load-interrupter or isolating switch in an AC or DC power circuit when this device is electrically operated or has electrical accessories, such as an auxiliary switch or magnetic lock.
90	**Regulating Device** – A device that functions to regulate a quantity, or quantities, such as voltage, current, power, speed, frequency, temperature, and load, at a certain value or between certain (generally close) limits for machines, tie lines, or other apparatus.
91	**Voltage Directional Relay** – A relay that operates when the voltage across an open circuit breaker or contactor exceeds a given value in a given direction.
92	**Voltage and Power Directional Relay** – A relay that permits or causes the connection of two circuits when the voltage difference between them exceeds a given value in a predetermined direction and causes these two circuits to be disconnected from each other when the power flowing between them exceeds a given value in the opposite direction.
93	**Field Changing Contactor** – A device that functions to increase or decrease in one step the value of field excitation on a machine.
94	**Tripping or Trip-Free Relay** – A device that functions to trip a circuit breaker, contactor, or equipment, to permit immediate tripping by other devices, or to prevent immediate reclosure of a circuit interrupter in case it should open automatically even though its closing circuit is maintained closed.
95 96 97	Used only for specific applications on individual installations where none of the assigned numbered functions from 1 to 94 is suitable.

APPENDIX 26306-23B

Typical Manufacturer Drawings

NOTE: The drawing package associated with *Section 2.2.5* can be viewed using the QR code provided here.

APPENDIX 26309-23A

Motor Circuit Calculation Worksheet

MOTOR CIRCUIT CALCULATION WORKSHEET

For standard single-speed, 3ϕ, 60 Hz continuous-duty squirrel-cage induction, synchronous, and wound-rotor motors (without external resistors) rated 600V or less with separate overcurrent devices and controllers.

MOTOR NAMEPLATE DATA

	Motor 1	Motor 2	Motor 3	Motor 4	Motor 5
TYPE OF MOTOR					
Rated volts					
Rated full-load amps (FLA)[1]					
Rated horsepower (hp)[2]					
Design code letter (B, C, or D)					
Thermally protected (yes or no)					
Service factor					
Temperature rise					

[1] If rated in kVA, convert kVA to VA, then use VA divided by rated voltage × 1.73 to determine current.
[2] If not rated, use the rated FLA to find the horsepower that corresponds with the current in *NEC Table 430.250*, interpolating as necessary.

BRANCH CIRCUIT CALCULATIONS

(Controller for each motor is marked at no less than the hp of the motor and with integral selectable overload relays)

Calculations	*NEC*® Reference	Motor 1	Motor 2	Motor 3	Motor 4	Motor 5
1. Lowest temperature rating of termination provisions or devices in circuit[1]	*NEC Section 110.14(C)*					
2. Select FLA from table for motor type, voltage, and hp	*NEC Table 430.248, 430.249, or 430.250*					
3. Select conductor size from table using FLA × 125%[2]	*NEC Table 310.16*					
4. Select overcurrent protective device (OCPD) using FLA x percentage from *NEC Table 430.52*[3]	*NEC Table 430.52* *NEC Section 240.6(A)* *NEC Section 430.52(C)(1) & (2)*					
5. Select equipment grounding conductor size from *NEC Table 250.122* using OCPD value	*NEC Table 250.122*					
6. Select overload protection size for motors not thermally protected[4]	*NEC Section 430.32(A)(1) & (B)(1)*					

[1] If marked for 100A or less, for use of No.14 through No. 1 AWG conductors, or if unknown, use 60°C. For Design Code B, C, or D motors use 75°C. For termination provisions or devices rated for more than 100A or AWG conductor sizes larger than 1 AWG, use 75°C.
[2] Use the ampacity column for the temperature entered on line 1 to select the conductor size from the table. Higher temperature rated conductors for the same conductor size may be used if desired.
[3] If the result is not a standard size, use the next standard size specified in *NEC Table 240.6(A)*. Values can be increased as specified in *NEC Section 430.52(C)(1)(b)(2)*, but cannot exceed the manufacturer's maximum overcurrent device rating per *NEC Section 430.52(C)(2)*.
[4] For motor service factors of 1.15 or greater or marked with a temperature rise of 40°C or less, multiply the nameplate FLA by 125%. For all other motors, multiply the nameplate FLA by 115%.

FEEDER CIRCUIT CALCULATIONS (MORE THAN ONE MOTOR)

Determine	Using Data from Branch Circuit Calculations	Result
Conductor size [*NEC Table 310.16*][1]	From line 2 above, select the largest motor FLA and multiply by 125%, then add the remaining FLAs. Use the total to determine conductor size from *NEC Table 310.16*.	
Overcurrent protective device (OCPD)[2]	From line 4 above, select the largest OCPD and add the remaining motor FLAs from line 2 to determine the total for the feeder circuit OPCD per *NEC Table 240.6(A)*.	
Equipment grounding conductor from *NEC Table 250.122*	Add all OCPDs from line 4 above to determine the total current rating required for the conductor. Use the total to determine conductor size from *NEC Table 250.122*.	

[1] If termination provisions or devices in a feeder circuit are marked for 100A or less, for use of No. 14 through No. 1 AWG conductors, or if unknown, use 60°C column. For termination provisions or devices rated for more than 100A or AWG conductor sizes larger than 1 AWG, use 75°C.
[2] If the total is not a standard OCPD, use the next smaller size listed in *NEC Table 240.6(A)*.

ADDITIONAL RESOURCES

26301-23 Load Calculations – Branch and Feeder Circuits

National Electrical Code® Handbook, Latest Edition. Quincy, MA: National Fire Protection Association.

26302-23 Motors: Conductor Selection and Calculations

National Electrical Code® Handbook, Latest Edition. Quincy, MA: National Fire Protection Association.

26303-23 Practical Applications of Lighting

Controlling LEDs, a Technical white paper, May 2014, P/N 367-2035 REV C. Download from: **http://www.lutron.com/TechnicalDocumentLibrary/367-2035_LED_white_paper.pdf**.

Lighting Handbook. New York, NY: Illuminating Engineering Society of North America (IESNA), Latest Edition.

National Electrical Code® Handbook, Latest Edition. Quincy, MA: National Fire Protection Association.

Wi-Fi® is a registered trademark of the Wi-Fi Alliance®, **www.wi-fi.org**.

26304-23 Hazardous Locations

Code Digest, Latest Edition. Syracuse, NY: Crouse-Hinds.

National Electrical Code® Handbook, Latest Edition. Quincy, MA: National Fire Protection Association.

26305-23 Overcurrent Protection

National Electrical Code® Handbook, Latest Edition. Quincy, MA: National Fire Protection Association.

NFPA 70E®, *Standard for Electrical Safety in the Workplace®*, Latest Edition. Quincy, MA: National Fire Protection Association.

26306-23 Distribution Equipment

National Electrical Code® Handbook, Latest Edition. Quincy, MA: National Fire Protection Association.

NFPA 70B, *Standard for Electrical Equipment Maintenance*, Latest Edition. Quincy, MA: National Fire Protection Association.

NFPA 70E®, *Standard for Electrical Safety in the Workplace®*, Latest Edition. Quincy, MA: National Fire Protection Association.

26307-23 Transformers

National Electrical Code® Handbook, Latest Edition. Quincy, MA: National Fire Protection Association.

26308-23 Commercial Electrical Services

National Electrical Code® Handbook, Latest Edition. Quincy, MA: National Fire Protection Association.

26309-23 Motor Calculations

National Electrical Code® Handbook, Latest Edition. Quincy, MA: National Fire Protection Association.

26310-23 Voice, Data, and Video

ANSI/TIA Telecommunication Building Wiring Standards, Latest Edition. Englewood, CO: Global Engineering Documents.

Cabling: The Complete Guide to Copper and Fiber-Optic Networking, 2014. Hoboken, NJ: John Wiley and Sons, Inc.

Cisco Home Technology Integration Fundamentals and Certification, 2005. Upper Saddle River, NJ: Pearson Education, Inc.

National Electrical Code® Handbook, Latest Edition. Quincy, MA: National Fire Protection Association.

NFPA 1221, *Standard for the Installation, Maintenance, and Use of Emergency Services Communications Systems*, Latest Edition. Quincy, MA: National Fire Protection Association.

26311-23 Motor Controls

Electrical Motor Controls for Integrated Systems, 5th Edition. Gary Rockis and Glen Mazur. Orland Park, IL: American Technical Publishers, Inc., 2013.

IEC Standard 60947-4-1, Low-Voltage Switchgear and Controlgear, Contactors and Motor Starters. Available from **https://webstore.iec.ch/home.**

National Electrical Code® Handbook, Latest Edition. Quincy, MA: National Fire Protection Association.

NEMA Standard ICS 1, Industrial Control and Systems: General Requirements, Latest Edition. Available from the National Electrical Manufacturers Association (NEMA) at **www.nema.org**.

NEMA Standard ICS 2, Controllers, Contactors and Overload Relays Rated 600V, Latest Edition. Available from the National Electrical Manufacturers Association (NEMA) at **www.nema.org**.

NFPA 70B®, *Recommended Practice for Electrical Equipment Maintenance®*, Latest Edition. Quincy, MA: National Fire Protection Association.

GLOSSARY

Actuator: The mechanism of a switch or switch enclosure that moves to operate the contacts.

American National Standards Institute/Telecommunications Industry Association (ANSI/TIA): ANSI oversees the development of voluntary consensus standards. TIA is a trade association representing the global information and communications technology industries. TIA is accredited by ANSI to develop voluntary industry standards that address telecommunications products, such as *ANSI/TIA 568*.

Ampacity: The maximum current that a conductor can carry continuously without warming beyond its temperature rating.

Ampere squared seconds: A unit of measure for heat energy developed within a circuit during a fuse's clearing of a fault. It can be expressed as *melting* I^2t, *arcing* I^2t, or the sum of them as *clearing* I^2t. I stands for the effective let-through current and t stands for time to open in seconds.

Ampere turns: The product of the current (in amperes) times the number of turns in a transformer coil.

Amperes interrupting capacity (AIC): The maximum short circuit current that a circuit breaker or fuse can safely interrupt.

Appliance: Equipment typically manufactured in standardized sizes and types that are installed as a unit to perform one or more specific functions, such as clothes washers and dryers, blenders, toasters, and similar appliances.

Appliance branch circuits: Branch circuits supplying energy to outlets to which only appliances are connected.

Arcing time: The amount of time passing from the instant a fuse link has melted until the overcurrent event is interrupted.

Attenuation: The reduction of a signal due to system losses.

Autotransformer: A transformer in which primary and secondary connections are made to a single winding.

Auxiliary contacts: An added set of contacts that are mechanically connected to a contactor, either opening or closing as the contactor itself moves. Auxiliary contacts can be normally open (NO) or normally closed (NC).

Backbone: A high-capacity network link between two sections of a network.

Ball mill: A grinding unit used to grind and/or blend rocks and minerals for use in manufacturing or other processes.

Bandwidth: The range of frequencies that will pass through a particular portion of a system.

Basic impulse level (BIL): A means of expressing the maximum impulse voltage surge that electrical insulation can withstand without damage; also referred to as the *basic insulation level*.

Branch circuit: The circuit conductors connecting the final overcurrent protective device and the outlets or other loads it protects.

Bushing: A hollow electrical insulator, usually made from porcelain, that provides safe passage for a conductor through a conductive material, such as the case of a distribution-class transformer or circuit breaker, preventing electrical contact.

Cam: A machine component that applies a force to the actuator of a limit switch, causing the actuator to move.

Capacitance: The storage of electricity in a capacitor that produces an opposition to voltage change. The unit of measurement is the *farad* (*F*) or *microfarad* (μF).

Clearing time: The total time between the beginning of an overcurrent event and the complete opening of the circuit by an overcurrent protective device operating within its rated voltage. Clearing time is the sum of the melting time and arcing time.

Closed-circuit television (CCTV): The use of video cameras to transmit a signal to one or more television monitors.

Coaxial cable: A type of cable that has a center conductor separated from the outer conductor by a dielectric. Coaxial cable is used to transmit very small currents or high-frequency signals and is commonly used in closed-circuit television (CCTV) systems.

Cold sequence metering: A service installation where the meter disconnect is installed before the metering for power to be disconnected to allow for safe maintenance of the metering.

Continuous load: A load where the maximum anticipated current is expected to continue for at least three hours.

Cross-connect: A facility for connecting cable runs, subsystems, and equipment using patch cords or jumper cable.

Current-limiting device: A device designed to clear a short circuit in less than one half cycle. This limits the instantaneous peak let-through current to a value substantially less than would pass if the device were replaced with a solid conductor of equal impedance.

Dashpot: A mechanical device that resists movement to cushion or dampen motion, such as the action of a shock absorber. A typical dashpot slows and controls mechanical movement by forcing air or liquid through an orifice with a fixed or adjustable diameter.

Daylight harvesting: Making use of available sunlight in an area while maintaining required illumination levels using lighting control systems to dim or switch electric lighting as needed to supplement the natural light; also referred to as *sunlight harvesting*.

Deadband: The difference between the setpoint at which a switch activates as input increases and the point at which the switch resets to the former position when the input decreases.

Delay skew: A change of timing or phase in a transmission signal.

Delta-connected: A three-phase transformer connection in which the terminals are connected in a triangular shape like the Greek letter delta (Δ).

Demand factors: The ratio of the maximum anticipated demands of certain electrical loads compared to the actual connected load.

Device: Components other than conductors with the primary purpose of carrying or controlling electrical power, but that do not consume significant electrical energy.

Dielectric constant: The ratio of the capacitance of a capacitor with a given dielectric to the capacitance of an identical capacitor using only air as a dielectric.

Diffused lighting: Light that has been filtered or scattered, without the normal reflective characteristic of direct light.

Distributed antenna system (DAS): A network of antennas throughout a building that serves as a repeater to remedy areas of poor coverage inside the building.

Distribution system equipment: Switchboard equipment downstream from the service-entrance equipment.

Distribution transformers: Transformers that are used for reducing the voltage and transferring power from a primary distribution circuit, usually originating at a substation, to the secondary distribution circuit. Distribution transformers are typically rated between 5 kVA and 500 kVA.

Dropout voltage: The voltage at which the armature return spring of a relay or contactor overpowers the operating coil's weakened magnetic field, allowing it to return to its de-energized state.

Eddy currents: Currents induced in the body of a metallic material by an oscillating magnetic field.

Equal level far end crosstalk (ELFEXT): A value derived by subtracting the attenuation of the interfering pair from the far end crosstalk (FEXT) that it has caused in the interfered pair.

Ethernet: A local area network (LAN) developed by Xerox Corporation that uses a bus or star topology. Ethernet is currently the most popular type of LAN.

Explosion proof: Designed and constructed to withstand an internal explosion without allowing an external explosion or fire.

Fast-acting fuse: A fuse that opens very quickly in response to overloads and short circuits and is not designed to withstand temporary overload events associated with common inductive loads.

Feeder: A set of conductors originating at a main distribution center that supply one or more secondary distribution centers, one or more branch circuit distribution centers, or any combination of the two.

Firestopping: A material or mechanical device used to block openings in walls, ceilings, or floors to prevent the passage of fire or smoke.

Flux: A measure of the electric field passing through a surface or area.

Footcandle (fc): Unit of measure for light reaching a surface. One lumen falling on one square foot of surface produces an illumination level of one footcandle. The international unit for illumination levels is the *lux*, which equates to one lumen per square meter. A footcandle is equal to approximately 10 lux.

Fresnel spotlights: Focusable spotlights that can be adjusted using a control knob on the back of the light from a narrow, focused beam to a wide beam.

General-purpose branch circuits: Branch circuits that supply two or more outlets or receptacles for lighting and appliances.

Gigabit per second (Gbit/s): An extremely fast rate of data transfer roughly equal to 1,000,000,000 bits per second.

Hazardous locations: Locations where ignitable vapors, dust, or fibers can cause a fire or explosion when exposed to a source of ignition.

Hermetic: Isolated and sealed airtight.

Individual branch circuits: Branch circuits that power only one electrical load, such as a clothes dryer or motor.

Induction: The production of an electromotive force (emf) in an electrical conductor when placed in a changing magnetic field.

Inductive loads: Electrical loads that consume significantly more current, referred to as *inrush current*, when first energized. The current rapidly declines to the normal, anticipated load current.

Infrared (IR): A portion of the light spectrum that is not visible with a wavelength beyond 690 nanometers. IR radiation is emitted by certain LEDs.

Innerduct: A type of high-density polyethylene (HDPE) conduit that is available in both corrugated and smooth types and is often placed inside a larger raceway to run communications cable.

Jogging: Intentional, repeated closure of a circuit used to start and stop a motor to initiate small, incremental movements in the driven load; also referred to as *inching*.

Line-powered sensor: A three-wire sensor that draws its operating current, referred to as burden current, directly from the line. The sensor's operating current does not flow through the load.

Load-powered sensor: A two-wire sensor in series with the load that draws its operating current, referred to as *leakage* or *residual current*, through the load.

Magnetic field: The area around a magnet in which the effect of the magnet can be detected.

Maintained contact switch: A switch with contacts that remain in the operating position when the actuating force is removed, with other provisions to reset the switch to its normal position.

Megabits per second (Mbit/s): A rate of data transfer roughly equal to 1,000,000 bits per second.

Melting time: The amount of time required for a fuse link to melt during an overcurrent event.

Metal-clad switchgear: A subset of metal-enclosed switchgear, with the internal components also enclosed in metal individually. Each component may also be individually insulated and grounded.

Metal-enclosed switchgear: Switchgear that is enclosed on all sides, except for small ventilation and/or viewport openings that are commonly required.

Momentary contact switch: A switch with contacts that return from the operating position to their normal position when the actuating force is removed.

Motor control circuit: A typically low-current electrical circuit that controls the operation of a contactor or motor starter but does not carry the current powering the motor.

Multioutlet assembly: A type of surface mounted, flush mounted, or freestanding raceway designed to hold conductors and receptacles.

Mutual induction: The condition of voltage generation in a second conductor caused by current flow in another conductor.

Near end crosstalk (NEXT): a measurement of the interference (crosstalk) between two wire pairs.

Operator: A switch component that is pressed, pulled, or rotated by the person operating the circuit to activate the switch contacts.

Outlet: A point in a branch circuit to which a load device can be connected.

Outside plant (OSP): Refers to all the communications equipment (cable, conduit, cabinets, and hardware) located between switching facilities or between a switching facility and a customer.

Overcurrent: Current that exceeds the rating of a conductor or a connected load, resulting from a ground fault, short circuit, or an overload condition.

Overload: Any overcurrent condition that exceeds the normal full-load current of a circuit.

Patch cord: A connecting cable used with cross-connect equipment and work area cables. It has three conductive portions separated by insulators.

Peak let-through: The instantaneous value of peak current allowed to pass through a current-limiting fuse when operating within its current-limiting range. Represented by the variable I_p.

Pickup voltage: The minimum voltage applied to a relay or contactor coil that will allow it to energize. The device is not likely to operate below this voltage due to the weakness of the magnetic field.

Pilot devices: Devices such as electrical push buttons and switches that influence the operation of related relays and contactors.

Plugging: Braking a three-phase motor by reversing the phase sequence of the power supply to the motor to achieve both rapid stops and quick reversal of rotation.

Power loss: Power expended without accomplishing useful work.

Power transformers: A term often used to identify larger transformers used at transmission and distribution substations, as opposed to smaller transformers found in electrical subsystems and control systems.

Proximity sensors: Sensing switches that detect the presence or absence of an object using an electronic noncontact sensor.

Reactance: The opposition to AC due to capacitance and/or inductance.

Receptacle outlet: A power outlet location where one or more receptacles are connected.

Receptacles: Points of connection installed at outlets that allow for the quick connection of an electrical load device with a compatible configuration. A *duplex receptacle* contains two such receptacles.

Rectifiers: Devices used to change alternating current to direct current.

Root-mean-square (rms): The multiplication of the square root of the mean (average) value by the squared function of the instantaneous values throughout the period. The rms value of an AC voltage or current is the effective value of the voltage or current.

Seal-in voltage: The minimum voltage applied to a relay or contactor coil that will maintain a sufficient magnetic field to keep the device energized The seal-in voltage is usually less than the pickup voltage.

Sealing compound: A material poured into an electrical fitting as a liquid that cures and hardens to positively prevent the passage of air, gases, and vapors.

Sealoff fittings: Fittings required in conduit systems to prevent the passage of gases, vapors, or fire from one portion of an electrical installation to another through the conduit. Also referred to as *sealing fittings* or *conduit seals*.

Service: Conductors and equipment for delivering electric energy from the serving utility to the wiring system of the premises served.

Service conductors: The conductors between the point of termination of the overhead service drop or underground service lateral and the main disconnecting device in the building or on the premises.

Service drop: The overhead conductors between the utility electric supply system and the service point.

Service entrance: All components between the point of termination of the overhead service drop or underground service lateral and the building's main disconnecting device, except for metering equipment.

Service equipment: The necessary equipment, usually consisting of a circuit breaker or switch along with fuses and their accessories, located near the entrance point of supply conductors to a building and intended to constitute the main control and cutoff means for the electric supply to the building.

Service factor: A number by which the horsepower rating of a motor is multiplied to determine the maximum safe load a motor can be expected to carry continuously at its rated voltage and frequency.

Service lateral: The underground conductors between the utility main electric supply system and the service point.

Service-entrance equipment: Equipment located at the service entrance of a structure that provides overcurrent protection to the feeder and service conductors, as well as a means of disconnecting the feeders from the energized service equipment.

Setpoint: The desired value to which a switching device is set or adjusted to respond.

Short circuit current: An overcurrent condition that greatly exceeds the normal current-carrying capacity of a circuit. This occurs when the current leaves the normal current-carrying path of the circuit, bypassing the load and returning to the source.

Single phasing: A condition that occurs when one phase of a three-phase system is lost, leaving loads energized by only the two remaining phases. Single phasing results in unbalanced loads in polyphase motors. Unless protective devices respond, it will cause overheating and failure.

Spacing criterion (SC): The ratio of spacing to mounting height that will produce uniform lighting when the luminaire is used. Exceeding the recommended spacing criterion results in uneven lighting on the work plane.

Sulfur hexafluoride (SF_6): A stable, synthetic, odorless gas that is used as an insulator in electrical apparatus.

Surface tracking: The formation of conductive, carbonized paths along the surface of insulators and similar components that provide a path for current from high-voltage sources.

Switchboard: A large single panel, frame, or assembly of panels on which switches, fuses, buses, and instruments are mounted.

Switchgear: A general term describing a collection of switching and interrupting devices that can also include control, instrumentation, metering, protective, and regulating devices.

Telecommunications main grounding busbar (TMGB): A copper busbar typically located where the telecommunications cables enter the building. It serves as the grounding point for the telecommunications systems/equipment and is connected to the main electrical ground using an appropriately sized copper conductor.

Tinsel cords: Cables typically used for telephones and similar applications that offer excellent flexibility but low current-carrying capacity.

Torque: A twisting force that produces rotary motion, typically measured in foot-pounds or inch-pounds of force.

Turns: The basic coil element in a transformer that forms a single conducting loop comprised of one insulated conductor.

Turns ratio: The ratio of the number of turns in the transformer primary winding compared to that in the secondary winding. In current transformers, it is the ratio of the turns in the secondary winding to the primary winding.

Utilization equipment: Equipment that consumes electrical energy for a wide variety of purposes.

Volatile: Describes a substance that evaporates (vaporizes) readily at room temperatures.

Voltage rating: The maximum voltage a fuse can be applied to and safely interrupt an overcurrent event.

Wi-Fi®: Acronym for *wireless fidelity*; the technology allowing communication via radio signals over a local or wide-area network equipped with a wireless access point or the internet.

Work plane: The distance above the floor where a task is to be performed. In offices and schools, for example, the distance is usually 30" (750 mm).